VOLUME 49

Advances in
CHROMATOGRAPHY

VOLUME 49

Advances in
CHROMATOGRAPHY

EDITORS:

ELI GRUSHKA
Hebrew University of Jerusalem
Jerusalem, Israel

NELU GRINBERG
Boehringer-Ingelheim Pharmaceutical, Inc.
Ridgefield, Connecticut, U.S.A.

CRC Press
Taylor & Francis Group
Boca Raton London New York

CRC Press is an imprint of the
Taylor & Francis Group, an **informa** business

CRC Press
Taylor & Francis Group
6000 Broken Sound Parkway NW, Suite 300
Boca Raton, FL 33487-2742

First issued in paperback 2019

© 2011 by Taylor and Francis Group, LLC
CRC Press is an imprint of Taylor & Francis Group, an Informa business

No claim to original U.S. Government works

ISBN-13: 978-1-4398-4091-7 (hbk)
ISBN-13: 978-0-367-38302-2 (pbk)

Visit the Taylor & Francis Web site at
http://www.taylorandfrancis.com

and the CRC Press Web site at
http://www.crcpress.com

Contents

Contributors

Friedrich Altmann
Department of Chemistry
University of Natural Resources and
    Life Sciences
Vienna, Austria

Emidio Camaioni
Dipartimento di Chimica e Tecnologia
    del Farmaco
University of Perugia
Perugia, Italy

Teresa Cecchi
Itis Montani
Fermo, Italy

Corina Flangea
Department of Chemical and Biological
    Sciences
Aurel Vlaicu University of Arad
Arad, Romania

Bernard Fried
Department of Biology
Lafayette College
Easton, Pennsylvania

Jun Haginaka
School of Pharmacy and
    Pharmaceutical Sciences
Mukogawa Women's University
Nishinomiya, Hyogo, Japan

Toshihiko Hanai
Health Research Foundation
Institut Pasteur
Kyoto, Japan

Kyu-Bong Kim
Department of Pharmaceutical
    Engineering
Inje University
Gimhae, Gyeorgnam, South Korea

Hwa Jeong Lee
Division of Life and Pharmaceutical
    Sciences
Ewha Women's University
Seoul, South Korea

Qing Liao
Department of Chemistry and Chemical
    Biology
Northeastern University
Boston, Massachusetts

Antonio Macchiarulo
Dipartimento di Chimica e Tecnologia
    del Farmaco
University of Perugia
Perugia, Italy

Maura Marinozzi
Dipartimento di Chimica e Tecnologia
    del Farmaco
University of Perugia
Perugia, Italy

Benedetto Natalini
Dipartimento di Chimica e Tecnologia
   del Farmaco
University of Perugia
Perugia, Italy

Roberto Pellicciari
Dipartimento di Chimica e Tecnologia
   del Farmaco
University of Perugia
Perugia, Italy

Krystyna Pyrzynska
Laboratory for Flow Analysis and
   Chromatography
Department of Chemistry
University of Warsaw
Warsaw, Poland

Andreas M. Rizzi
Institute of Analytical Chemistry
University of Vienna
Vienna, Austria

Roccaldo Sardella
Dipartimento di Chimica e Tecnologia
   del Farmaco
University of Perugia
Perugia, Italy

Joseph Sherma
Department of Chemistry
Lafayette College
Easton, Pennsylvania

Terrence P. Tougas
Analytical Department
Boehringer–Ingelheim Pharmaceuticals
Ridgefield, Connecticut

Paul Vouros
Department of Chemistry and Chemical
   Biology
Northeastern University
Boston, Massachusetts

Alina D. Zamfir
Department of Chemical and Biological
   Sciences
Aurel Vlaicu University of Arad
Arad, Romania

1 Retention Mechanism for Ion-Pair Chromatography with Chaotropic Reagents
From Ion-Pair Chromatography toward a Unified Salt Chromatography

Teresa Cecchi

CONTENTS

ABSTRACT

The breakthrough of chaotropic mobile phase modifiers in reversed-phase high-performance liquid chromatography (RP-HPLC) is due to their strong potential to provide adequate retention of ionic analytes without the blamed semipermanent modification of the chromatographic packing often connected to the use of classical ion-pair reagents. The lack of a physicochemical framework that is able to unify eclectic experimental evidence concerning the use of a wide gamut of ionic additives in RP-HPLC is the primary motive force for recent theoretical efforts to model their behavior. The time-honored solvophobic theory cannot properly explain salt effects. Its theoretical basis was recently questioned by breaking experimental evidence at variance with the textbook knowledge of ionic solutions interfaces; meanwhile, a recently proved extended thermodynamic approach to ion-pair chromatography (IPC) is challenged by the breakthrough of neoteric ionic additives whose behavior questions the rigidity of previous retention schemes and bridges salting chromatographic phenomena to IPC. Building on these research needs, the aims of this review are (1) to illustrate a comprehensive theory of analyte retention in the presence of any kind of electrolytes (hydrophobic ions, chaotropes, kosmotropes, ionic liquids [ILs]) to capture and rationalize the main salting effects and to support their strong practical impact for the separation of organic and inorganic ions, ionogenic, neutral, and zwitterionic analytes; (2) to explain why ion-specific salting chromatographic effects that represent a diachronic scientific consideration were not satisfactorily explained in the rubric of the solvophobic theory; and (3) to highlight the eligibility of chromatography as a basic technique that is able to clarify the currently hotly debated behavior of ions at water interfaces. The practical impact of chaotropic chromatography will also be detailed, and urgent research needs and suggestions will be illustrated.

1.1 INTRODUCTION

Chromatography is probably the most widely used technique in the modern analytical laboratory since it is able to simultaneously separate and quantitate analytes in a few minutes and possibly in a few seconds with good sensitivities and selectivities. A wide range of very different samples can be analyzed using several chromatographic strategies. High-performance liquid chromatography (HPLC) is the modern translation of column liquid chromatography invented by Tsvet in 1901. This technique eclipsed other branches of chromatography: experienced practitioners are able to exploit its formidable potential using hydroorganic eluents and hydrocarbonaceous-bonded stationary phases. The majority of separations are performed by reversed-phase high-performance liquid chromatography (RP-HPLC).

One of the major challenges for chromatographers is the separation of ionic analytes for which the use of RP-HPLC is somewhat constrained since they are hardly retained on the apolar stationary phase. pH adjustment of the mobile phase to suppress analyte ionization is right and proper only if the analytes to be separated have similar pKas that lie in the pH range of stability of the chromatographic packing (usually 2–8 for silica-based RP columns). Ion exchange chromatography is an alternative separative strategy for ionic compounds: since it is based on the electrostatic interaction of charged analytes with the oppositely charged groups grafted onto the stationary phase, its selectivity for organic ions is limited since the interactions of the hydrophobic moiety of the charged compounds is not fully exploited. The influence of inorganic salts used as mobile phase additives on RP-HPLC of neutral and ionogenic analytes has been studied in depth since this technique's infancy and represents a diachronic scientific consideration. However, inorganic salts were effectively used only in the realm of salting-out chromatography of neutral analytes.

The drawbacks of ionic suppression and ion exchange chromatography led to the development of ion-pair chromatography (IPC) [1,2], which classically used large organic ionic additives to obtain adequate retention of oppositely charged ionic analytes. Polarizable hydrophobic ionic additives, such as alkylsulfonates or alkylammonium ions, were named ion-pair reagents (IPRs) since they were supposed to be effective in enhancing oppositely charged analyte retention because they form ion pairs with the analyte in bulk eluent. The technique was known as ion-pair chromatography, but the first name was probably "soap chromatography" [3] since many IPRs are actually tensioactive compounds that were easily predicted to adsorb onto the apolar stationary phase/bulk eluent interface. Many other names were used to denote this separation strategy [3–11]; they alternatively emphasize a particular phenomenon in the chromatographic system that includes the ionic additive. The early description of the retention mechanism focused only on a partial interaction and gradually broadened its interest to cover the multifaceted interactions in both the mobile and stationary phases. Although the most suitable name of this technique still seems to be *ion-interaction chromatography* [6,12], it is recognized worldwide as IPC [1]. For brevity, the following section omits description of old partial retention models.

1.2 CHRONOLOGICAL DESCRIPTION OF THE RETENTION MODELING IN IPC

The rationalization of the IPC retention is a nontrivial task, and historically two classes of retention models were used to explain and predict it. According to the first stoichiometric models, in the chromatographic system the main equilibria are those illustrated in Figure 1.1: the adsorption of the analyte (E) onto the stationary phase hydrocarbonaceous ligand site (L); the adsorption of IPR (H) onto L; the ion-pair formation in the mobile phase (EH); the ion-pair formation in the stationary phase

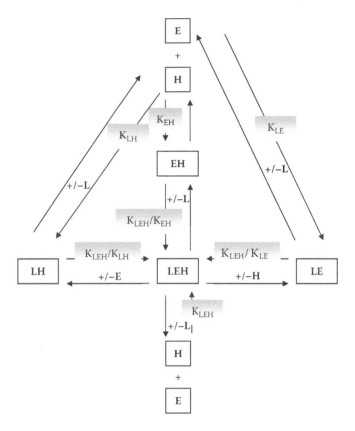

FIGURE 1.1 Comprehensive retention mechanism of IPC. *Ion-Pair Chromatography and Related Techniques*, 1st Edition by T. Cecchi. Copyright 2010 by Taylor & Francis Group. Reproduced with permission of Taylor & Francis Group LLC.

(EHL); and the displacement of H by E. The equilibrium constants for these equilibria are, respectively, K_{LE}, K_{LH}, K_{EH}, K_{LEH}, and K_{LE}/K_{LH}.

Since all stoichiometric models suggest the same structure of the adsorbed complex, they yield the same retention equation [11]. The nontrivial drawback shared by all stoichiometric models is that they neglect the experimentally proven establishment of the stationary phase electrostatic potential [13] due to the adsorption of the lipophilic IPR ions; it was proven [12] and will be explained subsequently that the stoichiometric constants used to describe these equilibria are not actually constant in the presence of the stationary phase-bulk eluent electrified interface: electrochemical potentials should replace chemical potentials to obtain meaningful thermodynamic expressions. This issue absolutely prejudices the foundations of all these models in physical chemistry. Stoichiometric models can be credited only with a descriptive value regarding the ion-pair retention mechanism. In this context, it is important to underline that the most comprehensive stoichiometric model was put forth by Horvath and coworkers [5] in the framework of the solvophobic theory, and this emphasizes the weakness of this long-established model regarding the chromatography of ionogenic analytes.

All nonstoichiometric models disagree with the stoichiometric hypothesis of an electroneutral stationary phase; they acknowledge the electrical double layer (DL) concept [12–21] and ascribe its development at the stationary phase interface to the higher adsorbophilicity of the IPR compared with that of its counterion. Figures 1.2 and 1.3 outline the DL concept according to the Gouy–Chapman and Stern–Gouy–Chapman models, respectively. Early electrostatic versions of the nonstoichiometric model [13,14] are only concerned with the interaction of the charged analyte with the surface potential; since analyte ions and IPR ions are oppositely charged, the electrical force is attractive, which results in the increased retention the chromatographer searches for. Unfortunately, they disregard the ion-pair equilibria in the mobile phase. Recently, these interactions in the mobile phase were experimentally demonstrated [22–35] beyond any dispute, for both inorganic and organic ion pairs, via a number of different analytical techniques.

It follows that electrostatic nonstoichiometric models [13,14] are not able to rationalize experimental evidence related to ion-pair formation. It was recently demonstrated that they misinterpret retention data. A new comprehensive retention model of IPC was put forth by Cecchi and coresearchers [12]. It is a thermodynamic model because it correctly takes into account all equilibria detailed in Figure 1.1 by using electrochemical potentials to obtain the related equilibrium constants. The model is quite complex even from the algebraic point of view, but the claims for a leading-edge theory, epistemologically speaking, are sustained by the following arguments [12,15–21,34]:

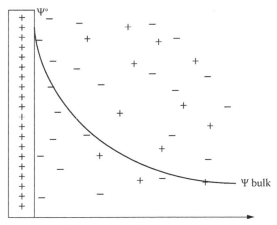

Distance from the stationary phase

FIGURE 1.2 Gouy–Chapman double layer model. *Ion-Pair Chromatography and Related Techniques*, 1st Edition by T. Cecchi. Copyright 2010 by Taylor & Francis Group. Reproduced with permission of Taylor & Francis Group LLC.

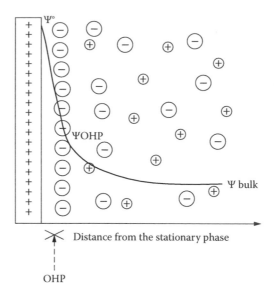

FIGURE 1.3 Stern–Gouy–Chapman model, in which only the OHP is indicated. *Ion-Pair Chromatography and Related Techniques*, 1st Edition by T. Cecchi. Copyright 2010 by Taylor & Francis Group. Reproduced with permission of Taylor & Francis Group LLC.

1. It is possible to comprehend previous theories of IPC: (a) if the surface potential is deliberately and mistakenly equated to zero, the model reduces to the stoichiometric model of IPC: the latter takes the counter ions necessary to ensure the balance of electrical neutrality into consideration, in open conflict with the experimental proof of the development of a stationary phase surface potential [13]; and (b) if the pairing equilibria are deliberately and mistakenly neglected, the model reduces to the law sanctioned by the pure electrostatic approach [13]. It can be concluded that the theory improves on previous outstanding retention models that can both be obtained as limiting cases of the new theory.

2. The model is inclusive: the retention equations quantitatively explain the IPC retention pattern of charged, multiply charged, neutral, and zwitterionic solutes and the adjustment of retention via the manipulation of the organic modifier concentration and of the ionic strength of the mobile phase.

3. The model, in the absence of the ion-pair reagent, mathematically gives the well-known relationships of RP-HPLC for the influence of the organic modifier or the ionic strength in the eluent.

4. The model explains experimental evidence related to ion pairing in the mobile phase that is obviously in open contrast with the genuine electrostatic retention model's predictions, such as (a) different analyte retention dependence on the adsorbed ion pair reagent, for different IPRs; (b) the influence of the IPR concentration on the ratio of the retention of two different analytes; and (c) the influence of the analyte hydrophobicity on the

retention pattern. Since all these issues are related to the ion specificity of both the IPR and the analytes, they cannot be modeled by simple electrostatic interactions that depend only on the ionic charges.

5. The model quantitatively explains that the lower the likelihood of ion-pairing interactions in the mobile phase, the better the genuine electrostatic retention model's predictions are—that is, (a) at high methanol percentages in the eluent (because in this case retention is basically electrostatic since hydrophobic ion-pairing is strongly dampened down); (b) if the chain length of the analyte is shorter, since in this case the solute architecture approaches the point charge description adopted by the electrostatic theories; (c) for analytes and IPRs in the same charge status because in this case ion-pairing interactions do not obviously apply.

6. Fitting parameters estimates, whose physical meaning is clear, are reliable since they compare well with the literature's chromatographic and nonchromatographic estimates.

1.3 RENEWED INTEREST IN INORGANIC IONIC ADDITIVES IN THE RP-HPLC OF IONOGENIC ANALYTES

1.3.1 INTRODUCTION AND USE OF INORGANIC IONIC ADDITIVES AS ELUENT MODIFIER

As pointed out already, the use of simple inorganic ions (not lipophilic and adsorbophilic like the organic ionic IPRs) as eluent additives to increase the RP-HLPC retention of ionic analytes was quickly obscured by the popularity and efficacy of IPRs. Their employment, described for the first time by Shepard and Tiselius in 1949 [36], was typically and ordinarily restricted to the "salting-out chromatography" of neutral molecules. This technique was subsequently described as *hydrophobic chromatography* or *hydrophobic adsorption chromatography* (HIC) [37] and *hydrophobic interaction chromatography* [38]. Nowadays, this versatile and efficient strategy is successfully used for the analytical or industrial separation and purification of proteins on weakly hydrophobic matrices [39].

1.3.2 THEORETICAL PREDICTION AND PRACTICAL ASSESSMENT OF IONIC PRESENCE AT THE CHROMATOGRAPHIC INTERFACE

While the adsorption of classical IPRs onto the apolar stationary phase was deeply and experimentally assessed [13] and was soon taken into account by model makers [12,15–21,34], the adsorption of inorganic salts from water solutions onto apolar surfaces is a current paradox for physical chemistry. Textbook knowledge about aqueous interfaces rationalizes the surface tension (σ) increase of aqueous salt solution as the consequence of the absence of ionic species at the interface, according to the Gibbs adsorption isotherm:

$$d\sigma = -\Sigma\ \Gamma_i\ d\mu_i \qquad (1.1)$$

since

$$\mu_i = \mu_i^\circ + RT \ln a_i \tag{1.2}$$

it follows that

$$\Gamma_i = -(1/RT)(\delta\sigma/\delta\ln a_i) \tag{1.3}$$

where Γ_i is the surface excess of the solute at the Gibbs dividing surface—that is, the location in the interfacial region where the excess of the solvent is zero (the interface can be arbitrarily located anywhere in that region, but in this case the mathematical handling of the expression is better)—and μ_i is the chemical potential of each component of the system. Onsanger and Samaras [40] justify the absence of ions at the interface as a result of their repulsion by a fictitious image charge. All charged compounds near any dielectric boundary would experience the image charge repulsion. For a chromatographic interface the following pictorial description would fit: since the water-rich eluent is strongly dipolar, it solvates the ions well. They feel an attraction to the bulk eluent, due to the dipole-ion interactions, that is not counterbalanced by a corresponding attraction to the hydrocarbonaceous apolar stationary phase. The inorganic ions, at variance with organic IPRs, lack a large hydrophobic moiety able to stack to the apolar stationary phase via Van der Waals interactions; hence, they should be repelled from the chromatographic interface. This repulsion was considered by Onsanger and Samaras identical to what would occur if a mirror image of the same sign was on the other side on the dielectric boundary. Since exclusion of ions at the surface involves a negative entropy of mixing variation, the deformation of the surface would be more difficult, which in turn increases interface tension. This entropically driven increase of surface tension was usually studied for the water–air interface. Unfortunately, this classical theory of colloid science does not accommodate "specific ion effects": the theory is flawed in that image charge repulsion is the same for all ions of the same valence. Ion-specific effects are ubiquitous and are commonly called Hofmeister effects. Hofmeister in 1888 [41] introduced the concept of ranking salts based on both cations and anions to explain their salting-out properties. Based on the position of an ion in the Hofmeister series, it is possible to foretell the relative effectiveness of salts in an enormous number of systems, so this series emphasizes a general underlying mechanism for salt effects. The rank of an ion was related to myriad chemical features: electrolyte activities, micelle and microemulsion structure, wettability, protein precipitation and unfolding-denaturation, gas and organic compounds solubilities, atmospheric chemistry, surface tension increments, salting-in and salting-out of salt solutions, as well as ionic hydration.

Moving our horizon from analytical chemistry toward physical chemistry and computational chemistry, we note that electrostatic forces, as described within the classical model of a charged sphere in a dielectric continuum, are important but do not provide the whole picture. Polarization, dispersion, and solvation effects all must be taken into account to answer theoretical questions with atomic resolutions at the angstrom scale. Recently, molecular dynamics (MD) computations [42–44] of the simple air–water interface definitively demonstrated that, at variance with the Onsanger–Samaras pictorial representation, ions do accumulate at interfaces. At variance with bulk solution, at the surface the environment is asymmetric, and the large solvent electric field polarizes the ion; if the ion is sufficiently soft, this polarization stabilization may compensate much of the electrostatic penalty due to a partial loss of solvation out of the bulk solution [42–44]. Additional experimental evidence that the cation and the anion of a salt partition differently between the surface and the bulk solution, and anions actually accumulate (even if to a much lower extent compared with surfactants) at the surface was also provided by surface spectroscopic measurement and electrospray ionization mass spectrometry [45–48].

An electrical double layer and a surface electrostatic potential is easily predicted to develop at the solution interface as the result of any difference in the degree of adsorption or repulsion from the surface. How these findings can be reconciled with (a) the ion-free Onsanger–Samaras surface and (b) the contrast between the experimentally demonstrated surface tension increase that is tantamount to the negative adsorption expected from the Gibbs adsorption isotherm and a surface accumulation of ions is a matter of debate.

Consider the following:

1. The electrical double layer gives rise to an additional influence on the surface tension via the Lippman–Helmoltz equation [49].

$$(\delta\sigma/\delta\Phi)_{T,P,n} = -q \tag{1.4}$$

   where q is the surface charge density, but this effect was neglected by the Onsanger–Samaras theory.

2. Moreover, since a double layer develops, the chemical potential of each charged species should be replaced by its electrochemical potential to correctly account for the transfer toward an electrified region.

3. It was also observed that since the surface excess is an integral quantity over the whole interfacial region, it can also be negative if surface accumulation of the salt is overwhelmed by subsurface depletion [42].

4. At high salt concentration, ion pairing of inorganic electrolytes may dampen the importance of the repulsive interaction toward a low dielectric constant phase more than the simple effect of the ionic atmosphere (the Onsanger–Samaras theory takes the latter into account via the Debye–Hukel factor and the inverse Debye length) since an ion may not behave independently of its companion. Although Hofmeister originally ranked salts acknowledging that there is an inherent specificity of the ion pair contributed by the salt, the standard Hofmeister ion specific series currently used was soon extrapolated [50].

5. The Gibbs–Onsanger–Samaras model is at variance not only with an interface devoid of ions but also with the Jones–Ray effect investigated in the 1930s: some salts did not produce the predicted monotonic increase in

surface tension but exhibited a minimum in the surface tension at milimolar concentrations. Recent explorations verified the Jones–Ray effect for alkali iodide and potassium ferrocyanide solutions by ultraviolet (UV) femtosecond second-harmonic generation (SHG) experiments [51], which revealed an initial surface ionic enhancement followed by a depletion at higher concentrations.

The driving force behind chromatographic interface adsorption has many origins, depending on the specificities of the absorbate and surface. In the majority of cases, covalent interactions between the surface and the solute molecule are absent, and the adsorption is governed by nonbonding forces even if electrostatic forces play a crucial role. The interface-driving polarization effects that were demonstrated to control ion specific effects at the water–air interface were never satisfactorily studied via molecular dynamics simulations for chromatographic surfaces in contact with eluents containing ionic additives and represent an urgent research need. The recently studied ionic adsorption at the dodecanol–water interface [48] and at a hydrophobic self-assembled alkane monolayer [52] could be a good starting point. In this context it is interesting to observe that RP-HPLC, which makes use of ionic additives in the mobile phase, is a perfect and new tool to test the adsorption of various electrolyte solutions components onto nonpolar solid surfaces. This is a process of outstanding importance in colloid science, physical chemistry, and biology, and chromatography is the most eligible technique to complement computational studies. Recently the adsorption of inorganic ions onto reversed-phase packing was experimentally demonstrated via measurement of the adsorption isotherm: the surface excess of inorganic electrolyte at the chromatographic interface [53–56] can be modeled by the same kind of adsorption isotherm—the Freundlich one—that also holds true for classical IPRs. Again, such adsorption of charged spheres at the dielectric boundary is in stark contrast to the Onsanger–Samaras theory and should encourage a better representation of the physical phenomenon.

It is interesting to observe that anions usually show a stronger surface affinity than cations [42,43,52,57]. This is not surprising if one takes into account that the polarizability of the cation is much lower than that of an anion; it was also suggested that the thermodynamics of transferring similar ions of opposite charges from the bulk to the surface are dominated by different interactions [57]. This observation links to the stronger influence of the anion (a) on the Hofmeister effect [57], (b) on the surface tension increase [57], and (c) on the concept of specific adsorption ("chemisorption") of anions at Hg electrodes [58]. Noteworthy, the chromatographic breakthrough of inorganic additives was due just to the "chaotropic" character of the salt anion. What is chaotropicity?

The key to understanding the Hofmeister effect is related to the ion-induced change of the thermodynamics of the interfacial water layers. A structural approach to describing ionic hydration is to provide information on the way water molecules are arranged around the ion and to distinguish between kosmotropes and chaotropes [1]. If the charge to radius ratio of the ion is sufficiently high (e.g., fluoride), the ion force field may be strong enough to break the hydrogen bonded structure of water and to order solvent molecules through its charge: these ions are known as structure making or kosmotropes. Usually, the energy of ion–dipole attractive interactions overbalances the losses of lattice and water structure energies, and a net hydrophilic hydration energy results. These ions are classical salting-out agents. Large hydrophobic ions (classical IPRs, e.g., tetrabuthylammonium) are often structure making due to a different reason, related to their hydrophobic hydration. The alkyl chains do not interact favorably with water molecules and force them to form strongly hydrogen bonded ice-like cages into the water structure beyond them; the lives of the ice-like clusters are increased because structure disruption by thermal agitation is prevented by their long chains, and the hydrogen bond network is actually fortified. Lipophilic kosmotropic ions and strongly hydrated kosmotropic ions have the opposite effect on the surface tension of water, since the former tends to accumulate at the interface and lower the surface tension, in stark contrast with the latter. Conversely, they both salt out neutral molecules, and this effect increases slightly with size [59]. Conversely, the behavior of large polarizable inorganic ions is often chaotropic; they are prone only to break the water structure since the ion–dipole interactions are not strong enough to order solvent molecules around the ion.

1.3.3 WEAKNESS OF THE SOLVOPHOBIC THEORY AND EARLY ATTEMPTS TO RATIONALIZE THE BEHAVIOR OF THE INORGANIC ADDITIVE IN RP-HPLC

The behavior of inorganic additive in salt-mediated separations was tentatively rationalized within the framework of the theory that, during the last decades, performed most efforts to capture the main physical effects responsible for the RP-HPLC retention: the solvophobic theory by Horvath and coworkers [60,61]. This theory was widely and successfully used to explain the mechanism of uncharged analyte RP-HPLC retention. The driving force for retention is the tendency of water to reduce the nonpolar surface areas of the solutes via association with the stationary phase ligands. This is tantamount to saying that water is what pushes solutes in the stationary phase rather than their mutual affinity. Retention was postulated to depend on the free energy of creation of a cavity in the mobile phase of such a size that it would be able to accommodate the solute. In this context it is important to remember that chromatographic retention is based on the transfer of the analyte from the mobile to the stationary phase. The solvophobic theory only underlines the role played by the cavity in the mobile phase and does not take into account its destruction on analyte retention on the stationary phase and its consequent modification. The retention behavior of analytes not strongly affected by polar interactions can be satisfactorily elucidated within the hermeneutics of the solvophobic theory; for them it is still the most rigorous and comprehensive RP-HPLC theoretical framework. Several theories based on various partition models put forward during the last two decades [62] proved not to be competitive for the quantitative interpretation of RP-HPLC data [61]. In spite of this the authors of the solvophobic theory had to realize that, for polar and ionized solutes, retention is affected considerably by polar interactions, and explanation of their retention behavior in the rubric of the solvophobic theory in its present state is out of reach; it would be facilitated by incorporating additional terms in the solvophobic framework that quantify electrostatic interactions of polar moieties [63].

Actually, although the solvophobic theory usually works well with uncharged analytes, the limits of the solvophobic theory regarding the rationalization of the RP-HPLC retention of ionogenic analytes in the presence of ionic additives in the eluent are due to a main crucial and false assumption: the adsorption of an ionized solute is solely determined by solvophobic interactions; that is, no ionic and hydrogen bonding interactions occur between the solute and the stationary phase [60]. Recent experimental proofs demonstrate that this conjecture is false. Sound experimental breaking results indicate that chaotropic inorganic ions accumulate at the apolar chromatographic interface [53–56], in stark disagreement with common solvophobic theory wisdom actually based on assumptions that parallel those of the Onsanger–Samaras theory [40]. This chromatographic outcome parallels mounting experimental and computational evidence for the presence of ions at other water interfaces.

The increase of the electrolyte solution surface tension with increasing electrolyte concentration was classically viewed as the result of a depletion of electrolyte ions at the solution surface, according to the Gibbs adsorption equation. The ion-free and not electrified stationary phase concept, described in Section 3.2, was adopted by the solvophobic theory, and this allows one to rationalize the two big failures of this theory when it was applied to chromatographic systems with inorganic ionic additives.

First, the solvophobic model was not able to quantitatively describe the retention of ionogenic analytes in the presence of simple inorganic ions as a function of the eluent ionic strength. The experimental proof of the theory emphasized a strong scatter about the theoretical curve, and the fitting was not even attempted [60]. Moreover, the final relationship put forth by the solvophobic theory to describe the influence of ionic strength (I) on analyte retention (k) in the absence of IPRs [60]

$$\ln k = \ln k° + \alpha(BI^{1/3} + CI) + \beta I$$

cannot take into account the experimental evidence of charged analyte exclusion [64]; a negative k can on the converse be easily explained by the electrostatic repulsion experienced by an analyte similarly charged to the electrified stationary phase due to the adsorption of the ionic modifier. Furthermore, the declared and deliberated carelessness for polar interactions with the stationary phase caused the solvophobic theory to fail many times even to explain the salting-out chromatography of neutral analytes [65–68].

Second, when the solvophobic approach tried to model the influence of the eluent ionic strength due to inorganic "inert salts" (not tensioactive as IPRs) on retention under IPC condition—that is, in the presence of IPRs—the IPC retention equation was modified via the introduction of additional stoichiometric constants to take the analyte–salt pairing constant and the analyte–salt adsorption constant into account. A quite empirical expression is introduced only for high salt and IPR concentration, and a quantitative test of this equation was never attempted [5].

The main problem of the solvophobic theory in the rationalization of analyte retention in the presence of inert salts or surface active IPRs or both is that it completely disregards the adsorption of the ionic additives (in the ionic form) onto the apolar stationary phase and the modification of its electrostatic potential and chemical nature as well as the decrease of the available surface area. In this context, it is important to observe that the demonstrated adsorption of a wide range of ionic additives onto the apolar stationary phase questions the foundations of the solvophobic theory, which is constructed on the assumption that the retention is a function of the eluent surface tension.

1.3.4 Transition State of IPC: From IPRs to Chaotropic Additives

Chaotropes are recently rehabilitated as useful additives in liquid chromatography of ionogenic analytes since they are very effective in enhancing oppositely charged analyte retention. Chromatographically speaking, it is astonishing to see that chaotropic ions may mimic the role played by lipophilic kosmotropic IPRs and eventually replace them, since they have opposite effects on the water structure and on surface tension. What common feature do they share that makes them similarly useful in enhancing the RP-HPLC of oppositely charged analytes? In the present view, this characteristic is that they have superior adsorbophilicity compared with that of hard, kosmotropic inorganic ions. Their surface affinity is due to their loose hydration and consequent lower hydration energy loss on adsorption and their high polarizability. Their adsorption results in an electrified interface that is able to provide increased retention for oppositely charged solutes. In this respect, it is spectacular to see that anions can be ranked according to their chromatographic effectiveness to provide adequate retention of positively charged analytes according to the classical Hofmeister scale

$$PO_4^{-3} < SO_4^{-2} < H_2PO_4^- < HCOO^- < CH_3SO_3^- < Cl^- < Br^- < NO_3^-$$

$$< CF_3COO^- < BF_4^- < ClO_4^- < PF_6^-$$

and in agreement with MD simulations estimates of the ionic adsorbophilicity. The rank of an ion in the Hofmeister series was actually demonstrated to be a measure of its tendency to accumulate at the stationary phase in RP-HPLC [56]. In this instance, chaotropicity mimics hydrophobicity; dispersive forces that are responsible for the ion polarizability are similar to $\pi-\pi$ interactions. It should also be considered that since ion pairing of organic analytes in the eluent usually involves hydrophobic and dispersive interactions it is clear that chaotropes are expected to be good pairing counterions, such as classical IPRs.

Nowadays the common practice of IPC points toward its transition state. Chaotropic salts achieved good and increasing recognition among chromatographers during the past decade as neoterics ionic eluent modifiers. If one takes into account that anions are more important than cations [42,43,52,57] to determine salt features, as illustrated previously, it is easy to understand why chaotropic chromatography is devoted to the chromatographic separations of positively charged analytes. The classical distinction between classical hydrophobic IPRs and "inert electrolytes" is labile because the latter may mimic the role played by the former depending on their chaotropicities. ILs, whose importance in many fields of chemistry is undisputed, have just appeared as possible chromatographic additives, but their full potential was not fully exploited, since they were typically used only to prevent adverse silanophilic interactions or to mimic the role of the organic modifier. A viable revival of the first simple ionic inorganic additives (usually kosmotropes) and the already acclaimed breakthrough chaotropes and ionic liquids may bridge salting chromatographic phenomena to IPC. In addition, surface inactive salts not only can be used in salting-out chromatography but can also be exploited to depress and fine-tune the analyte retention under IPC conditions: inorganic ions similarly charged to the analyte would compete with it in pairing equilibria, thereby reducing its retention.

The influence of the Hofmeister rank on ionic adsorption at a solid surface is surely related to the ion-induced change of the thermodynamics of the interfacial water layers; however, it is still not well understood, so understanding the Hofmeister effects is an urgent research need that challenges model makers to put forth a unified model of salt chromatography.

1.4 STOICHIOMETRIC APPROACHES TO CHAOTROPIC CHROMATOGRAPHY

Few attempts toward a general quantitative approach of the problem involved in salt-mediated chromatographic separations were recently made. These attempts are described below.

The large number of optimization variables (pH, nature of the ionic additive, organic modifier and ionic additive concentration, stationary phase packing) and the wide differences in the analyte nature and charge status (nonionogenic, ionizable, ionic, and zwitterionic compounds) make rationally selecting the best experimental conditions for adequate resolution in reasonable run time a nontrivial task for the chromatographer. Using salt additives in liquid chromatography calls for a unifying theoretical modeling to translate IPC into a mature and general technique.

A model is the description and interpretation of a phenomenon. Theoretical models, at odds with empirical models, follow from known theoretical principles; they convert raw data to meaning and predict new results. If sound outcomes of experiments are at variance with theoretical predictions, the theory needs to be improved. The theoretical weakness of the time-honored solvophobic theory is evident in the presence of ionic additives. Salt effects question the rigidity of previous retention schemes and have to be treated within a unifying theoretical framework. Unfortunately, the theoretical efforts to model chaotropic chromatography are fraught with the same flaws that were typical of pioneering stoichiometric approaches to IPC. The dominance of the ion pairing in solution was claimed by many authors, in agreement with a common wisdom that rules out the presence of ions at a water dielectric boundary, especially because chaotropic ions are much less hydrophobic than classical lipophilic IPRs. Ion-pairing reactions were supposed to require the exclusion of water molecules between the partner ions. Chaotropes are supposed to give strong pairing interactions since they are weakly hydrated, and increased interaction of the neutral complex with the hydrophobic stationary phase is regarded as the only cause of analyte retention increase in chaotropic chromatography. This view was shared by different research groups [69–75]. Kazakevich and coworkers [75], in particular, emphasize the importance of the degree of analyte solvation: the added chaotropic additives disrupt the solvation shell via ion pairing; this way the analyte hydrophobicity improves, and enhanced RP retention can be attained. The analyte solvation–desolvation equilibrium is expressed by an equilibrium constant that corresponds to the ion-pairing equilibrium constant K_{EH}, illustrated in Figure 1.1, between the analyte and the chaotrope (H; we use the same notation for the chaotrope and the IPR in Figure 1.1 since their role is similar), since the desolvated analyte exists only as an ion-associated complex with the chaotropic additive. θ represents the fraction of solvated unpaired analyte that increases if the ion-pairing constant is lower. The model equates the analyte retention factor to the sum of the retention factor of the free solvated analyte (k_s) multiplied by the solvated fraction and the retention factor of the unsolvated ion-paired analyte (k_{us}) multiplied by its fraction $(1 - \theta)$, that is,

$$k = k_s \theta + k_{us}(1 - \theta) \tag{1.5}$$

This approach is regrettably stoichiometric since it takes for granted that the value of the equilibrium constants remain the same upon addition of the ionic additive in the mobile phase. Convincing experimental proofs [53–56] definitively demonstrate the adsorbophilicity of chaotropic ions onto the reversed-phase packing. It is then clear that k_s is not constant as it was presumed, but it is surface-potential modulated. With increasing chaotrope concentration in the mobile phase its surface excess increases, according to its adsorption isotherm, and the surface potential is stronger. It turns out that the analyte is more strongly electrostatically attracted toward the stationary phase than what the model prescribes. In the following, we will provide the thermodynamic relationship that parallels Equation (1.5). The neglect of the chaotrope adsorption leads to the following drawbacks.

Since it is clear that

$$\theta = (K_{EH}[H] + 1)^{-1} \tag{1.6}$$

analyte retention is predicted to increase with increasing [H] and show an asymptote at $k = k_{us}$. The typical plot [75] can be observed in Figure 1.4. The presence of asymptote implies that the model will never be able to predict the retention maxima (which are sometimes observed in IPC). Moreover, if $\theta = 1$ and ion-pairing interactions are not conceivable—as in the case of analytes in the same charge status of H—the model cannot predict the experimental retention decrease with increasing [H] [53–54]. Conversely, Equation (1.5) erroneously predicts their retention to be constant with $k = k_s$, which represents analyte retention in the absence of the chaotrope. The model makers fitted Equation (1.5) to experimental results, but k_s, which could better have been obtained from experimental results, was considered a fitting parameter, thereby adding a degree of freedom to the fitting procedure. In addition, its estimates were not compared with its physical values, which were not reported [75]. Furthermore, the estimates of the ion-pairing constants were at variance with the chaotropic scale [76]; in this context it is important to emphasize that enough

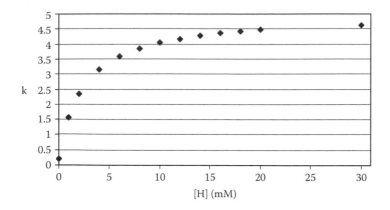

FIGURE 1.4 Retention pattern of an analyte oppositely charged to the chaotropic ion according to Equation (1.5), typical of a stoichiometric model [75] with $K_{EH} = 0.4$ mM^{-1}, $k_s = 0.2$, $k_{us} = 5$.

processing of large quantities of data with a sufficient number of fitting parameters may lead to any deduction [77], hence the comments on the estimates of the equation parameters and their physico-chemical interpretation are crucial to determine the adequacy of any theoretical model. Unfortunately, that is where the retention modeling was left.

Carr and coworkers revisited the stoichiometric model to extend its predictions to chaotropic chromatography [22]. The presence of ionized silanols led the authors to presume that the negative charges on the stationary phase prevent the anionic additive absorption. The developed stoichiometric retention equation (Equation (1) of [22]) is not actually new since it parallels the one developed by Horvath [5] within the framework of the solvophobic theory, so the same criticism as that illustrated previously applies to Carr's model as well. The authors could have tested this hypothesis via the measurement of the chaotrope adsorption via the breakthrough method; during the discussion, they actually endorsed the performance of such an experiment. Interestingly, capillary electrophoresis (CE) was used to obtain nonchromatographic estimates of ion-pairing constants. The authors had to admit that the errors are bigger and the fit is poorer at lower pH since the suppression of ionic ionization of silanols facilitates the additive adsorption onto the chromatographic packing. This effect was taken into account by another stoichiometric equation (Equation (5) of [22]), but it was not tested. They unfortunately did not fit that equation to experimental results. Similarly, another equation that accounted for silanophilic interactions was developed, but, again, it was not used for retention results-fitting.

1.5   DEVELOPMENT OF THE FIRST THERMODYNAMIC THEORETICAL MODEL OF SALT CHROMATOGRAPHY

Only one recent study goes beyond the stoichiometric approach for retention modeling in the presence of chaotropic additives. The classical IPC thermodynamic model developed by Cecchi and coworkers proved to be able to comprehend the laws sanctioned by previous theories of IPC that are only limiting cases of the new theory and to be evidence based since it explains experimental behaviors that cannot be explained by other models. In addition, this model gives estimates of chemical parameters such as pairing constants that were validated via extrachromatographic means. The theory was recently suited to chaotropic chromatography to take into account its peculiarities. This extended thermodynamic approach to IPC was developed for classical IPRs that are large, adsorbophilic organic ions with inorganic counterions. Cecchi and coworkers cultivated the theory [53,54] they developed for IPC to make it able to deal with the retention behavior of all kinds of analytes in the presence of neoteric mobile phase ionic additives such as chaotropes. The model captures and rationalizes the main physical salting effects. The retention mechanism is constructed on experimental evidence that understands salt chromatography as a range of chromatographic modes: chaotropic chromatography, salting-out chromatography, and ion-pair chromatography are limiting examples in the multifaceted multiplicity of phenomena involved when a mobile phase containing an electrolyte is used in RP-HPLC.

To develop a comprehensive theoretical description of salt effects in liquid chromatography, the steps outlined in the subsequent sections should be followed.

1.5.1 MEASUREMENT OF THE ADSORPTION OF THE IONIC ADDITIVE AND ITS DESCRIPTION VIA THE ADSORPTION ISOTHERM

The adsorption of a number of different ionic mobile phase additives that ranges from classical hydrophobic IPRs to chaotropes and ionic liquids can be studied in a number of different stationary phases (classical silica-based hydrocarbonaceous reversed-phase material, polymeric reversed-phase material, porous graphitized carbon [PGC] stationary phases) via the measurement of the amount of additive adsorbed onto the stationary phase. Each specific ionic additive adsorption can be experimentally quantified via the breakthrough method of frontal chromatography at increasing eluent concentrations of the additive [12].

The adsorption of the ionic additive is described via its adsorption isotherm (i.e., the relationship that algebraically relates the amount of adsorbed additive to its concentration in the mobile phase) to take into account the modification of the stationary phase in salt chromatography because of the development of a surface charge density as a consequence of the adsorption of the ionic additive. Since the chaotropic ion is more adsorbophilic than its counterion, the Langmuir isotherm does not hold because the electrostatic surface potential that develops is of the same sign as the chaotropic ion and runs counter to its further adsorption. Surface coverage is less than that predicted in the absence of the surface potential. A potential modified Langmuir adsorption isotherm can be easily obtained from the thermodynamic equilibrium constant (K_{LH}) for adsorption of the ionic additive onto the stationary phase-free ligand site (L) [78]

$$[LH] = \frac{K_{LH}[L]_T \exp(-z_E F\Psi^\circ/RT)}{\left(1 + K_{LH}[H]\exp(-z_H F\Psi^\circ/RT)\right)} \tag{1.7}$$

where [LH] is the surface concentration of H—that is, the specific surface excess (μmol/m^2) of the additive—[H] is its eluent concentration, K_{LH} is the thermodynamic equilibrium constant for its adsorption onto the stationary phase-free ligand site (L), and $[L]_T$ is the total ligand sites concentration, or the monolayer capacity (μmol/m^2) of the column. This equation is usually linearized under the assumption that as [H] approaches zero the denominator approaches one. Actually, this approximation is seldom chromatographically meaningful [14]. If its linearization is acceptable, and if the surface potential is high [17,78,79] it reduces to a Freundlich adsorption isotherm

$$[LH] = a\,[H]^b \tag{1.8}$$

If these conditions are not fulfilled, its use is semiempirical, and a and b are two constants that can be experimentally evaluated. The worth of the expression is due to its simplicity end effectiveness in modeling the adsorption of lipophilic, classical IPRs, and chaotropic ionic additives [1,12,15–21,34,53,54].

1.5.2   Calculation of the Surface Potential

The development of an electrostatic potential difference (Ψ°) between the surface and the bulk solution is quantified as a function of the stationary phase charge density via the Gouy–Chapman equation; the estimate of the charge density has to account for the different adsorbophilicities of the anion and cation of the additive. If the surface potential determining (the most adsorbophilic) ion is H, its electrical charge is z_H, and its counterion is negligibly adsorbophilic, the surface charge is $[LH]|z_H|F$; hence, the absolute value of the surface potential is given by [12].

$$\Psi^\circ = \frac{2RT}{F} \ln\left\{ \frac{[LH]\cdot|z_H|F}{\left(8\varepsilon_0\varepsilon_r RT \sum_i c_{0i}\right)^{\frac{1}{2}}} + \left[\frac{\left([LH]\cdot z_H F\right)^2}{8\varepsilon_0\varepsilon_r RT \sum_i c_{0i}} + 1 \right]^{\frac{1}{2}} \right\} \tag{1.9}$$

where ε_0 is the electrical permittivity of vacuum, ε_r is the dielectric constant of the mobile phase, R and F are the gas and Faraday constants, respectively, T is the absolute temperature, and Σc_i is the mobile phase concentration (mM) of singly charged electrolyte ion. Ψ° is positive or negative, according to the charge status of the adsorbed ion.

The supposition that the counterion of the IPR is not adsorbophilic is correct for classical tensioactive IPRs, such as sodium dodecanesulfonate or tetrabutylammonium chloride, since the adsorbophilicity of the small inorganic counterions (Na$^+$ and Cl$^-$, respectively) is negligible compared with that of the large organic potential determining ions (dodecanesulfonate and tetrabutylammonium, in that order). If the ionic modifier is NaClO$_4$ in chaotropic chromatography, the adsorbophilicity of the anion is stronger than that of the cation, and a negative sign of the surface potential is expected. However, the latter cannot be completely neglected [14], and this should be taken into account including the surface excess of Na$^+$ in the surface charge density calculation (see, e.g., Equation (A11) in [80]). Since the amount of Na$^+$ that is paired to the adsorbed ClO$_4^-$ is not easily measurable, we may expect that the surface excess of ClO$_4^-$ is a fraction of the total adsorbed NaClO$_4$ that can be determined via the adsorption isotherm.

1.5.3 ACCOUNTING FOR THE ION PAIRING INTERACTIONS IN THE MOBILE PHASE AND ELECTROSTATIC INTERACTIONS WITH THE STATIONARY PHASE

The model now needs to consider the experimental proof [22–35] of the effectiveness of ion pairing in the eluent. The multiplicity of phenomena the analytes undergoes in the chromatographic system, outlined in Figure 1.1, is described at a thermodynamic level via thermodynamic equilibrium constants. They can all be derived from the condition of equilibrium expressed as a function of the electrochemical potential (μ) of each species involved in the equilibrium to take into account the electrification of the surface on ionic additive adsorption. Since a thermodynamic equilibrium constant has the general expression

$$K = \exp(-\Delta\mu°/RT) \qquad (1.10)$$

where $\Delta\mu°$ represents the standard electrochemical potential difference for the considered equilibrium, the following equations are easily obtained for each thermodynamic equilibrium constant (a represents the activity of each species) [12]:

$$K_{LE} = \frac{a_{LE}}{a_L \cdot a_E} \exp\left(\frac{z_E F \Psi°}{RT}\right) \qquad (1.11)$$

$$K_{LH} = \frac{a_{LE}}{a_L \cdot a_H} \exp\left(\frac{z_H F \Psi°}{RT}\right) \qquad (1.12)$$

$$K_{EH} = \frac{a_{EH}}{a_E \cdot a_H} \qquad (1.13)$$

$$K_{LEH} = \frac{a_{LEH}}{a_L \cdot a_E \cdot a_H} \qquad (1.14)$$

Interestingly, stoichiometric constants that parallel the thermodynamic Equations (1.11) and (1.12) and that are typical of stoichiometric models lack the exponential term, since they disregard the experimental proof of the development of a stationary phase electrostatic potential. This way they are not constant upon addition of the ionic additive. Conversely, the exponential term in Equations (1.11) and (1.12) describes the fact that the electrostatic potential favors (runs counter to) the adsorption of the ionic species of opposite (similar) sign.

Introducing the expressions for the equilibrium constant as a function of the additive concentration into the analyte retention factor (k) that recognizes the multiplicity of the analyte chemical species (see Figure 1.1) in the system

$$k = \phi \frac{[LE] + [LEH]}{[E] + [EH]} \tag{1.15}$$

leads to an expression that relates the latter to the mobile phase or stationary phase composition. If the activity coefficient ratio is considered constant, the final algebraic expression (the retention equation), ready to be validated, is

$$k = \Phi \, [L] \cdot (K_{EL} \exp(-z_E F \Psi^\circ / RT) \, \theta + (K_{EHL}/K_{EH})(1 - \theta)) \tag{1.16}$$

where θ is given by Equation (1.6), and [L] represents the amount of free sites on the stationary phase ($\mu mol/m^2$).

1.5.4   COMPARISON OF CECCHI'S RETENTION EQUATION WITH STOICHIOMETRIC RETENTION EQUATIONS FOR CHAOTROPIC CHROMATOGRAPHY

Equation (1.16) matches the stoichiometric Equation (1.5) put forth by Kazakevich and coworkers; however, at odds with that expression, it is able to predict (a) the retention decrease of analytes similarly charged to the chaotropic reagent (in this case z_E and Ψ° are of the same sign and the analyte is electrostatically repelled by the electrified surface) and (b) retention maxima due to adsorption competition for available ligand sites.

Equation (1.16) is algebraically equivalent to the retention equation of the thermodynamic model of classical IPC by Cecchi and coworkers [12,53,54] on the basis of Equation (1.6):

$$k = \phi[L]_T \frac{K_{LE} \dfrac{\gamma_L \gamma_E}{\gamma_{LE}} epx\left(-z_E F \Psi^\circ / RT\right) + K_{LEH} \dfrac{\gamma_L \gamma_E \gamma_H}{\gamma_{LEH}} [H]}{\left(1 + K_{EH} \dfrac{\gamma_E \gamma_H}{\gamma_{EH}} [H]\right)\left(1 + K_{LH} \dfrac{\gamma_L \gamma_H}{\gamma_{LH}} [H] epx\left(-z_H F \Psi^\circ / RT\right)\right)} \tag{1.17}$$

where γ represents the activity coefficient for each species.

Equation (1.17) describes (a) the electrostatic interaction of an analyte with the surface potential (the first term in the numerator), (b) ion-pair formation at the stationary phase (second term in the numerator), (c) ion-pair formation in the mobile phase (left factor of the denominator), and (d) the competition between the analyte and the IPR for the available ligand sites that decline with increasing IPR concentration (right-hand factor of the denominator). Equation (1.17) matches the stoichiometric equation of the retention model described by Carr and coworkers [22 Eq.

5] unfortunately, they did not test it except in Equation (1.17) there is a potential modulation of the terms that are related to the transfer of a charged species onto the electrified stationary phase surface. If the activity coefficient ratios in Equation 1.17 are constant it is equivalent ot Equation 1.16.

1.5.5 INTRODUCTION OF THE POTENTIAL EXPRESSION IN THE RETENTION EQUATION TO OBTAIN THE RELATIONSHIP TO BE TESTED

When the G–C expression for the potential and the Freundlich equation for the adsorption isotherm of the ionic additive are introduced in Equation (1.17), the following practical retention equation is obtained [53,54]:

$$k = \frac{c_1 \left\{ a[\mathrm{H}]^b f + \left[\left(a[\mathrm{H}]^b f \right)^2 + 1 \right]^{\frac{1}{2}} \right\}^{\pm 2|z_E|} + c_2[\mathrm{H}]}{\left(1 + c_3[\mathrm{H}]\right)\left\{ 1 + c_4[\mathrm{H}]\left\{ a[\mathrm{H}]^b f + \left[\left(a[\mathrm{H}]^b f \right)^2 + 1 \right]^{\frac{1}{2}} \right\}^{-2} \right\}} \tag{1.18}$$

where

$$f = \frac{|z_\mathrm{H}|\mathrm{F}}{8\varepsilon_0\varepsilon_r \mathrm{RT} \sum_i c_{0i}} \tag{1.19}$$

f (m²/mol) can be evaluated from the eluent composition and operative temperature.

In the exponent of the first term of the numerator, the plus (minus) sign applies for oppositely (similarly) charged analytes and IPR, as expected on the basis of the electrostatic behavior. When Equation (1.18) is fitted to experimental data in classical IPC systems, excellent results are obtained [12,15–21,34,53,54].

The retention of a wide gamut of properly designed test solutes (nonionic, ionizable, ionic, and zwitterionic solutes) can now be studied as a function of both the mobile phase and stationary phase concentration of the ionic additive; each analyte retention factor at a given mobile phase composition is a data point. Data points are modeled according to the developed retention equation via the fitting of experimental results.

Since c_1 represents k_0 (that is, analyte retention without the ionic additive in the mobile phase), it can be experimentally obtained. Thus, the adjustable variables are only $c_2 - c_4$, and since they have a clear physical meaning [12,53,54] their estimates can be commented upon. The right factor of the denominator can often be considered not very different from one, especially if the ionic surface potential determining additive is not very strongly adsorbophilic like chaotropic reagents. Hence, c_4 will not be included in the fitting; c_2 and c_3 are related to ion-pair equilibrium constants. The values of these parameters, already estimated by the model from the fitting of retention data in various chromatographic systems, have been very reasonable and agree with chromatographic and nonchromatographic estimates [1,12,15–21,34,53,54]. This strongly endorses the model.

The progress from theoretical issues to application enables the chromatographer to perform educated guesses. An a priori retention prediction on changing the value of an eluent parameter is possible, and the commonly used trial-and-error procedures (i.e., one-at-a-time changing of parameters, without regard to parameter interactions) may be avoided. Theory-driven optimization finds the unique combination of values of the adjustable parameters that gives the best performance possible for a set of solutes.

A slight modification of the retention equation enables the model to predict capacity factor of zwitterionic solutes [15,16,53,54]; their electrical dipole is electrically pushed toward the interphase, where the field is stronger. The electrical dipole of the zwitterion can be easily considered to be equivalent to a fractional electrical charge whose magnitude can be estimated by the model.

As explained already, the majority of experimental studies concerning chaotropes are in regards to oppositely charged analytes, since the practical goal of chaotropic chromatography is to increase ionic analyte retention; although the study of the retention of analytes in the same charge status as the chaotrope is a major theoretical target, it was very seldom experimentally studied [81]. The Cecchi's model tried to address this issue and also focused on a model analyte similarly charged to the chaotropic reagent.

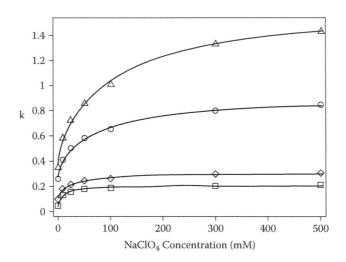

FIGURE 1.5 Dependence of the retention of positively charged analytes on the eluent concentration of $NaClO_4$, fitted via Equation (1.18). Squares: octopamine. Diamonds: 3-hydroxytyramine hydrochloride. Circles: atenolol. Triangles: N,N′-dimethylbenzylamine. (Reprinted from Cecchi, T. and Passamonti, P., *Journal of Chromatography A,* Vol. 1216, 1789–1797, 2009. With permission.)

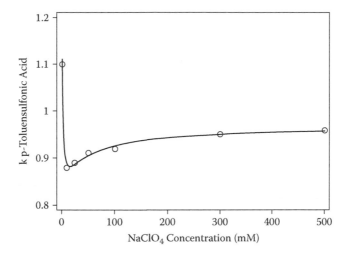

FIGURE 1.6 Dependence of the retention of p-toluensulfonate on the eleuent concentration of NaClO$_4$, fitted via Equation (1.18). (Reprinted from Cecchi, T. and Passamonti, P., *Journal of Chromatography A*, Vol. 1216, 1789–1797, 2009. With permission.)

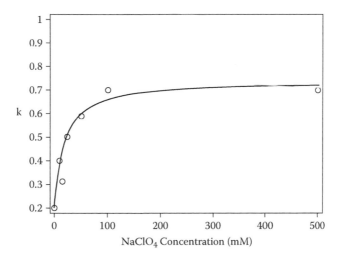

FIGURE 1.7 Dependence of the retention of 1-butyl-3-methylimidazolium cation on the eluent concentration of NaClO$_4$, fitted via Equation (1.18). (Reprinted from Cecchi, T. and Passamonti, P., *Journal of Chromatography A*, Vol. 1216, 1789–1797, 2009. With permission.)

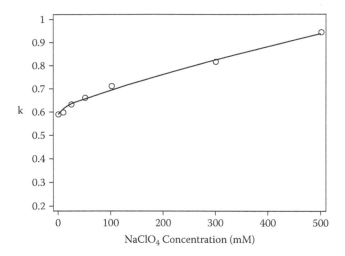

FIGURE 1.8 Dependence of the retention of 3-(4-tert-butyl-1-pyridino)-1-propanesulfonate on the eluent concentration of NaClO₄, fitted via Equation (1.18). (Reprinted from Cecchi, T. and Passamonti, P., *Journal of Chromatography A,* Vol. 1216, 1789–1797, 2009. With permission.)

Figures 1.5 through 1.8 illustrate that the model benchmarks well against the influence of a classical chaotropic ion, such as perchlorate, on the retention of positively and negatively charged analytes as well as a zwitterion and a model ionic liquid. Figure 1.6 definitely demonstrates the inability of the stoichiometric approach (Eq. 1.5) to model the retention of analytes similarly charged to the chaotrope, since in this case constant retention would be predicted.

1.6 PRACTICAL IMPACT OF CHAOTROPIC CHROMATOGRAPHY

While traditional IPRs tend to stick so strongly to the stationary phase that its pristine properties may be irreversibly impaired when their presence in the eluent is discontinued, chaotropic additives are readily soluble in the mobile phase; hence, they can be easily removed from the column.

Chaotropic salts proved particularly useful in the RP-HPLC of positively charged basic compounds [69–71,74–76,81–87], and this is particularly interesting for the pharmaceutical industry since about 80% of drugs are characterized by basic functional groups. The retention activities of selected alkaloids were studied using chaotropic additives whose behaviors agreed with their ranks in the Hofmeister series [71].

RP-HPLC chaotropic separations of basic samples are appealing not only because these additives are able to provide adequate retention for the otherwise scarcely retained analytes but also because they positively influence the selectivity, efficiency, and repeatability [71,76,88], probably because silanophilic interactions are shielded. Interestingly, weak chaotropes such as acetate [84,89–91] and phosphate, which are usually regarded as a kosmotropic ion [69,82,92], also proved to be useful, thereby demonstrating how fine-tuned features may be provided by an ionic additive. Peptides separation currently relies on the use of perfluorinated carboxylic acids [93] and small chaotropes [91,94].

The retention enhancement of an analyte usually increases with (a) increasing chaotropic concentration, as described by Equation (1.18), (b) increasing its hydrophobicity [71,95], or (c) increasing the additive chaotropicity [96]; for example, perchlorate is better than dihydrogen phosphate for the separation of β-blockers [69] and better than trifluoroacetate for the retention of peptides [74]. These effects were related to the dominance of ion-pairing interactions in solution [71–73], but it is clear that they can be better rationalized within the framework of the extended thermodynamic approach [53–54]. Even if the large use of anionic chaotropes obscured the potential of polarizable cations, chaotropic sulfonium and phosphonium reagents showed single selectivity toward polarizable anions [70,97], we believe that the full exploitation of chaotropic cations is not out of reach.

Retention maxima, often observed at high concentrations of traditional IPRs, are usually missing in the retention plot of charged analytes as a function of the additive concentration, probably because a chaotropic ion has no critical micellar concentration. In addition, since its hydrophobicity is low, adsorption competitions for available ligand sites on the stationary phase are easily predicted to be negligible.

ILs have recently gained popularity in a variety of fields since they show an exclusive and spectacular broad range of physico-chemical properties. Many ILs show significant ecotoxicities even if they were considered environmentally friendly. The exponentially increasing use of ILs brought the focus of interest in their analytical control, and chaotropic IPRs such as $NaPF_6$ or $NaClO_4$ proved to be able to enhance their peak shapes and increase their retention, even if it was not possible to differentiate cations associated with different anions [98]. Their adjustable features are fascinating separation scientists, who are increasingly considering them not only as simple analytes but also as putative ionic additives with strong practical impact. If its concentration is lowered to the millimolar range, an IL is a mobile phase modifier, which is very useful to suppress deleterious effects of silanophilic interactions [99–104]. It was soon clear that they also affect analyte retention via their molecular architecture and concentration [100,103], similar to classical IPRs and chaotropic inorganic ions. In IL-mediated RP-HPLC separations, both the anion and its counterion contribute to solute retention and improvement of peak shape. Since both the IL anion and cation can adsorb on a hydrophobic stationary phase according to their own hydrophobicity and following a lyotropic series directly related to their alkyl chain lengths [88,105,106], the stationary phase will show an electrostatic potential difference whose sign depends on the IL partner that shows the strongest adsorbophilic attitude toward the RP material (potential determining ion). In this case the same reasoning that led to Equation (1.18) can be repeated here, thereby demonstrating how wide is the gamut of chromatographic salt-mediated situations that can be modeled by a single retention equation.

It was reported that the replacement of NaCl by NaBF4 increases retention and slightly improves peak shapes of cationic analytes. If NaCl is replaced by the butyl-3-methyl imidazolium (BMIM) chloride IL, a 30% decrease in retention factor of the positively charged analytes associated with a remarkable peak shape improvement is observed, due to charge–charge repulsion between a positively electrified stationary phase and the analytes; this in turn results in a better peak shape, since the analyte cation is largely retained by hydrophobic fast interactions. When BMIM BF4 IL replaces NaCl, both the cation and anion of the IL adsorb on the C18 surface, and the ionic interactions take place simultaneously so that global retention depends on the extent to which one interaction is stronger than the other [105]. The theory thereby elegantly explains seemingly contradictory results about the retention modulation via the use of an IL in the eluent: overall retention of the analyte was reported to decrease [107], increase [105,108], or remain almost constant [103]; the effect was demonstrated to depend on the specific IL concentration in the mobile phase [109]. The synergistic contributions of both cationic and anionic components of ILs let chromatographers predict their fast breakthrough in salt-mediated separations. Their main drawback is probably the nonvolatility of ILs that may create condensation and pollution in the ionization sources of LC-MS hyphenated techniques and their slight UV absorbance.

1.7 URGENT RESEARCH NEEDS AND SUGGESTIONS

The state of the art detailed in the present review is a motivating force suggesting to separation scientists the directions to arrive at a physicochemical framework that is able to unify eclectic experimental evidence concerning the use of a wide gamut of ionic additives in RP-HPLC.

First, molecular dynamics simulation of the chromatographic interphase and chromatographic measurement of the ionic adsorption could be jointly used to clarify the currently hotly debated behavior of ions at water interfaces. Chromatography can be recommended as a technique able to cross-validate the computationally studied accumulation of ions at the water–apolar medium interface.

Second, the predictive abilities of theoretical modeling and artificial intelligence should be thoroughly compared, since few studies concern the potential of the latter [110–112]. Artificial neural network (ANN) programs are able to learn by introducing new knowledge to a computer and developing a representation of information (expert system). ANNs are an artificial intelligence tool (along with genetic algorithms and induction) and may be used as a "soft" retention modeling because they do not follow from theoretical principles translated to formulas as "hard" models do. ANNs can be used to investigate the effects of several input factors (components of the eluent) on the chromatographic response surface of analyte retention in the design space. The comparison between the theory-driven and ANN optimization should indicate the eligibility of "soft" and "hard" modeling as a function of the number of data.

Third, the importance of the dielectric constant of the medium is beyond dispute in salt chromatography and has been long overlooked. Since the dielectric constant of aqueous electrolyte solutions is usually lower than that of water, one would predict electrolytic solutions to mimic the behavior of organic solvents of lower dielectric constant or superheated water. Conversely, organic neutral analytes are usually salted out by electrolyte solutions, and their solubility is poorer than that in pure water at the same temperature. The salting-out properties of electrolyte solutions seem to be at variance with the salt dielectric decrement. It has to be taken into account that electrolyte solutions are not molecular liquids and that the presence of electrolytes alters the water structure and reduces the number of water molecules available to react to the electric field. The relationship between the Hofmeister rank of a salt (i.e., related to its salting characteristic and particularly to its chromatographic effectiveness as ionic additive) and its specific dielectric decrement should be investigated and modeled via both chromatography and dielectric relaxation spectroscopy (DRS). This technique, however, shows great potential in elucidating the pairing equilibria in solution. It can be recommended as a promising tool to complement chromatographic retention data modeling with an extrachromatographic validation since, although the theoretical modeling of retention gives chromatographic estimates of the pairing equilibrium constants, DRS may offer a cross-validation of the results in an interdisciplinary way. The mobile phase has to be stimulated by an electromagnetic field applied over the microwave region, and the dielectric response of the sample has to be measured. The complex dielectric response is frequency sensitive and dependent on the square of the dipole moment of each species in solution. Both solvent molecules and dipolar ion pairs contribute to the signal. Additional contributions arise from the polarizability of the species and from the ions. These components are more important at higher and lower frequencies, respectively. Since under the effect of the applied field there is ionic migration, the ionic conductivity term must be deconvolved from the dipolar response at lower frequencies. What sets DRS apart from other methods [35,113,114] to detect ion-pairing equilibria is its great sensitivity to very weakly associated ion pairs and its ability to distinguish between contact and solvent-separated ion pairs. For the latter, the dipole is higher, and the DRS sensitivity is better. DRS will shed light on the evidence of ion pairing in just the aqueous electrolyte solutions used as chromatographic eluent. This issue is hotly debated. One merit of the retention model by Cecchi and coworkers was to emphasize the peculiarities of the hydrophobic ion-pairing concept [1,114] and to demonstrate that the widely accepted common wisdom provided by Bjerrum's theory is at variance with experimental evidence concerning the influence of the medium dielectric constant and the possibility for univalent ions to pair in water solutions [18]. Again, it is instructive to observe that Bjerrum's theory, erroneously, did not allow for interactions other than electrostatic. It is interesting to extend the modeling of ion-pairing constants to electrolytes of chromatographic interest but different from the classical hydrophobic IPRs. This is particularly important since current views in the literature range from a position that there is no ion association whatsoever in most aqueous electrolyte solutions to the claim that even for NaCl(aq) ions are "inseparable."

Fourth, since the main hindrance to gradient IPC once was the slow desorption of classical IPRs from the chromatographic bed with increasing organic modifier percentage in the eluent, the use of chaotropic additives that are not as adsorbophilic as the earlier tensioactive IPRs should pave the way to a customary use of the gradient organic modifier elution and possibly to the chaotropic additive concentration gradient. It is clear that these operative procedures need to be properly modeled from a theoretical point of view.

Furthermore, the description of the influence of the analyte nature of retention has to be thoroughly considered. For IPC system it was only attempted [115] the authors realize that the charge may be too simple as a solute descriptor for ion-pair effect because it does not contain the hydrophobic effect, they also neglected ion pairing in the eluent at variance with large experimental evidence. They did not devise better descriptors. Quantitative structure retention relationships (QSRRs) should be used to identify the properties of analytes that control their retention. The retention factors of a set of model solutes should be measured at certain experimental conditions, and multiple regression analysis should be used to correlate k with several solute molecular descriptors such as physicochemical, topological, geometrical, and electronic factors.

Ionic additives can also be exploited to discriminatingly tune the retention and hence the resolution among difficult-to-separate neutral polar analytes, such as isomers, via the interaction of their dipole with the electrified stationary phase [15,16] or via adsorption competition between the solute and the ionic additive [19]. This new idea stems from the observation of a small retention decrease with increasing ionic additive concentration in the eluent for small neutral molecules; this repeatable experimental evidence was at variance with the textbook knowledge about classical IPC but was properly rationalized within the framework of the extended thermodynamic model of IPC [12]. It is now time to exploit this effect from a practical point of view and to compare it with the salting-in and -out effect strictly related to the Hofmeister series. Simple and accurate prediction of salting coefficients (Setchenov constants) from chromatographic retention data should spur further theoretical development since they were modeled by various theoretical treatments [1] unfortunately their predictions are only fairly accurate. In the treatment by Debye and McAulay [116], salting effects are related to the change of the relative permittivity of the medium when electrolytes and nonelectrolytes are added to a pure solvent. Salting-in would be obtained for neutral solutes more polar than water. Long and McDevit's thermodynamic "electrostriction" theory [117] tries to calculate it from the excess work necessary to put an element of volume of the neutral solute into the electrolyte solution instead of into water; the compression of the solvent in the presence of ions (solvent electrostriction) results in a higher internal pressure that reduces the available space for the nonelectrolyte, pushing it out of the liquid phase; salting-it would be obtained for electrolytes that give a negative electrostriction. At variance with electrostatic theories, treatments based on scaled particle theory do not consider electrostatic interactions and structural modification since the ionic solution is regarded as another kind of solvent and the Setchenov coefficient is related to the free-energy change for the formation of a cavity large enough to hold the nonelectrolyte and for the nonelectrolyte introduction into the cavity. The model by Masterton and Lee [118] is suitable only for nonpolar nonelectrolytes and must be customized for polar nonelectrolytes, but in this case the complexity of the representation increases considerably. Statistical mechanical treatments fail to address the dispersion and polarization interactions that are particularly important when hydrophobic reagents such as IPRs and chaotropes are dealt with. The quantitative treatment of these interactions was introduced by Bockris, Bowler-Reed, and Kitchener [119]; their work is important for explaining anomalous salting-in when the simple electrostatic theory would predict salting-out. By this scenario, it is clear that chromatographic information on salting effects are particularly interesting and welcome also because the lower surface affinity of the cation compared with that of the chaotropic anion parallels its lesser effect on the Setchenov constant, thereby showing a thriving research field. The possibility to arrange a gamut of chaotropicity for a number of different ionic additives from retention data of model analytes is also promising. Actually, even if there are many indicators of chaotropicity (e.g., the viscosity B coefficient of the Dole–Jones equation that describes change in viscosity upon salt addition, or the hydration number that can be obtained via DRS), there are discrepancies among literature results, probably because the rank of a single ion is erroneously deduced from the rank of the salt to which it belongs. Judiciously chosen additives with common anions or common cations can help overcome this question via the deconvolution of the anion and cation effects. As detailed in Seciton 1.3.2., MD simulations of the RP-eluent interface in the presence of ionic additives, represent a thriving research field.

Finally, on the basis of breaking preliminary experimental evidence, the stronger adsorbophilicity of the anion compared with that of the cation on an RP packing can be used to obtain a nonelectroneutral solution. Feeding a generic $A^+ B^-$ salt solution in frontal chromatography mode, if B^- tends to adsorb onto the stationary phase, the column effluent is B^- free and positively charged as long as the column stationary phase is saturated. This will open a prophetic scenario for chemistry.

1.8 CONCLUSIONS

The development of salt chromatography should reconcile ion-pair effects and salt effects that were contrasted for decades as opposite phenomena in RP-HPLC. It can become an intriguing separation mode of HPLC, with a strong practical impact on the separation of complex mixtures of polar, ionic, and ionogenic species.

REFERENCES

1. Cecchi, T. 2009. *Ion-pair chromatography and related techniques.* Boca Raton: Taylor and Francis Group.
2. Cecchi, T. Ion pairing. *Crit. Rev. Anal. Chem.* 2008, 38, 1–53.
3. Knox, J.H. and Laird, G.R. Soap chromatography: a new high performance chromatographic technique for separation of ionizable materials: dyestuff intermediates. *J. Chromatogr.* 1976, 122, 17–34.
4. Kraak, J.C., Jonker, K.M., and Huber, J.F.K. Solvent-generated ion exchange systems with anionic surfactant for rapid separation of amino acids. *J. Chromatogr.* 1977, 142, 671–688.
5. Horvath, C., Melander, W., Molnar, I., and Molnar, P. Enhancement of retention by ion-pair formation in liquid chromatography with nonpolar stationary phases. *Anal. Chem.* 1977, 49, 2295–2305.
6. Bidlingmeyer, B.A., Deming, S.N., Price, W.P. Jr., Sachok, B., and Petrusek, M. Retention mechanism for reversed-phase ion-pair liquid chromatography. *J. Chromatogr.* 1979, 186, 419–434.
7. Ghaemi, Y. and Wall, R.A. Hydrophobic chromatography with dynamically coated stationary phases. *J. Chromatogr.* 1979, 174, 51–59.
8. Melander, W.R. and Horvath, C. Mechanistic study of ion pair reversed-phase chromatography. *J. Chromatogr.* 1980, 201, 211–224.

9. Fornstedt, T. Peak distortion effects of suramin due to large system peaks in bioanalysis using ion-pair adsorption chromatography. *J. Chromatogr. B.* 1993, 612, 137–144.

10. Mulholland, M., Haddad, P.R., and Hibbert, D.B. Expert systems for ion chromatographic methods using dynamically coated ion-interaction separation *J. Chromatogr.* 1992, 602, 9–14.

11. Knox, J.H. and Hartwick, R.A. Mechanism of ion pair liquid chromatography of amines, neutrals, zwitterions and acids using anionic hetaerons. *J. Chromatogr.* 1981, 204, 3–21.

12. Cecchi, T., Pucciarelli, F., and Passamonti, P. Extended thermodynamic approach to ion interaction chromatography. *Anal. Chem.* 2001, 73, 2632–2639.

13. Cantwell, F.F.P. 1979. Mechanism of chromatographic retention of organic ions on a noionic adsorbent. *Anal. Chem.* 51, 623–632.

14. Bartha, A. and Ståhlberg, J. Electrostatic retention model of reversed-phase ion-pair chromatography. *J. Chromatogr. A.* 1994, 668, 255–284.

15. Cecchi, T. and Cecchi, P. The dipole approach to ion interaction chromatography of zwitterions: use of a potential approximation to obtain a simplified retention equation. *Chromatographia* 2002, 55, 279–282.

16. Cecchi, T., Pucciarelli, F., and Passamonti, P. The dipole approach to ion interaction chromatography of zwitterions. *Chromatographia* 2001, 54, 38–44.

17. Cecchi, T. Extended thermodynamic approach to ion-interaction chromatography: a thorough comparison with the electrostatic approach and further quantitative validation. *J. Chromatogr. A.* 2002, 958, 51–58.

18. Cecchi, T., Pucciarelli, F., and Passamonti, P. Extended thermodynamic approach to ion-interaction chromatography: influence of organic modifier concentration. *Chromatographia* 2003, 58, 411–419.

19. Cecchi, T., Pucciarelli, F., and Passamonti, P. Ion interaction chromatography of neutral molecules. *Chromatographia* 2000, 53, 27–34.

20. Cecchi, T., Pucciarelli, F., and Passamonti, P. Extended thermodynamic approach to ion interaction chromatography: a mono- and bivariate strategy to model the influence of ionic strength. *J. Sep. Sci.* 2004, 27, 1323–1332.

21. Cecchi, T., Pucciarelli, F., and Passamonti, P. Ion interaction chromatography of zwitterions: fractional charge approach to model the influence of the mobile phase concentration of the ion-interaction reagent. *Analyst* 2004, 129, 1037–1042.

22. Dai, J. and Carr, P.W. Role of ion pairing in anionic additive effects on the separation of cationic drugs in reversed-phase liquid chromatography. *J. Chromatogr. A.* 2005, 1072, 169–184.

23. Mbuna, J. et al. Capillary zone electrophoretic studies of ion association between inorganic anions and tetraalkylammonium ions in aqueous dioxane media. *J. Chromatogr. A.* 2005, 1069, 261–270.

24. Dai, J. et al. Effect of anionic additive type on ion pair formation constants of basic pharmaceuticals. *J. Chromatogr. A.* 2005, 1069, 225–234.

25. Steiner, S.A., Watson, D.M., and Fritz, J.S. Ion association with alkylammonium cations for separation of anions by capillary electrophoresis. *J. Chromatogr. A.* 2005, 1085, 170–175.

26. Motomizu, S. and Takayanagi, T. Electrophoretic mobility study of ion–ion interactions in an aqueous solution. *J. Chromatogr. A.* 1999, 853, 63–69.

27. Takayanagi, T., Wada, E., and Motomizu, S. Electrophoretic mobility study of ion association between aromatic anions and quaternary ammonium ions in aqueous solution. *Analyst* 1997, 122, 57–62.

28. Takayanagi, T., Wada, E., and Motomizu, S. Separation of divalent aromatic anions by capillary zone electrophoresis using multipoint ion association with divalent quaternary ammonium ions. *Analyst* 1997, 122, 1387–1392.

29. Takayanagi, T., Tanaka, H., and Motomizu, S. Ion association reaction between divalent anionic azo dyes and hydrophobic quaternary ammonium ions in aqueous solution. *Anal. Sci.* 1997, 13, 11–18.

30. Popa, T.V., Mant, C.T., and Hodges, R.S. Capillary electrophoresis of amphipathic *a*-helical peptide diastereomers. *Electrophoresis* 2004, 25, 94–107.

31. Popa, T.V., Mant, C.T., and Hodges, R.S. Ion interaction capillary zone electrophoresis of cationic proteomic peptide standards. *J. Chromatogr. A.* 2006, 1111 192–199.

32. Popa, T.V. et al. Capillary zone electrophoresis of *a*-helical diastereomeric peptide pairs with anionic ion pairing reagents. *J. Chromatogr. A.* 2004, 1043, 113–122.

33. Mbuna, J. et al. Evaluation of weak ion association between tetraalkylammonium ions and inorganic anions in aqueous solutions by capillary zone electrophoresis. *J. Chromatogr. A.* 2004, 1022, 191–200.

34. Cecchi, T. Influence of the chain length of the solute ion: a chromatographic method for the determination of ion pairing constants. *J. Sep. Sci.* 2005, 28, 549–554.

35. Barthel, J., Hetzenauer, H., and Buchner, R. Dielectric relaxation of aqueous electrolyte solutions II: ion pair relaxation of 1:2, 2:1, and 2:2 electrolytes, *Ber. Bunsen Ges. Phys. Chem.* 1992, 96, 1424–1432.

36. Shepard, C.C. and Tiselius, A. *Discussions of the Faraday Society,* Vol. 7 (275). London: Hazell Watson and Winey, p. 275.

37. Er-el, Z., Zaidenzaig, Y., Shaltiel, S., Er-El, Z., and Shaltiel, S. Hydrophobic chromatography in the resolution of the interconvertible forms of glycogen phosphorylase. *FEBS Lett.* 1974, 40, 142–145.

38. Hofstee, B. H. Protein binding by agarose carrying hydrophobic groups in conjunction with charges. *Biochem. Biophys. Res. Commun.* 1973, 50, 751–757.

39. Mahn, A., Lienqueo, M.E., and Salgado, J.C. Methods of calculating protein hydrophobicity and their application in developing correlations to predict hydrophobic interaction chromatography retention. *J. Chromatogr. A*, 2009, 1216, 1838–1844.

40. Onsager, L. and Samaras, N.N.T. The surface tension of Debye-Hückel electrolytes. *J. Phys. Chem.* 1934, 2, 528–36.

41. Hofmeister, F. Zur Lehre von der Wirkung der Salze. *Arch. Exp. Pathol. Pharmakol.* 1888, 24, 247–260.

42. Jungwirth, P. Ions at aqueous interfaces. *Faraday Discuss.*, 2009, 141, 9–30.

43. Jungwirth, P. and Tobias, D. Specific ion effects at the air/water interface. *J. Chem. Rev.* 2006, 106, 1259–1281.

44. Petersen, P.B. and Saykally, R.J. On the nature of ions at the liquid water surface. *Annu. Rev. Phys. Chem.* 2006, 57, 333–364.

45. Liu, D., Ma, G., Levering, L.M., and Allen, H.C. Vibrational spectroscopy of aqueous sodium halide solutions and air–liquid interfaces: Observation of increased interfacial depth. *J. Phys. Chem. B* 2004, 108, 2252–2260.

46. Ghosal, S., Hemminger, J.C., Bluhm, H., Mun, B.S., Hebenstreit, E.L.D., Ketteler, G., Ogletree, D.F., Requejo, F.G., and Salmeron, M. Electron spectroscopy of aqueous solution interfaces reveals surface enhancement of Halides. *Science* 2005, 307, 563–566.

47. Cheng, J., Vecitis, C.D., Hoffmann, M.R., and Colussi, A.J. Experimental anion affinities for the air/water interface *J. Phys. Chem. B* 2006, 110, 25598–25602.

48. Onorato, R.M., Otten, D.E., and Saykally, R.J. Adsorption of thiocyanate ions to the dodecanol/water interface characterized by UV second harmonic generation. *PNAS*, 2009, 106, 15176–15180.

49. Shafer, K.L., Pérez Masiá, A., Jüntgen, H. Z. Untersuchungen über Grenzflächen-effekte kapillar-inactive wässiger Lösungen *Z. Elektrochem.* 59, 425–434, 1955.

50. Lillie, R.S. The influence of electrolytes and of certain other conditions on the osmotic pressure of colloidal solutions. *Am. J. Physiol.*, 1907, 20, 127–169.

51. Petersen, P.B. and Saykally, R.J. Adsorption of ions to the surface of dilute electrolyte solutions: the jones-ray effect revisited. *J. Am. Chem. Soc.* 2005, 127 (44), 15446–15452.
52. Horinek, D., Serr, A., Bonthuis, D.J., Boström, M., Kunz, W., and Netz. R.R. Molecular hydrophobic attraction and ion-specific effects studied by molecular dynamics. *Langmuir* 2008, 24, 1271–1283.
53. Cecchi, T. and Passamonti, P. Retention mechanism for ion-pair chromatography with chaotropic additives. *J. Chromatogr. A.* 2009, 1216, 1789–1797.
54. Cecchi, T. and Passamonti, P. Erratum to "Retention mechanism for ion-pair chromatography with chaotropic reagents" [*J. Chromatogr. A* 1216, 1789–1797, 2009] *J. Chromatogr. A.* 2009, 1216(26), 5164.
55. Kazakevich, I.L. and Snow, N.H. Adsorption behavior of hexafluorophosphate on selected bonded phases. *J. Chromatogr. A.* 2006, 1119, 43–50.
56. Kazakevich, Y.V., LoBrutto, R., and Vivilecchia, R. Reversed-phase high-performance liquid chromatography behavior of chaotropic counter anions. *J. Chromatogr. A.* 2005, 1064, 9–18.
57. Pegram, L., Record, Jr., M., and Hofmeister, M.T. Salt effects on surface tension arise from partitioning of anions and cations between bulk water and the air-water interface. *J. Phys. Chem. B* 2007, 111, 5411–5417.
58. Conway, B.E. The solvation factor in specificity of ion adsorption at electrodes. *Electrochim. Acta*, 1995, 40, 1501–1512.
59. Treiner, C. and Chattopadhyay, A.K. The setchenov constant of benzene in non-aqueous electrolyte solutions. Alkali-metal halides and aliphatic and aromatic salts in methanol at 298.15 K. *J. Chem. Soc. Farad. T. 1* 1983, 79(12), 2915–2927.
60. Horvath, C.S., Melander, W., and Molnár, I. Liquid chromatography of ionogenic substances with nonpolar stationary phases. *Anal. Chem.* 1977, 49, 142–154.
61. Melander, W. and Horvath, C. Salt effects on hydrophobic interaction precipitation and chromatography of proteins: an interpretation of the lyotropic series. *Arch. Biochem. Biophy.*, 1977, 183, 200–215.
62. Dill, K.A. The mechanism of solute retention in reversed phase liquid chromatography. *J. Phys. Chem.* 1987, 91, 1980–1988.
63. Vailaya, A. and Horvath, C. Solvophobic theory and normalized free energies of nonpolar substances in reversed-phase chromatography. *J. Phys. Chem. B*, 1997, 101, 5875–5888.
64. Jandera, P., Churáček, J., and Bartosŏvá, J. Reversed-phase liquid chromatography of aromatic sufonic and carboxylic acids using inorganic electrolyte solutions as mobile phase. *Chromatographia*, 1980, 13, 485–492.
65. Gelsema, W.J., Brandts, P.M., De Ligny, C.L., Theeuwes, A.G.M., and Roozen, A.M.P. Hydrophobic interaction chromatography of aliphatic alcohols and carboxylic acids on octyl-sepharose CL-4B: mechanism and thermodynamics. *J. Chromatogr.* 1984, 295, 13–29.
66. Brandts, P.M., Gelsema, W.J., and De Ligny, C.L. Influence of additives to the eluent on the hydrophobic interaction chromatography of simple componds IV. Influence of electrolytes on the retention of normal alkanols at 25°C. *J. Chromatogr.* 1988, 437, 337–350.
67. Diogo, M.M., Queiroz, J.A., and Prazeres, D.M.F. Hydrophobic interaction chromatography of homo-oligonucleotides on derivatized Sepharose CL-6B Application of the solvophobic theory. *J. Chromatogr. A*, 2002, 944, 119–128.
68. Heron, S. and Tchapla, A. Description of retention mechanism by solvophobic theory. Influence of organic modifiers on the retention behaviour of homologous series in reversed-phase liquid chromatography. *J. Chromatogr.* 1991, 556, 219–234.

69. Hashem, H. and Jira, T. Effect of chaotropic mobile phase additives on retention behaviour of *b*-blockers on various reversed-phase high-performance liquid chromatography columns. *J. Chromatogr. A.* 2006, 1133, 69–75.

70. Harrison, C.R, Sader, J.A., and Lucy, C.A. Sulfonium and phosphonium, new ion-pairing agents with unique selectivity toward polarizable anions. *J. Chromatogr. A.* 2006, 1113, 123–129.

71. Flieger, J. Effect of chaotropic mobile phase additives on the separation of selected alkaloids in reversed-phase high-performance liquid chromatography. *J. Chromatogr. A* 2006, 1113, 37–44.

72. Gritti, F. and Guiochon, G. Effect of the ionic strength of salts on retention and overloading behaviour of ionizable compounds in reversed-phase liquid chromatography I: XTerra-C18. *J. Chromatogr. A.* 2004, 1033, 43–55.

73. Gritti, F. and Guiochon, G. Role of the buffer in retention and adsorption mechanism of ionic species in reversed-phase liquid chromatography I: analytical and overloaded band profiles on Kromasil-C18. *J. Chromatogr. A.* 2004, 1038, 53–66.

74. Shibue, M., Mant, C.T., and Hodges, R.S. The perchlorate anion is more effective than trifluoroacetate anion as an ion-pairing reagent for reversed-phase chromatography of peptides. *J. Chromatogr. A.* 2005, 1080, 49–57.

75. LoBrutto, R., Jones, A., and Kazakevich, Y.V. Effect of counter-anion concentration on retention in high performance liquid chromatography of protonated basic analytes. *J. Chromatogr. A.* 2001, 913, 189–196.

76. Jones, A., LoBrutto, R., and Kazakevich, Y.V. Effect of the counter-anion type and concentration on the liquid chromatography retention of β-blockers. *J. Chromatogr. A.* 2002, 964, 179–187.

77. Lavine, B. and Workman, J. Chemometrics. *Anal. Chem.* 2006, 78, 4137–4145.

78. Cecchi, T. Use of lipophilic ion adsorption isotherms to determine the surface area and the monolayer capacity of a chromatographic packing, as well as the thermodynamic equilibrium constant for its adsorption. *J. Chromatogr. A.* 2005, 1072, 201–206.

79. Davies, J.T. and Rideal, E.K. 1961. *Adsorption at liquid interfaces—Interfacial phenomena.* New York: Academic Press, Ch. 4. pp. 154–216.

80. Lu, P., Zou, H., and Zhang, Y. The retention equation in reversed phase ion pair chromatography. *Mikrochim. Acta* 1990, III, 35–53.

81. Loeser, E. and Drumm, P. Investigation of anion retention and cation exclusion effects for several C18 stationary phases. *Anal. Chem.* 2007, 79, 5382–5391.

82. Basci, N.E. et al. Optimization of mobile phase in the separation of β-blockers by HPLC. *J. Pharm. Biomed. Anal.* 1998, 18, 745–750.

83. LoBrutto, R. et al. Effect of the eluent pH and acidic modifiers in high-performance liquid chromatography retention of basic analytes. *J. Chromatogr. A.* 2001, 913, 173–187.

84. Gritti, F. and Guiochon, G. Effect of the pH, concentration and nature of the buffer on the adsorption mechanism of an ionic compound in reversed-phase liquid chromatography II: analytical and overloaded band profiles on Symmetry-C18 and Xterra-C18. *J. Chromatogr. A.* 2004, 1041, 63–75.

85. Courderot, C.M. et al. Chiral discrimination of dansyl-amino-acid enantiomers on teicoplanin phase: sucrose–perchlorate anion dependence. *Anal. Chim. Acta* 2002, 457, 149–155.

86. Pilorz, K. and Choma, I. Isocratic reversed-phase high-performance liquid chromatographic separation of tetracyclines and flumequine controlled by a chaotropic effect. *J. Chromatogr. A.* 2004, 1031, 303–305.

87. Crespi, C.L., Chang, T.K., and Waxman, D.J. CYP2D6-dependent bufuralol 1′-hydroxylation assayed by reverse-phase ion-pair high-performance liquid chromatography with fluorescence detection. *Methods Mol. Biol.* 2006, 320, 121–125.

88. Pan, L. et al. Influence of inorganic mobile phase additives on the retention, efficiency and peak symmetry of protonated basic compounds in reversed-phase liquid chromatography. *J. Chromatogr. A.* 2004, 1049, 63–73.

89. Rodriguez-Nogales, J.M., Garcia, M.C., and Marina, M.L. Development of a perfusion reversed-phase high performance liquid chromatography method for the characterisation of maize products using multivariate analysis. *J. Chromatogr. A.* 2006, 1104, 91–99.

90. Castro-Rubio, A. et al. Determination of soybean proteins in soybean–wheat and soybean–rice commercial products by perfusion reversed phase high-performance liquid chromatography. *Food Chem.* 2007, 100, 948–955.

91. Fanciulli, G. et al. Quantification of gluten exorphin A5 in cerebrospinal fluid by liquid chromatography–mass spectrometry. *J. Chromatogr. B.* 2006, 833, 204–209.

92. Mant, C.T. and Hodges, R.S. Context-dependent effects on the hydrophilicity and hydrophobicity of side-chains during reversed-phase high-performance liquid chromatography: implications for prediction of peptide retention behaviour. *J. Chromatogr. A.* 2006, 1125, 211–219.

93. Kim, J. et al. Phosphopeptide elution times in reversed-phase liquid chromatography. *J. Chromatog. A.* 2007, 1172, 9–18.

94. Wang, X. and Carr, P.W. Unexpected observation concerning the effect of anionic additives on the retention behavior of basic drugs and peptides in reversed-phase liquid chromatography. *J. Chromatog. A.* 2007, 1154, 165–173.

95. Hashem, H. and Jira, T. Retention behaviour of *b*-blockers in HPLC using a monolithic column. *J. Sep. Sci.* 2006, 29, 986–994.

96. Roberts, J.M. et al. Influence of the Hofmeister series on retention of amines in reversed-phase liquid chromatography. *Anal. Chem.* 2002, 74, 4927–4932.

97. Shapovalova, E.N. et al. Ion-pair chromatography of metal complexes of unithiol in the presence of quaternary phosphonium salts. *J. Anal. Chem.* 2001, 56, 160–165.

98. Ruiz-Angel, M.J. and Berthod, A. Reversed-phase liquid chromatography analysis of alkyl-imidazolium ionic liquids II: effects of different added salts and stationary phase influence. *J. Chromatog. A.* 2008, 1189, 476–482.

99. Kaliszan, R. et al. Suppression of deleterious effects of free silanols in liquid chromatography by imidazolium tetrafluoroborate ionic liquids. *J. Chromatogr. A.* 2004, 1030, 263–271.

100. Marszałł, M.P. and Kaliszan, R. Application of ionic liquids in liquid chromatography. *Crit. Rev. Anal. Chem.* 2007, 37, 127–140.

101. Marszałł, M.P., Bączek, T., and Kaliszan, R. Evaluation of the silanol-supressing potency of ionic liquids. *J. Sep. Sci.* 2006, 29, 1138–1145.

102. Marszałł, M.P., Bączek, T., and Kaliszan, R. Reduction of silanophilic interactions in liquid chromatography with the use of ionic liquids. *Anal. Chim. Acta* 2005, 547, 172–178.

103. Ruiz-Angel, M.J., Carda-Broch, S., and Berthod A. Ionic liquids versus triethylamine as mobile phase additives in the analysis of β-blockers. *J. Chromatogr. A.* 2006, 1119, 202–208.

104. Waichigo, M.M., Riechel, T.L., and Danielson, N.D. Ethylammonium acetate as a mobile phase modifier for reversed phase liquid chromatography. *Chromatographia* 2005, 61, 17–23.

105. Berthod, A., Ruiz-Angel, M.J., and Huguet, S. Nonmolecular solvents in separation methods: dual nature of room temperature ionic liquids. *Anal. Chem.* 2005, 77, 4071–4080.

106. Gritti, F. and Guiochon, G. Retention of ionizable compounds in reversed-phase liquid chromatography: effect of ionic strength of the mobile phase and the nature of salts used on overloading behavior. *Anal. Chem.* 2004, 76, 4779–4789.

107. He, L. et al. Effect of 1-alkyl-3-methylimidazolium-based ionic liquids as the eluent on the separation of ephedrines by liquid chromatography. *J. Chromatogr. A.* 2003, 1007, 39–45.

108. Tang, F. et al. Determination of octopamine, synephrine and tyramine in Citrus herbs by ionic liquid improved "green" chromatography. *J. Chromatogr. A.* 2006, 1125, 182–188.

109. Jin, C.H., Polyakova, Y., and Kyung, H.R. Effect of concentration of ionic liquids on resolution of nucleotides in reversed-phase liquid chromatography. *Bull. Kor. Chem. Soc.* 2007, 28, 601–606.

110. Sacchero, G. et al. Comparison of prediction power between theoretical and neural-network models in ion-interaction chromatography. *J. Chromatogr. A.* 1998, 799, 35–45.

111. Marengo, E. et al. Optimization by experimental design and artificial neural networks of the ion-interaction reversed-phase liquid chromatographic separation of 20 cosmetic preservatives. *J. Chromatogr. A.* 2004, 1029, 57–65.

112. Marengo, E., Gennaro, M.C., and Angelino, S. Neural networks and experimental design to investigate the effects of five factors in ion-interaction high-performance liquid chromatography. *J. Chromatogr. A.* 1998, 799, 47–55.

113. Buchner, R. and Hefter G. Interactions and dynamics in electrolyte solutions by dielectric spectroscopy. *Phys. Chem. Chem. Phys.*, 2009, 11, 8984–8999.

114. Spickermann, C., Thar, J., Lehmann, S.B.C., Zahn, S., Hunger, J., Buchner, R., Hunt, P.A., Welton, T., and Kirchner, B. Why are ionic liquid ions mainly associated in water? A Car–Parrinello study of 1-ethyl 3 methyl-imidazolium chloride water mixture. *J. Chem. Phys.* 2008, 129, 104–505.

115. Li, J. Prediction of internal standards in reversed-phase liquid 1. chromatography IV: Correlation and prediction of retention in reversed-phase ion-pair chromatography based on linear solvation energy relationships. *Anal. Chim. Acta* 2004, 522, 113–126.

116. Debye, P., McAulay, J. Z. Das Elektrische Feld Der Ionen Und Die Neutralsalzwirking (The electric field of ions and the action of neutral salts). *Physik* 1925, 26, 22–29.

117. Long, F.A. And McDevit, W.F. Activity coefficients of nonelectrolyte solutes in aqueous salt solutions. *Chem. Rev.* 1952, 51, 119–169.

118. Masterton, W.L. and Lee, T.P. Salting coefficients from scaled particle theory. *J. Phys. Chem.* 1970, 74, 1776–1782.

119. Bockris, J.O., Bowler-Reed, J., and Kitchener, J.A.The salting-in effect. *Trans. Faraday Soc.* 1951, 47, 184–192.

2 Mechanistic Aspects of Chiral Recognition on Protein-Based Stationary Phases

Jun Haginaka

CONTENTS

2.1 INTRODUCTION

In 1973, Stewart and Doherty [1] separated Trp enantiomers on a bovine serum albumin (BSA)-immobilized agarose gel. Since then a lot of chiral stationary phases (CSPs) based on a protein or glycoprotein have been developed and used for enantioseparations of a wide variety of compounds in liquid chromatography (LC). These CSPs have enantioselectivity for a wide range of compounds because of multiple binding sites on the surface of a chiral selector or multiple binding interactions between a chiral selector and selectant [2–6]. Protein-based CSPs so far developed have included albumins such as BSA [7], human serum albumin (HSA) [8], and serum albumin from other species [9]; glycoproteins such as α_1-acid glycoprotein

(AGP) [10], ovomucoid from chicken egg whites (OMCHI) [11], ovoglycoprotein from chicken egg whites (OGCHI) [12] (now termed chicken AGP, or cAGP [13]), avidin [14], and riboflavin binding protein [15]; enzymes such as trypsin [16], α-chymotrypsin [17], cellobiohydrolase (CBH) [18], lysozyme [19], pepsin [20], glucoamylase [21], and penicillin G-acylase (PGA) [22]; antibodies [23]; and other proteins such as ovotransferrin (or conalbumin) [24], streptavidin [25], and fatty acid binding protein [26].

Antibodies are glycoproteins, which are produced by the immune system of vertebrates in response to antigens, invading pathogenic microorganisms or nonself biological compounds [27]. Antibody-based CSPs have been introduced [23,27]. Since antibodies could be raised by an enantiomer, the elution order of the enantiomer is predictable on antibody-based CSPs in contrast to conventional CSPs based on a protein or glycoprotein. Furthermore, membrane-bound proteins such as receptors and ion channels, whose fragments were obtained from cell lines, could be investigated as CSPs [28,29]. Physical properties of these proteins and glycoproteins are shown in Table 2.1.

This review chapter mainly deals with recent progresses in mechanistic aspects of chiral recognition on protein-based stationary phases.

2.2   BASIC PRINCIPLES OF CHIRAL RECOGNITION ON A PROTEIN

2.2.1   CHIRAL RECOGNITION MODEL

In his recent review article [30], Bentley attempted to explain chiral recognition mechanism of a substrate or drug on a protein (receptor or enzyme) historically, focused on specific interactions such as binding, combinations, or associations among the involved components. A three-point attachment (TPA) model was first proposed by Easson and Stedman [31] to explain stereoselective bindings of drug enantiomers to the receptor. When the three groups (a, b, and c) of one enantiomer bind to the receptor surface at specific sites (A, B, and C), respectively, the TPA model is shown in Figure 2.1a. However, in the case of the other enantiomer an equivalent binding via the same three specific sites could not be attained, as shown in Figures 2.1b and 2.1c. Ogston [32] independently proposed a TPA model to account for stereoselective conversion of a prochiral reactant to a chiral product by the enzyme. Furthermore, the model contained the implicit assumption that the substrate could reach the receptor or enzyme only from the surface (Figures 2.1a, 2.1b, and 2.1c) but not from the interior or back (Figure 2.1d). The basic idea of the TPA model could be well received in the field of receptor–ligand interactions (receptor-histamine [33] and receptor-sweetener [34]) and chiral chromatography [35–37].

There has been much controversy over the TPA model. Wilcox et al. [38] pointed out that, in addition to three distinct and specific points of interaction between an enzyme and a substrate, some other condition such as steric hindrance, directed forces, or another point of interaction was required for stereoselective enzyme reactions. This is the first proposal of four-contact point (FCP) model for chiral recognition in the enzyme or protein. For each contact point, two atoms (or groups of atoms)—one from the protein site and the other from the substrate or drug—are

TABLE 2.1

Physical Properties of Proteins or Glycoproteins Used for CSPs

Protein	Molecular Mass	Carbohydrate Content (%)	Isoelectric Point	Origin
Albumins				
Bovine serum albumin (BSA)	66000	—[a]	4.7	Bovine serum
Human serum albumin (HSA)	65000	—	4.7	Human serum
Glycoproteins				
α_1-Acid glycoprotein (AGP)	41000–43000	45	2.7–3.8	Human serum
	33000	35	2.7–3.8	
Ovomucoid (OMCHI)	28000	30	4.1	Egg white
Ovoglycoprotein (OGCHI) (chicken AGP (cAGP))	30000	25	4.1	Egg white
Avidin	66000	7	10	Egg white
Riboflavin binding protein	32000–36000	14	4	Egg white
Enzymes				
Cellulase				Fungus
Cellobiohydrolase I (CBH I, Cel7A)	65000	6	3.9	
Cellobiohydrolase II (CBH II, Cel6A)	53000		5.9	
Cellobiohydrolase 58 (CBH 58, Cel6B)	60000		3.8	
Lysozyme	14300	—	10.5	Egg white
Pepsin	34600	—	<1	Porcine stomach
Glucoamylase				Fungus
Glucoamylase G1	94000	30–35		
Glucoamylase G2	85000			
Penicillin G-acylase (PGA)	85000	—	8.1	Bacterium
Antibody (Immunoglobulin G)	150000	2–3	6–8	Vertebrate
Others				
Ovotransferrin (conalbumin)	77000	2.6	6.1	Egg white
Streptavidin	53000	—	5	Bacterium
Fatty acid binding protein	14000	—	9.0	Chicken liver

[a] No sugar moieties.

involved. Hence, an FCP model requires the participation of eight atoms or groups of atoms. However, these contact points are either binding or repulsive. On the other hand, Sokolov and Zefirov [39] proposed a two-point attachment model for an enzyme reaction. Two binding interactions of a prochiral substrate with the enzyme result in an unequal reactivity of two identical prochiral groups of the substrate. If the groups are sterically hindered to different extents, enantiospecific enzyme

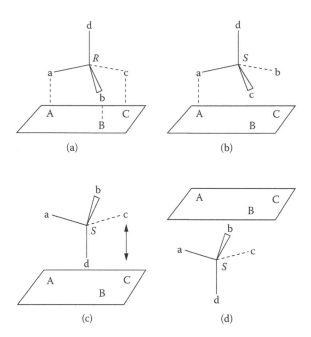

FIGURE 2.1 The Easson–Stedman model. For the purpose of making *RS* configurational assignments, it is assumed that the priority sequence is a > b > c > d. The binding sites for a, b, and c are represented as A, B, and C. In the Easson–Stedman model (a), the *R* enantiomer can bind at all three sites and would be assumed to be the physiologically active material. However, the *S* enantiomer is limited to a single contact point (b). An alternative possibility (c) for the *S* enantiomer is ruled out because of steric hindrance by the d group. The distances a–A, b–B, and c–C (indicated by the double arrow) are too large to permit binding. Furthermore, the approach of the *S* enantiomer from the interior (d) is not allowed. (Adapted from Bentley, R., *Arch Biochem Biophys*, 414, 2, 2003. With permission.)

reactions occur, resulting in stereoselective conversion of a prochiral reactant to a chiral product. Davankov [40] defined a three-point interaction (TPI) rule that, at minimum, three configuration-dependent active points of a selector molecule should interact with three complementary and configuration-dependent active points of an enantiomer molecule. Furthermore, the TPI rule is different from the TPA model in that the former considers repulsive interactions as another point of interaction.

Topiol and Sabio [41] claimed FCP interactions for chiral recognition, based on a theoretical analysis of interactions between two asymmetric tetrahedrons. Furthermore, the FCP model was discussed for a protein (enzyme or receptor) interaction with the substrate or drug by Bentley [42]. Recently, Mesecar and Koshland [43] proposed a four-location (FL) model, which is the same as an FCP model. As shown in Figure 2.2, the three groups (a, b, and c) of different enantiomers bind to the same protein location at specific sites (A, B, and C), respectively, whereas the d groups interact at different positions (D' and D''). The TPA model works only under the assumption that the two enantiomers can approach a flat protein surface only from the top. Therefore, a fourth location is essential to distinguish the two

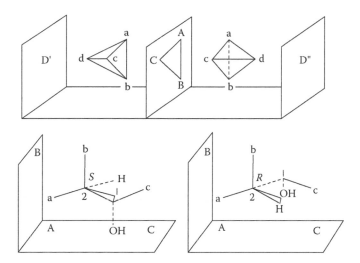

FIGURE 2.2   Models for IDH. The top line shows the model proposed by Mesecar and Koshland [43]. It is essentially the original Easson–Stedman model but allowing each enantiomer to approach from opposed directions and with the fourth defining contact points, D′ and D″. In the bottom line, (1R, 2S) and (1S, 2R) enantiomers of isocitrate can fit into the same site and can approach this site from the same direction. In this model, a and c = COOH and b = CH₂COOH. Two necessary fourth contact points for the OH groups at C-1 determine the actual binding. (From Bentley, R., *Arch Biochem Biophys*, 414, 8, 2003. With permission.)

enantiomers in the actual protein structure. Mesecar and Koshland [43] found that this model could work in isocitrate dehydrogenase (IDH), whose substrate is the (1R, 2S) form (D-isocitrate) in the presence of Mg^{2+}, while the (1S, 2R) enantiomer (L-isocitrate) binds in the absence of Mg^{2+} without any catalytic activities. The two isocitrate enantiomers approached the active site from the same direction. As shown in Figure 2.2, only binding positions of two carboxylic acids are planar, whereas that of the third carboxylic acid is on a second plane at right angles to the first. The fourth contact point is, of course, necessary in each case for locating the hydroxy groups. This situation could be attained by the presence of a CH_2 (or CHX) group between the chiral center and one of the binding groups. However, it is noted that isocitrate has two stereocenters. In the case of phenylalanine ammonia-lyase, L-Phe is the substrate, but D-Phe is an inhibitor. Both enantiomers could bind at the same active site with only minor bond angle changes, as shown in Figure 2.3. Since Phe, having only one stereocenter, has a CH_2 group between the chiral center and one of the binding groups, it could be considered that the fourth contact point is necessary as in isocitrate enantiomers [30].

Recently, Sundaresan and Abrol [44,45] proposed a stereocenter-recognition (SR) model, which provides the minimum number of substrate locations interacting with protein (enzyme or receptor) sites. This model is based on the structural consideration that all substrate stereoisomers would approach the same binding surface of the receptor from the same direction, while allowing for the possibility that

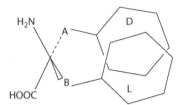

FIGURE 2.3 Enantiomer superposition of L- and D-phe. In this composite structure, the C_α–NH_2 and C_α–COOH are drawn with NH_2 above the plane of the paper and COOH below this plane. These positions are identical for both enantiomers. For L-phe, A = H and B = CH_2, and the phenyl group is represented by the hexagon labeled L. For D-phe, A = CH_2 and B = H with the phenyl hexagon identified as D. In both cases, the three groups, NH_2, COOH, and C_6H_5, occupy essentially the same positions. (From Bentley, R., *Arch Biochem Biophys*, 414, 5, 2003. With permission.)

different substrate stereoisomers can bind in different conformations. This assumption is reasonable, because substrate stereoisomers interact with the pocket or cleft on the receptor surface. Sundaresan and Abrol [44,45] defined a substrate location as a functional group or groups attached to a stereocenter in the substrate and receptor sites as either specific functional groups in the receptor or the contour of a large part of the receptor surface. It is noted that a substrate location may interact with multiple receptor sites as shown in Figure 2.4 [44] or that multiple substrate locations may interact with a single receptor site. The SR model for a molecule with two stereocenters is as follows. The minimum number of substrate locations required for the simplest case of a molecule with two adjacent stereocenters is illustrated in Figures 2.5 and 2.6 [45]. Interactions with only three substrate locations, either in the (3-0) or (2-1) distributions on the two stereocenters resulted in no recognition of one

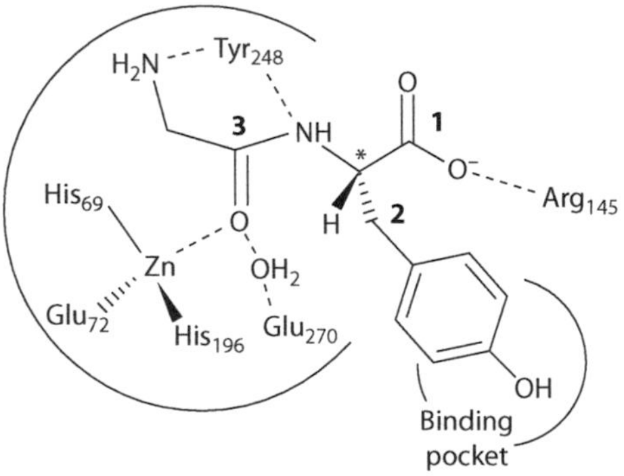

FIGURE 2.4 Multiple interactions between receptor sites and a substrate location. In the binding of carboxypeptidase A to glycyl-L-tyrosine (an asterisk marks the chiral center), the free amino group of glycine and the amide functionality together constitute a single substrate location, interacting with multiple sites (Tyr248, Glu270 via a bound H_2O molecule and His69, Glu72, His196 via the Zn^{2+} ion) distributed over a large part of the enzyme's binding surface. (Adapted from Sundaresan, V. and Abrol, R., *Protein Sci*, 11, 1132, 2002. With permission.)

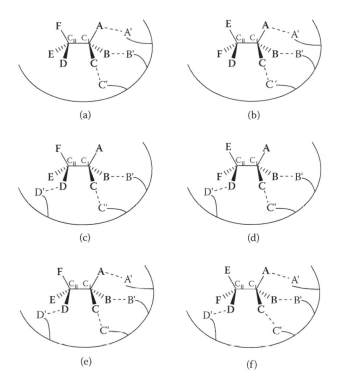

FIGURE 2.5   Molecules with two stereocenters. I. The two stereocenters in the substrate molecule are marked C_I and C_{II}. The notation (x–y) refers to the number of distinct substrate locations on each stereocenter, that is, x locations on C_I and y locations on C_{II}. (a, b) Interactions of a receptor with three locations in a (3-0) distribution of stereocenters would allow two different stereoisomers to interact in a similar fashion. (c, d) Interactions of a receptor with three locations in a (2-1) distribution would also allow different stereoisomers to interact similarly. (e, f) Interactions with four locations in a (3-1) distribution likewise fails to prevent similar binding of the two stereoisomers. Therefore, all three distributions of substrate locations would result in decreased stereoselectivity. It follows that the (0-3), (1-2), and (1-3) distributions of substrate locations would also fail for the same reasons. In this figure and the next, the dotted lines drawn between substrate locations and corresponding receptor sites are not necessarily attachments. They simply represent any kind of interaction between receptor sites and substrate locations. (From Sundaresan, V. and Abrol, R., *Chirality*, 17, S34, 2005. With permission.)

stereoisomer (Figures 2.5a through 2.5d). The (3-1) distribution of substrate locations (Figures 2.5e and 2.5f) interacting with the receptor resulted in no recognition as well. A minimum of four substrate locations in a (2-2) distribution over the two stereocenters (Figures 2.6a through 2.6d) would need to interact with a receptor. For a substrate that has two stereocenters, the (2-2) configuration of substrate locations implies that two locations are directly attached to each of the two stereocenters (Figure 2.7a) [45]. In addition, all locations on C_{II} effectively act together as the third location on CI (Figure 2.7b) as they are attached indirectly to C_I via C_{II}. Similarly, all locations on C_I effectively act together as the third location on C_{II} (Figure 2.7c).

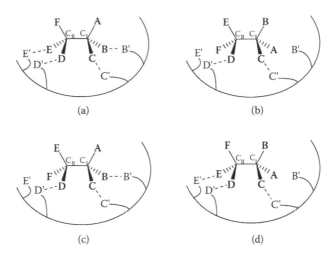

FIGURE 2.6 Molecules with two stereocenters. II. The two stereocenters in the substrate molecule are marked C_I and C_{II}. Interactions of a receptor with four locations in a (2-2) distribution on the two stereocenters in the substrate are shown in (a). The enantiomer (b) and the diastereomers (c) and (d) of this molecule would be able to interact through only two or three out of these four locations. This ensures complete recognition of all four stereoisomers of the substrate. (From Sundaresan, V. and Abrol, R., *Chirality*, 17, S34, 2005. With permission.)

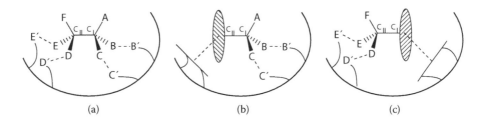

FIGURE 2.7 Three effective locations per stereocenter. (a) Four locations in a molecule with two stereocenters interacting with a receptor in a (2-2) distribution. The two stereocenters are marked C_I and C_{II}. (b) Interactions with locations on C_{II} can be grouped together, resulting in an interaction with one effective location with respect to C_I. Along with the two locations on C_I, this gives three effective locations for C_I. (c) Interactions with locations on C_I can be grouped together, resulting in an interaction with one effective location with respect to C_{II}, giving a total of three effective locations for C_{II}. (Adapted from Sundaresan, V. and Abrol, R., *Chirality*, 17, S35, 2005. With permission.)

Thus, a receptor has to effectively interact with at least three locations on each stereocenter in the substrate [45]. This SR model could be expanded to chiral recognition of a compound with three stereocenters. It was concluded that stereoselectivity toward a substrate with N stereocenters in a linear structure involves a minimum of $N + 2$ substrate locations. This means that effectively at least three locations per stereocenter should interact with one or more receptor sites [45].

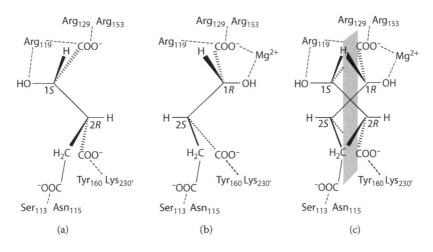

FIGURE 2.8 Isocitrate and its interactions with IDH. (a) Schematic view of four noncoplanar locations in L-isocitrate [(1S, 2R)], in a (2-2) distribution, and interacting with IDH residues in the absence of Mg^{2+} (–OH and –COO^- on the 1S stereocenter and –CH_2COO^- and –COO^- on the 2R stereocenter). Only the important enzyme residues interacting with the substrate are shown here. (b) In the presence of Mg^{2+}, the same four locations in D-isocitrate [(1R, 2S)] interact with the metal ion and enzyme residues. (c) Superposition of D- and L-isocitrate bound conformations, in the presence and absence of Mg^{2+}, respectively. A hypothetical mirror-plane (gray), determined by three of the four locations, relates the two superposed isocitrate molecules. The fourth location (i.e., the –OH groups of D- and L-isocitrate) lies on opposite sides of this mirror plane. (From Sundaresan, V. and Abrol, R., *Chirality*, 17, S36, 2005. With permission.)

With regard to IDH, in the FL model isocitrate could be considered as a single asymmetric tetrahedron. Because isocitrate has two stereocenters, Mg^{2+}-dependent inversion of stereoselectivity in IDH is better explained by the SR model. Irrespective of the metal ion, the same enzyme residues interact with the –COO^- group attached to one stereocenter and the –COO^- and –CH_2COO^- groups attached to the other stereocenter of D- and L-isocitrate, as shown in Figures 2.8a and 2.8b [45]. However, in the presence of Mg^{2+} the metal ion coordinates with the –OH group of D-isocitrate, whereas in the absence of Mg^{2+} a hydrogen bonding interaction with an Arg residue is possible for the –OH group of L-isocitrate. In both cases, IDH interacts with a total of four noncoplanar substrate locations, in a (2-2) distribution on the two stereocenters of isocitrate: that is, –OH and –COO^- on one stereocenter; and –CH_2COO^- and –COO^- on the second stereocenter. Therefore, the enzyme binds selectively to L-isocitrate in the absence of Mg^{2+}, whereas it binds selectively to D-isocitrate in the presence of Mg^{2+}. Superposition of the bound structures shows that the conformations of D- and L-isocitrate bear a mirror image relationship to each other (Figure 2.8c). However, it should be noted that both D- and L-isocitrate bind to the same residues in the enzyme, within the same binding cavity, and not on two opposite sides of an enzyme surface. Furthermore, this model is applicable to chiral stationary phases: the interaction between a receptor and a substrate would be equivalent to that between a chiral selector and a chiral selectant. Receptor sites would then correspond to interacting functional groups in the chiral selector and substrate locations to those in the chiral selectant.

As pointed out by Sundaresan and Abrol [45], the SR model as well as the TPI model present a static view of chiral recognition, without taking into account the dynamic process. Wainer and his group [46] proposed a dynamic model of chiral recognition as the following four steps: (1) formation of selector–selectant complex (tethering); (2) positioning of selectant–selector to optimize interactions (conformational adjustments); (3) formation of secondary interactions (activation of the complex); and (4) expression of the molecular fit. Furthermore, they confirmed this proposal by chiral recognition of dextromethorphan (DM) and levomethorphanon (LM) on $\alpha_3\beta_4$ nicotinic acetylcholine receptor ($\alpha_3\beta_4$ nAChR) [47] and chiral recognition of (R)- and (S)-verapamil on human organic cation transporter 1 (hOCT1)[48].

2.2.2 THERMODYNAMICS OF SELECTOR–SELECTANT COMPLEX

Several researchers emphasized that the formation of diastereomeric complexes with differing thermodynamic stabilities was the fundamental prerequisite and that an interaction-based model was not needed [30].

The relationship of the observed enantioselectivity (α) and differences in the Gibbs free energy change ($\Delta\Delta G^0$) of the complexes formed between the two enantiomers and the chiral selector can be defined as

$$\Delta\Delta G^0 = -RT \ln \alpha \qquad (2.1)$$

where R is the gas constant, and T is the temperature of experimental conditions in Kelvin.

The Gibbs free-energy change (ΔG^0) of each enantiomer–selector complex is composed of enthalpic and entropic contributions and can be defined as

$$\Delta G^0 = \Delta H^0 - T\Delta S^0 \qquad (2.2)$$

where ΔH^0 is the change in enthalpy, and ΔS^0 is the change in entropy associated with the formation of the enantiomer–selector complex. By combining Equations (2.1) and (2.2), the following equation could be obtained:

$$\ln \alpha = -\frac{1}{T}\frac{\Delta\Delta H^0}{R} + \frac{\Delta\Delta S^0}{R} \qquad (2.3)$$

and, by plotting ln α against 1/T, the corresponding enthalpic and entropic contributions ($\Delta\Delta H^0$ and $\Delta\Delta S^0$) to enantioselectivity (α) could be estimated. Therefore, two chiral recognition mechanisms—enthalpic driven and entropic driven—could be considered.

It was pointed out that the distinction between enthalpy-driven and entropy-driven molecular chiral recognition is only a reflection of the relative contributions of these parameters to a single process and that interaction-based models, which concentrate on enthalpy and measure the relative stabilities of the selector–selectant complexes, have obscured this fact [49].

2. 3 CHIRAL RECOGNITION MECHANISM ON PROTEIN-BASED STATIONARY PHASES

2.3.1 HUMAN SERUM ALBUMIN

HSA, which has a molecular mass of 67 kDa, consists of 585 amino acid residues [2]. CSPs based on HSA were first bound to diol-silica particles by Domenici et al. [8]. A variety of weakly acidic and neutral compounds—which include 2-arylpropionic acid derivatives such as naproxen, flurbiprofen, ibuprofen, ketoprofen, and fenoprofen; reduced folates such as leucovorin and 5-methyltetrahydrofolate; and benzodiazepines such as oxazepam, lorazepam, and temazepam—are resolved on HSA-based CSPs [3]. HSA involves closely related proteins with serum albumins from other mammalian species. However, the binding stereoselectivity of the 2,3-benzodiazepine, tofisopam, in human is opposite to that in all other species (bovine, dog, horse, pig, rabbit, and rat). In the binding of 1,4-benzodiazepines, dog albumin is very similar to HSA [9].

Stereoselective binding characteristics of HSA have been thoroughly examined. Stereoselective binding of drugs on HSA occurs principally at two major binding sites: warfarin (Wf)-azapropazone site (site I); and indole-benzodiazepine site (site II) (see Figure 2.9) [50]. He and Carter determined the three-dimensional structure of HSA, which shows that the binding sites I and II are located in hydrophobic

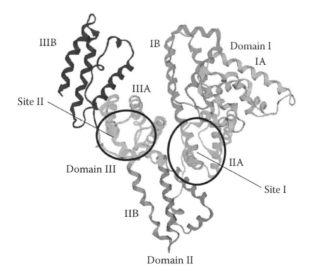

FIGURE 2.9 Crystal structure of HSA. The subdivision of HSA into domain (I–III) and subdomains is indicated, and approximate locations of Sudlow's site I and site II are shown. (From Chuang, V.T.G. and Otagiri, M., *Chirality*, 18, 160, 2006. With permission.)

cavities in subdomains IIA and IIIA, respectively [51]. In addition, the crystal structure of HSA-myristate complexed with the (R)- and (S)-Wf was determined [52]. The
structures confirm that Wf binds to subdomain IIA in the presence of fatty acids
and reveals the molecular details of the protein–drug interaction. Figure 2.10 shows
structural details of the interaction of the (R)- and (S)-enantiomers of Wf with HSA

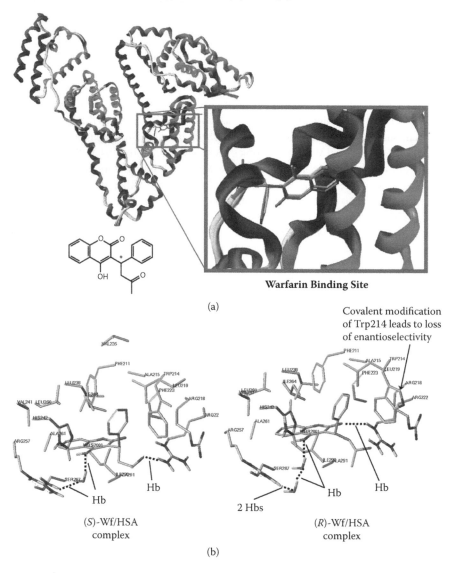

Warfarin Binding Site

(a)

Covalent modification
of Trp214 leads to loss
of enantioselectivity

(S)-Wf/HSA
complex

(R)-Wf/HSA
complex

(b)

FIGURE 2.10   (*See color insert following page 152.*) X-ray crystal structures of HSA-Wf
complexes. (a) Superimposed complexes, (R)-Wf (magenta), (S)-Wf (cyan). (b) Active site
with binding modes. (For interpretation of the references to color in this figure legend, the
reader may refer to color insert.) (From Lämmerhofer, M., *J Chromatogr A*, 1217, 840, 2010.
With permission.)

[53]. The (R)- and (S)-enantiomers bind in the pocket in almost identical conformations, and the main difference in the drug occurs in the conformations in the acetonyl group and in hydrogen bonding interactions that are formed between Arg222 residue and the carbonyl of the coumarin ring for (R)-enantiomer and of the acetonide for (S)-enantiomer [53]. However, the two enantiomers of Wf adopt very similar conformations when bound to the protein and make many of the same specific contacts with amino acid side chains at the binding site, thus accounting for the relative lack of stereospecificity of the HSA-Wf interaction [52,53].

2.3.2   α₁-Acid Glycoprotein

AGP (α orosomucoid (ORM)), a member of the lipocalin family, is a 41–43-kDa glycoprotein with an isoelectric point (pI) of 2.8–3.8. The peptide moiety is a single chain of 183 amino acids for human AGP. The carbohydrate content represents 45% of the molecular weight attached in the form of five to six highly sialylated complex-type N-linked glycans [54]. Recently, it was reported that the average molecular mass of AGP could be ca. 33 kDa [55] by matrix-assisted laser-desorption ionization time-of-flight mass spectra and that the sugar content of AGP could be estimated to be about 35% [55]. Besides the high heterogeneity of glycans, the protein part has also been found to show polymorphism. The variants are encoded by two different genes: the F1 and S variants are encoded by the alleles of the same gene, whereas the A variant is encoded by a different gene [56]. There is a difference of at least 22 amino acid residues between the F1–S (ORM 1) and A (ORM 2) variants, whereas F1 and S forms differ only in a few residues.

CSPs based on AGP were developed by Hermansson [10]. In general, AGP was bound to silica gels via its amino groups. A wide range of basic, neutral, and acidic enantiomers was separated using AGP columns [3,57,58]. It was thought that drug binding to AGP occurred at a single hydrophobic pocket or cleft within the protein domain of the molecule [59]. Alternatively, more than one binding site was presented [59]. It is thought that the hydrophobic, electrostatic, and hydrogen bonding interactions could play important roles in the retentivity and enantioselectivity of a solute on an AGP column [57].

Selective binding of coumarin enantiomers (Wf, phenprocoumon, and acenocoumarol) to human AGP genetic variants was investigated. All investigated compounds bound stronger to ORM 1 than to ORM 2 [56]. ORM 1 and human native AGP preferred the binding of (S)-enantiomers of Wf and acenocoumarol, whereas no enantioselectivity was observed in phenprocoumon binding. Furthermore, a new homology model of AGP was built, and the models of ORM 1 and ORM 2 suggested that the binding cavity, including Trp122, for ORM1 was the same as that for ORM2. In addition, the difference in binding to AGP genetic variants could be caused by steric factors: ORM 2 formed a smaller, more hydrophobic cavity compared with ORM 1 [56]. Figure 2.11 shows comparison of dockings of (R)- and (S)-acenocoumarol to ORM 1 and ORM 2 models. Dockings to ORM 1 resulted in a much lower intermolecular energy than dockings to ORM 2, suggesting that although binding to both variants is possible, ORM 1 binding is more favorable. Energy differences between

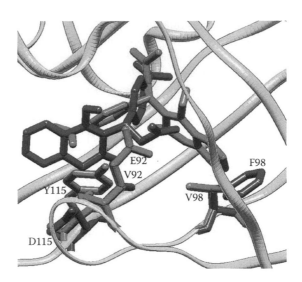

FIGURE 2.11 *(See color insert following page 152.)* Comparison of dockings to ORM 1 and ORM 2 models. ORM 1 side chains and (*S*)-acenocoumarol docked to ORM 1 are indicated with blue color, ORM 2 side chains and (*S*)-acenocoumarol docked to ORM 2 with red color. (For interpretation of the references to color in this figure legend, the reader is referred to the Web version of the original article.) (From Hazai, E. et al., *Bioorg Med Chem*, 14, 1964, 2006. With permission.)

(R)- and *(S)*-enantiomers are not significant and show a slight preference for the *(S)*-enantiomer in the case of both ORM 1 and ORM 2 [56].

Ligand-binding properties of AGP were investigated by using circular dichroism (CD) methods [60]. The induced CD spectra of drug–AGP complexes were observed with many classes of drugs. Results of additional CD experiments performed by using recombinant AGP mutants showed no changes in the ligand binding ability of Trp122Ala in sharp contrast with the Trp25Ala, which was unable to induce extrinsic CD signal with either ligand. These findings suggest that, likely via π–π stacking mechanism, Trp25 is essentially involved in the AGP binding of drugs studied [60].

2.3.3 OVOMUCOID AND OVOGLYCOPROTEIN (CHICKEN α_1-ACID GLYCOPROTEIN)

The CSP based on OMCHI was developed by Miwa et al. [11]. It was used for the resolution of acidic, basic, and neutral enantiomers in bulk drugs and formulations [61–63] and for the assay of enantiomers in biological fluids [64]. Various ovomucoids such as ovomucoid from turkey egg whites (OMTKY) [65] and OMCHI [66] exist as three tandem, independent domains [67]. Each domain and combination domains, first and second, second and third domains, were isolated, purified, and characterized [65]. Furthermore, columns were made with purified OMTKY and OMCHI domains to test chiral recognition properties [65,68]. The third domain of OMTKY and OMCHI consisted of glycosylated (OMTKY3S and OMCHI3S) and unglycosylated domains (OMTKY3 and OMCHI3). The third domains of the OMTKY and OMCHI domains were found to be enantioselective to at least two classes of

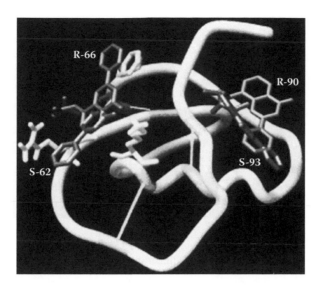

FIGURE 2.12 *(See color insert following page 152.)* Molecular modeling simulation of U-80,413 enantiomers bound to OMTKY3. The white curved graphic represents the protein backbone. Selected protein side chains are also shown in white. Ligands are labeled according to their *R* or *S* chirality and numbered according to their position among the 100 lowest energy-minimized binding orientations. (For interpretation of the references to color in this figure legend, the reader is referred to the original article.) (From Pinkerton, T.C. et al., *Anal Chem*, 67, 2365, 1995. With permission.)

compounds: benzodiazepines and 2-arylpropionic acid derivatives. Glycosylation of the third domain did not affect chiral recognition.

The chiral recognition mechanism of OMTKY3 was elucidated by using NMR measurement and molecular modeling and ligand docking [65]. On the surface of OMTKY3, there are two distinct binding sites (nonselective and enantioselective binding sites) as shown in Figure 2.12. In the former, hydrophobic interactions mainly work for the binding, whereas in the latter hydrophobic, electrostatic, and hydrogen bonding interactions play important roles. The enantioselective binding model for *(R)*- and *(S)*-U-80413, which is one of 2-arylpropionic acid derivatives with OMTKY3 is shown in Figure 2.13 [65]. One can see similarities and differences in orientation and intermolecular interactions between *(R)*- and *(S)*-U-80413. The carboxy groups of each enantiomer engage in electrostatic interactions with the positive charge on Arg21. The carbonyl group on U-80413's central ring shares a hydrogen bond with NH_3^+ group of Lys34. The distinguishing difference between the enantiomers is the proximity of the phenyl group of *(R)*-U-80413 and Phe53.

Recently, a new protein from chicken egg whites, which is termed OGCHI, was isolated and characterized [12]. In addition, it was found that 10% of OGCHI was included in crude OMCHI preparations [12]. The OGCHI column gave much more excellent chiral recognition ability than the crude OMCHI column. Moreover, OMCHI and OGCHI columns were made from isolated, pure proteins and were compared with regard to chiral recognition abilities. It was found [12] that chiral

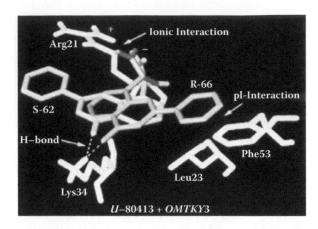

FIGURE 2.13   (*See color insert following page 152.*) U-80413 enantiomers in the enantiose-lective binding site of OMTKY3. (From Pinkerton, T.C. et al., *Anal Chem*, 67, 2366, 1995. With permission.)

recognition ability of OMCHI reported previously [11] came from OGCHI and that OMCHI had no appreciable chiral recognition ability.

A cDNA clone encoding OGCHI was isolated, and the amino acid sequence of OGCHI was clarified [13]. OGCHI consisted of 203 amino acids including a predictable signal peptide of 20 amino acids. The mature OGCHI showed 31–32% identities to rabbit and human AGPs. Thus, OGCHI should be the chicken AGP (cAGP), a member of the lipocalin family. Furthermore, the recombinant cAGP was prepared by the *Escherichia coli* expression system, and its chiral recognition ability was confirmed by CE. Since proteins expressed in *Escherichia coli* are not modified by any sugar moieties, this result shows that the protein domain of the cAGP is responsible for the chiral recognition [13]. cAGP consists of 183 amino acid residues and has only one Trp residue at the 26 position [13]. The Trp26 residue was modified with 2-nitrophenylsulfenyl chloride, and chiral separations of neutral, acidic, and basic compounds were examined on cAGP and Trp-modified cAGP columns [69]. Chiral separations of propranolol, alprenolol, and oxprenolol were lost on the Trp-modified cAGP column, whereas chlorpheniramine, ketoprofen, and benzoin were still enantioseparated on the Trp-modified cAGP column despite lower enantio-selectivity than that on the cAGP column. These results suggest that the Trp26 residue could be responsible for chiral recognition of these compounds. Competition studies using *N,N*-dimethyl-*n*-octylamine (DMOA) as a competitor indicated that propranolol, alprenolol, and oxprenolol competed with DMOA on a single binding site near the Trp26 region and that further bindings of chlorpheniramine, ketoprofen, and benzoin occurred at the secondary binding site in a noncompetitive fashion with DMOA [69]. Furthermore, ligand-binding properties of cAGP were investigated by using CD methods. Analysis of the extrinsic CD spectra with the study of the Trp26-modified protein and CD displacement experiments revealed that a single Trp26 residue of cAGP conserved in the whole lipocalin family is part of the binding site. In addition, it showed that it is essentially involved in the ligand-binding process via π-π stacking interaction resulting in the appearance of strong induced CD bands due to the nondegenerate intermolecular exciton coupling between the π-π* transitions of the stacked indole ring–ligand chromophore [70].

2.3.4 CELLOBIOHYDROLASE

Cellulases are cellulose-degradating enzymes. Fungus *Trichoderma reesei* produces four major cellulases: two cellobiohydrolases, CBH I (Cel7A, 64 kDa, p*I* 3.9) and CBH II (Cel6A, 53 kDa, p*I* 5.9); and two endoglucanases, EG I (Cel7B, 55 kDa, p*I* 4.5) and EG II (Cel6B, 48 kDa, p*I* 5.5) [71,72]. They are all acidic glycoproteins and have a common structural organization with a binding domain connected to the rest of the enzyme (i.e., the core) through a flexible arm [73]. The core is enzymatically active. CSPs based on CBH I (Cel7A), which can resolve acidic, basic, and uncharged racemates into their enantiomers, have been most extensively investigated among CSPs based on cellulases [72]. Especially, higher enantioselectivity was obtained for the separation of β-blocking agents such as propranolol, oxprenolol, and metoprolol [74]. Cellobiohydrolase 58 (CBH 58, Cel7D) from *Phanerochaete chrysosporium*, which is the counterpart to Cel7A from *Trichoderma reesei*, was immobilized on silica. CBH 58 (Cel7D, 60 kDa, p*I* 3.8), similarly to CBH I (Cel7A), was an excellent chiral selector for β-blockers and even expresses a broader enantioselectivity for other basic compounds [75].

The three-dimensional structure of the catalytic domain of Cel7A was elucidated by x-ray crystallography [76]. Furthermore, the catalytic domain of Cel7A was co-crystallized with the (*S*)-propranolol, and the structure of the complex was determined [77]. As shown in Figure 2.14, (*S*)-propranolol binds at the active site. The catalytic residues Glu212 and Glu217 make tight salt links with the secondary amino group of (*S*)-propranolol. The oxygen atom attached to the chiral centre of (*S*)-propranolol forms hydrogen bonds to the nucleophile Glu212 and to Gln175, whereas the aromatic naphthyl moiety stacks with the indole ring of Trp376.

Only the Asp214Asn of the Cel7A-mutant CSPs retained enantioselectivity, whereas the enantioselectivity was completely lost for the Glu212Gln and Glu217Gln columns [77]. The loss of enantioselectivity was accompanied by the loss of catalytic activity for the mutants. These results revealed that carboxy functions of Glu212 and Glu217 were essential for catalysis as well as chiral recognition and that exchange of either one impaired both the activity and chiral recognition. Asp214 appears to be involved in the chiral recognition but is of less importance [77]. Cocrystallization of Cel7A with the (*R*)-propranolol failed. It was supposed that (*R*)-enantiomer could interact with Cel7A in the same binding site with (*S*)-enantiomer, resulting in a poorer fit of the (*R*)-enantiomer due to steric hindrance with the hydroxy group, rather than weaker binding due to the loss of hydrogen bonding [77].

It is interesting that at mobile phase pH 5.5 the retention time of the less-retained (*R*)-propranolol enantiomer decreases with increasing temperature on the CBH I (Cel7A) column, whereas that of the (*S*)-enantiomer increases, causing a large increase of the enantioseparation factor when the temperature is raised from 5 to 45°C as shown in Figure 2.15 [72]. Thermodynamic studies indicate that the interaction

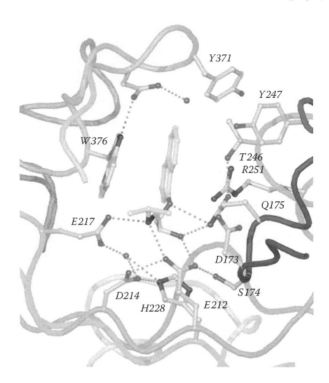

FIGURE 2.14 (*See color insert following page 152.*) Ball and stick representation of the active site with the (*S*)-propranolol molecule (green C atoms), protein residues (light-brown C) with atoms within 4 Å distance from the ligand, and two water molecules (magenta). Probable hydrogen bonds are indicated with blue dots and relevant parts of the C_α backbone are shown, color-ramped from blue at residue 171 to red at the C terminus, residue 434. The main interactions are the hydrophobic stacking of the naphthyl group with Trp376, the bidentate salt-link interaction between the positively charged secondary amine and the catalytic residues Glu212 and Glu217, and hydrogen bonding of the chiral hydroxy group with Gln175 and Glu212. (From Ståhlberg, J. et al., *J Mol Biol*, 305, 83, 2001. With permission.)

enthalpy and entropy on nonspecific binding sites were −1.1 kcal mol^{-1} and +0.1 cal mol^{-1} K^{-1}, respectively. For specific binding sites, the values were −1.9 kcal mol^{-1} and −2.6 cal mol^{-1} K^{-1}, respectively, for (*R*)-propranolol, and +1.6 kcal mol^{-1} and −11.6 cal mol^{-1} K^{-1}, respectively, for (*S*)-propranolol. The adsorption of the more retained enantiomer, (*S*)-propranolol, is entropy driven and endothermic, whereas that of *(R)*-enantiomer is enthalpy driven and exothermic [78]. This is why unusual chiral separation of propranolol on the CBH I (Cel7A) column occurs.

2.3.5 PENICILLIN G-ACYLASE

PGA from *Escherichia coli* catalyzes the hydrolysis of penicillin G to phenylacetic acid and 6-aminopenicillanic acid and is composed of two nonidentical subunits of 23 and 62 kDa [79]. PGA was immobilized via its amino and carboxy group [22]. The

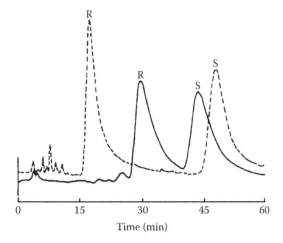

FIGURE 2.15   Separation of (*R,S*)-propranolol at 10 and 45°C. Mobile phase: acetate buffer pH 5.5, I = 0.02. Solid-line chromatogram at 10°C; dotted-line chromatogram at 45°C. (From Isaksson, R. et al., *Trends Anal Chem* 13, 435, 1994. With permission.)

former was attained using amino-silica gels activated by *N,N'*-disuccinimidylcarbonate (DSC) and epoxy-silica gels, and the latter using amino-silica gels with 1-ethyl-3-(3'-dimethylaminopropyl)carbodimide and *N*-hydroxysulfosuccinimide. The best results in terms of bound amount of PGA, enzymatic activity, and enantioselectivity were obtained with PGA immobilized on epoxy-silica particles. Enantiomers of 2-arylpropionic acid derivatives such as ketoprofen, suprofen, and fenoprofen and phenoxypropionic acids such as 2-(4-phenoxyphenoxy)propionic acid and 2-(4-benzylphenoxy)propionic acid were separated on PGA-based CSPs [22].

Chiral recognition mechanism on PGA was investigated using molecular modeling and docking studies [80,81]. As depicted in Figure 2.16a, the associated binding mode for (*S*)-2-(4-chlorophenyl)-2-phenoxyacetic acid is characterized by the presence of numerous H-bond interactions involving the carboxy group with both the backbone NH and the OH group of SerB386 [80]. An additional H-bond is established between the ether oxygen of the ligand and the ArgB263 side chain. Moreover, an ionic interaction is formed between the negatively charged carboxy group and the positively charged *N*-terminal SerB1. A set of charge–transfer interactions is also found between the ligand phenoxy and phenyl moieties and the PheB24, PheA146, PheB71, and PheB256 aromatic rings. Regarding (*R*)-2-(4-chlorophenyl)-2-phenoxyacetic acid, the calculated posing is similar to the one found for the (*S*)-enantiomer, even if the H-bond interaction between the ether oxygen and ArgB263 side chain is absent and the phenoxy moiety is placed in the same position occupied by the phenyl ring in the (*S*)-enantiomer (Figure 2.16b) [80].

2.3.6   ANTIBODY

In 1982, Mertens et al. [23] raised polyclonal antibodies for (+)- and (–)-abscisic acid (ABA) in rabbit sera and prepared their immunoglobulin G (IgG) fractions. Then,

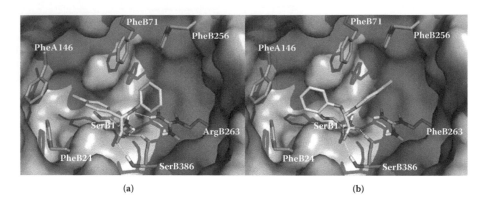

(a)　　　　　　　　　　　　　　　　　　　(b)

FIGURE 2.16　(*See color insert following page 152.*) Binding mode of (*S*)-2-(4-chlorophenyl)-2-phenoxyacetic acid (a) and (*R*)-2-(4-chlorophenyl)-2-phenoxyacetic acid (b) within PGA. For clarity reasons, only interacting residues are displayed. Hydrogen bonds between ligand and protein are shown as dashed yellow lines. Ligand (white) and interacting key residues (orange) are represented as stick models; the protein is a light gray Connolly surface. (From Lavecchia, A. et al., *J Mol Graph Mod*, 25, 777, 2007. With permission.)

the IgG antibodies bound to cyanogen bromide-activated Sepharose 4B were used for the isolation of ^{3}H–(+)-ABA as well as ^{3}H–(–)-ABA [23]. Similarly, an antibody for (+)-ABA was produced using ABA-4'-*p*-aminobenzoyl-hydrazone coupled to keyhole limpet hemocyanin (KLH) as an immunogene [82]. The obtained monoclonal antibodies were bound to cyanogen bromide-activated Sepharose 4B. The target enantiomer was retained by the antibody CSPs and eluted second, whereas the opposite enantiomer eluted with the void volume. The major disadvantage of an antibody column is that for elution of a more retained enantiomer a strong mobile phase is required and that it is unstable for repeated uses.

Recently, Hofstetter and his group raised stereospecific antibodies against a broad class of substances, α-amino acids, and applied them to enatioseparations of α-amino acids [83–86]. *p*-Amino-D- and L-Phe, respectively, were coupled to KLH or BSA via the *p*-amino group by diazotization of its Tyr residue, and the resulting conjugates, *p*-azo-D-Phe-KLH or -BSA and *p*-azo-L-Phe-KLH or -BSA, were used as the immunogenes for rabbits. The produced monoclonal antibodies were bound to Sepharose 4B [83], POROS-OH (poly(styrene-divinylbenzene) porous perfusion beads) [84], and silica gels [85,86] activated by DSC. Routine enantioseparations of α-amino acids could be achieved on antibodies bound to the latter two supports under true HPLC conditions. Figure 2.17 shows enantioseparations of Tyr, Phe, and *p*-aminophenylalanine on a monoclonal antibody column against L-amino acid [84]. The obtained column gave the enantioselectivity for a wide variety of α-amino acids: the higher one for aromatic and bulky side-chain amino acids; and the lower one for aliphatic amino acids [84]. Using an anti-D-amino acid antibody as a chiral selector, the L-enantiomers eluted with the void volume, whereas the D-enantiomers eluted second. The enantioselective binding of D-amino acids on the anti-D-amino acid antibody was investigated using molecular modeling and ligand docking [86]. The

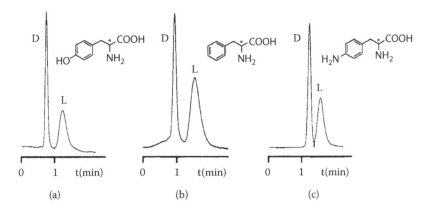

FIGURE 2.17   Separations of D,L-Tyr (a), D,L-Phe (b) and D,L-*p*-aminophenylalanine (c) on anti-L-amino acid column at flow rates of 4 (a, b) and 2 mL/min (c). (From Hofstetter, O. et al., *Anal Chem*, 74, 2124, 2002. With permission.)

results indicate that in addition to four hydrogen bonds, formed between amino acid residues in the binding site and the ligand, a number of hydrophobic interactions are involved in the formation of the antibody–ligand complex, as shown in Figure 2.18 [86]. The aromatic side chain of the ligand interacts with Trp and Tyr residues in the binding site through π–π stacking.

Recombinant antibody fragments (antigen binding fragment, Fab) have been prepared for finrozole (Figure 2.19), which has two chiral centers, using genetic engineering techniques [87]. The obtained Fab-fragments, which have a molecular weight of about 30 kDa, were bound to chelating sepharose fast flow loaded with copper ions. Furthermore, single mutants (Tyr96Val) or double mutants (Tyr96Val/Trp33Ala mutants) of the Fab-fragments were prepared to increase the lifetime of the antibody fragment columns [88]. Decreasing the affinity by one or more orders of magnitude resulted in enabling the use of a lower concentration of organic solvent for elution. The crystal structures of this enantioselective antibody Fab-fragment are determined in the absence or the presence of the hapten [89]. The hapten molecule was tightly bound in a deep cleft between the light and heavy chains of the Fab-fragment. From the complex structure, it was also possible to describe the molecular basis for enantioselectivity and to deduce the absolute configurations of all the four different stereoisomers (a–d) of finrozole. The antibody fragment selectively bound the SR (a) enantiomer from the racemic mixture of a- and d-enantiomers. Asp95 and Asn35 of the H-chain in the antibody seem to provide this specificity through hydrogen bonding interactions [89], as shown in Figure 2.20.

2.3.7   Nicotinic Acetylcholine Receptor and Human Organic Cation Transporter

Recently, membrane-bound proteins such as receptors, ion channels, and drug transporters, have been incorporated into chromatographic systems [28,29]. In this approach, cellular membrane fragments obtained from cell lines expressing a target

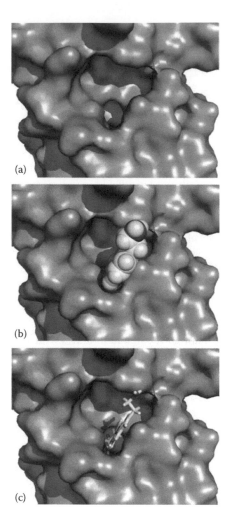

FIGURE 2.18 *(See color insert following page 152.)* Surface contour images of the modeled anti-D-amino acid structure (a) without the ligand, (b) with the docked ligand in a spherical representation, and (c) with the docked ligand in a stick representation. (From Ranieri, D.I. et al., *Chirality*, 20, 565, 2008. With permission.)

protein were used to create cellular membrane affinity chromatography (CMAC) columns, which were used to study ligand binding to the target proteins as well as the functional consequences of the binding interactions [28,29]. Furthermore, by using the resulting CMAC column chiral separations of some racemates were attained with poor column efficiencies.

In molecular modeling and docking studies of $\alpha_3\beta_4$ nAChR with DM and LM, the lowest energy docked conformations of the DM and LM complexes were both located at the Val/Phe ring and involved the insertion of the hydrophobic portion of both molecules into the hydrophobic cleft found at this position, as shown in Figures 2.21a

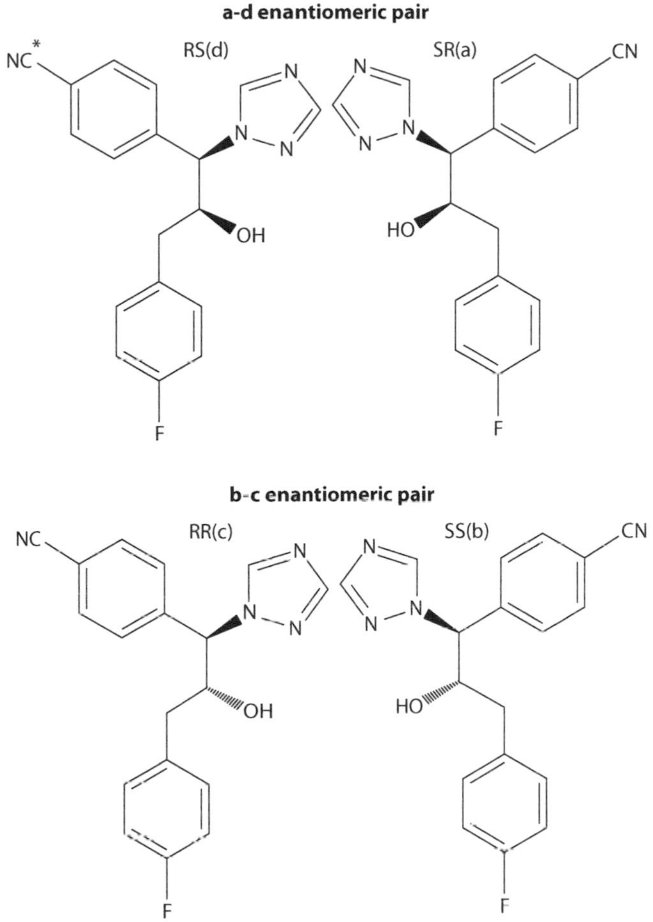

FIGURE 2.19   The four stereoisomers (a–d) of the hapten 4′-[3-(4-fluorophenyl)-2-hydroxy-1-(1,2,4-triazol-1-yl)-propyl]benzonitrile. The position of the linker, which was used to conjugate a protein for the development of the original monoclonal antibodies through immunization, is marked by an asterisk in the d-enantiomer form. (From Parkkinen, T. et al., *J Mol Biol*, 357, 472, 2006. With permission.)

and 2.21b [47]. The mirror image relationship between the two enantiomers and their lack of conformational mobility produce two unique orientations. In the case of DM (Figure 2.21a), the bridgehead nitrogen atom of the docked molecule is oriented toward Ser residues located on the α_3 helix at position 8 (Ser ring). With LM, the bridgehead nitrogen atom was pointing away from the two helices forming the α_3 and β_4 subunits (Figure 2.21b). The orientation of DM increases the probability of hydrogen bond formation between the bridgehead nitrogen and a hydroxy moiety on Ser residues, whereas the orientation of LM reduces this probability.

From the viewpoint of the dynamic chiral recognition mechanism, the first step is the insertion of the hydrophobic moiety of the methorphan molecule into the

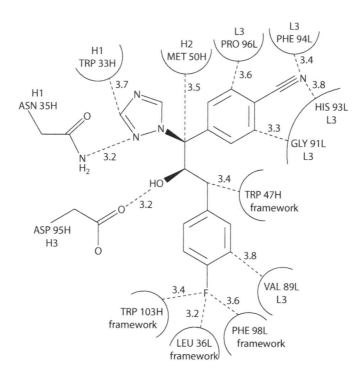

FIGURE 2.20 The binding of the a-(*SR*)-enantiomer of the hapten and its contacts (in Å) with the ENA11His Fab-fragment. The locations of amino acid residues in four CDR loops (L3, H1, H2, and H3) and in the framework area of variable domains are shown. (From Parkkinen, T. et al., *J Mol Biol*, 357, 475, 2006. With permission.)

hydrophobic pocket on the nAChR, which occurs at the same rate and with the same probability for both DM and LM. This initial binding interaction of DM and LM with the hydrophobic pocket is not enantioselective, but it tethers the molecules to the receptor. The second step is the configurationally defining hydrogen bonding interactions. These two steps in the binding process are interconnected and produce a dynamic chiral recognition mechanism.

With regard to the modeling of stereoselective binding to hOCT1, the resulting model contained a positive ion interaction site, a hydrophobic interaction site, and two hydrogen-bond acceptor sites [48]. When (*R*)-verapamil was fit to the proposed pharmacophore, all the relevant functional groups of the molecule matched the hypothesis (Figure 2.22a), whereas (*S*)-verapamil could be mapped to only three of the model feature sites (Figure 2.22b). These data suggest that for (*R*)- and (*S*)-verapamil, chiral recognition is a multistep process involving an initial tethering of the selectant to the selector, most probably occurring at the positive ion interaction site, followed by conformational adjustments that produce the optimum interactions. This process results in a distribution of selectant–selector complexes of varying relative stabilities and the observed enantioselectivity.

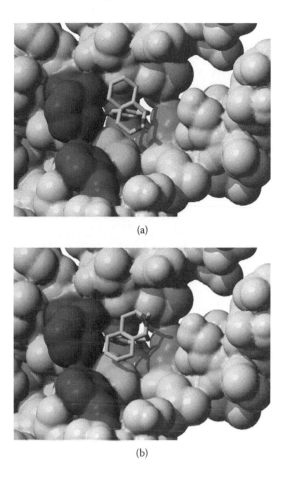

(a)

(b)

FIGURE 2.21 *(See color insert following page 152.)* The most stable docked orientations of (a) DM and (b) LM complexes with the model of the central lumen of the $\alpha_3\beta_4$ nAChR. Hydrophobic clefts formed within the channel are shown in detail. Residues forming the cleft are color coded Phe, blue; Val, green; and Ser, orange. (From Jozwiak, K. et al., *J Chromatogr B*, 797, 203, 2008. With permission.)

2.4 CONCLUSIONS

Many CSPs based on a protein or glycoprotein have been developed and used for enantioseparations of a wide variety of compounds in LC. These CSPs have the enantioselectivity for a wide range of compounds because of multiple binding sites on the surface of a chiral selector or multiple binding interactions between a chiral selector and selectant. Chiral recognition mechanism on a protein has been elucidated using spectroscopy, molecular modeling and ligand docking, or x-ray crystallography. In the static viewpoint, chiral recognition on a protein or protein-based stationary

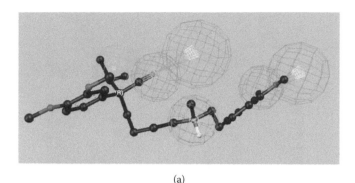

(a)

(b)

FIGURE 2.22   (*See color insert following page 152.*) The fit of verapamil enantiomers in the proposed pharmacophore, where (a) is the mapping of (*R*)-verapamil and (b) is the mapping of (*S*)-verapamil. Nonpolar hydrogen atoms have been omitted for clarity. (From Moaddel, R. et al., *Br J Pharmacol*, 151, 1311, 2007. With permission.)

phases could be explained based on interaction models such as the TPI rule and SR model. A dynamic model for chiral recognition on a protein or protein-based stationary phases is also proposed. Both static and dynamic models could well explain chiral recognition of enantiomers on a protein or protein-based stationary phases.

2.5   LIST OF ABBREVIATIONS

ABA	abscisic acid
AGP	α_1-acid glycoprotein
Ala	alanine
Asn	asparagine
Asp	aspartic acid
Arg	arginine
BSA	bovine serum albumin

cAGP	chicken α_1-acid glycoprotein
CBH I, Cel7A	cellobiohydrolase I from *Trichoderma reesei*
CBH II, Cel6A	cellobiohydrolase II from *Trichoderma reesei*
CBH 58, Cel7D	cellobiohydrolase 58 from *Phanerochaete chrysosporium*
CD	circular dichroism
cDNA	complementary DNA
CDRs	complementarity determining regions
CE	capillary electrophoresis
CMAC	cellular membrane affinity chromatography
CSP	chiral stationary phase
DM	dextromethorphan
DMOA	*N,N*-dimethyl-*n*-octylamine
DSC	*N,N′*-disuccinimidylcarbonate
EG I, Cel7B	endoglucanase I from *Trichoderma reesei*
EG II, Cel6B	endoglucanase II from *Trichoderma reesei*
Fab	antigen binding fragment
FCP	four-contact point
FL	four-location
Gln	glutamine
Glu	glutamic acid
hOCT1	human organic cation transporter 1
HPLC	high-performance liquid chromatography
HSA	human serum albumin
IDH	isocitrate dehydrogenase
IgG	immunoglobulin G
KLH	keyhole limpet hemocyanin
LC	liquid chromatography
LM	levomethorphanon
Lys	lysine
nAChR	nicotinic acetylcholine receptor
NMR	nuclear magnetic resonance
OGCHI	ovoglycoprotein from chicken egg whites
OMCHI	ovomucoid from chicken egg whites
OMCHI3	unglycosylated, third domain of ovomucoid from chicken egg white
OMCHI3S	glycosylated, third domain of ovomucoid from chicken egg white
OMTKY3	unglycosylated, third domain of ovomucoid from turkey egg white
OMTKY3S	glycosylated, third domain of ovomucoid from turkey egg white
ORM	orosomucoid
ORM1	F1-S variant of orosomucoid
ORM2	A variant of orosomucoid
PGA	penicillin G-acylase

Phe   phenylalanine
p*I*   isoelectric point
RfBP   riboflavin binding protein
Ser   serine
SR   stereocenter-recognition
TPA   three-point attachment
TPI   three-point interaction
Trp   tryptophan
Tyr   tyrosine
Val   valine
Wf   warfarin

REFERENCES

1. Stewart, KK, Doherty, RF. 1973. Resolution of DL-tryptophan by affinity chromatography on bovine-serum albumin-agarose columns. *Proc. Nat. Acad. Sci., USA* 70: 2850–2852.
2. Allenmark, S. 1991. *Chromatographic Enantioseparation: Methods and Applications*, 2nd edn. New York: Ellis Horwood.
3. Wainer, IW. Ed. 1993. *Drug Stereochemistry: Analytical Methods and Pharmacology*, 2nd edn. New York: Marcel Dekker.
4. Gubitz, G, Schmid, MG. Eds. 2004. *Chiral Separations: Methods and Protocols*. Totowa: Humana Press.
5. Haginaka, J. 2001. Protein-based chiral stationary phases for HPLC enantioseparations. *J Chromatogr A* 906: 253–273.
6. Haginaka, J. 2008. Recent progresses in protein-based chiral stationary phases for enantioseparations in liquid chromatography. *J Chromatogr B* 875: 12–19.
7. Allenmark, S, Bomgren, B, Borén, H. 1983. Direct liquid chromatographic separation of enantiomers on immobilized protein stationary phases. III. Optical resolution of a series of *N*-aroyl D,L-amino acids by high-performance liquid chromatography on bovine serum albumin covalently bound to silica. *J Chromatogr A* 264: 63–68.
8. Domenici, E, Bertucci, C, Salvadori, P, Felix, G, Cahagne, I, Motellier, S, Wainer, IW. 1990. Synthesis and chromatographic properties of an HPLC chiral stationary phase based upon human serum albumin. *Chromatographia* 29: 170–176.
9. Fitos, I, Visy, J, Simonyi, M. 2002. Species-dependency in chiral-drug recognition of serum albumin studied by chromatographic methods. *J Biochem Biophys Methods* 54: 71–84.
10. Hermansson, J. 1983. Direct liquid chromatographic resolution of racemic drugs using α_1-acid glycoprotein as the chiral stationary phase. *J Chromatogr A* 269: 71–80.
11. Miwa, T, Ichikawa, M, Tsuno, M, Hattori, T, Miyakawa, T, Kayano, M, Miyake, Y. 1987. Direct liquid chromatographic resolution of racemic compounds. Use of ovomucoid as a column ligand. *Chem Pharm Bull* 35: 682–686.
12. Haginaka, J, Seyama, C, Kanasugi, N. 1995. The absence of chiral recognition ability in ovomucoid: Ovoglycoprotein-bonded HPLC stationary phases for chiral recognition. *Anal Chem* 67: 2539–2547.
13. Sadakane, Y, Matsunaga, H, Nakagomi, K, Hatanaka, Y, Haginaka, J. 2002. Protein domain of chicken α_1-acid glycoprotein is responsible for chiral recognition ability *Biochem Biophys Res Commun* 295: 587–590.
14. Miwa, T, Miyakawa, T, Miyake, Y. 1988. Characteristics of an avidin-conjugated column in direct liquid chromatographic resolution of racemic compounds. *J Chromatogr A* 457: 227–233.

15. Mano, N, Oda, Y, Asakawa, N, Yoshida, Y, Sato, T. 1992. Development of a flavoprotein column for chiral separation by high-performance liquid chromatography. *J Chromatogr A* 623: 221–228.
16. Thelohan, S, Jadaud, P, Wainer, IW. 1989. Immobilized enzymes as chromatographic phases for HPLC: the chromatography of free and derivatized amino acids on immobilized trypsin. *Chromatographia* 28: 551–555.
17. Wainer, IW, Jadaud, P, Schombaum, GR, Kadodkar, SV, Henry, MP. 1988. Enzymes as HPLC Stationary phases for chiral resolutions: initial investigations with α-chymotrypsin. *Chromatographia* 25: 903–907.
18. Erlandsson, P, Marle, I, Hansson, L, Isaksson, R, Petterson, C, Petterson, G. 1990. Immobilized cellulase (CBH I) as a chiral stationary phase for direct resolution of enantiomers. *J Am Chem Soc* 112: 4573–4574.
19. Haginaka, J, Murashima, T, Seyama, C. 1994. Separation of enantiomers on a lysozyme-bonded silica column. *J Chromatogr A* 666: 203–210.
20. Haginaka, J, Miyano, Y, Saizen, Y, Seyama, C, Murashima, T. 1995. Separation of enantiomers on a pepsin-bonded column. *J Chromatogr A* 708: 161–168.
21. Nyström, A, Strandberg, A, Aspergren, A, Behr, S, Karlsson, A. 1999. Use of immobilized amyloglucosidase as chiral selector in chromatography. Immobilization and performance in liquid chromatography. *Chromatographia* 50: 209–214.
22. Massolini, G, Calleri, E, de Lorenzi, E, Pregnolato, M, Terreni, M, Félix, G, Gandini, C. 2001. Immobilized penicillin G acylase as reactor and chiral selector in liquid chromatography. *J Chromatogr A* 921: 147–160.
23. Mertens, R, Stuning, M, Weiler, EW. 1982. Metabolism of tritiated enantiomers of abscisic acid prepared by immunoaffinity chromatography. *Naturwissenschaften* 69: 595–597.
24. Mano, N, Oda, Y, Asakawa, N, Yoshida, Y, Sato, T. 1992. Conalbumin-conjugated silica gel, a new chiral stationary phase for high-performance liquid chromatography. *J Chromatogr A* 603: 105–109.
25. Ravelet, C, Michaud, M, Ravel, A, Grosset, C, Villet, A, Peyrin, E. 2004. Streptavidin chiral stationary phase for the separation of adenosine enantiomers. *J Chromatogr A* 1036: 155–160.
26. Massolini, G, de Lorenzi, E, Calleri, E, Bertucci, C, Monaco, HL, Perduca, M, Caccialanza, G, Wainer, IW. 2001. Properties of a stationary phase based on immobilised chicken liver basic fatty acid-binding protein. *J Chromatogr B* 751: 117–130.
27. Hofstetter, H, Hofstetter, O. 2005. Antibodies as tailor-made chiral selectors for detection and separation of stereoisomers. *Trends Anal Chem* 24: 869–879.
28. Jozwiak, K, Hernandez, SC, Kellar, KJ, Wainer, IW. 2003. Enantioselective interactions of dextromethorphan and levomethorphan with the $\alpha_3\beta_4$-nicotinic acetylcholine receptor: comparison of chromatographic and functional data. *J Chromatogr B* 797: 373–379.
29. Moaddel, R, Patel, S, Jozwiak, K, Yamaguchi, R, Ho, PC, Wainer, IW. 2005. Enantioselective binding to the human organic cation transporter-1 (hOCT1) determined using an immobilized hOCT1 liquid chromatographic stationary phase. *Chirality* 17: 501–506.
30. Bentley, R. 2003. Diastereoisomerism, contact points, and chiral selectivity: a four-site saga. *Arch Biochem Biophys* 414: 1–12.
31. Easson, LH, Stedman, E. 1933. Studies on the relationship between chemical constitution and physiological action. V. Molecular dissymmetry and physiological activity. *Biochem J* 27:1257–1266.
32. Ogston, AG. 1948. Interpretation of experiments on metabolic processes, using isotopic tracer elements. *Nature* 162: 963.

33. Nederkoorn, PHJ, van Gelder, EM, Donne-Op den Kelder, G, Timmerman, H. 1996. The agonistic binding site at the histamine H2 receptor. II. Theoretical investigations of histamine binding to receptor models of the seven α-helical transmembrane domain. *J Comput Aid Mol Des* 10: 479–489.
34. Suami, T, Hough L. 1993. Molecular mechanisms of sweet taste 3: aspartame and its non-sweet isomers. *Food Chem* 46: 235–238.
35. Dalgliesh, CE. 1952. The optical resolution of aromatic amino acids on paper chromatograms. *J Chem Soc*: 3940–3952.
36. Davankov, VA, Kurganov, AA. 1983. The role of achiral sorbent matrix in chiral recognition of amino acid enantiomers in ligand-exchange chromatography. *Chromatographia* 17: 686–690.
37. Pirkle, WH, Welch, CJ, Hyun, MH. 1983. A chiral recognition model for the chromatographic resolution of *N*-acylated 1-aryl-1-aminoalkanes. *J Org Chem* 48: 5022–5026.
38. Wilcox, PE, Heidelberger, C, Potter, VR. 1950. Chemical preparation of asymmetrically labeled citric acid. *J Am Chem Soc* 72: 5019–5024.
39. Sokolov, VI, Zefirov, NS. 1991. Enantioselectivity in 2-point binding: The model of rocking tetrahedron. *Dokl Akad Nauk SSSR* 319: 1382–1383.
40. Davankov, VA. 1997. The nature of chiral recognition: Is it a three-point interaction? *Chirality* 9: 99–102.
41. Topiol, S, Sabio, M. 1989. Interactions between eight centers are required for chiral recognition. *J Am Chem Soc* 111: 4109–4110.
42. Bentley, R. 1995. Chirality in biology. In *Encyclopedia of Molecular Biology and Molecular Medicine, vol. 1*, ed. Myers, RA., 332–347. Weinheim: VCH.
43. Mesecar, AD, Koshland Jr, DE. 2000. A new model for protein stereospecificity. *Nature* 403: 614–615.
44. Sundaresan, V, Abrol, R. 2002. Towards a general model for protein–substrate stereoselectivity. *Protein Sci* 11: 1330–1339.
45. Sundaresan, V, Abrol, R. 2005. Biological chiral recognition: The substrate's perspective. *Chirality* 17: S30–S39.
46. Booth, T, Wahnon, D, Wainer, IW. 1997. Is chiral recognition a three-point process? *Chirality* 9: 96–98.
47. Jozwiak, K, Ravichandran, S, Collins, JR, Wainer, IW. 2004. Interaction of noncompetitive inhibitors with an immobilized $\alpha_3\beta_4$ nicotinic acetylcholine receptor investigated by affinity chromatography, quantitative-structure activity relationship analysis, and molecular docking. *J Med Chem* 47: 4008–4021.
48. Moaddel1, R, Ravichandran, S, Bighi1, F, Yamaguchi1, R, Wainer, IW. 2007. Pharmacophore modelling of stereoselective binding to the human organic cation transporter (hOCT1). *Br J Pharmacol* 151: 1305–1314.
49. Jozwiak, K, Moaddel, R, Ravichandran, S, Plazinska, A, Kozaka, J, Patel, S, Yamaguchi, R, Wainer, IW. 2008. Exploring enantiospecific ligand–protein interactions using cellular membrane affinity chromatography: Chiral recognition as a dynamic process. *J Chromatogr B* 875: 200–207.
50. Chuang, VTG, Otagiri, M. 2006. Stereoselective binding of human serum albumin. Chirality 18:159–166.
51. He, XM, Carter, DC. 1992. Atomic structure and chemistry of human serum albumin. *Nature* 358: 209–215.
52. Petitpas, I, Bhattacharya, AB, Twine, S, East, M, Curry, S. 2001. Crystal structure analysis of warfarin binding to human serum albumin. *J Biol Chem* 276: 22804–22809.
53. Lämmerhofer, M. 2010. Chiral recognition by enantioselective liquid chromatography: Mechanisms and modern chiral stationary phases. *J Chromatogr A* 1217: 814–856.
54. Fournier, T, Medjoubi-N, N, Porquet, D. 2000. Alpha-1-acid glycoprotein. *Biochim Biophys Acta* 1482: 157–171.

55. Haginaka, J, Matsunaga, H. 2000. Separation of enantiomers on HPLC chiral stationary phases based on human plasma α_1-acid glycoprotein: effect of sugar moiety on chiral recognition ability. *Enantiomer* 5: 37–45.

56. Hazai, E, Visy, J, Fitos, I, Bikádi, Z, Simonyi, M. 2006. Selective binding of coumarin enantiomers to human α_1-acid glycoprotein genetic variants. *Bioorg Med Chem* 14: 1959–1965.

57. Hermansson, J. 1989. Enantiomeric separation of drugs and related compounds based on their interaction with α_1-acid glycoprotein. *Trends Anal Chem* 8: 251–259.

58. Makamba, H, Andrisano, V, Gotti, R, Cavrini, V, Felix, G. 1998. Sparteine as mobile phase modifier in the chiral separation of hydrophobic basic drugs on an α_1-acid glycoprotein column. *J Chromatogr A* 818: 43–52.

59. Kremer, JM, Wilting, J, Janssen, LH. 1988. Drug binding to human alpha-1-acid glycoprotein in health and disease. *Pharmacol Rev* 40: 1–47.

60. Zsila, F, Iwao, Y. 2007. The drug binding site of human α_1-acid glycoprotein: Insight from induced circular dichroism and electronic absorption spectra. *Biochim Biophys Acta* 1770: 797–809.

61. Irdale, J, Aubry, AF, Wainer, IW. 1991. The effects of pH and alcoholic organic modifiers on the direct separation of some acidic, basic and neutral compounds on a commercially available ovomucoid column. *Chromatographia* 31: 329–334.

62. Haginaka, J, Seyama, C, Yasuda, H, Takahashi, K. 1992. Investigation of enantioselectivity and enantiomeric elution order of propranolol and its ester derivatives on an ovomucoid-bonded column. *J Chromatogr A* 598: 67–72.

63. Kirkland, KM, Neilson, KL, McCombs, DA. 1991. Comparison of a new ovomucoid and a second-generation α_1-acid glycoprotein-based chiral column for the direct high-performance liquid chromatography resolution of drug enantiomers. *J Chromatogr A* 545: 43–58.

64. Oda, Y, Asakawa, N, Kajima, T, Yoshida, Y, Sato, T. 1991. On-line determination and resolution of verapamil enantiomers by high-performance liquid chromatography with column switching. *J Chromatogr A* 541: 411–418.

65. Pinkerton, TC, Howe, WJ, Urlich, EL, Comiskey, JP, Haginaka, J, Murashima, T, Walkenhorst, WF, Westler, WM, Markley, JL. 1995. Protein-binding chiral recognition of HPLC stationary phases made with whole, fragmented and third domain turkey ovomucoid. *Anal Chem* 67: 2354–2367.

66. Haginaka, J. 2002. LC packing materials for pharmaceutical and biomedical analysis. *Chromatography* 23: 1–11.

67. Kato, I, Schrode, J, Kohr, WJ, Laskowski, Jr, M. 1987. Chicken ovomucoid: determination of its amino acid sequence, determination of the trypsin reactive site, and preparation of all three of its domains. *Biochemistry* 26: 193–201.

68. Haginaka, J, Seyama, C, Murashima, T. 1995. Retentive and enantioselective properties of ovmucoid-bonded silica columns. Influence of protein purity and isolation method. *J Chromatogr A* 704: 279–287.

69. Matsunaga, H, Haginaka, J. 2006. Investigation of chiral recognition mechanism on chicken α_1-acid glycoprotein using separation system. *J Chromatogr A* 1106: 124–130.

70. Zsila, F, Matsunaga, H, Bikádi, H, Haginaka, J. 2006. Ligand-side chain intermolecular exciton circular dichroism spectra reveal the essential role of the conserved tryptophan residue in the molecular recognition properties of the lipocalin member chicken α_1-acid glycoprotein. *Biochim Biophys Acta* 1760: 1248–1273.

71. Uzcategui, E, Ruiz, A, Montesino, R, Johansson, G, Petterson, G. 1991. The 1,4-β-D-glucan cellobiohydrolases from *Phanerochaete chrysosporium*. I. A system of synergistically acting enzymes homologous to *Trichoderma reesei*. *J Biotecnol* 19: 271–286.

72. Isaksson, R, Pettersson, C, Pettersson, G, Jönsson, S, Ståhlberg, J, Hermansson, J, Marle, I. 1994. Cellulases as chiral selectors. *Trends Anal Chem* 13: 431–439.

73. Tomme, P, van Tilbeurgh, H, Pettersson, G, van Damme, J, Vandekerckhove, J, Knowles, J, Teeri, T, Claeyssens, M. 1988. Studies of the cellulolytic system of *Trichoderma reesei* QM 9414. Analysis of domain function in two cellobiohydrolases by limited proteolysis. *Eur J Biochem* 170: 575–581.

74. Marle, I, Erlandsson, P, Hansson, L, Isaksson, R, Pettersson, C, Pettersson, G. 1991. Separation of enantiomers using cellulase (CBH I) silica as a chiral stationary phase. *J. Chromatogr A* 586: 233–248.

75. Henriksson, H, Munoz, IG, Isaksson, R, Pettersson, G, Johansson, G. 2000. Cellobiohydrolase 58 (P.c. Cel 7D) is complementary to the homologous CBH I (T.r. Cel 7A) in enantioseparations. *J. Chromatogr. A* 898: 63–74.

76. Divne, C, Ståhlberg, J, Reinikainen, T, Ruohonen, L, Pettersson, G, Knowles, JKC, Teeri, TT, Jones, TA. 1994. The three-dimensional structure of the catalytic core of cellobiohydrolase I from *Trichoderma reesei. Science* 265: 524–528.

77. Ståhlberg, J, Henriksson, H, Divne, C, Isaksson, R, Pettersson, G, Johansson, G, Jones, TW. 2001. Structural basis for enantiomer binding and separation of a common β-blocker: Crystal structure of cellobiohydrolase Cel7A with bound (*S*)-propranolol at 1.9 Å resolution. *J Mol Biol* 305: 79–93.

78. Fornstedt, T, Sajonz, P, Guiochon, G. 1999. Thermodynamic study of an unusual chiral separation. Propranolol enantiomers on an immobilized cellulase. *J Am Chem Soc* 119: 1254–1264.

79. Barbero, JL, Buesa, JM, de Buitrago, GG, Méndez, E, Pérez-Aranda, A, Carcía, JL. 1986. Complete nucleotide sequence of the penicillin acylase gene from *Kluyveru cifrophila. Gene* 49: 69–80.

80. Lavecchia, A, Cosconati, S, Novellino, E, Calleri, E, Temporini, C, Massolini, G, Carbonara, G, Fracchiolla, G, Loiodice, F. 2007. Exploring the molecular basis of the enantioselective binding of penicillin G acylase towards a series of 2-aryloxyalkanoic acids: A docking and molecular dynamics study. *J. Mol. Graph. Mod.* 25: 773–783.

81. Temporini, C, Calleri, E, Fracchiolla, G, Carbonara, G, Loiodice, F, Lavecchia, A, Tortorella, P, Brusotti, G, Massolini, G. 2007. Enantiomeric separation of 2-aryloxyalkyl- and 2-arylalkyl-2-aryloxyacetic acids on a Penicillin G Acylase-based chiral stationary phase: Influence of the chemical structure on retention and enantioselectivity. *J Pharmaceut Biomed Anal* 45: 211–218.

82. Quarrie, SA, Galfre, G. 1985. Use of different hapten-protein conjugates immobilized on nitrocellulose to screen monoclonal antibodies to abscisic acid. *Anal Biochem* 151: 389–399.

83. Hofstetter, O, Hofstetter, H, Wilchek, M, Schurig, V, Green, BS. 2000. Production and applications of antibodies directed against the chiral center of α-amino acids. *Int J BioChromatogr* 5: 165–174.

84. Hofstetter, O, Lindstrom, H, Hofstetter, H. 2002. Direct resolution of enantiomers in high-performance immunoaffinity chromatography under isocratic conditions. *Anal Chem* 74: 2119–2125.

85. Franco, EJ, Hofstetter, H, Hofstetter, O. 2006. A comparative evaluation of random and sitespecific immobilization techniques for the preparation of antibody-based chiral stationary phases. *J Sep Sci* 29: 1458–1469.

86. Ranieri, DI, Corgliano, DM, Franco, EJ, Hofstetter, H, Hofstetter, O. 2008. Investigation of the stereoselectivity of an anti-amino acid antibody using molecular modeling and ligand docking. *Chirality* 20: 559–570.

87. Nevanen, TK, Söderholm, L, Kukkonen, K, Suortti, T, Teerinen, T, Linder, M, Söderlund, H, Teeri, TT. 2001. Efficient enantioselective separation of drug enantiomers by immobilised antibody fragments. *J Chromatogr A* 925: 89–97.

88. Nevanen, TK, Hellman, ML, Munck, N, Wohlfahrt, G, Koivula, A, Söderlund, H. 2001. Model-based mutagenesis to improve the enantioselective fractionation properties of an antibody. *Protein Eng* 16: 1089–1097.
89. Parkkinen, T, Nevanen, TK, Koivula, A, Rouvinen, J. 2006. Crystal structures of an enantioselective Fab-fragment in free and complex forms. *J Mol Biol* 357: 471–480.

3 Mechanistic Aspects and Applications of Chiral Ligand-Exchange Chromatography

Benedetto Natalini, Roccaldo Sardella,
Antonio Macchiarulo, Maura Marinozzi,
Emidio Camaioni, and Roberto Pellicciari

CONTENTS

3.1 INTRODUCTION

Different from other chromatographic techniques, the interaction between the chiral selector and the enantiomer in chiral ligand-exchange chromatography (CLEC) does not take place in direct contact. The interaction is mediated by a central metal ion that as a Lewis acid simultaneously coordinates the two species through dative bonds with the following formation of a mixed ternary complex. The absence of a close contact between the selector and the analyte and the presence of a metal cation in the chromatographic environment contribute to the difficulty of the rationalization of the chromatographic event, especially when computational approaches are employed for this study. Among the enantiomer chromatographic separation techniques, ligand-exchange has been exploited in all the main ways of accomplishment: chiral mobile phase (CMP), covalently bound chiral stationary phase (B-CSP), and coated chiral stationary phase (C-CSP). Although the first, commercially available chiral columns were those based on B-CSP, soon after followed by the others, the mechanistic aspects dealing with the chiral recognition and the chromatographic separation process in CLEC are among the less studied ones, most probably because of the complexity of the theoretical models of reference. More recently, this difficulty has however spurred different computational approaches with the aim of clarifying the interaction mechanism and indicating the main parameters that describe the analyte responsible for the enantiorecognition process.

This chapter illustrates the state-of-the-art in this field by highlighting selected examples from the recent literature. For more comprehensive discussion [1,2] and early approaches [3], the reader is referred elsewhere.

3.2 THERMODYNAMIC ENANTIOSELECTIVITY IN CLEC SYSTEMS

One of the attracting features of CLEC is the possibility of selecting among a wide number of discriminating agents either acting as mobile phase additives or expressing their attitude to resolve enantiomers as part of a more complex stationary phase. A brief description of the fundamental set of equilibria that turn up in the chromatographic medium will aid understanding of the main aspects governing the enantiorecognition process in CLEC. Moreover, for the sake of clarity, the two main chromatographic modalities (CSP and CMP approaches) will be separately treated.

3.2.1 CHIRAL STATIONARY PHASES (CSPs)

In the case of a chiral selector C, covalently grafted (B-CSP) on the stationary phase, or dynamically coated (C-CSP) on a suitable material (generally through hydrophobic interactions), the reversible incorporation of the analyte A into the ternary complex AMC^s may happen by direct interaction of A^m with MC^s or, alternatively, through a two-step process as formulated in

$$\text{Mobile phase} \quad A^m$$
$$\updownarrow \qquad\qquad\qquad\qquad (3.1)$$
$$\text{Stationary phase} \quad A^s + MC^s \xrightarrow{\;K^s_{AMC}\;} AMC^s$$

where M indicates the central metal ion, and the superscripts m and s refer to the location of the species in the mobile and stationary phase, respectively.

Although a complex series of dissociation and association equilibria take place inside the column, the fundamental processes responsible for both retention and chiral recognition in all CSP-CLEC media are, however, contained in Equation (3.1) [4,5]. In other words, the enantiorecognition event that materializes in such environments exclusively relies on the different participation of the two enantiomers in the formation of the corresponding diastereomeric ternary complex. The retention of each enantiomer (expressed as the retention factor k) can be put into relation with its adsorption and subsequent (or even contemporary) complexation through

$$k_A = \varphi \cdot \frac{[A^s] + [AMC^s]}{[A^m]} \tag{3.2}$$

where φ is the phase ratio. By assuming as K^s_{AMC} the equilibrium constant of the complexation process in Equation (3.2), the k_A values can be computed as

$$k_A = \varphi \cdot \frac{[A^s] + K^s_{AMC}[A^s][MC^s]}{[A^m]}$$

$$= \varphi \cdot \frac{[A^s]}{[A^m]} \cdot (1 + K^s_{AMC}[MC^s]) \tag{3.3}$$

$$= \underline{k}_A (1 + K^s_{AMC}[MC^s])$$

where $\underline{k}_A$ is the retention factor of the solute A in the absence of any reaction of complexation, $[MC^s]$ corresponds to the concentration of the chiral sorption sites, and K^s_{AMC} represents the formation constant of the analyte–enantioresolving agent adduct on the stationary phase. From these formulations, the enantioselectivity of the column can be expressed as

$$\alpha = \frac{k_{A(R)}}{k_{A(S)}} = \frac{\underline{k}_{A(R)}}{\underline{k}_{A(S)}} \cdot \frac{1 + K^s_{A(R)MC}[MC^s]}{1 + K^s_{A(S)MC}[MC^s]}$$

$$= \frac{1 + K^s_{A(R)MC}[MC^s]}{1 + K^s_{A(S)MC}[MC^s]} \tag{3.4}$$

For all the chromatographic systems that operate in CSP mode, the enantioselectivity α approaches to the value of $K^s_{A(R)MC}/K^s_{A(S)MC}$ when the nonspecific adsorption of the solute enantiomers onto the multicomposed stationary phase is negligible. For all the enantioselective chromatographic systems, the *chromatographic (apparent) enantioselectivity, α*, is always lower than the *thermodynamic (true) enantioselectivity, α^** [6]. Without entering into detail, several adsorption sites can be classified as nonspecific in a CLEC medium. Among these, worth mentioning are the underivatized silanols, the solid sorbent material, and the alkyl chains in reverse-phased (RP)-coupled systems.

3.2.2 Chiral Mobile Phases (CMPs)

Two distinct situations can be described in the case of chiral species employed as additive to the eluent (CMP approach): that is, dealing with the discriminating agent exclusively located into the eluent and the one it partitions between the two chromatographic phases [5]. If compared with the previously given CSP systems, unremarkable differences characterize the theoretical treatment of the former CMP environments. The overall complexation equilibria can be schematized as

$$
\begin{array}{ll}
\text{Mobile phase} & A^m + MC^m \xrightarrow{\quad K^m{}_{AMC}\quad} AMC^m \\[4pt]
& \qquad\qquad \updownarrow \\[4pt]
\text{Stationary phase} & A^s
\end{array}
\tag{3.5}
$$

and, in analogy to the previous case

$$
\begin{aligned}
k_A &= \varphi \cdot \frac{[A^s]}{[A^m]+[AMC^m]} = \varphi \cdot \frac{[A^s]}{[A^m]+K^m{}_{AMC}[A^m][MC^m]} \\[6pt]
&= \frac{k_A}{1+K^m{}_{AMC}[MC^m]}
\end{aligned}
\tag{3.6}
$$

$$
\begin{aligned}
\alpha &= \frac{k_{A(R)}}{k_{A(S)}} = \frac{\underline{k}_{A(R)}}{\underline{k}_{A(S)}} \cdot \frac{1+K^m{}_{A(R)MC}[MC^m]}{1+K^m{}_{A(S)MC}[MC^m]} \\[6pt]
&= \frac{1+K^m{}_{A(R)MC}[MC^s]}{1+K^m{}_{A(S)MC}[MC^s]}
\end{aligned}
\tag{3.7}
$$

In "pure" CMP contexts, both chromatographic phases can have a prominent role in the α^* value. Moreover, a reversal in the enantiomer elution order (and, in turn, in the sign of the enantioselectivity) takes place in such systems if compared with the previously debated case.

More intricate is the instance of environments dealing with a chiral selector that dwells on both phases. The occurring complexation events can be reasonably summarized as

$$
\begin{array}{llll}
\text{Mobile phase} & A^m + MC^m & \xrightarrow{\quad K^m{}_{AMC}\quad} & AMC^m \\[4pt]
& \;\updownarrow \quad\; \updownarrow & & \;\updownarrow \\[4pt]
\text{Stationary phase} & A^s + MC^s & \xrightarrow{\quad K^s{}_{AMC}\quad} & AMC^s
\end{array}
\tag{3.8}
$$

so that

$$
k_A = \varphi \cdot \frac{[A^s] + [AMC^s]}{[A^m] + [AMC^m]} = \varphi \cdot \frac{[A^s] + K^s_{AMC}[A^s][MC^s]}{[A^m] + K^m_{AMC}[A^m][MC^m]}
$$
$$
= \underline{k}_A \cdot \frac{1 + K^s_{AMC}[MC^s]}{1 + K^m_{AMC}[MC^m]}
$$

(3.9)

It can be effortlessly deduced that a stronger complexation on the stationary phase accompanies a higher enantiomer retention. Conversely, all the conditions promoting the complexation into the eluent do contribute to decreasing the k_A value. Consequently, the enantioselectivity turns out to be a more complex function of the processes in the two phases:

$$
\alpha = \frac{k_{A(R)}}{k_{A(S)}} = \frac{\underline{k}_{A(R)}}{\underline{k}_{A(S)}} \cdot \frac{(1 + K^s_{A(R)MC}[MC^s])}{(1 + K^m_{A(R)MC}[MC^m])} \cdot \frac{(1 + K^m_{A(S)MC}[MC^m])}{(1 + K^s_{A(S)MC}[MC^s])}
$$
$$
= \frac{(1 + K^s_{A(R)MC}[MC^s])}{(1 + K^s_{A(S)MC}[MC^s])} \cdot \frac{(1 + K^m_{A(S)MC}[MC^m])}{(1 + K^m_{A(R)MC}[MC^m])}
$$

(3.10)

By assuming higher formation constants of the ternary complexes, the α value can be roughly calculated through the ratio

$$
\alpha = \frac{K^s_{A(R)MC}}{K^m_{A(R)MC}} \cdot \frac{K^m_{A(S)MC}}{K^s_{A(S)MC}} = \alpha *^s / \alpha *^m
$$

(3.11)

As evident from Equation (3.11), the selectivity in the two phases oppositely affects the chromatographic enantioselectivity. This can be logically explained as follows: whereas in a pure CSP system the less strongly retained enantiomer is, in principle, eluted earlier than its specular, an opposite situation occurs for pure chiral selector-containing eluents (CMPs). For the latter, the higher the complex stability, the faster its elution through the column. Due to the number of complexation equilibria from Equation (3.8), a more pronounced variation of the overall chromatographic performances (mainly in terms of retention, enantioselectivity, and resolution factors) on even slight modifications of the experimental conditions (e.g., pH, type, and content of cupric salt and organic modifier) can occur in CMP systems. Although this higher flexibility can appear as univocally advantageous, the column performance comes to be often rather difficult to predict.

3.3 KINETIC ENANTIOSELECTIVITY

Examples of kinetic enantioselectivity have also been encountered for particular CLEC systems [7]. Theoretically, it represents the difference in terms of interaction rate of a chiral discriminating agent with the two enantiomers to be separated.

Formally, this can be explained starting from the classical concept of thermodynamic enantioselectivity, α^*. For the generic reaction scheme reported in Equation (3.12), this last parameter can be expressed as the ratio of the stability constants of the two diastereomeric adducts (K^{RS} and K^{SS}).

$$CuA_2 + CuB_2 \underset{k_{-1}}{\overset{k_1}{\longleftrightarrow}} 2CuAB$$

$$\alpha^* = K^{RS}/K^{SS}$$

(3.12)

(Note the chiral selector bearing an S configuration).

Moreover, both stability constants (K) can be defined as the ratio of the rate constants of the forward (k_1) and reversed (k_{-1}) reactions according to

$$K = k_1/k_{-1}$$

(3.13)

Thereby, the thermodynamic enantioselectivity can be seen as the ratio of the kinetic enantioselectivities of the forward and reversed reactions (α_1 and α_{-1}, respectively).

$$\alpha^* = \frac{K^{RS}}{K^{SS}} = \frac{\dfrac{k_1^{RS}}{k_{-1}^{RS}}}{\dfrac{k_1^{RS}}{k_{-1}^{SS}}} = \frac{\alpha_1}{\alpha_{-1}}$$

(3.14)

It is thus possible to state that thermodynamic enantioselectivity cannot exist without the kinetic enantioselectivity.

3.4   RECIPROCITY OF CHIRAL RECOGNITION IN CLEC SYSTEMS

Pirkle indicated the recourse to the reciprocity rule as highly profitable for the design of new enantioresolving agents [8]. However, as clearly stated by Pirkle et al. for immobilized selectors, the employment of a ligand A to resolve the enantiomer pair B can also be energetically not equivalent than the reversed situation. Among the reasons mostly accountable for this discrepancy, the chemistry of the ligand immobilization and the occurrence of secondary interactions (by both the ligand and the analyte enantiomers) with the underlying chromatographic support as well as of specific either repulsive or attractive contacts were remarked upon. As a consequence of the possible nonreciprocal behavior, they suggested paying special attention when evaluating chromatographically derived models. Interesting cases of deviation from the reciprocity rule were also reported by Davankov et al. [5] for some proline derivatives, chromatographed in CLEC systems. Before illustrating the more striking example of this phenomenon, it is necessary to recall inherent theoretical aspects. First, consider the species A_S employed in a CMP system for the separation of the racemic mixture B_{RS}. The two diastereomeric complexes A_SCuB_R and A_SCuB_S are expected to be differently retained in an achiral medium if they are characterized by a different thermodynamic stability and partitioning between the chromatographic phases. The separation factor α_1 should be in principle identical to that (α_2) calculated for a system in which the chiral selector B_S is able to separate the A_{RS} racemate. Indeed, in the latter system the ternary complex A_SCuB_S is the same as in the former, whereas the assembly A_RCuB_S is the antipode of A_SCuB_R. For the series of CMP-CLEC systems investigated by Davankov, there were many instances in which the α_1 value was different from α_2. The singular case of the two species (S)-allo-hydroxyproline [(S)-aHyp] and N-benzyl-(S)-allo-hydroxyproline [(S)-BzlaHyp] alternatively employed as analyte and selector, was particularly highlighted [5]. While an α_1 value of 0.33 emerged when the (S)-aHyp was adopted to resolve the BzlaHyp enantiomers (elution order $k_R < k_S$), α_2 turned out to be 1.21 in the opposite situation (elution order $k_S < k_R$). A different density of the RP-18 chains coating during the column conditioning was invoked to explain this intriguing outcome. Indeed, the hydrophobic $Cu[(S)\text{-BzlaHyp}]_2$ was deemed to cover the reversed-phase material more densely than the more hydrophilic $Cu[(S)\text{-aHyp}]_2$. This led to a different mechanism of chiral recognition, which reflected both in a different α value and in an inverted enantioselectivity.

3.5 SECONDARY EQUILIBRIA IN CLEC SYSTEMS

A resuming scheme for the equilibria that take place in most CLEC systems can be written as

$$
\begin{array}{ll}
\text{Mobile phase} & A^m + M^m + C^m \leftrightarrow AMC^m \\
& \updownarrow \quad \updownarrow \quad \updownarrow \quad\quad \updownarrow \\
\text{Stationary phase} & A^s + M^s + C^s \leftrightarrow AMC^s
\end{array}
\tag{3.15}
$$

On considering the number of secondary processes, "apparent" equilibrium constants can serve to formalize the intricate pattern of events in a more realistic way. The engagement of these thermodynamic parameters is however rational only in cases when the chiral selector concentration C is assumed to be constant. In such cases, and for the two chromatographic phases,

$$
K^m_{app} = K^m[M^m][C^m]
\tag{3.16}
$$

and

$$
K^s_{app} = K^s[M^s][C^s]
\tag{3.17}
$$

The formation of a ternary mixed complex unit can also materialize following several other routes than those illustrated in Equation (3.15). In other words, the equilibria formalized by Equation (3.15) do not represent the complete group of species present into a CLEC environment operating in accordance with the CMP mode. Among these, some of the more plausible are reported in the following list of equilibria in which all constants relate to the mobile phase:

$$A + M + C \leftrightarrow AMC, \quad K_1 = [AMC]/[A][M][C] \tag{3.18}$$

$$AM + C \leftrightarrow AMC, \quad K_2 = [AMC]/[AM][C] \tag{3.19}$$

$$A + MC \leftrightarrow AMC, \quad K_3 = [AMC]/[A][MC] \tag{3.20}$$

$$AM + CM \leftrightarrow AMC + M, \quad K_4 = [AMC][M]/[AM][CM] \tag{3.21}$$

$$AM + C_2M \leftrightarrow AMC + CM, \quad K_5 = [AMC][CM]/[AM][C_2M] \tag{3.22}$$

$$A_2M + CM \leftrightarrow AMC + AM, \quad K_6 = [AMC][AM]/[A_2M][CM] \tag{3.23}$$

$$A_2M + C_2M \leftrightarrow 2AMC, \quad K_7 = [AMC]^2/[A_2M][C_2M] \tag{3.24}$$

As an approximation, for each of them, a comparable stability of the complexes formed by the chiral selector and the two enantiomers was assumed. This assumption was done for the otherwise more intricate situation occurring in the case of complexes endowed with markedly different stabilities. Indeed, for such situations, additional equilibria, like the displacement of the chiral selector by the solute, should be taken into account [9]. Moreover, to simplify the following debated theoretical model (reducing, in turn, the number of equilibria to be considered), both C and A are assumed to be added to the system as C_2M and A_2M, respectively. Thereby, free-metal ions and enantiomers exclusively stem from the relative complex dissociation. The sophistication of the CLEC system can be furthermore reduced by considering the complex distribution as a function of the eluent pH as reported in Figure 3.1.

For the plot shown in Figure 3.1, a model system constituted of valine and serine is considered. The distribution is calculated based on the stability constants of the complexes reported in the literature [10,11]. Without any specification, one of the two amino acids can be thought of as the chiral discriminating agent, and the other represents the analyte. Being available in the literature, the values of the stability constants can be exploited to estimate the concentration distribution as a function of the eluent pH. The amount of both the free chiral selector and the free enantiomer is practically negligible in the pH range generally adopted for CLEC studies.

Owing to their presence in the eluent as protonated form or complexed species, Equations (3.18)–(3.20) do not appreciably contribute to the enantiomer separation. The importance of Equilibrium (3.21) comes up for pH values where both the free-metal ions and the mixed ternary complexes are simultaneously present at a sufficient concentration. Such pH values do not exist in practice. Indeed, while a relevant concentration of free metal can be found at pH values lower than 4, these acidic conditions do not allow any complexation event. On the other hand, significant amounts of ternary complex are produced for pH values higher than 5. In such cases,

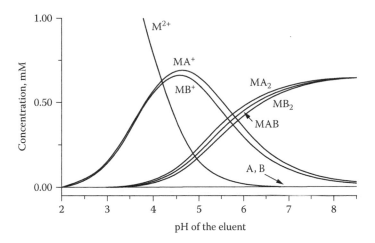

FIGURE 3.1 Distribution of complexes formed in a solution containing (*S*)-valine (A, 2mM), (*S*)-serine (B, 2mM), and copper(II) ions (4 mM) depending on the pH. (From Kurganov, A. A., *J. Chromatogr. A* 906, 51–71, 2001. With permission.)

the processes described by Equations (3.22)–(3.24) are the prevailing ones inside the chromatographic medium. If comparing Equilibria (3.22) and (3.23), involving the bis-complexes of the solute, with Equilibrium (3.21) dealing instead with its mono-complex, the first two should have a greater influence at higher pH. Conversely, Equilibrium (3.21) should reveal itself as prominent in the more acidic pH range. Because the concentration of chiral selector and metal ions are constant during the chromatographic runs, the concentration of mono- and bis-complexes of the former also remain constant. Consequently, the complexity of Equilibria (3.18)–(3.24) can be markedly reduced through the introduction of the apparent equilibrium constants

$$Eq.(18) \Rightarrow K_{1app} = K_1[M][C], \quad A \leftrightarrow AMC \tag{3.25}$$

$$Eq.(19) \Rightarrow K_{2app} = K_2[C], \quad AM \leftrightarrow AMC \tag{3.26}$$

$$Eq.(20) \Rightarrow K_{3app} = K_3[CM], \quad A \leftrightarrow AMC \tag{3.27}$$

$$Eq.(21) \Rightarrow K_{4app} = K_4[CM]/[M], \quad AM \leftrightarrow AMC \tag{3.28}$$

$$Eq.(22) \Rightarrow K_{5app} = K_5[C_2M]/[CM], \quad AM \leftrightarrow AMC \tag{3.29}$$

$$Eq.(23) \Rightarrow K_{6app} = K_6[CM], \quad A_2M \leftrightarrow AMC + AM \tag{3.30}$$

$$Eq.(24) \Rightarrow K_{7app} = K_7[C_2M], \quad A_2M \leftrightarrow 2AMC \tag{3.31}$$

Indeed, Equilibria (3.18)–(3.22) can be considered as conventional isomerization types and therefore are instrumental for the easy calculation of the enantiomer retention. On the contrary, while Equilibrium (3.30) is of an association type, (3.31) is of the dimerization type.

The secondary equilibria of the dimerization type have been established generating relevant variations on the overall enantiorecognition mechanism even with slight modifications of the apparent equilibrium constants, sample amount, and enantiomer ratio in the sample [3]. This represents the main distinctive feature of CLEC systems in which only secondary equilibria of the isomerization type take place. However, even into a CLEC environment, the occurrence of the isomerization type interactions is much more frequent. Worth mentioning also is the peak distortion (in terms of both fronting and tailing phenomena) one can observe in the presence of such multiple solute-chiral selector interactions. Interesting simulation studies on the impact exerted by the secondary dimerization type contacts on the chromatographic behavior were exhaustively described by Kurganov.

3.6 RELEVANT FACTORS INFLUENCING THE CHROMATOGRAPHIC BEHAVIOR OF AMINO ACIDS IN CLEC

As discussed so far, all the retention processes in CLEC systems are based on a combination of complex-forming equilibria and hydrophobic interactions. Consequently, the factors that in principle should mainly control both the retention and the enantioselectivity are the pH and the ionic strength of the eluent (namely, the cupric concentration), the concentration and type of the organic modifier, and the nature of the central metal ion and of the copper(II) counter ion. However, the column temperature and the eluent flow rate should also be considered in this scenario. In the following discussion, a basic elucidation of the role played by each of the aforementioned parameters on the mechanism of chiral recognition as a whole is reported.

3.6.1 ELUENT pH

Eluent pH is of utmost importance since it governs the retention of the chelating species in all the CLEC environments, irrespective of the adopted operation mode. Based on thermodynamic investigations carried out on systems with two different amino acids and Cu(II) ions, the formation of the mixed assemblies is favored at pH values around the neutrality [12]. Moreover, it was reported for CSP systems that at pH $\geq$ 7.0 the formation of the ternary (stationary) complexes is favored over that of the binary (mobile) adducts [13]. This is also in line with what Galaverna claimed as the noticeable influence of the eluent pH on the type and amount of specific bi- and tri-component adducts within specific pH ranges and in CMP systems in which the selector is present both in solution and on the alkyl chains [9,14,15]. In some cases, both retention and enantioselectivity were observed to be enhanced on an increase of the eluent pH [12,13,16]. This experimental outcome was explained with an increasing presence of the bidentate coordination by the chelating compounds

[16]. However, in other cases, the increase in retention did not parallel the trend for the enantioselectivity [17,18]. The same evidence turned out for the resolution factor and the column efficiency did not reveal a univocal trend [17,19].

3.6.2 METAL ION CONCENTRATION

A progressive increase in the central metal ion concentration generally causes a continuous decrease of the enantiomer retention [12,16,20,21]. For systems dealing with N-alkyl-L-hydroxyprolines (N-alkyl-L-Hyp) as C-CSPs, Davankov et al. [16] explained this behavior through the effect played by the amount of the copper(II) acetate [Cu(AcO)$_2$] on the equilibrium between the two mono(amino acidato)copper complexes (i.e., [CuA]$^+$AcO$^-$ and [Cu(N-alkyl-L-Hyp)]$^+$AcO$^-$) and the bis(amino acidato)copper complex (i.e., [Cu(A)(N-Alkyl-L-Hyp)]). The cited equilibrium can be formalized as

$$[Cu(A)(N-alkyl-L-Hyp)] + Cu(AcO)_2 \leftrightarrow$$
$$\leftrightarrow [Cu(A)]^+ AcO^- + [Cu(N-alkyl-L-Hyp)]^+ AcO^-$$

(3.32)

An increase in the cupric salt concentration shifts the equilibrium to the right, thus reducing the ternary mixed complex stability and, in turn, the enantiomer retention. This explanation can be extended to all CLEC systems. The Cu(II) salt content modification does not generally affect the selectivity to a remarkable extent [12,16,17,20,21]. Indeed, the Cu(II) concentration mainly controls the extent of the ternary mixed assembly formation without influencing the stereochemically dependent interactions between ligands that are responsible for the enantioselectivity [22].

3.6.3 TYPE AND AMOUNT OF ORGANIC MODIFIER

The presence of an organic modifier in the eluent, even in low percentages generally leads to a decrease in the enantiomer retention and enantioselectivity as well [12,21,23]. A major impact on the retention is felt by the stronger retained enantiomer due to the higher contribution by the hydrophobic contacts in stabilizing the ternary mixed complex in which it is contained [12,21,23]. In fact, for C-CSP type systems, the organic modifier molecules strongly affect all the hydrophobic interactions between the sample side-chain residual and the lipophilic bed from the chiral selector anchor and the alkyl chains of the column. Nonetheless, the perturbing effect is expected to be less pronounced when organic modifiers endowed with a higher polar character are engaged. This aspect was experimentally confirmed by comparing the chromatographic outcomes from acetonitrile (MeCN) and methanol (MeOH) separately used as the organic modifier into the eluent [12,23]. Indeed, MeCN (less polar) containing mobile phases generally produced a lower retention than that from the equally composed MeOH carrying ones [12,23].The previous assumption was further corroborated when the performances from the use of MeOH and 2-propanol were compared [21]. Because the latter is less polar, it more strongly affects the enantiomer retention. A reversal of the elution order with different MeCN percentage into the eluent was also reported by the same authors ([21]; see Section 3.8). The use of a low percentage of organic solvent can be useful to drastically reduce the run time for the most retained samples [17]. However, in C-CSP environments this parameter needs to be used with special care because it could drastically compromise the column stability (detachment of the enantioresolving agent).

3.6.4   TYPE OF METAL ION

Only a few studies dealing with a comparison of the chromatographic performances deriving from the use of different central cations—such as Cu(II), Ni(II), Zn(II), Co(II), Fe(II), Mn(II), Cd(II)—have been carried out so far. Most of them rely on the ligand-exchange capillary electrophoresis technique and were directed to different classes of chemicals: free [25] and derivatized [26] amino acids; sugars [27]; and quinolones [28]. The data from the literature clearly indicate the absence of a common trend in terms of chromatographic performances. In other words, the enantioseparation extent strictly relates to the property of the mixed ternary complex as a whole. As far as the liquid chromatography is concerned, the literature data claim the Cu(II) providing the best performances for its ability to undergo profitably thermodynamically stable and kinetically labile complexes [2,29,30] and have to be considered as the preferential cation for the analysis on amino acids [31].

3.6.5   TYPE OF COPPER SALT

The problem of the copper(II) counter ion effect has been very scarcely investigated. However, it has been observed that the physico-chemical nature of the Cu(II) anion exerts an utmost influence on the between-ligand interactions [17,19,32,33]. Among the chromatographic parameters, the resolution factor (R_S) underwent the most pronounced variation. The nitrate anion proved to afford the best chromatographic performances in almost all the CLEC systems investigated so far in this respect [17,33]. Section 3.8 treats the proposed explanation of this outcome in detail..

3.6.6   COLUMN TEMPERATURE

The formation of the diastereomeric ternary complexes is markedly influenced by the column temperature. In fact, the multispecies assemblies are increasingly stabilized by decreasing the temperature of the environment where the ligand-exchange process takes place. Moreover, for all the CLEC systems where the chiral selector exclusively resides onto the stationary phase (CSPs), the formation of the more stable ternary mixed complex—namely, that bearing the more retained enantiomer—is more stabilized than that carrying its specular. Consequently, an improvement of the chromatographic performance has to be expected as far as the enantioselectivity (α) is concerned. The almost generally accepted explanation for this behavior accounts for a variation in the rate of equilibrium for the formation of the ternary complexes. An alternative explanation for the change in the chromatographic performance was proposed by Sanaie and Haynes [34,35] as discussed in Section 3.10.1.

They claimed that the lifetime of the transient diastereomeric complexes were longer at low temperatures, which leads to an increase in the retention and enantioselectivity accompanied with a reduction in column efficiency. What is lost in terms of column efficiency, often measured as the number of the theoretical plates N, stems from a broadening of the enantiomer peaks. Conversely, the lifetime of the transient ternary complexes progressively diminishes as the column temperature increases, thus resulting into a fast transfer kinetics and sharper peaks. The derived higher-resolution factors, R_S, can be explained on this basis [23]. The very intriguing case of an improvement in the enantiomer separation on an increase in the column temperature was also reported [36]. The unusual case was diagnosed during the separation of the N-benzylproline enantiomers on a cross-linked polystyrene based chiral stationary phase containing the proline as the selector. The entropic control of the enantioseparation was explained as a result of the different axial accommodation of a solvent molecule in the *homo*-chiral and *hetero*-chiral adsorbates. Indeed, the solvent molecule complexation was deemed to be sterically hindered in the latter, thus favoring the elution of the former complex.

3.6.7 FLOW RATE

For all the chromatography-based enantioseparation approaches, a modification of the eluent flow rate can result as decisive in tuning the overall performance. This effect is particularly evident in CLEC where the ligand-exchange processes occur with slow kinetics [4,13,37]. Generally, low flow rates turn into higher-resolution factors, R_S [17,37], which accompany increased column efficiencies (i.e., higher number of theoretical plates, N). Indeed, the ligands undergo a complete exchange when low flow rates are fixed [37]. Worth noting is that a variation in the eluent velocity does not modify the strength of the stereoselective contacts for the two enantiomers to a different extent, which accounts for the enantioseparation (α) being almost always unaffected.

3.7 COVALENTLY BOUND CHIRAL STATIONARY PHASES (B-CSPS)

As detailed before, the enantioseparation process that realizes a CLEC environment relies on the reversible complex formation between doubly charged cations (usually Cu) and bi- (or even multi-) chelating compounds. The ligand exchange as the driving force of the enantioseparation accomplishment was clearly demonstrated by Galaverna et al. [9]. In a study in which the authors employed the fluorescent L-tryptophanamide (L-Trp-NH$_2$) as the resolving agent in the CMP mode and the solely fluorescent-based detector, a unique peak appeared on the chromatographic trace after an enantiomeric mixture was injected. By taking into account the following set of equilibria

$$Cu(Trp-NH_2)_2 + D - A \leftrightarrow Cu(Trp-NH_2)(D-A) + Trp-NH_2$$
$$Cu(Trp-NH_2)_2 + L - A \leftrightarrow Cu(Trp-NH_2)(L-A) + Trp-NH_2$$
(3.33)

the aforementioned peak is shown to be related to the free L-Trp-NH$_2$, which was displaced by the sample molecule (irrespective of its absolute configuration) from the ternary mixed complexes. The absence of the other two peaks (each related to one of the mixed ternary complexes) can be explained on the basis of the nonfluorescent feature of all cupric containing binary and ternary complexes. As expected, three signals (one for the L-Trp-NH$_2$ and two for the diastereomeric complexes) were detected when an ultraviolet (UV)-Vis-based detector was used.

Interestingly, the "three-point interaction model" (mostly known as the "three-point" rule), which theorizes the minimum requisites for the enantiorecognition success, can be applied as well to CLEC systems. Accordingly, by considering amino acidic chiral selectors and analytes, two points of the intermolecular interaction relate with the coordination bonds to the central metal ion through their amino and carboxylic moieties, and the third stereoselective contact takes place instead between the side-chain residual from the enantiomer analyte and a specific portion of the resolving agent. Depending on the physico-chemical character of these regions, such a third interaction can either stabilize or destabilize the mixed complex. It should be pointed out that both attractive and repulsive interactions within the diastereomeric adducts are productive with respect to the chiral recognition accomplishment.

In the presence of amino acidic bifunctional chelators, the two groups are equatorially disposed around the metal ion in an alternating manner. One or more solvent molecules can complete the first solvation sphere of the central cation. This is especially the case when the selector or the analyte lack further electron-donating groups endowed with coordination capability.

Without recurring to thermodynamic formalism, as a general rule for all the CSP contexts, the chromatographic enantioseparation can be observed when even a subtle difference in the relative stability (free energy of formation) of the adsorbates occurs. In other words, to get a successful enantiorecognition, the chiral discriminating entity must adequately perceive the difference in the spatial configuration of its partners thus giving rise to differently stable diastereomeric complexes. The different stability of the two diastereomeric ternary complexes cannot alone explain the achievement of the enantioseparation in a medium that operates in CMP mode. However, since the theory for CMP systems is more intricate than that for the CSP ones, some mechanistic aspect dealing with the latter environment will first be discussed. Moreover, since Marchelli et al. [2] clearly discussed the CLEC chromatographic behavior of diverse chelating dansyl-derivative compounds, the following discussion will be exclusively focused on underivatized species (with a particular regard to the amino acids).

In the frame of the following discussion, a series of examples mainly dealing with proline-based chiral selectors will be treated. Among the reasons accounting for the preferential employment of configurationally rigid ligands, the generally large enantioselectivity is of primary importance. Conversely, unfavorable steric interactions are often furnished by flexible ligands in this respect. Moreover, for each of the examined CLEC systems, particular attention will be given to the molecular basis of the observed elution order.

Also, the role played by specific portions of the nonchiral matrix of different sorbents in the selector–selectand interactions within each of the two diastereomeric transient ternary complexes taking form in each specific CSP-CLEC setting is of

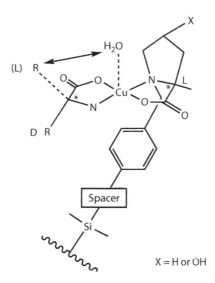

FIGURE 3.2 Influence of one axially coordinated water molecule on the overall stability of the two ternary complexes carrying a proline-based chiral selector, Cu(II), and L- or D-amino acid. (Adapted from Davankov, V. A. and Kurganov, A. A., *Chromatographia* 17, 686–90, 1983. With permission.)

great importance. Through the collection of exhaustive examples, Davankov and Kurganov [38] introduced and emphasized the concept of the *cooperation* of achiral elements on the overall ligand-exchange-based recognition mechanism. By taking into consideration the molecular assembly depicted in Figure 3.2, one can readily figure out the effect provided by these adjunctive exogenous elements.

In a general system in which the L-proline or the L-4-hydroxyproline are grafted onto a cross-linked polystyrene matrix, the aromatic ring of the more extended N-benzyl-containing structure (Figure 3.2) tends to be projected toward the axial coordinative position of the central metal thus facilitating the attractive interaction with the hydrophobic side-chain residual R of a generic D-enantiomer [38,39].

The deriving configurationally driven stabilization enhances the D-amino acid retention. On the contrary, the R group of the L-amino acid realizes a destabilizing interaction with a water molecule that can plausibly be coordinated in the upper axial coordination site of the Cu(II). This achiral species is instead absent between the benzyl group of the selector and the R group of the D-enantiomeric form, thus allowing the favorable, between-ligand hydrophobic interactions. The observed elution order $k_L < k_D$ for hydrophobic amino acids in such a CLEC system can be explained on this basis. An inversion of the elution order (namely, $k_D < k_L$) turns up in the case of trifunctional ligands as histidine, *allo*-hydroxyproline (L-4-hydroxyproline), and aspartic acid. For all these compounds, the additional hydrophilic moiety is able to replace the axially coordinated water molecule in the L-isomer, thus giving rise to a stronger sorption complex [7,38] (Figure 3.3a). Conversely, a repulsive interaction characterizes the *hetero*-chiral complex (Figure 3.3b). No difference in the elution order between bifunctional and trifunctional analytes is instead observed

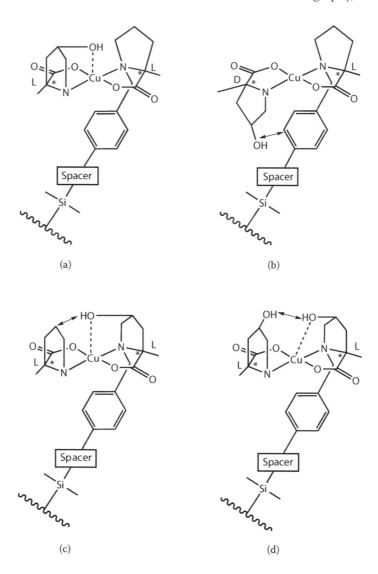

FIGURE 3.3 Schematic illustration for the chiral discrimination of enantiomers of bifunctional and trifunctional amino acids, and resins incorporating L-proline and L-*allo*-hydroxyproline as chiral selectors. (Adapted from Davankov, V. A., *Enantiomer* 5, 209–23, 2000. With permission.)

when trifunctional chiral selectors are exploited. This is the case of *allo*-hydroxy-L-proline-based enantioresolving agents which, independently of the chelating attitude of analytes, accomplish a higher destabilization in the *homo*-chiral complexes (Figure 3.3c and 3.3d).

A completely reversed situation appeared when matrices carrying electron-donating groups are adopted. In these instances, the additional chelating moieties can occupy the lower axial position of the first Cu(II) coordination sphere.

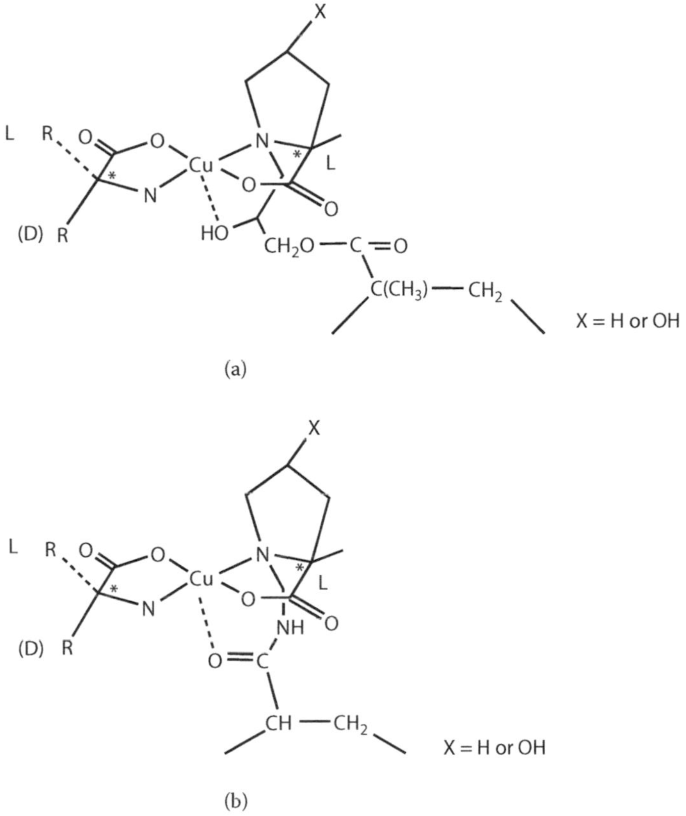

FIGURE 3.4 Nonbonding interactions between the side group R of D-enantiomers of amino acids and functional groups of chiral selectors, which are coordinated in axial position of Cu(II) ions. (From Davankov, V. A., *Chirality* 9, 99–102, 1997. With permission.)

Exemplary are the cases of poly(2,3-epoxycyclopropyl methacrylate)-based [38,39] (Figure 3.4a) or polyacrylamide-based sorbents [38–40] (Figure 3.4b) functionalized with L-proline or L-4-hydroxyproline units. In the *hetero*-chiral multicomponent assembly, the axial coordination of the hydroxy or carbonyl group from the initial sorbent matrix can exert a destabilizing effect that turns in the faster elution of the D-enantiomer (namely, $k_D < k_L$).

Both the models drawn in Figures 3.2 and 3.4 aid in understanding why the sign of enantioselectivity inverts when one of the selector (comprising also the underlying matrix)–selectand interactions is repulsive rather than attractive.

Theoretically, a reversal of the enantiomer retention is also expected by inverting the absolute configuration of the selector molecule. Although partially different chiral selectors were employed, this was observed by Hyun [23,41] during the elution of a series of α- and β-amino acids on two CSPs obtained by covalently bonding the species (S)-N,N-carboxymethyl undecyl leucinol monosodium salt [23] and (R)-N,N-carboxymethyl undecyl phenylglycinol mono sodium salt [41] onto silica gel. Besides

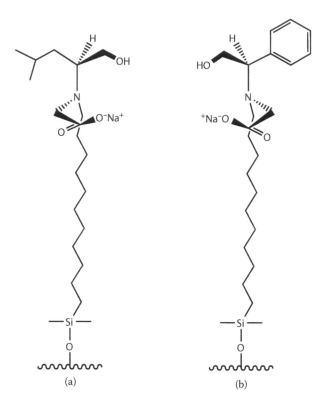

FIGURE 3.5 The proposed structures for (a) (*S*)-N,N-carboxymethyl undecyl leucinol and (b) (*R*)-N,N-carboxymethyl undecyl phenylglycinol monosodium salts. (From Hyun, M. H. et al., *J. Chromatogr. A* 950, 55–63, 2002. With permission.)

being different in terms of absolute configuration at the asymmetric carbon, the two selectors also vary in terms of steric bulkiness and electron density of their side-chain residual (isobutyl vs. phenyl group). The reversal of the enantiomer elution order observed with the two chiral discriminating species unambiguously indicates an identical recognition mechanism. Figure 3.5 shows the proposed structures of the leucinol and phenylglycinol-based CSPs allowing the observed elution orders to be rationalized [23]. The hydroxyl group of the chiral selector is oriented in a way that promotes its axial coordination to the central cupric cation.

In the *hetero*-chiral diastereomeric complex, the alkyl group at the chiral center of the L-enantiomer (Figure 3.5a) undergoes a stabilizing lipophilic interaction with the tethering C-11 chain on the CSP. This turns into an energetically more stable diastereomeric adduct. A similar interaction with the alkyl group at the chiral center of the D-enantiomer (Figure 3.5b) would require the inversion of the configuration of the tertiary amino group of the CSP. As a result, since this inversion is unfeasible, the D-enantiomer with its R moiety projected toward the bulky of eluent, results less retained. Interestingly, the length of the tethering chain was established to play a decisive part in addressing the chiral recognition mechanism as a whole. Indeed, for L-4-hydroxyproline-based chiral selectors anchored to the silica bed via N-alkyl linear groups, a progressive inversion of the elution order turned out on an increase of the number of N-methylene units from one to eight. Likewise, regarding the case of the leucinol-based device, hydrophobic interactions between the R side-chain residual of the D-enantiomer and the N-alkyl chain of the chiral selector enhanced the affinity between the two partners that tuned into a reversal of the previous elution order (i.e., $k_L < k_D$ in place of $k_D < k_L$).

The higher flexibility stemming from an N-alkyl chain elongation, which markedly reduces the steric factors, can account for the discussed outcomes.

A modification of the "Davankov rule" was proposed by [42] in a study in which a series of α-hydroxy acids were enantioresolved on an L-hydroxyproline-based chiral selector. While for the major part of the tested compounds the (S)-isomers were eluted before their (R)-speculars ($k_S < k_R$), a reversal of the elution order ($k_R < k_S$) was instead encountered for the remaining few species. The experimental observations cannot be explained on the basis of the models developed by Davankov [38].

Indeed, by relying on these models, the more bulky R group from the (R)-hydroxy acid (D-hydroxy acid) side-chain would be oriented toward the more hindered side (Figure 3.6a). This, in turn, would result in a destabilization of the corresponding diastereomeric complex and thereby in its lower retention. Conversely, in another structure in which the alcoholic hydroxy and the amino groups are arranged in a *cis*-configuration as in Figure 3.6b, then the R residual from the (R)-enantiomer side-chain would be projected toward the less hindered side. In such an instance, a stabilizing selector–selectand interaction would come out, thus providing for a higher retention of the resulting enantiomer. Analogously drawn models could also serve to furnish a tentative explanation of the reversed elution order ($k_R < k_S$) for the species endowed with a further hydrophilic functional group on their side chain (Figures 3.7a and 3.7b). For these analytes, additional interactions of the polar side-chain groups stabilized the complexes carrying the (S)-enantiomer.

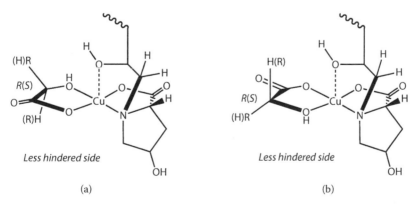

(a)   (b)

FIGURE 3.6 (a) A possible structure of a mixed ternary complex containing an α-hydroxy acid. The structure is obtained by a simple substitution of the amino group by a hydroxyl group in the Davankov model. (b) A modification of the Davankov model of ternary complex with an α-hydroxy acid. The OH oxygen atom is *cis*-positioned with respect to the hydroxyproline nitrogen atom. (From Chilmonczyk, Z. et al., *Chirality* 10, 821–30, 1998. With permission.)

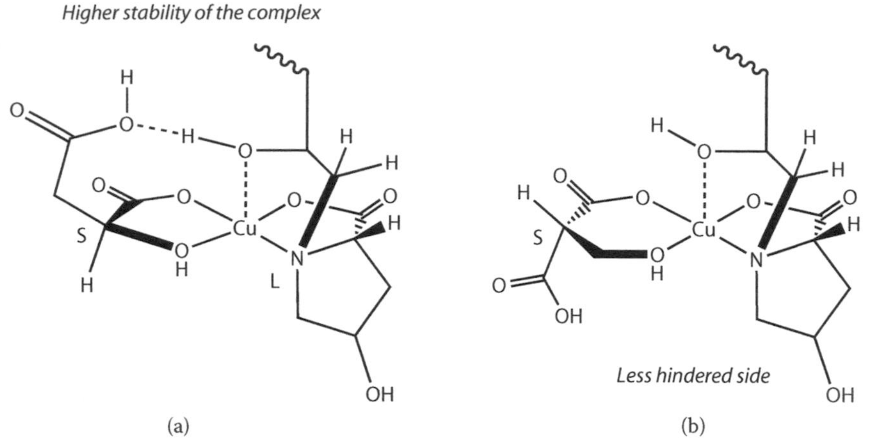

Higher stability of the complex

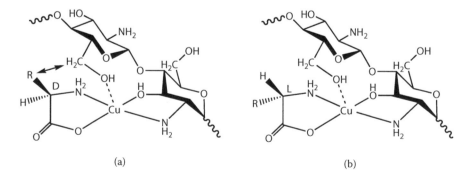

FIGURE 3.7 (a) A possible structure of a ternary complex with an α-hydroxy acid containing three functionalities suited to a complexation. (b) A possible structure of a ternary complex with an α-hydroxy acid containing one of a chelating group located in the β position with respect to a chiral center. (From Chilmonczyk, Z. et al., *Chirality* 10, 821–30, 1998. With permission.)

Owing to its excellent binding capacity for heavy metal ions, Liu et al. [43] successfully exploited chitosan as the chiral discriminating agent for α-amino acids and α-hydroxycarboxylic acids. For the former class of compounds the authors assumed an OH group on the adjacent glucosamine unit being involved in the formation of the diastereomeric ternary complexes (Figures 3.8a and 3.8b). The lower stability in the complex bearing the D-enantiomer analyte was explained on the basis of a steric hindrance between the R substituent in the analyte and the hydroxymethyl group on the chiral selector, since the latter was able to weaken the corresponding complex (Figure 3.8a).

FIGURE 3.8 The proposed structures of the ternary complex formed from the fixed ligand, chitosan CSP, Cu(II), and (a) D-amino acid or (b) L-amino acid. (From Liu, Y., Zou, H., and Haginaka, J., *J. Sep. Sci.* 29, 1440–6, 2006. With permission.)

3.8 COATED CHIRAL STATIONARY PHASES (C-CSPS)

The influence of achiral regions (within the column) on the overall enantiorecognition mechanism is especially evident when a stationary phase resulting from the dynamic adsorption of the selected enantioresolving agent onto suitable materials is engaged. Since the coated chiral stationary phases (C-CSPs) doubtless represent the most adopted CLEC approach, a selection of the most representative and meaningful examples will be presented in the following discussion. In this frame, for all the illustrated cases, the models proposed by the authors to explain the relative enantiorecognition event will be debated.

A milestone in CLEC literature is Davankov's [16] dealing with the enantioseparation of unmodified α-amino acids through hydrophobically impregnated RP-18 stationary phases with N-alkyl-L-hydroxyproline units. In this study, the performances from differently extended alkyl groups as $n-C_7H_{15}-$, $n-C_{10}H_{21}-$, and $n-C_{16}H_{33}-$ were explored. Although this represents the first study with this chromatographic approach, generally accepted principles on the mechanism of chiral recognition are reported there. Consequently, this work can be reasonably adopted to provide the theoretical background for further applications that will be reported in the following discussion. For the first time, it was claimed that highly performing and permanently stable C-CSPs were attainable by simply stratifying suitable amino acids derivatives (acting as the chiral selectors) onto conventional octadecylsilica-based surfaces. As the mobile phase, a simple Cu(II) containing solution (with the sulphate salt as the cupric source) was engaged. A detailed mechanism of retention was proposed with the N-alkyl chain of each selector unit lying parallel to the C-18 chains. In such a way, the most stable fixation of the enantioresolving compound onto the achiral surface was achieved. Once coordinated to the central cupric cation, the N-atom of the chiral selector becomes asymmetrical assuming an (*S*)-configuration. Accordingly, the combination with the (*S*)-configuration of the α-C-atom forced the hydroxypyrrolidine ring in an almost parallel disposition with respect to the equatorial plane of the ternary assembly (Figure 3.9). When a D-amino acid was present in the ternary complex, its side-chain residual resulted to be projected toward the octadecyl chains, without distorting the conformation of the sorption complex (Figure 3.9a). As expected, an increase in the hydrophobic character of the aforementioned residual (from the methyl in the alanine to the n-butyl group in the Nor-leucine) resulted in a higher retention, due to the outstanding effect of superimposed hydrophobic interactions in this CLEC system. Interestingly, the authors also described the possibility for the α-radical from L-amino acids to experience hydrophobic contacts with the C-18 chains as well (Figure 3.9b). In this distorted arrangement of the dative bonds, the side-chain residual tended to be equatorially located just "leaning" on the hydrophobic bed. However, it was clearly emphasized that those two types of hydrophobic interactions were adequately different from an energetic standpoint. Moreover, the discrepancy between the two enantiomers becomes much more evident on an increase in the number of the side-chain carbon atoms. Worth mentioning is the case of proline. Extraordinary α values (up to 16) were observed for this compound. This exceptional behavior was explained basing on its rigidity that precludes the pyrrolidine ring of the L-enantiomer to be profitably interacting with the C-18

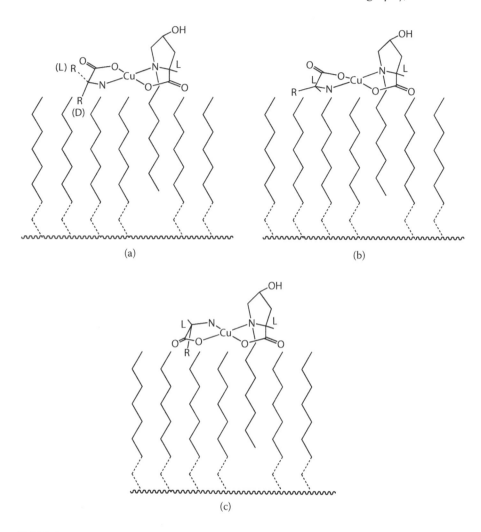

FIGURE 3.9 The proposed structure of mixed-ligand sorption complexes containing one C$_7$-L-Hyp ligand and a (a) D- (or L-) amino acid with planar chelate ring, (b) L-amino acid with puckered chelate ring conformation, (c) L-amino acid in *cis*-position to C$_7$-L-Hyp ligand. Reprinted with permission from Davankov, V. A. et al., *Chromatographia* 13, 677–85, 1980. With permission.)

chains. Thereby, for such analytes, the ternary complex with the L-enantiomeric form was particularly unstable with a consequent particularly low retention. On the contrary, the D-enantiomer underwent strong interactions with the hydrophobic bed, thus stabilizing the corresponding mixed complex. As expected, the introduction of a polar moiety in the analyte side chain produced a decrease in terms of retention and a reduction of the overall chromatographic performance by the mixed stationary phase. In Figure 3.9c, the case of a *cis*-arrangement of the N and O atoms around the Cu(II) cations is depicted. If drawn in this way, then no noticeable difference in terms of stability could be presumed for the two diastereomeric complexes. Moreover, a very scanty chance to have *cis*-arranged complexes of this kind was expressed. Such a disposition, indeed, is characterized by a low stability.

The inversion of the elution order for the histidine (namely, $k_D < k_L$ in place of the "regular" $k_L < k_D$) was explained on the basis of the tridentate chelating ability of its L-enantiomer. Indeed, the imidazole nitrogen of the L-enantiomeric form was assumed to occupy the axial position in the first coordination sphere of Cu(II), in turn stabilizing the corresponding ternary complex. A steric hindrance prevents a similar behavior for the D-enantiomer. The multimodal complexation ability of histidine in Cu(II) containing environments is a matter of great interest. Among the studies treating this object, worth mentioning is the paper authored by Deschamps et al. [44].

The reliability of the chiral recognition model suggested by Davankov, which will be referred as the "Davankov model" in the following discussion, was furthermore demonstrated by several other authors. In this frame, work by Hyun [45,46] provided particularly interesting results, where the N-dodecyl (1S, 2R)-norephedrine derivative was coated onto a C-18-based column (Figure 3.10a). Based on the Davankov model, the almost total explanation of the observed elution order, and in turn of the enantiorecognition mechanism as a whole, could be done. D-enantiomers resulted more retained than their antipodes for all compounds bearing a simple α-alkyl substituent (namely, $k_L < k_D$ for, e.g., alanine, valine, leucine). Conversely, with two

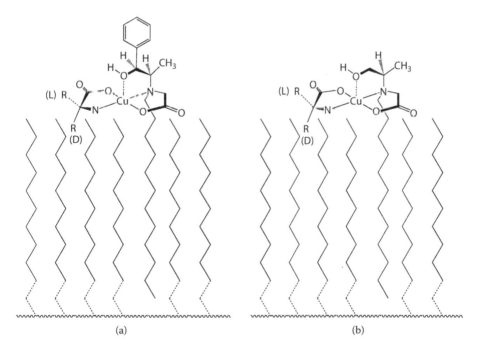

(a)   (b)

FIGURE 3.10 The proposed structures of the ternary complexes formed from the fixed ligand (a) (1S,2R)-N,N-carboxymethyl-N-dodecylnorephedrine monosodium salt and (b) (R)-N-carboxymethyl-N-dodecyl-2-amino-1-propanol monosodium salt, Cu(II), and a D- or L-amino acid. (From Hyun, M. H. et al., *J. Chromatogr. A* 684, 189–200, 1994. With permission.)

exceptions—glutamine and glutamic acid—L-enantiomers were eluted last for the species endowed with an extra hydrophilic group at the α-alkyl substituent (i.e., the elution order was $k_D < k_L$ for tyrosine, asparagine, serine, threonine, aspartic acid, and histidine). As pointed out in Figure 3.10a, the *trans*-configuration with a hydroxy group from the chiral selector axially coordinated to the central Cu(II), being energetically more favorable, was still assumed as the prevailing one. Without repeating the proposed reason accounting for the stronger retention of some D-enantiomers, it is more interesting to direct attention toward the higher stability observed for almost all the complexes containing L-enantiomers with an additional hydrophilic moiety in their side chain. The outcome was tentatively explained through two alternative ways. Indeed, the L-amino acid polar side-chain residual could either enter a hydrogen bond contact with the hydroxy group of the fixed ligand or displace it, thus occupying the axial coordination. Moreover, a destabilizing interaction between the reversed-phase material and the previously given side-chain hydrophilic α-radical in the D-enantiomers was contemporarily assumed. The controversial behavior observed with glutamine and glutamic acid is still unsolved.

An almost univocal elution order was instead observed by the same group of authors when the (R)- N-carboxymethyl-N-dodecyl-2-amino-1-propanol monosodium salt was engaged as the chiral discriminating agent [47]. Indeed, with only the exception of histidine, all the compounds underwent the same elution order, with the D-enantiomer being more retained than its specular. The elution order ($k_L < k_D$) for all the compounds having a simple α-alkylic side-chain residual was explained by the authors just based on the Davankov model. Moreover, the absence of the opposite elution order, which previously turned out for some of the investigated species, was also rationalized. The C-CSP-based on the (R)-2-amino-1-propanol (Figure 3.10b) was supposed to form more stable and tight ternary complexes with Cu(II), and the amino acids had a simpler α-alkyl substituent than those realized with the (1S, 2R)-norephedrine-based material. The lower stability of the adrenergic agent-based complexes was explained on the basis of a destabilizing interaction between the axially coordinated hydroxy group and the phenyl ring. The precarious stability of the axial coordination to the central cation by the chiral selector hydroxy group was invoked to explain its previously debated displacement by the polar functionality in the side-chain residual of some analyte. The already cited remarkable strength of the axial coordination by the hydroxy group from the chiral selector was deemed to prevent any displacement promoted by polar functionalities on the L-enantiomers. This turned out into a lower stability of the ternary complexes from L-enantiomers than that from D-enantiomers (which explains the observed elution order, $k_L < k_D$). The unusual behavior shown by L-histidine was explained on the basis of the relatively high basicity of the imidazole nitrogen enabling its preferential axial coordination (Figure 3.11a). In accordance with a well-known work from Gil-Av et al. [48], an alternative model for the chiral recognition of this species was brought to light (Figure 3.11b). However, both models supported the higher stability of the ternary complexes carrying the L-isomer of histidine. In the latter case, the α-amino group and the imidazole nitrogen underwent a planar coordination with the Cu(II) cation.

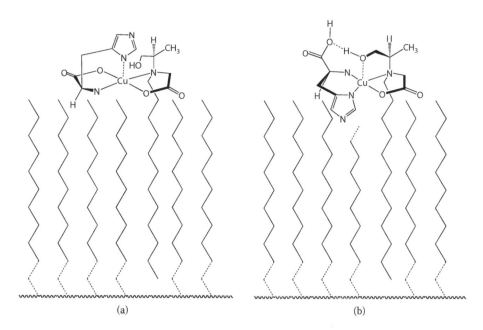

(a)   (b)

FIGURE 3.11 The proposed structures of the ternary complexes formed from the fixed ligand (R)-N-carboxymethyl-N-dodecyl-2-amino-1-propanol monosodium salt, Cu(II), and L-histidine. (From Hyun, M. H. et al., *J. Chromatogr. A* 684, 189–200, 1994. With permission.)

The importance of the axial hydroxy group in the overall chiral recognition mechanism was highlighted when the authors used a (R)-α-phenylethylamine-based selector [47] (Figure 3.12).

Through employing this species lacking from the hydroxy group and still adsorbed by means of the N-dodecyl chain, a completely opposite elution order than that observed with the (R)-2-amino-1-propanol-based selector (namely, $k_D <$ k_L instead of $k_L < k_D$) was achieved. The energetically plausible model accounting for this outcome was proposed as depicted in Figure 3.12. The authors pointed out the possibility of representing the model with the α-radical of the D-enantiomers protruded toward the hydrophobic bed. This would mean a stronger retention of the D-enantiomers over their speculars. According to the chromatographic results, this second model was however indicated as less stable and hence with a statistically lower chance of taking form. The possibility of drawing such multispecies assemblies in a different way directly stems from the assumed chirality of the N atom. As clearly stated by Davankov [29], its absolute configuration is not fixed and can invert during the formation of the ternary mixed complexes. Depending on this configuration, the side-chain residual from either the D- or L-enantiomer can be intercalated among the C-18 chains. However, as previously stated, energetic differences characterize the two resulting complexes, which means that a preferential arrangement has to be expected.

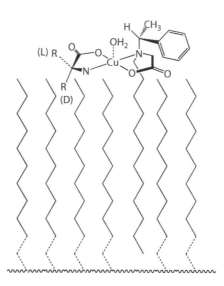

FIGURE 3.12 The proposed structures of the ternary complexes formed from the fixed ligand based on the (R)-α-phenylethylamine, Cu(II), and a D- or L-amino acid. (From Hyun, M. H. et al., *J. Chromatogr. A* 684, 189–200, 1994. With permission.)

Because of the previously claimed singular behavior from histidine as analyte, major attention has been given to the enantiorecognition mechanism promoted by histidine-based resolving agents. In this frame, a key contribution was afforded by Davankov et al. [22] with a study focused on the use of the N-decyl-L-histidine as the hydrophobically impregnated chiral selector for the resolution of α-amino acids. In a very direct way, the authors indicated the "structural ambiguity" of the sorption complexes from histidine-type fixed ligands as the main source for the generally observed low enantioselectivity. Among the reasons accounting for this unattracting feature was the possibility of a double configuration by the primary nitrogen atom after its alkylation was emphasized. Moreover, as already cited, the assumed configuration can even invert during the ternary diastereomer formation. However, it was claimed that the combination, or most properly the equilibrium, between these configurations can be of aid to fully rationalize the overall enantiorecognition mechanism with this material. The "glycine-like" (gly-like) coordination mode was made to correspond to the (R)-nitrogen. In this instance, the carboxy and amino groups from the selector and the selectand were placed on the main coordination square of copper in an alternated fashion (*trans*-configuration). In this scenario, the nitrogen atom from the imidazole ring resulted to be axially coordinated to the central metal cation (Figure 3.13a) and the equatorial plane in both diastereomeric assemblies lies perpendicularly to the C-18 alkyl chains. The opposed elution order for aspartic acid ($k_D < k_L$) was explained by supposing a stabilizing hydrogen bond between the imidazole ring of the fixed ligand and the β-hydroxy group in the L-enantiomer. Contemporarily, only weak hydrophobic interactions between the α-radical of the D-isomer and the reversed-phase material turned up.

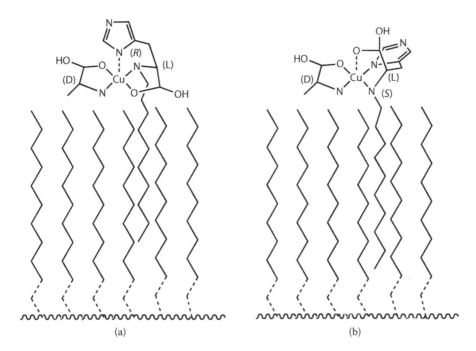

FIGURE 3.13 The proposed structures for the (a) glycine-type and (b) diamine-type ternary sorption complexes with the N-decyl-L-histidine as the chiral selector, Cu(II), and D-amino acid. (Adapted from Davankov, V. A., Bochkov, A. S., and Belov, Y. P., *J. Chromatogr.* 218, 547–57, 1981. With permission.)

In the second fashion of assembly, which seems to be preferred by some analytes and corresponds to the (*S*)-configuration of the N atom, the carboxy group from the selector resulted to be axially coordinated to the Cu(II) (Figure 3.13b). In this situation called "diamine-type" interaction, the nitrogens of the α-amino group and of the imidazole ring were located on the equatorial plane. Such a kind of coordination was deemed to be unfavorable to the chiral recognition accomplishment. Indeed, owing to the inability by the resolving agent to univocally direct its carboxy group into a *trans*-orientation, its α-amino group could occupy both the *cis*- and *trans*-position in the copper coordination square. This implied a higher stabilization of the L-isomers in the case of *trans*-structures and the opposite when the amino groups of the two partners experienced a *cis*-orientation.

Different from this α-amino modified chiral discriminating agent, Remelli et al. [12] obtained an effective coated phase relying on a N^τ-alkylated L-histidine-based material, the N^τ-n-decyl-L-histidine. Also, in this case the pH-dependent possibility to get two diverse coordination modalities by the chiral selector was underlined. While a gly-like type assembly (Figure 3.14a) was deemed as the preferential in the presence of rather acidic eluents, the one referred to here as the "histamine-like" (hm-like) coordination was inferred at pH ≥ 4.5 (Figure 3.14b). However, a *cis/trans* equilibrium often needs to be taken into account to explain the observed

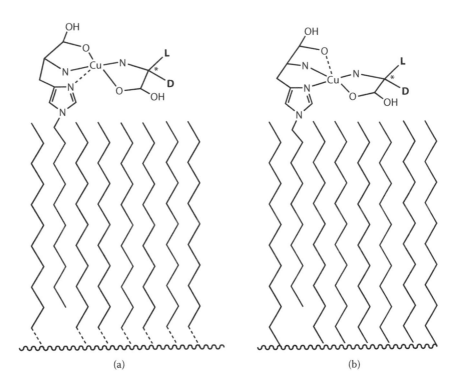

FIGURE 3.14 Structural hypotheses for the stationary Cu(II) complex of the chiral selector N^{τ}-n-decyl-L-histidine with an amino acid sample. The position of the side chain of the sample is indicated by the letters L or D. (a) represents the gly-like coordination mode, and (b) represents the hm-like mode. (Adapted from Remelli, M., Fornasari, P., and Pulidori, F., *J. Chromatogr. A* 761, 79–89, 1997. With permission.)

enantioselectivity for such a system. As already proposed by Davankov for the CLEC environment based on the employment of the N^{α}-n-decyl-L-histidine [22], in this case two different arrangements around the central cation also exist as the primary amino groups of the two partners in the *trans*-position in the gly-like structure (Figure 3.14a) and in *cis*-position in the hm-like one (Figure 3.14b). However, in both cases, the histidine acts as a tridentate ligand.

The different localization of the N-alkylation in the Davankov system compared with that of Remelli was called into play to explain the reversed elution order for aspartic acid (i.e., $k_L < k_D$ with the Remelli system instead of $k_D < k_L$ in the Davankov system) and diamino butyric acid (i.e., $k_D < k_L$ with the Remelli system instead of $k_L < k_D$ in the Davankov system). Among the reasons accounting for this dissimilarity, the different orientation of the sorption complexes was remarked by the authors. In both systems, the α-radical of D-amino acid enantiomers is projected toward the hydrophobic RP material to stabilize the interaction of all those samples endowed with an apolar side chain (which means L- before D- as the elution order). The inversion of the elution order (namely, $k_D < k_L$) for the investigated basic compounds (i.e., diaminobutyric acid, arginine, histidine, lysine, and ornitine) was rationalized for the model in Figure 3.14b by hypothesizing an attractive intermolecular electrostatic interaction between the protonated (at all the fixed pH) ω-group from the analyte and the negatively charged carboxylic group on the chiral selector amino acidic moiety. The elution order for acidic species like aspartic and glutamic acids (namely, $k_L < k_D$) was instead explained on the basis of an interaction between the positively charged imidazole nitrogen and the deprotonated carboxylic group in the D-analyte side-chain as shown in the *hetero*-chiral complex in Figure 3.14a.

In contrast to the presented cases, Galaverna et al. [49] found it highly profitable to engage amino acid amides as CLEC enantiodiscriminating agents. In particular, the three species N_2-octyl-(*S*)-phenylalaninamide (Noc-Phe-NH$_2$), N_2-dodecyl-(*S*)-phenylalaninamide (Ndo-Phe-NH$_2$), and N_2-octyl-(*S*)-norleucinamide (Noc-Nleu-NH$_2$) dynamically adsorbed onto an octadecyl silica gel column, enabling the chiral resolution of physico-chemically diverse amino acid racemates. An interesting interpretation of the observed elution order was advanced by attributing a central role to the chiral selector side-chain residual, as exemplified with Noc-Phe-NH$_2$ in Figure 3.15. Indeed, repulsive interactions between the benzyl moiety of the selector and the side-chain group (Figure 3.15a) of the investigated polar compounds (e.g., serine, glutamic acid, histidine, threonine) were invoked to account for the relative elution order ($k_S < k_R$). On the other hand, the same elution order observed for all the remaining samples endowed with a hydrophobic α-radical ($k_S < k_R$) was explained through the Davankov model of retention (Figure 3.15b).

In the frame of a study focused on the enantiomeric and diastereomeric separation of cyclic β-substituted α-amino acids by means of an N,S-dioctyl-D-penicillamine coated phase, a very interesting behavior was observed by Schlauch et al. [21]. Surprisingly, by means of this column, previously developed by Ôi et al. [50], an increase in the MeCN content turned out into an inversion of the elution order for

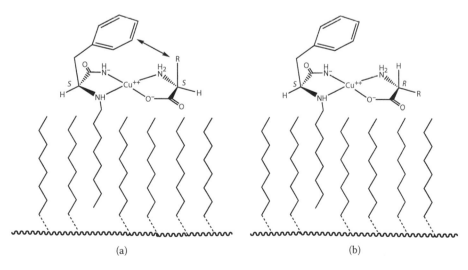

(a)   (b)

FIGURE 3.15 Structures of the proposed diastereomeric ternary complexes with the Noc-Phe-NH$_2$ as the chiral selector. (From Galaverna, G. et al., *Chirality* 8, 189–196, 1996. With permission.)

the enantiomers coming from *cis*-configured complexes of the 1-amino-2-hydroxy-cyclohexanecarboxylic acid and the enantiomers from *trans*-configured complexes of the 1-amino-2-methylcyclopentanecarboxylic acid when chromatographed at 20°C. The authors found the MeCN content to exert a pronounced influence on the thermodynamic parameters of retention and hence of the chiral discrimination as a whole [51]. Inherently, the higher the MeCN amount in the eluent, the lower the difference between the adsorption enthalpy ($\Delta\Delta H°$) and entropy ($\Delta\Delta S°$). Having the content of the organic modifier a prominent influence on the hydrophobic interactions, the authors pointed out the key role played by such a kind of contacts in the overall enantiorecognition process. Moreover, the dependence of the isoenantioselective temperature, T_{iso} from the experimental conditions was also highlighted.

Of particular interest, both in pure scientific and technological terms, is Wan's [52] work dealing with the use of L-phenylalanine-based enantioresolving agents, which were hydrophobically adsorbed onto a porous graphitic carbon-based material. In this frame, N-alkyl- as well as N-aralkyl-substituted L-phenylalanines were highly successful for the enantioseparation of nonpolar, polar, and ionisable amino acid enantiomer couples. Regarding enantioselectivity, the authors first remarked on the noticeable benefit stemming from the elimination of the nonstereoselective secondary interactions with the unalkylated silanols, with carbon-based material unavoidably present on all the conventional RP columns. Moreover, the authors remarked on the influence of the hydrophobic interactions between the anchor molecule and the enantiomers, representing the critical aspect of the overall chiral separation mechanism with this kind of C-CSPs. In fact, besides affecting the enantiomer retention, these interactions were also found to direct the elution order of enantiomers in the relative CLEC system. Indeed, the order of the enantiomer retention with N-alkylated L-phenylalanine coated phases was opposite in comparison with that observed in the presence of the N-aralkyl substituted, with only few exceptions: $k_D < k_L$ and $k_L < k_D$, respectively. The rationalization of the relative complex stability was attempted on the basis of structural models of the diastereomeric assemblies. The possibility for the analyte side chain to undergo contact with either the N-anchor or the phenyl ring of the chiral selector was considered of utmost importance. Indeed, these competitive interactions determine the elution order. Figure 3.16 shows the exemplary case of leucine enantiomers with two selected N-alkyl- and N-aralkyl-substituted L-phenylalanines—namely, the N-(n-heptyl)-L-phenylalanine and the N-(4-methoxybenzyl)-L-phenylalanine. The opposite elution order of amino acid enantiomers observed with the investigated chiral discriminating species was explained in line with the previously mentioned models.

With the selectors of the former group (Figures 3.16a and 3.16b), the interaction between the L-isomer alkyl side chain and the phenyl group on the selector was reputed to be most likely dominant because of the aromatic nature of the phenyl ring. This is in agreement with the assumption of the stronger character of the alkyl–aryl interactions if compared with the alkyl–alkyl ones. Conversely, the stronger stabilizing alkyl–aryl interaction took place with the D-enantiomer form when the N-aralkyl-substituted chiral discriminating agents were used (Figures 3.16c and 3.16d). Wan et al. [52] stated that the hydrophobic surface of graphite was an extension of the benzyl group or the N-substituent depending on the competitive interactions of

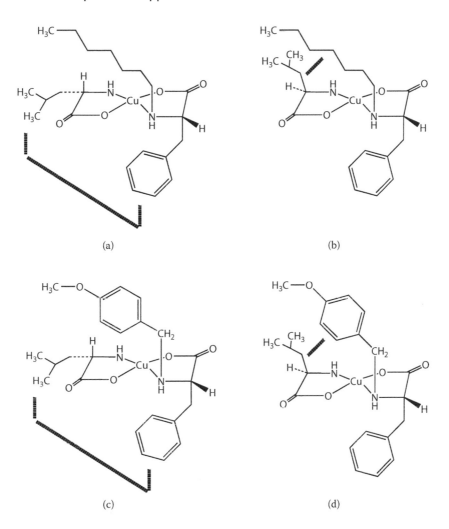

FIGURE 3.16 The proposed structure for the complexes formed between (a) L-leucine (analyte) and N-alkyl L-phenylalanine (selector); (b) D-leucine (analyte) and N-alkyl L-phenylalanine (selector); (c) L-leucine (analyte) and N-aralkyl L-phenylalanine (selector); and (d) D-leucine (analyte) and N-aralkyl L-phenylalanine (selector). Dotted lines indicate potential hydrophobic interactions that stabilize the complexes and thereby influence enantioselectivity. (From Wan, Q.-H. et al., *J. Chromatogr. A* 765, 187–200, 1997. With permission.)

the surface with the hydrophobic portion of the chiral selector. In a follow-up study, the same group of authors [53] evaluated the chromatographic performances afforded by some N-alkyl- and N-aralkyl-substituted L-prolines employed as chiral selectors adsorbed on porous graphitic carbon. Interestingly, the same elution order occurred with all the adopted chiral media, with the L-enantiomers of the selected amino acid analytes being less retained than their antipodes ($k_L < k_D$). Again, structural models were engaged to explain this dissimilar behavior in comparison with the previously

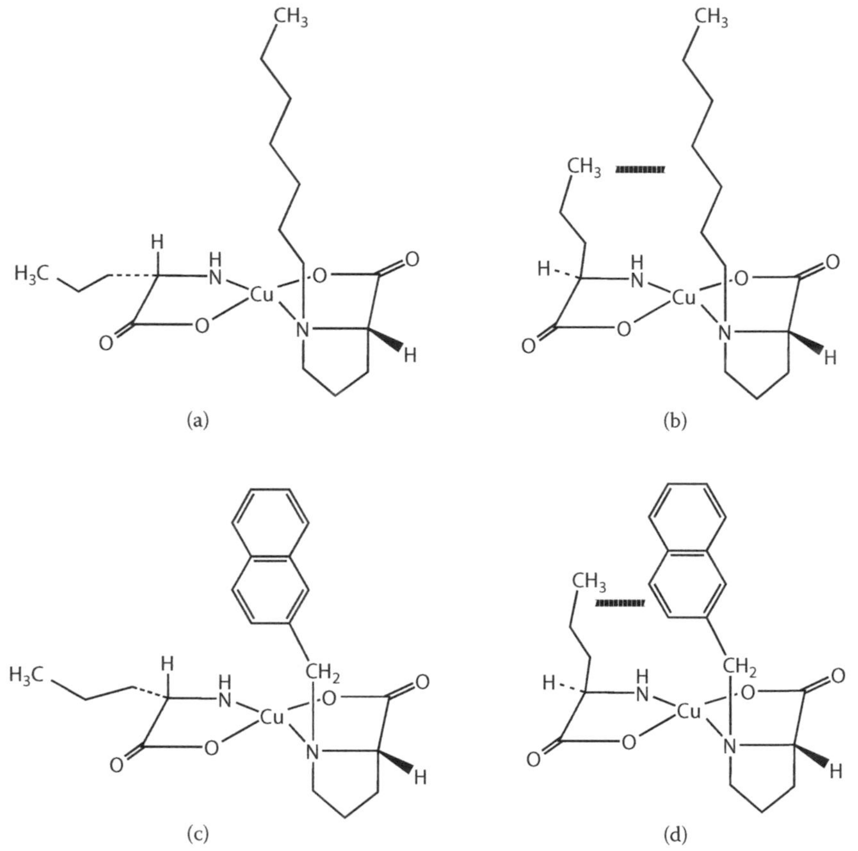

FIGURE 3.17 The proposed structure for the complexes formed between (a) L-norvaline (analyte) and N-alkyl L-proline (selector); (b) D-norvaline (analyte) and N-alkyl L-proline (selector); (c) L-norvaline (analyte) and N-aralkyl L-proline (selector); and (d) D-norvaline (analyte) and N-aralkyl L-proline (selector). Dotted lines indicate potential hydrophobic interactions that stabilize the complexes and thereby influence enantioselectivity. (From Wan, Q.-H. et al., *J. Chromatogr. A* 786, 249–57, 1997. With permission.)

debated case of L-phenylalanine derivatives. Again, exemplary diastereomeric complexes were considered (Figure 3.17).

In accordance with literature data [54], the complexes were drawn so that the N-substituent and the pyrrolidine ring are oriented in an opposite fashion with respect to the main coordination plane. In the two D-enantiomer analytes containing complexes (both with the illustrated N-alkyl and N-aralkyl chiral selectors), strong hydrophobic interactions between the analyte side chain and the N-anchor of the selector accounted for the observed elution order. In this frame, whereas the Nor-valine was engaged as the representative analyte, the N-(n-heptyl)-L-proline (Figures 3.17a and 3.17b) and the N-(2-naphtylmethyl)-L-proline (Figures 3.17c and 3.17d) were chosen as the N-alkyl- and N-aralkyl-substituted chiral selectors, respectively.

Natalini profitably engaged cysteine-based amino acids as dynamically coated chiral discriminating agents [19,32,55] for the enantiodiscrimination of several physico-chemically heterogeneous amino acid enantiomer couples. Particularly, in the three species S-benzyl-(R)-cysteine [(R)-SBC] [19], S-diphenylmethyl-(R)-cysteine [(R)-SDC] [55], and S-trityl-(R)-cysteine [(R)-STC] [32] the primary amino group of the amino acidic moiety was kept free of any derivatization, thus dedicating to the side chain the task of interacting with the RP-18 alkyl chains. Hence, the free amino acidic moiety represents the main point of diversity from the above discussed chiral selectors. Except for the case of histidine, aspartic acid, and 1-aminoindan-1,5-dicarboxylic acid (AIDA), the (S)-enantiomers were always retained more strongly than the (R) ones (i.e., $k_R < k_S$). Whereas for the case of histidine the reversed elution order ($k_S < k_R$) was explained in accordance with the Davankov model of retention, an adjunctive axial coordination by the side-chain hydroxy group, either direct or mediated by a water molecule, was instead inferred for the (R)-aspartic acid enantiomer. With the aid of 3-D structures (Figure 3.18), the controversial behavior provided by the AIDA enantiomers was clarified. From the structures in Figures 3.18a and 3.18b, a wider projection of the polar surface on the RP-18 hydrophobic monolayer by the ternary complex bearing the (S)-AIDA is evident. An opposite situation can be instead visualized for the tyrosine enantiomers (Figures 3.18c and 3.18d), thus providing for a "regular" elution order.

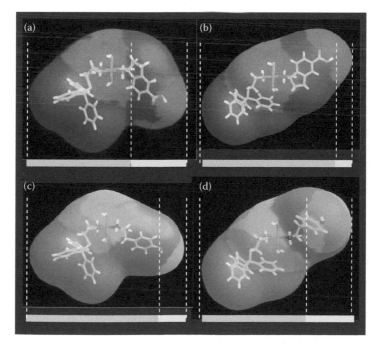

FIGURE 3.18 (*See color insert following page 152.*) The proposed structures for the ternary complexes and surface projections for S-trityl-(R)-cysteine as the chiral selector and (a) (S)-AIDA, (b) (R)-AIDA, (c) (S)-Tyr, and (d) (R)-Tyr.

Interestingly, Natalini provided evidence on the noticeable influence exerted by the type of the cupric anion on the enantioseparation of both natural and synthetic amino acids when chromatographed in the presence of the (R)-STC as the dynamically coated chiral selector [17]. Accordingly, when different cupric salts (i.e., acetate, bromide, chloride, formate, nitrate, perchlorate, sulfate, and triflate) were employed as the Cu(II) source into unbuffered systems, a noticeable gain mainly in terms of resolution factor (R_S) was achieved with the Cu(II) nitrate, whereas the acetate generally furnished almost the worst results in this respect [17]. The following equilibrium was tentatively used as an explanation of the intriguing outcomes:

$$[\text{selector} - \text{Cu(II)} - \text{analyte}] + \text{Cu(anion)}_2 \leftrightarrow$$
$$\leftrightarrow [\text{selector} - \text{Cu(II)}]^+ \text{anion}^- + [\text{Cu(II)} - \text{analyte}]^+ \text{anion}^-. \tag{3.34}$$

A competitive behavior by the copper(II) counter ion over each enantiomer toward the ternary complex formation was realistically hypothesized, which means a different rate for the ligand-exchange processes being accomplished. The highest R_S values from the nitrate salt were assumed to stem from its limited attitude to form complexes in an aqueous system with metal ions [56]. Conversely, the chelating character of the acetate accounts for the pronounced competition with the enantiomer analytes toward the complexation and hence of the low R_S values.

In the previous detailed discussion, any reference to the comparison of the enantiomer retention with structurally analogued C-CSPs has been deliberately avoided. Indeed, since it is very difficult to establish the exact entity of the surface coverage by a coated selector, all the interpretations are meaningless.

3.9 CHIRAL MOBILE PHASES (CMPS)

Regarding the CMP-CLEC approach, an important contribution was published by Weinstein in 1982 [57]. This paper reports the use of N,N-dialkylated L-valine- and L-alanine-based chiral additives for the enantiodiscrimination of underivatised amino acids. The author focused his attention mainly on the effect of the size of the N-alkyl group on the overall enantiorecognition mechanism. Weinstein underlined the hydrophobicity of all the investigated ligands being responsible for their relevant adsorption onto the RP-18 stationary phase. Thus, the chiral differentiation was stated to be principally referred to interactions with species adsorbed on the stationary phase. This assumption found support in the previous observation that stability constants of amino acid-copper(II) complexes are higher in organic solvents (likewise, the RP-18 environment can be considered) than in aqueous solution [58]. Some very interesting observations were made regarding the N-alkyl length. Accordingly, even small variations in the N-alkyl extension were found to be because of an inversion of the enantiomer elution order, thus revealing a strict incidence on the enantiorecognition mechanism. Indeed, by passing from the N,N-dimethylated to the N,N-diethylated ligands, the inversion of the elution order occurred for all the discriminated analytes (with the exception of histidine). Steric hindrance-based

argumentations were invoked to account for this experimental outcome. More pre-
cisely, this was explained simply by considering that, to keep the planarity safe in
all the Cu(II) complexes, subtle changes in the mixed complex structures led to
conformational modifications. These variations were surmised by the author to be
responsible for different selector–selectand association fashions. Particularly, the
modification of the steric barrier in the coordination sphere was used to explain these
results. A further reversal of the elution order was achieved in the case of the N,N-di-
n-propyl-based enantiodiscriminating agent. However, the enantiomer elution order
from the column no longer changed by passing to the n-butyl and n-pentyl groups
as the N-substituents. The previously cited controversial behavior from histidine
enantiomers with both the N,N-dimethyl and N,N-diethyl carrying selectors was still
ascribed to the tridentate coordinative fashion by this compound. However, its action
as a gly-like coordinant took place by further extending the N-alkyl residual length.
In this instance, the coordination by the imidazole nitrogen was reputed as sterically
prevented, and the regular elution order was accomplished by the enantiomers of
this analyte.

The use of chiral additives, which allowed the realization of mixed CLEC environments, was also experienced by Wernicke ([18]. Accordingly, the author reported
on the successful enantioseparation of underivatised amino acid enantiomers by
means of three chiral additives to the eluent: L-phenylalanine (L-Phe), N-methyl-
L-phenylalanine (N-me-L-Phe), and N,N-dimethyl-L-phenylalanine (N,N-dime-
L-Phe). Some attention was given both to the *trans*-configuration assumed by all
the complexes and to the role of the hydrophobic interactions within the employed
reversed-phase environment. Regarding the latter, for a homologous series of analytes (i.e., glycine, alanine, amino butyric acid, Nor-valine, and Nor-leucine), the
progressive increase of the methylene units in their side chain was followed by an
increase in terms of retention. Still, with all the three chiral phases, a higher retention was found for both enantiomers of Nor-valine with respect to that of valine, thus
underlying the stronger hydrophobic interaction between the RP-18 chains and the
normal chain residual (in the former). Less intense hydrophobic interactions instead
took place with the branched α-radicals of valine. A reduced bulk effect for Nor-
valine was claimed to rationalize this finding. The much more intricate and manifold
situation characterizing the leucine isomers, when eluted in the presence of the three
selectors, was deputed to adjunctive ligand-exchange equilibria, added to steric hindrance phenomena.

Another intriguing aspect is the absence of a progressive increment of retention with the chiral selector hydrophobicity. Indeed, with the employment of the
N-me-L-Phe in place of the less hydrophobic L-Phe, a general decrease of the analyte enantiomer permanence into the column emerged. The interference exerted
by the N-methyl group during the diastereomeric complex formation was invoked
to explain this outcome. In comparison with the mono-substituted ligand, an
increase in the enantiomer retention was instead revealed with the use of the N,N-
dimethylated chiral additive. This situation, found specifically in the L-enantiomeric
forms, accounted for the generally higher α values, with the ligand having the highest hydrophobicity in the series. A configuration-dependent effect allowed the complex bearing the L-enantiomeric form to undergo stronger hydrophobic contacts with

the RP-18 chains inside the column. Conversely, the occurrence of steric hindrance phenomena during the ternary complex formation explained the scarce retention of the assemblies carrying the D-enantiomers. The combination of the aforementioned two events was responsible for the relevant α values in the presence of the N-dime-L-Phe. The inverted elution order observed for serine, asparagines, and threonine ($k_L < k_D$ in place of $k_D < k_L$), when both the L-Phe and the more hydrophobic N-me-L-Phe were used, was explained by calling into play a third coordinative bond by their polar side-chain residual. Interestingly, the enantiomers of serine and threonine experienced the regular elution order in the presence of the N,N-dime-L-Phe. The steric hindrance exerted by the bulky $-N(CH_3)_2$ group, which avoided a further coordination on the Cu(II), was used to explain this controversial behavior. While the elution order of all the remaining analytes was fully explained with the Davankov model, the observed reversal enantiomer retention for histidine was still related to its pH dependent tridentate character.

A seminal work for understanding the basic principles governing the enantiorecognition as a whole in CMP-CLEC systems was provided by Galaverna [9,14,15]. In a very exhaustive way, the nature of all the species involved in the ligand-exchange processes was investigated by the authors. Additionally, they also furnished a deeper insight into the variation in the enantiorecognition mechanism rising from a progressive increase in the structural complexity of the employed chiral discriminating agent. This investigation was carried out by maintaining the L-phenylalanine building block as the common feature of all the investigated ligands, thus allowing a rational comparison among the cases. The authors relied on this species for the enantioseparation studies because of its hydrophobicity that enables, to some extent (depending on the adopted experimental conditions), its adsorption onto the RP-18 chains of the column employed as stationary phase. The use of chiral additives not exclusively resident into the eluent unavoidably generates very CLEC processes that are very complicated to formalize. Indeed, as already cited, depending on the selected experimental conditions, the chromatographic process responsible for the enantiorecognition accomplishment can mainly take place either on the stationary phase or in the bulk of the solution. Consequently, an intricate series of complexation equilibria turns up. In this scenario, Galaverna systematically investigated this subject, thus contributing to establish the relevance of specific complexation equilibria in the two chromatographic phases on the overall mechanism of enantioseparation. A selection of exemplary cases is presented in the following discussion, with consideration given to bi-, ter-, and tetradentate-type ligands from amino acid amides. In each of the presented cases, the complex formation equilibria are relevant in both the mobile and stationary phases, and partitioning of all species between the two phases must be taken into account to explain the enantiomeric discrimination. Whether a phase predominates over the other strictly depends on the structural feature of the initial binary complex as well as the partitioning of the ternary assemblies between the two phases, as already stated. The complexity of all the CLEC systems characterized by ubiquitarian chiral selectors was brought into play by the authors to explain the discrepancy that often occurs between the experimental results and the thermodynamic enantioselectivity evaluated in solution. As the simplest case, the use of the three ligands L-phenylalanine amide (L-Phe-A), N-methyl-L-phenylalanine amide (L-MePhe-A), and N,N-dimethyl-L-phenylalanine amide (L-Me$_2$Phe-A) will be illustrated [9].

Using Wernicke's work as a starting point [18], the different enantioselectivity profiles coming out from a progressive N-methylation were underlined by Galaverna et al. [9]. First, it was again pointed out how steric factors can markedly affect the selector–selectand association even when mediated by a central metal ion. Indeed, the pronounced steric hindrance in the N,N-dimethylated selector resulted in the best enantiodiscrimination ability toward the nonpolar amino acids. Interestingly, this selector resulted completely ineffective for the polar ones, in this regard. For the polar amino acidic species, a better compromise between steric hindrance and hydrophobicity was invoked to give a rationale for the enantioseparation ability by Phe-A and MePhe-A. Moreover, the same observed reversal in the elution order by passing from the polar ($k_L < k_D$) to the nonpolar ($k_D < k_L$) analytes stood for analogue interactions materializing during the selector–enantiomer association processes in the two systems. This clearly indicated that the presence of an N-methyl group alone to an extent still did not affect the principal interactions allowing the enantiorecognition accomplishment. Another major point emphasized throughout the work was the enantioselectivity being not necessarily ascribable to a single binary Cu(II)-ligand adduct but rather to the occurrence of a more complex environment characterized by the presence of several multispecies entities in equilibrium between each other. Accordingly, the Cu(II)-MePhe-A complexes identified within the studied pH range (5.0–7.5), by means of potentiometric titration and spectrophotometry, are displayed in Figure 3.19.

Experimental techniques such as potentiometry and spectrophotometry have been used on a regular basis by the same group to shed light on the enantiorecognition process. The usefulness of these techniques unequivocally demonstrates the necessity of not exclusively relying on chromatographic assays to interpret CLEC events. Indeed, focused parallel studies can provide for peculiar adjunctive elucidating information in this regard. The two diastereomeric structures displayed in Figure 3.20 and carrying the MePhe-A as the chiral selector allowed the authors to elucidate the previously cited elution order [9].

Accordingly, in the *homo*-chiral complex, the side chains from both the discriminating agent and the L-analyte enantiomer are oriented on the same side of the copper (II) coordination plane. In this way, strong interactions with the RP-18 chains are materialized in the case of nonpolar R residual (which emerged in the observed elution order $k_D < k_L$ for such species). Conversely, in the case of polar R radicals, less favorable interactions in the *homo*-chiral complexes took place, thus giving rise to the elution order $k_L < k_D$. The terdentate fashion of histidine was still remarked by Galaverna et al. [9], who also attributed an axial coordination ability to the hydroxy group of aspartic acid. The absence of this adjunctive point of attack to the Cu(II) by the glutamic acid was invoked to rationalize the lack of enantiodiscrimination for this analyte. Finally, the presence of the methyl group in the threonine side chain accounted for its enantioseparation that in contrast did not occur for the unhindered and too hydrophilic serine.

CuL^{2+}

CuL_2^{2+}

$CuLH_{-1}^{+}$

$CuL_2H_{-1}^{+}$

CuL_2H_{-2}

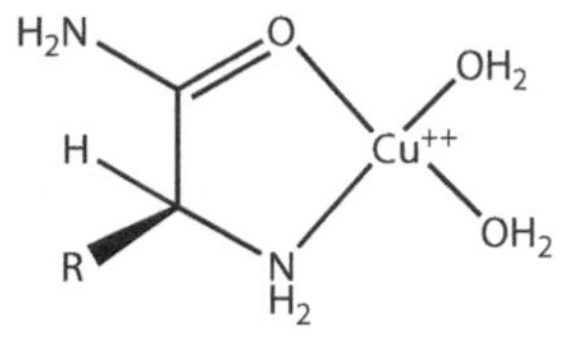

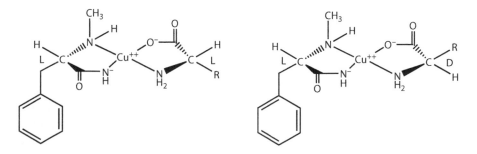

FIGURE 3.19 The proposed structures for the Cu(II)-MePhe-A complexes identified within the studied pH range (5.0–7.5) by means of potentiometric titration and spectrophotometry. (From Galaverna, G. et al., *J. Chromatogr. A* 657, 43–54, 1993. With permission.)

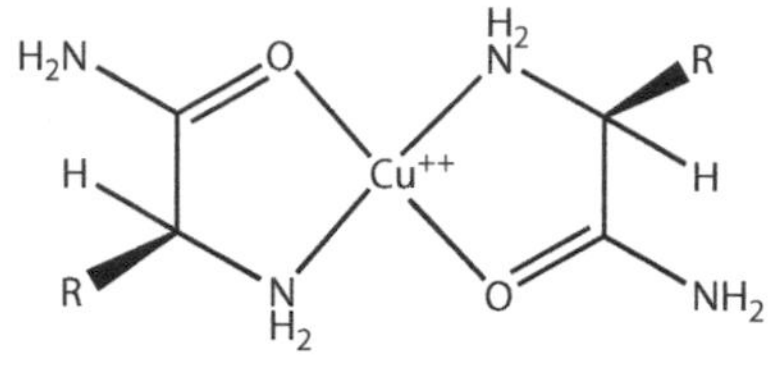

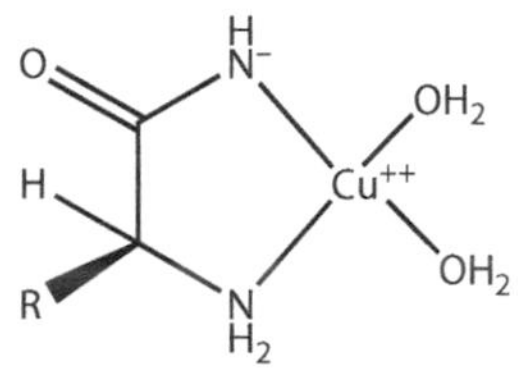

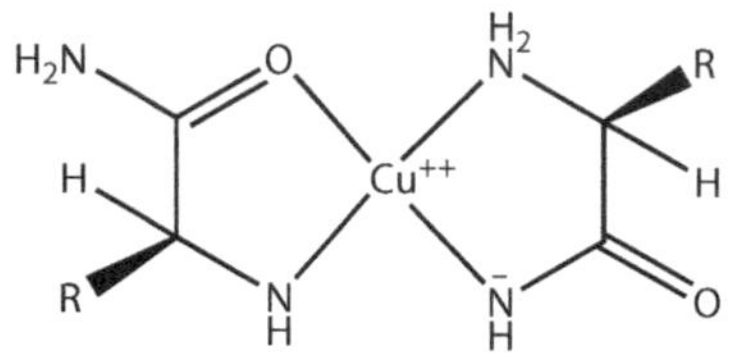

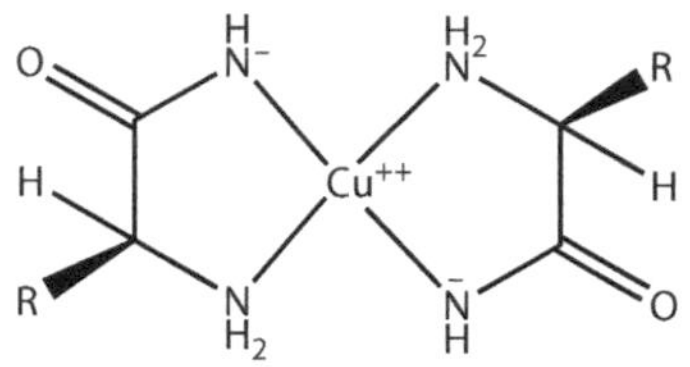

FIGURE 3.20 The proposed structures of the ternary complexes formed from the chiral additive to the eluent N-methyl-L-phenylalanine amide, Cu(II), and a D- or L-amino acid. (From Galaverna, G. et al., *J. Chromatogr. A* 657, 43–54, 1993. With permission.)

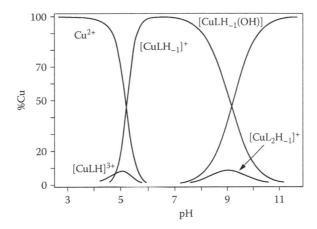

FIGURE 3.21   Specie distribution for Cu(II)/PheN-2 (1:2) system. $C_{Cu} = 1$ mM. (From Galaverna, G., Corradini, R., Dossena, A., et al., *J. Chromatogr. A* 829, 101–13, 1998. With permission.)

The effect deriving from an increased physico-chemical sophistication in the ligand structure was investigated by the same authors by employing 1-(N-L-phenylalanyl)-1,2-ethanediamine (PheN-2)- and 1-(N-Me-L-phenylalanyl)-1,2-ethanediamine (MePheN-2)-based selectors as terdentate ligands bearing a further amino or methylamino group [14]. Also, in this case all the ligands were designed by maintaining the L-phenylalanine portion that had proven to be beneficial in the previous study [9]. While few performances emerged for underivatised amino acid analytes, very satisfactory enantioseparations were achieved on their dansyl-derivatives (Dns-amino acids). Even if no definite trend in terms of α values was observed for this second group of compounds, by increasing the hydrophobicity of the chiral selector the same elution order ($k_D < k_L$) occurred. A crucial concept of the CLEC systems was the lack of a strict dependence of the enantioselectivity on the intrinsic chiral selector hydrophobicity. Similar to the previously debated case, by engaging the aforementioned ligands the authors showed how focused potentiometric measurements in solution can help tremendously to cast light on the relationship between the eluent pH and the produced binary complexes. The exemplary case of species distribution for the Cu(II)/PheN-2 (1:2) system is shown in Figure 3.21. The entire pH range generally covered in HPLC studies was investigated. With the aid of spectrophotometric analysis, the complete characterization of the prevalent binary adduct that took form with the PheN-2 and MePheN-2 selectors within the pH range investigated for this study was also furnished. The resulting assemblies were proposed as having a square planar structure in which the ligand generates two coplanar five-membered chelate rings with two amino and one deprotonated amide nitrogens. The fourth position was considered to be occupied by a water molecule.

Subsequent potentiometric studies undertaken to investigate the pH-dependent formation of ternary mixed complexes of PheN-2 and some of the free D- and L-amino acids (i.e., valine, glutamic acid, phenylalanine, proline, and tryptophan) allowed, *inter alia*, the feature of the prevailing diastereomeric complexes to be hypothesized

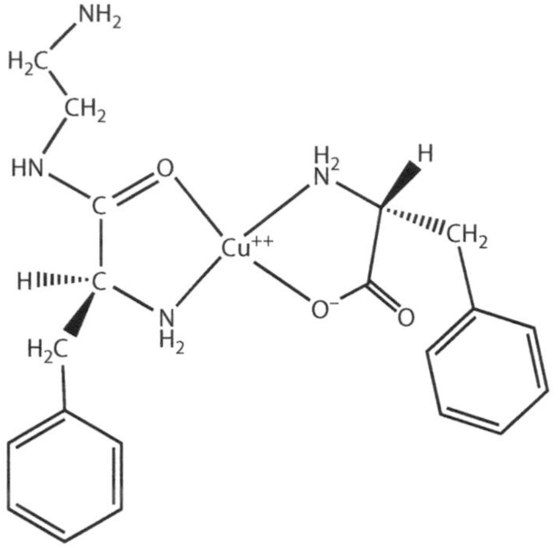

[CuLA]⁺

(a)

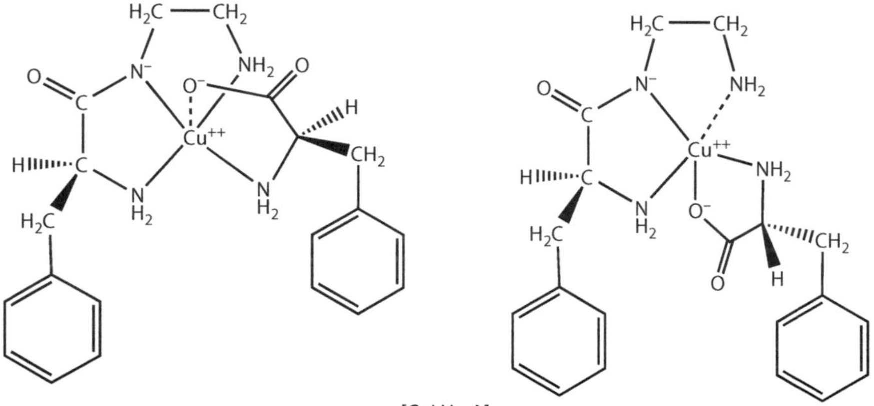

[CuLH₋₁A]

(b)

FIGURE 3.22 The proposed structures of the ternary complexes of Cu(II) with PheN-2 and D-Phe. (a) [CuLA]⁺; (b) [CuLH₋₁A]: the two different coordination modes at the apical positions are outlined. (From Galaverna, G., Corradini, R., Dossena, A., et al., *J. Chromatogr. A* 829, 101–13, 1998. With permission.)

for each analyte. As illustrated in Figure 3.22, in the presence of the D-phenylalanine as the analyte, the two prevailing ternary assemblies were assumed to be [CuLA]⁺ (Figure 3.22a) and [CuLH₋₁A] in the two interaction modes (Figure 3.22b). The latter, which is the main species (40% of the total copper concentration), was proposed to possess a tetragonal–pyramidal character with the amino acid analyte binding at the equatorial (–NH₂) and apical (–COO⁻) positions.

From this study, important information was gained: the enantiorecognition event can take place even when only one of the equatorial positions is occupied by the

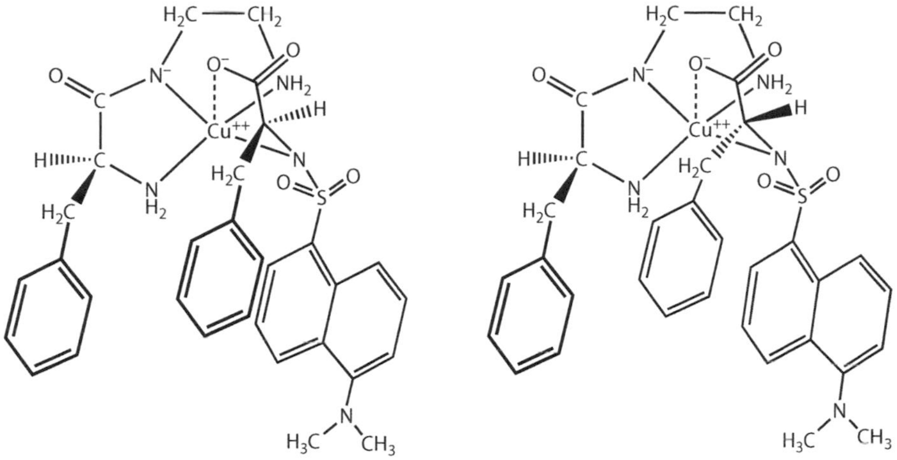

L-PheN-2/Cu$_{II}$/L-Dns-Phe   L-PheN-2/Cu$_{II}$/D-Dns-Phe

FIGURE 3.23 The proposed recognition models for Dns-Phe by L-PheN-2. (From Galaverna, G., Corradini, R., Dossena, A., et al., *J. Chromatogr. A* 829, 101–13, 1998. With permission.)

analyte enantiomer. Moreover, even if not completely explicatory, the difference in the stability constants calculated in solution for the two diastereomeric complexes of each analyte were still claimed to provide useful information on the potential enantioselective power of a definite chiral selector toward a specific species to be enantioresolved. However, as already pointed out, the discrepancies with the chromatographic results indicated that the thermodynamic selectivity in solution has only a partial effect on the overall discrimination. Indeed, the different affinity of the diastereomeric ternary complexes for the stationary phase plays a crucial role in CMP-CLEC environments. All the structural information gained through the previously mentioned potentiometric studies was, however, highly instrumental in drawing the binary diastereomeric complexes with the investigated Dns-amino acids (Figure 3.23). As evident, the chiral selector did not undergo any chelate ring modification after the coordination of the analyte. The same elution order for all the dansylated compounds ($k_D < k_L$) was attributed to the foremost role by the dansyl group over the analyte side chain.

Indeed, by considering the discriminating agent as (partially) adsorbed onto the alkyl chains, the dansyl portion from all the analytes did undergo aspecific hydrophobic contact with this surface so that the nature of the α-radical is irrelevant. The stronger retention of the L-enantiomers was explained invoking a better fit of their dansyl groups with the RP-18 chains.

As the last example of the interesting chiral discriminating agents developed by Galaverna, the fruitful employment of the two tetradentate (*S,S*)-N-N′-bis (phenylalanyl)ethanediamine (PheNN-2) and (*S,S*)-N-N′-bis(methylphenylalanyl) ethanediamine (Me$_2$PheNN-2) [15] is reported. Although L-phenylalanine is the highest structural complexity of this selector, it was still maintained as the chiral building block. Besides the relevant performances (mainly achieved with the latter selector) in enantiodiscriminating underivatised amino acids, a precious mechanistic information was also gained. Moreover, several assumptions already put forth in the previous studies [9,14] were further corroborated when these two ligands were used. First, the concept that the thermodynamic enantioselectivity (still measured in solution via potentiometric assays) does not necessarily parallel the chromatographic one was mentioned for the first time. In this frame, the outstanding effect exerted by the differential distribution of the two diastereomeric complexes between the two chromatographic phases on the overall enantiorecognition mechanism was still observed. Most importantly, the occurrence of different types of retention mechanisms being produced by changing the chiral selector concentration into the eluent was demonstrated. Indeed, both retention and enantioselectivity were observed to reach a maximum when the chiral selector concentration in the eluent was equal to 0.5 mM. In this circumstance, the enantiorecognition was assumed to take place mostly because of the adsorbed chiral selector ($Me_2PheNN-2$) onto the C-18-based material. The ability of the $Me_2PheNN-2$ to produce a fairly good enantiorecognition when exclusively resident onto the stationary phase was experimentally demonstrated. As before, the typology of the different binary complexes [Cu(II)-ligand] taking form in the correspondence of specific pH values was determined for the two ligands. Indeed, whereas with PheNN-2 the predominant species were found to be $[Cu_2L_2H_{-2}]$ (80% of total copper) in the pH range 5.5–7.5 (adopted for the chromatographic studies) and $[CuLH_{-2}]$ (100%) at pH > 7.5, in the presence of $Me_2PheNN-2$ the most abundant species were $[CuLH_{-2}]$ (100%) at pH > 6.5. This different distribution (in terms of both species typology and relative amount) also influenced the formation of the ternary mixed complexes for specific eluent pH values.

Ranging the pH studied between 5.5 and 7.5, Galaverna et al. [15] stated that for PheNN-2 (Figure 3.24a) the main selector form able to undergo ligand exchange was $[Cu_2L_2H_{-2}]^{2+}$. In this case, the two mixed assemblies $[CuLH_{-1}A]$ or $[CuLA]^+$ were inferred to be produced. Instead, in the case of $Me_2PheNN-2$ (Figure 3.24b), since $[CuLH_{-2}]$ was the major species present in solution, the occurrence of the ternary complexes $[CuLH_{-1}A]$ was hypothesized.

In $[CuLA^+]$, the copper(II) was made to coordinate an amino nitrogen and the carbonyl group of the selector on the equatorial plane, with the two other sites occupied into an alternating fashion by the amino and carboxylic moieties of the analytes. Concerning the $[CuLH_{-1}A]$, the ligand was assumed to chelate via an amino and a deprotonated amido nitrogen. In both models, the authors did not exclude an eventual axial interaction by the other amino group on the ligand. No interpretation of the observed elution order with $Me_2PheNN-2$ ($k_L < k_D$ for polar analytes and $k_D < k_L$ for the apolar ones) was proposed.

3.10 COMPUTATIONAL APPROACHES IN CHIRAL LIGAND-EXCHANGE CHROMATOGRAPHY

Diverse computational approaches have been successfully devoted to shedding light on the basic principles governing the chromatographic event in ligand-exchanging

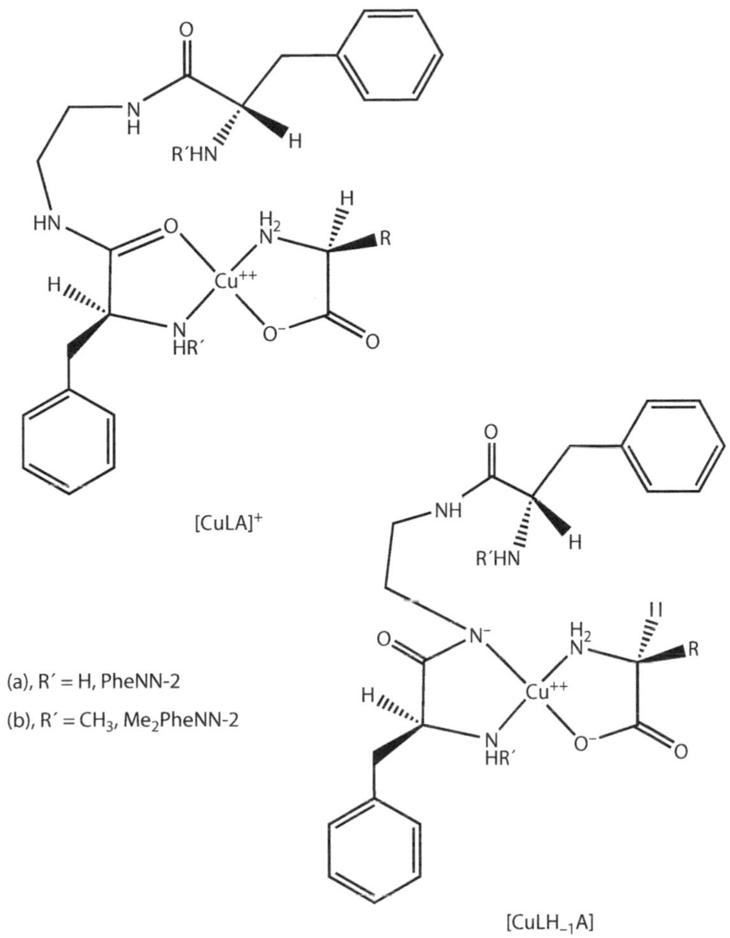

FIGURE 3.24 The proposed structures of the ternary complexes of Cu(II) with PheNN-2 and Me₂PheNN-2 and L-amino acids. (From Galaverna, G. et al., *J. Chromatogr. A* 922, 151–63, 2001. With permission.)

environments. In this section, we have divided these approaches into mathematical models, molecular modeling studies, and quantitative structure activity relationships (QSARs). Though each of the three computational approaches along with case studies is discussed, particular attention is directed to some recent applications in CLEC of QSAR studies.

3.10.1 Mathematical Models

In spite of increasing interest in the use of preparative-scale enantioseparative chromatography, lack of attraction to all its related approaches can be attributed mostly to its general cost limitations (viz. the preparation of suitable enantioresolving materials, the search for suitable experimental conditions, and the required technology).

Therefore, to make it more accessible, it is necessary to examine the method's essential operating conditions. In this scenario, the knowledge of the separation behavior in a specific chromatographic environment under increasing concentration of sample to be injected and operational condition variation can be helpful to reduce the number of unsuccessful efforts. Accordingly, to rely on adsorption isotherms studies aimed at predicting the chromatogram profiles is a matter of growing interest [59–61]. Although a large number of studies deal with the CLEC methodology on an analytical scale, only very few preparative-scale developments have been attempted. However, although only scarcely undertaken, the profitable isolation of pure enantiomeric forms in CLEC has been achieved through both the CMP [24,62,63] and the CSP approach [32,64–66], thus making the technique more versatile. Worth noting is the almost total absence of interest that emerges for any mathematical elaborations of the chromatographic data, like studies on adsorption isotherms. Indeed, apart from the exhaustive work of Kurganov et al. [61] on the unusual peak profile under overloaded conditions, only a small number of contributions quantitatively treating this aspect can be found in the recent literature.

Among these, owing to the systematic study described, two papers by Kostova [59,60] are worth mentioning. Kostova and Bart monitored the variation in the chromatographic behavior in different injected concentrations of physico-chemically diverse racemic amino acidic analytes on a conventional RP-18 column impregnated with the N-hexadecyl-L-hydroxyproline as the enantioresolving agent. In one of the two studies [59], a detailed comparison of three different methods (perturbation, adsorption, and desorption) was done with the intent of determining the most suitable approach for measuring the relative adsorption isotherms for the employed CLEC environment. In a follow-up study [60], once identified, the more appropriate isotherm equations were applied to mathematically simulate the preparative separation behavior both at the adsorption and desorption conditions and in the same chromatographic system. Interestingly, depending on the nature of the investigated sample, a different method can be suggested as the preferential tool for the measurement of its adsorption isotherms. Moreover, although the applied mathematical model provides a very good prediction of the desorption profiles, a sample concentration and chemical dependence affected some of the adsorption profiles.

The recorded outcomes, which were modeled via the multicompetitive Langmuir equation [67], can be successfully used to predict the chromatographic behavior of other structurally related compounds to be enantioresolved on a preparative scale. The attractive feature of the multicompetitive Langmuir equation is that it accounts for the influence of the adsorption of one enantiomer over its specular and vice versa.

Sanaie and Haynes [68], whose aim was to model the L-DOPA enantioseparation process by using the chiral selector N-octyl-3-octylthio-D-valine hydrophobically stratified onto a common RP-18 stationary phase, pointed out, *inter alia*, the noticeable contribution provided by adsorption isotherm studies in this respect. As before, a quantitative model endowed with predictive ability becomes of key importance both to economize and to speed up the preparative-scale enantioisolation of this anti-Parkinson's disease compound. Accordingly, through the use of an original model for the enantiomer transport and elution, which puts together the chromato-graphic theory for the reaction–diffusion with that of multiple chemical equilibria [34], Sanaie and Haynes [68] proved that the overall chromatographic process was successfully predicted and, in turn, advantageously modified. The model sheds light on the correlation between the series of chemical equilibria that materialize into all CLEC-based media and the variation in band profiles and band separation after changing the experimental conditions. Accordingly, through the use of an excellent adherence between experimental and model elution profiles in the presence of spe-cific operating conditions was obtained.

A detailed explanation of the multiple chemical equilibria approach theory is given by Sanaie and Haynes [34]. Keeping in mind that in CLEC even subtle changes in the experimental conditions can drastically affect the overall chromatographic behavior, they noted the necessity of using a model encoding for all the chemical equilibria taking place inside the column (to consider in combination with the equi-librium formation constants of the ternary complexes that take place at the stationary phase). Such a model is highly instrumental in rationalizing the separation process by considering together the elution band profiles and the temporal and spatial changes in the distribution of equilibrium complexes as the species are flowed along the chro-matographic medium. Particularly, in the work the model was developed to fur-nish an interpretation and a simulation of the elution profiles of racemic valine and phenylalanine under mobile phase composition: pH and temperature variation. The enantioseparation study was carried out using an L-3-hydroxyproline-based mate-rial as the enantioresolving compound. The set-up of a robust model relies on the availability of standard thermodynamic databases [69,70] reporting both protonation constant for single stereocenter enantiomers as well as formation constants for com-plexes carrying transition metal cation. The goodness of the model in terms of both retention times and peak profile prediction is confirmed by comparing the predicted and experimental traces. It is significant to note the reduced number of experiments required for employing the mathematical model.

Another remarkable work relying on computer-assisted calculations was car-ried out by the same research group [35]. Focused efforts were made in this case to define the impact of the temperature on the chemical equilibria in a system bearing the L-hydroxyproline as the enantiodiscriminating agent and the D,L-proline as the species to be resolved. Interestingly, Sanaie and Haynes (still aided by potentio-metric titration experiments) attributed the variation of the chromatographic trace profile (i.e., the peak retention time and shape) with increasing temperatures, to a modification in the solute diffusion through the pores of the stationary phase. This was reputed to markedly affect the solute adsorption on the ligand-exchange mate-rial. In other words, the change in the enantioseparative performance that generally turns out with a variation of the column temperature was related to changes in the occurring chemical equilibria rather than to the previously supposed increment in the kinetics of the ternary complex formation [23]. Thereby, increasing the column temperature was observed to unavoidably reduce the partitioning of the enantiomer between the two chromatographic phases.

3.10.2 MOLECULAR MODELING STUDIES

The influence of the solvent composition on the enantiorecognition mechanism was studied by Koska et al. [71]. Interestingly, for the first time, the role played by different water–MeOH-containing mobile phases on the chemical equilibria that took place in a specific CLEC environment was studied by means of molecular mechanic (MM) calculations. In this frame, whereas L-proline and L-hydroxyproline were chosen as the chiral discriminating agents, the enantiomers of valine, leucine, and phenylalanine represented the investigated analytes. Potentiometric titrations combined with the multiple chemical equilibria model allowed the determination of the protonation constants as well as the formation constants for all the mono and bis binary complexes and for each *homo-* and *hetero*-chiral ternary assembly. In addition to these thermodynamic studies, MM simulations represented a valid tool to understand at a molecular level the contribution of the direct participation of the solvent molecules on the ternary complex stability and, in turn, on the mechanism of the occurring enantiorecognition process. Sanaie and Haynes claimed such simulations as highly beneficial for further applications because they afford the possibility of predicting changes in the ligand enantioselectivities with the solvent composition. The reduction of the solvent dielectric constant with an increasing amount of MeOH was employed to explain the higher equilibrium formation constants for the ternary arrangements. Moreover, the aforementioned higher stability stemming from the axial coordination of MeOH molecules was accounted for. Worth noting is the different alteration of the chemical equilibria and of the enantioselectivity, as a result of the analyte physico-chemical nature. Indeed, while the calculated enantioselectivity for the phenylalanine resulted, in solution, practically unaffected when the MeOH concentration in the eluent was increased, a different situation was revealed for both valine and leucine. The phenylalanine enantiomers' insensitivity to the solvent modification was attributed to an axial coordination of its phenyl ring to the central Cu(II). Very interestingly, leucine enantiomers underwent an inversion of enantioselectivity when a 20% MeOH containing solution was adopted. This evidence doubtless stated for an influent contribution of the analyte side chain on the ternary complexes' stability (and more generally on the selector–selectand way of associating with each other). The previously mentioned increased equilibrium formation constants of the ternary complexes in the presence of MeOH did not, however, undergo a parallel increase in their concentration. Indeed, it was observed that the concentration of the exemplarily studied L-leucine:Cu(II):L-hydroxyproline system underwent a sensitive reduction in the presence of the organic modifier. Contemporarily, a significant increase in the stability of the bis binary complexes was recorded. Thereby, in fact, inspecting the measured equilibrium formation constants was found to not only predict the effect of the organic modifier content on the CLEC event and, in particular, on the amount of the different species for definite experimental conditions. In this frame, MM calculations were highly fruitful in establishing the equilibrium structures for both bis binary and ternary complexes. The estimation was done on the basis of the strain energies of the minimized complexes. For this study, L-hydroxyproline was selected to act as the chiral discriminating agent. The results obtained by considering the systems either in the vacuum or in the presence of solvent molecules were compared. In the latter case, the different behavior from a pure aqueous and a binary (water–MeOH) solution was also investigated. This comparison served to establish, in a more realistic way, the contribution of different compounds as a component in the complexes. A general stabilizing effect was observed when the axial solvent molecules (water or MeOH) were added within the complexes. However, since this effect was experienced by both the *homo*- and *hetero*-chiral complexes, the enantiodiscriminating effect was reduced. Interestingly, from MM calculations, although a *trans*-equatorial arrangement resulted more stable for leucine and valine, the *cis*-disposition around the central cation was instead found to provide a higher stabilization for the phenylalanine enantiomers.

A remarkable reduction of the strain energies (which means a higher stability) was diagnosed by replacing the water molecules with MeOH. A better packing was used to explain, for leucine, the lower energy values. However, the unprofitable orientation of the leucine enantiomer side chains limits the enantioselective interactions. The presence of MeOH molecules was found to direct in both diastereomeric complexes the leucine enantiomer side chain away from the metal ion center, thus limiting specific interactions enabling the enantiorecognition. Moreover, the inability of MeOH to undergo differential hydrophobic interactions with the leucine α-radical was emphasized as well. A relevant reduction in the enantioselectivity (still measured as the difference of energy between the minimized diastereomeric complexes) was also observed for the phenylalanine by passing from vacuum to the solvent-containing system. This outcome was attributed to the absence of a discriminating axial interaction that took place in the presence of water and McOH molecules. Interestingly, from the MM simulation, the presence of water molecules tended to destabilize the *hetero*-chiral complex, thus producing an inversion in the enantioselectivity. In the presence of MeOH, the differential π-cation interactions were furthermore reduced, thus turning out into a higher reduction of enantioselectivity.

Interestingly, the difference in the calculated ternary complex strain energies of the *hetero*- and *homo*-chiral complexes for leucine and phenylalanine was found to be in qualitative agreement with the experimental (chromatographic) values obtained with a 40% MeOH-containing mobile phase. This observation allowed the authors to point out the determining effect on the enantioselectivity by the direct participation of the solvent molecules in the complexes.

Another work relying on the use of MM calculations was carried out by Natalini et al. [72]. In this study, eight diastereomeric couples of copper complexes with N,N-dimethyl-(S)-phenylalanine and the enantiomeric amino acid were built and optimized using the Universal Force Field [73]: a dielectric constant of 80 simulated the water environment. For the resulting global minimum conformation of each complex, the water-accessible surface was calculated and analyzed by computing, in particular, the copper ion area exposed to the solvent. As a result, it was found that the elution order in the eight diastereomeric couples is independent of the amount of total water-accessible surface area of the complex because it is strictly correlated with the solvent-accessible area on copper ion. This observation was interpreted with the presence of a relationship between the elution order and the water coordination capability on the copper ion in the formation of the mixed ternary complexes.

Starting from the observation that the Davankov model, which is still widely accepted to explain the elution order of the amino acid enantiomers, does not always prove effective in the case of α-hydroxy acids, Chilmonczyk et al. [42] proposed a new one. Major details of this second model are presented in Section 3.7. To validate their innovative model, Chilmonczyk et al. engaged a series of semiempirical quantum mechanical calculations combined with studies based on density functional theory [74]. To mimic the L-hydroxyproline-based stationary phase, the species hydroxyproline bound to the dihydroxy aliphatic chain simulating the silica support was selected. However, as Chilmonczyk et al. declared, the proposed theoretical model revealed itself too crude for furnishing an explanation of the observed elution order and, in turn, of the overall enantiodiscrimination mechanism. The limited validity of the model was mainly ascribed to an unsuitable representation of the stationary phase, the solvent, and the ligand-exchange process, which are actually involved in the first Cu(II) coordination sphere.

3.10.3 QUANTITATIVE STRUCTURE ACTIVITY RELATIONSHIP (QSAR) STUDIES

Nowadays, computers and computational methods permeate many aspects of drug discovery ranging from pharmacodynamics to pharmacokinetics [75]. One of the main goals of such methodologies is the ability to quantitatively relate the structure of chemical compounds to their activity or chemical physical properties with the aim of accelerating the process of bringing new chemical entities from bench to bedside. This ability is the result of studies combining molecular modeling approaches and QSARs.

The various aspects of molecular modeling and QSAR approaches may fill a textbook in itself, so any attempt to cover the topic in a single section is merely brushing the surface. However, it is worth appreciating that both molecular modeling and QSAR approaches may prove useful when applied to study the molecular mechanisms occurring in CLEC.

This section focuses on some basic aspects of QSAR methodologies as applied to CLEC, beginning with a brief overview of the QSAR concept and then considering some approaches in the context of QSAR model development that can be summarized as follows: (1) computational representations of the molecules; (2) derivation of molecular descriptors; and (3) statistical methods used to generate the QSAR model. The conclusion contains some illustrations of QSAR approaches that have been used to shed light on the molecular mechanisms occurring in CLEC.

The QSAR methodology was introduced by Hansch in the first half of the 1960s [76,77] and is based on the assumption that the difference in structural properties of compounds accounts for their difference in biological activities. In particular, the biological activity is influenced by structural changes that impact both the interactions of the molecule (ligand) at the biological target (receptor) and the transport mechanisms of the molecule to the target tissue. These structural changes usually involve three major types of properties: electronic, steric, and hydrophobic [78]. In a very similar way, we may think that electronic, steric, and hydrophobic properties may also affect the separation behaviors of molecules in CLEC by impacting both the interactions of the molecule (analyte) with the stationary phase of the column and the stability in the eluent of the complex formed between the molecule and the selector.

The electronic, steric, and hydrophobic properties are usually described by molecular descriptors that are generated from various computational representations of the molecules. These may include 2-D chemical graphs or 3-D molecular geometries that are obtained using different methods and computational tools. 2-D chemical graphs are intuitive, understandable, and straightforward for presenting configuration and information of molecules in planar pages [79]. As a result, however, 2-D molecular graphics are more graphic representation than geometrical calculation tools. 3-D molecular geometries are representations of the molecules that result from MM [80], quantum mechanic (QM), and semiempirical calculations [81], to a different extent of accuracy. MM calculations treat molecules as collections of atoms and are based on the laws of classical physics, dealing with electronic interactions by highly simplified approximations such as Coulomb's law. QM calculations are more time-consuming and require great computational cost but afford the highest level of accuracy providing information about both nuclear position and electronic distribution. Using corrections and approximations derived from experimental data, semiempirical methods are a trade-off between the fast but moderately accurate MM and the highly accurate but slow QM calculations.

Depending on the way molecules are represented, 2-D and 3-D molecular descriptors can be calculated that lead to the terms of 2-D- or 3-D-QSAR, respectively. Among 2-D descriptors, the E-state indices are widely known descriptors that were developed to define an atom- or group-centered numerical code to represent the molecular structure [82]. Accordingly, the E-state indices are numerical values that encode information about the electronic and steric properties of atoms in a molecule. They reflect the electronegativity of specific, proximal, and distal atoms and their topological state. The topological state of a given atom, in particular, is based on the graph distance between it and the other atoms. The electronegativity aspect is based on an intrinsic state plus perturbation due to intrinsic state differences between atoms in the molecule. Three types of indices are generally used: the number of specific atom types (N-xxx); the presence/absence (0/1) of specific atom types (I-xxx); and the sum of electrotopological state indices of specific atom types (S-xxx). Symbols in the xxx types are related to the connectivity of the atom and can be single bond (s); double bond (d); triple bond (t); or aromatic bond (a). E-state indices have been used to study chiral recognition to derive relationships between structural features of the analytes and the separation ability (SA) with different albeit closely related selectors (Figure 3.25) [19].

In contrast to 2-D descriptors, 3-D descriptors depend on the molecular geometry resulting from the configurational and conformational aspects of the molecule. Among many others, these include Jurs and Shadow descriptors [83,84]: 4-carboxy-5-methyl-thien-2-yl-glycine (5-MATIDA) and asparagine have been taken as model compounds to exemplify the graphic representation (Figures 3.26 and 3.27). Jurs descriptors capture the shape and electronic information of the molecules by mapping

$$SA = 1.661 + 0.029 * (N\text{-}aaCH) - 0.467 * (N\text{-}sNH_2) - 0.331 * (I\text{-}sCH_3)$$

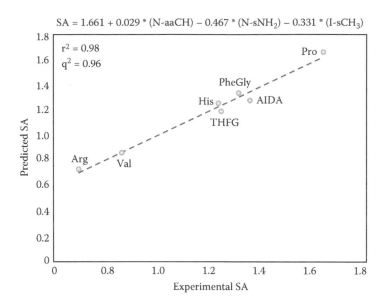

FIGURE 3.25 E-state indices relating the structural features of the analytes and the SA with different, albeit closely related, selectors. Plot of experimental versus predicted SA. (Adapted from Natalini, B. et al., *Chirality* 18, 509–18, 2006. With permission.)

atomic partial charges on solvent-accessible surface areas of individual atoms. Shadow descriptors encode the geometric arrangement of the molecular shape of the molecules by aligning the principal moments of inertia with the X, Y, and Z axes and then projecting the molecular surface on the three mutually perpendicular planes (XY, YZ, and XZ; Figures 3.26 and 3.27). Jurs and Shadow descriptors have been instrumental to relate hydrophobic, bulky, and electronic properties to the separation behavior of analytes in specific chromatographic environments [32,85,86].

In QSAR studies applied to CLEC, the target property is expressed with chromatographic parameters such as the elution order, separation ability, and separation factor (α), to name but a few. Hence, the target property can be defined either as continuous range of values or chromatographic behavioral classes. In this latter case, numbers encode the class membership of analytes: a value of zero for undiscriminated enantiomeric couples (separation factor $\alpha = 1$); and a value of one for discriminated enantiomeric couples (separation factor $\alpha > 1$).

Depending on how the target property is defined, the QSAR study is referred to as continuous property or classification analysis. To relate the target property—that is, the dependent variable—to the molecular descriptors, which are the independent variables, statistical methods are applied leading to the development of the QSAR model. During this procedure, the selection of the most explanatory independent variables is carried out to achieve the most significant correlation between the descriptor values and the target property.

The most used mathematical technique in QSAR analysis is multiple regression analysis (MRA) [87]. The principle of MRA is to model a quantitative dependent variable, the SA, through a linear combination of quantitative explanatory variables,

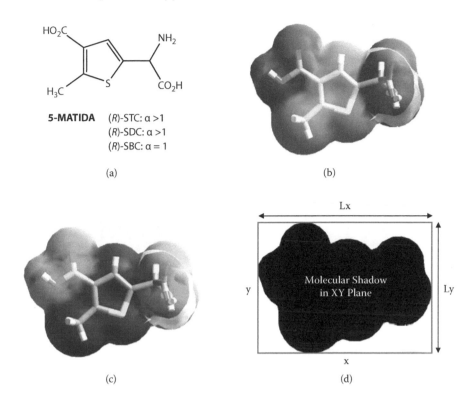

5-MATIDA　(R)-STC: α >1
　　　　　(R)-SDC: α >1
　　　　　(R)-SBC: α = 1

(a)

(b)

(c)

(d)

FIGURE 3.26 *(See color insert following page 152.)* Jurs and Shadow descriptors calculated on 5-MATIDA. (a) Chemical structure of 5-MATIDA and diverse separation factor according to the chiral selector used. (b) Representation of some Jurs descriptors on 5-MATIDA: partial positive surface area (PPSA, sum of blue areas) and partial negative surface area (PNSA, sum of red areas). (c) Representation of some Jurs descriptors on 5-MATIDA: total hydrophobic surface area (TASA, sum of brown areas). (d) Representation of some Shadow descriptors on 5-MATIDA: area of the molecular shadow in the xy plane (Sxy, black area). Length of molecule in the x dimension (Lx). Length of molecule in the y dimension (Ly).

namely, the molecular descriptors. The analysis yields an equation that is used to infer the quantitative relationships between the structure and the target chromatographic parameter of the compounds.

While the major advantage of the method is its computational simplicity, which makes it possible to easily interpret the resulting equation, its limitation is mainly linked to the ratio of objects and independent variables (Obs/Vi) that, as a rule of thumb, leads to a large risk in chance correlation if below a value of five [88]. The explanatory variables that highly correlate with the dependent variable are generally selected using a stepwise procedure. This process starts by incrementally adding the independent variables with the largest contribution to the model on the basis of "Student's t" statistic and a cutoff of probability for entry criterion. After the addition of the first three variables to the model, the impact of removing each new variable added to the model is evaluated using the "Student's t" statistic and a cutoff of probability of removal criterion. Then, a cross-validation protocol is used to

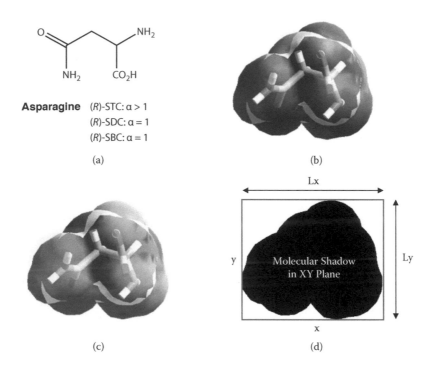

FIGURE 3.27 (*See color insert following page 152.*) Jurs and Shadow descriptors calculated on Asparagine. (a) Chemical structure of Asparagine and diverse separation factor according to the chiral selector used. (b) Representation of some Jurs descriptors on Asparagine: partial positive surface area (PPSA, sum of blue areas) and partial negative surface area (PNSA, sum of red areas). (c) Representation of some Jurs descriptors on Asparagine: total hydrophobic surface area (TASA, sum of brown areas). (d) Representation of some Shadow descriptors on Asparagine: area of the molecular shadow in the xy plane (Sxy, black area). Length of molecule in the x dimension (Lx). Length of molecule in the y dimension (Ly).

determine the statistical significance of the QSAR equation providing a predictive index (R_{XV}^2, q^2).

Regarding a classification analysis, recursive partitioning is a statistical method that predicts class membership for compounds in a given dataset [89,90]. The core of recursive partitioning is the construction of a decision tree, in which the data are organized (partitioned) into nodes (leaves) along branches. The construction of such a tree is carried out by analyzing a training set of compounds for which the target property has been used to define the class membership. The decision tree is then used to classify compounds with unknown target property, thereby making predictions of class membership. In the decision tree, nodes are questions that are posed incrementally on the molecular descriptors to split the training set into its classes of target property. Ideally, a single question node would perfectly divide the training set according to the class membership of the compounds. If this is not the case, then a set of question nodes is constructed to separate the training set in the cleanest way. The statistical significance of the question node is calculated on a quantitative basis by computing *p*-values, which discern the quality of a split relative to a random event.

The degree of impurity in the classes generated by a question node is related to the number of misclassified compounds. Generally, this is measured using two parameters: the entropy; and the Gini index. Given a value of impurity at the initial question node, incremental question nodes are then chosen on the molecular descriptors that minimize the weighted average of the impurity of the resulting classes of target property. At the end of this recursive process, the ensemble of question nodes constitutes the decision tree.

A key aspect of the decision trees concerns the level of complexity of the question nodes that may lead to overfit of the training set, thereby limiting the predictive applicability of decision trees to external test set. To avoid such problems, diverse approaches are available [90–92]. One approach is to stop the insertion of question nodes when the impurity of the resulting classes is below a given cutoff. An alternative way is to proceed with the insertion of question nodes until the tree is completed and no class can be further subdivided. In this case, however, to avoid the overfitting problem, the resulting tree is pruned by collapsing internal question nodes so that the classification error is reduced on external training examples. Once the decision tree is constructed, a cross-validation test should be run on leave-out training examples to assess the predictive applicability. Hence, decision trees to model and interpret the separation of aminoacidic enantiomers in CLEC have been applied. In particular, the analysis consisted of correctly assigning each enantiomeric couple of amino acids of the training set to a class of target property of separation factor ($\alpha = 1$, class 1; or $\alpha > 1$, class 2) on the basis of 3-D molecular shape descriptors (Jurs and Shadow descriptors; Figures 3.26 and 3.27) [83,84]. In this framework, the first effort was directed to study the separation behavior of a set of 24 amino acids with (S)-(-)-α, α-di(2-naphthyl)-2-pyrrolidine methanol [(S)-DNP] as the selector [86]. This model identified the separation likeness of the analytes with (S)-DNP as linked to their relative hydrophobic surface area (TASA). In particular, we observed, as rule of thumb, that a good separation was more likely when the analyte got more than 95.9 Å^2 of TASA (Figure 3.28).

In a following study, decision trees were applied to study the molecular shape properties affecting the separation of a set of aminoacidic enantiomer couples with three cysteine-based, dynamically coated selectors: S-benzyl-(R)-cysteine [(R)-SBC]; S-diphenylmethyl-(R)-cysteine [(R)-SDC]; and S-trityl-(R)-cysteine [(R)-STC] [32,85]. As a result, it was found that two different descriptors were the final splitting nodes with the use of (R)-SBC as selector, depending on the isomer of the training set (Figure 3.29). Thus, whereas the fractional negatively charged partial surface area (FNSA-3, ratio between the sum of the product of solvent-accessible surface area and partial charge for all negatively charged atoms and the total molecular solvent-accessible surface area) was identified as the discriminating descriptor for the (R)-enantiomers of the training set, the relative polar surface area (RPSA; the total polar surface area divided by the total molecular solvent-accessible surface area) was selected as the splitting node of the (S)-enantiomers. Accordingly, polar interactions affect the discriminating activity of (R)-SBC selector, accounting for the contribution to the separation process of the hydrogen-bonding ability of the sulfur atom and free silanols. The predictive applicability of these decision trees was also satisfactorily assessed using an external test set of amino acids.

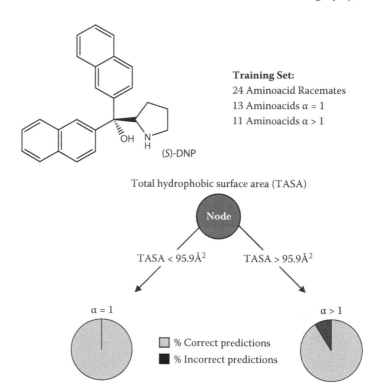

FIGURE 3.28 Partition tree predicting the separation behavior of a set of 24 amino acids with (S)-(-)-α,α-di(2-naphthyl)-2-pyrrolidine methanol [(S)-DNP] as selector.

Conversely, the total hydrophobic surface area (TASA) was identified as a suitable terminal splitting node for explaining the observed chromatographic behavior when the selector endowed with the intermediate lipophilicity [(R)-SDC] was employed (Figure 3.30). Thus, generally speaking, (R)- and (S)-aminoacidic enantiomers characterized by TASA > 133.8 Å^2 and 121.9 Å^2, respectively, were able to undergo separation with (R)-SDC. The selection of TASA as a question node of this decision tree was interpreted on the basis of the influence exerted by the interaction between the hydrophobic, wide electron-rich aromatic system of (R)-SDC and the total hydrophobic surface area of the analyte in stabilizing the diastereomeric complexes in the stationary phase and, consequently, in determining a separating or nonseparating behavior of each enantiomer couple.

Concerning the decision tree constructed on the separation behavior that was observed when the most lipophilic and aromatic (R)-STC was used (Figure 3.31), the partial positive surface area (PPSA-1) explained the observed separation behaviors of the training set. Since PPSA encodes the sum of the solvent-accessible surface areas of all the atoms with partial positive charges, the model suggests that the π/cation interaction between the analytes and (R)-STC is crucial to the enantiomeric separation process. Although the comparison with (S)-DNP may suffer the use of a reduced training set, (R)-STC proved the most performing selector among the four ones hitherto studied when attempting to separate amino acids. The added value

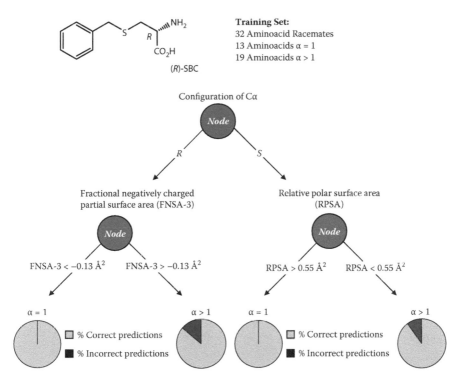

FIGURE 3.29 Partition tree predicting the separation behavior of a set of 32 amino acids with S-benzyl-(R)-cysteine [(R)-SBC] as selector.

of (R)-STC is its versatility in separating different classes of amino acids, including both hydrophobic and aromatic residues as well as polar amino acids [32,55]. This feature of (R)-STC is likely to be associated to its large number of aromatic moieties that may exploit three different types of interaction according to the type of amino acid bound to the complex when driving the enantiomeric separation process. Thus, while π/cation interaction between the analytes and (R)-STC is crucial to the enantiomeric separation process of mostly polar residues, hydrophobic and aromatic interactions may play roles in affecting the separation process of hydrophobic and aromatic amino acids, respectively.

3.11 CONCLUSION

CLEC is the first LC method with the ability to thoroughly separate enantiomers [7]. This separation process is the result of a complex equilibrium of factors that arise both from primary equilibria of interactions between the metal-mediated chelate and the stationary phase and from secondary equilibria of interactions occurring inside the chromatographic column. Due to the selected experimental conditions (e.g., pH, temperature, metal concentration, amount of organic modifier), in particular, secondary equilibria may occur that account for an intricate enantiorecognition mechanism. Thus, making the difference from any other chromatographic system

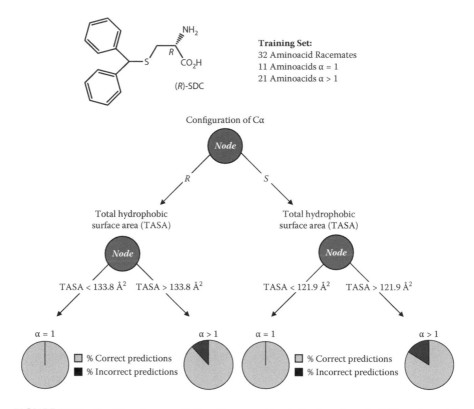

Training Set:
32 Aminoacid Racemates
11 Aminoacids α = 1
21 Aminoacids α > 1

FIGURE 3.30 Partition tree predicting the separation behavior of a set of 32 amino acids with S-diphenylmethyl-(R)-cysteine [(R)-SDC] as selector.

involving selector-analyte interactions, secondary equilibria are of great importance in CLEC since they carefully control and directly cause the chiral recognition and the separation selectivity.

Although it was clear from the beginning that a thorough understanding of primary and secondary equilibria taking place in CLEC is a very challenging task, the last decade has seen major progress in the application of computational approaches to their study for the rationalization of the chromatographic event. Grouping these approaches in mathematical models, molecular modeling, and QSAR studies, some general aspects of their theoretical background and some of the recently reported applications have been discussed.

In tandem with the development of new chiral selectors for CLEC methodology, these approaches provide pivotal contributions to clarify the mechanistic aspects and to understand the main equilibria influencing the separation process in CLEC. Although future work is expected to improve the predictive power of such computational approaches, they will ultimately prove useful (1) to identify the proper selector for a specific separation or (2) to design a new, suitable chiral discriminating agent in accordance with the suggested indications.

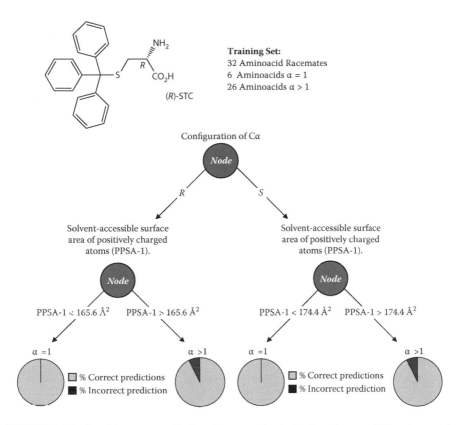

FIGURE 3.31 Partition tree predicting the separation behavior of a set of 32 amino acids with S-trityl-(R)-cysteine [(R)-STC] as selector.

3.12   LIST OF ABBREVIATIONS

$\Delta\Delta H°$	difference between the adsorption enthalpy
$\Delta\Delta S°$	difference between the adsorption entropy
α	enantioselectivity
(R)-SBC	S-benzyl-(R)-cysteine
(R)-SDC	S-diphenylmethyl-(R)-cysteine
(R)-STC	S-trityl-(R)-cysteine
(S)-aHyp	(S)-allo-hydroxyproline
(S)-BzlaHyp	N-benzyl-(S)-allo-hydroxyproline
(S)-DNP	(S)-(-)-α, α-di(2-naphthyl)-2-pyrrolidine methanol
5-MATIDA	4-carboxy-5-methyl-thien-2-yl-glycine
AIDA	1-aminoindan-1,5-dicarboxylic acid
Arg	arginine
B-CSP	covalently bound chiral stationary phase

C-CSP	coated chiral stationary phase
CLEC	chiral ligand-exchange chromatography
CMP	chiral mobile phase
CSP	chiral stationary phase
Dns	dansyl
FNSA-3	fractional negatively charged partial surface area
gly-like	glycine like
His	histidine
hm-like	histamine-like
I-xxx	presence/absence (0/1) of specific atom types
L-Me$_2$Phe-A	N,N-dimethyl-L-phenylalanine amide
L-MePhe-A	N-methyl-L-phenylalanine amide
L-Phe	L-phenylalanine
L-Phe-A	L-phenylalanine amide
L-Trp-NH$_2$	L-tryptophanamide
Lx	length of molecule in the x dimension
Ly	length of molecule in the y dimension
Me$_2$PheNN-2	(S,S)-N-N′-bis(methylphenylalanyl)ethanediamine
MeCN	acetonitrile
MeOH	methanol
MePheN-2	1-(N-Me-L-phenylalanyl)-1,2-ethanediamine
MM	molecular mechanic
MRA	multiple regression analysis
N	number of the theoretical plates
N,N-dime-L-Phe	N,N-dimethyl-L-phenylalanine
N-alkyl-L-Hyp	N-alkyl-L-hydroxyprolines
Ndo-Phe-NH$_2$	N$_2$-dodecyl-(S)-phenylalaninamide
N-me-L-Phe	N-methyl-L-phenylalanine
Noc-Nleu-NH$_2$	N$_2$-octyl-(S)-norleucinamide
Noc-Phe-NH$_2$	N$_2$-octyl-(S)-phenylalaninamide
N-xxx	number of specific atom types
PheGly	phenylglycine
PheN-2	1-(N-L-phenylalanyl)-1,2-ethanediamine
PheNN-2	(S,S)-N-N′-bis(phenylalanyl)ethanediamine
PNSA	partial negative surface area
PPSA-1	partial positive surface area
QM	quantum mechanic
QSAR	quantitative structure activity relationship
RP	reversed-phase
RPSA	relative polar surface area
R$_S$	resolution factor
SA	separation ability
S-xxx	sum of electrotopological state indices of specific atom types
Sxy	area of the molecular shadow in the xy plane.

TASA	total hydrophobic surface area
THFG	2-(2′-tetrahydrofuranyl)glycine
T_{iso}	isoenantioselective temperature
Tyr	tyrosine
Val	valine

REFERENCES

1. Gübitz, G. and Schmid, M. G. 2007. Chiral separation by ligand exchange. In: *Chiral separation techniques,* 3rd ed., ed. G. Subramanian, 155–179. Weinheim: Wiley-VCH.
2. Marchelli, R., Corradini, R., Galaverna, G., Dossena, A., Dallavalle, F., and Sforza, S. 2007. Enantioselective separation of amino acids and hydroxy acids by ligand exchange with copper(II) complexes in HPLC (chiral eluent) and in fast sensing systems. In: *Chiral separation techniques,* 3rd ed., ed. G. Subramanian, 301–331. Weinheim: Wiley-VCH.
3. Kurganov, A. A. 2001. Chiral chromatographic separations based on ligand exghange. *J. Chromatogr. A* 906:51–71.
4. Davankov, V. A. 1994. Chiral selectors with chelating properties in liquid chromatography: fundamental reflections and selective review of recent developments. *J. Chromatogr. A* 666:55–76.
5. Davankov, V. A., Kurganov, A. A., and Ponomareva, T. M. 1988. Enantioselectivity of complex formation in ligand-exchange chromatographic systems with chiral stationary and/or chiral mobile phases. *J. Chromatogr.* 452:309–16.
6. Götmar, G., Fornstedt, T., and Guiochon, G. 2000. Apparent and true enantioselectivity in enantioseparations. *Chirality* 12:558–64.
7. Davankov, V. A. 2000. 30 years of chiral lilgand exchange. *Enantiomer* 5:209–23.
8. Pirkle, W. H., Welch, C. J., and Lamm, B. 1992. Design, synthesis, and evaluation of an improved enantioselective naproxen selector. *J. Org. Chem.* 57:3854–60.
9. Galaverna, G., Corradini, R., de Munari, E., Dossena, A., and Marchelli, R. 1993. Chiral separation of unmodified amino acids by ligand-exchange high-performance liquid chromatography using copper(II) complexes of L-amino acid amides as additives to the eluent. *J. Chromatogr. A* 657:43–54.
10. Brookes, G. and Pettit, L. D. 1977. Complex formation and stereoselectivity in the ternary systems copper(II)–D/L-histidine–L-amino-acids. *J. Chem. Soc., Dalton Trans.* 1918–24.
11. Perrin, D. D. 1982. Stability constants of metal-ion complexes, 2nd Suppl. Part B, organic ligands. In: *IUPAC Chemical Data Series No. 22.* Oxford: Pergamon Press.
12. Remelli, M., Fornasari, P., and Pulidori, F. 1997. Study of retention, efficiency and selectivity in chiral ligand-exchange chromatography with a dynamically coated stationary phase. *J. Chromatogr. A* 761:79–89.
13. Remelli, M., Trombin, D., and Conato, C. 2002. Chiral ligand-exchange chromatography on an RP HPLC column coated with a new chiral selector derived from L-spinacine. *Chromatographia* 55:301–6.
14. Galaverna, G., Corradini, R., Dossena, A., et al. 1998. Chiral discrimination of Dns- and unmodified D,L-amino acids by copper(II) complexes of terdentate ligands in high-performance liquid chromatography. *J. Chromatogr. A* 829:101–13.
15. Galaverna, G., Corradini, R., Dallavalle, F., Folesani, G., Dossena, A., and Marchelli, R. 2001. Chiral separation of amino acids by copper(II) complexes of terdentate diamono-diamido-type ligands added to the eluent in reversed-phase high-performance liquid chromatography: a ligand exchange mechanism. *J. Chromatogr. A* 922:151–63.

16. Davankov, V. A., Bochkov, A. S., Kurganov, A. A., Roumeliotis, P., and Unger, K. K. 1980. Separation of unmodified α-amino acid enantiomers by reverse phase HPLC. *Chromatographia* 13:677–85.
17. Natalini, B., Sardella, R., and Pellicciari, R. 2005. O-benzyl-(*S*)-serine, a new chiral selector for ligand-exchange chromatography of amino acids *Curr. Anal. Chem.* 1:85–92.
18. Wernicke, R. 1985. Separation of underivatised amino acid enantiomers by means of a chiral solvent-generated phase. *J. Chromatogr. Sci.* 23:39–47.
19. Natalini, B., Sardella, R., Macchiarulo, A., and Pellicciari, R. 2006. Dynamic ligand-exchange chiral stationary phase from S-benzyl-(*R*)-cysteine. *Chirality* 18:509–18.
20. Hyun, M. H., Yang, D. H., Kim, H. J., and Ryoo, J. -J. 1994. Mechanistic evaluation of the resolution of α-amino acids on dynamic chiral stationary phases derived from amino alcohols by ligand-exchange chromatography. *J. Chromatogr. A* 684:189–200.
21. Schlauch, M., Volk, F. -J., Fondekar, K. P., Wede, J., and Frahm, A. W. 2000. Enantiomeric and diastereomeric high-performance liquid chromatographic separation of cyclic β-substituted α-amino acids on a copper(II)-D-penicillamine chiral stationary phase. *J. Chromatogr. A* 897:145–52.
22. Davankov, V. A., Bochkov, A. S., and Belov, Y. P. 1981. Ligand-exchange chromatography of racemates. XV. Resolution of α-amino acids on reversed-phase silica gels coated with N-decyl-L-histidine. *J. Chromatogr.* 218:547–57.
23. Hyun, M. H., Han, S. C., Lee, C. W., and Lee, Y. K. 2002. Preparation and application of a new ligand exchange chiral stationary phase for the liquid chromatographic resolution of α-amino acid enantiomers. *J. Chromatogr. A* 950:55–63.
24. Natalini, B., Marinozzi, M., Bade, K., Sardella, R., Thomsen, C., and Pellicciari, R. 2004. Preparative resolution of 1-aminoindan-1,5-dicarboxylic acid (AIDA) by chiral ligand-exchange chromatography. *Chirality* 16:314–7.
25. Qi, L., Han, Y., Zuo, M., and Chen, Y. 2007. Chiral CE of aromatic amino acids by ligand-exchange with zinc(II)-L-lysine complex. *Electrophoresis* 28:2629–34.
26. Yuan, Z., Yang, L., and Zhang, S. 1999. Enantiomeric separation of amino acids by copper(II)-L-arginine ligand exchange capillary zone electrophoresis. *Electrophoresis* 20:1842–5.
27. Hödl, H., Schmid, M. G., and Gübitz, G. 2008. Chiral separation of amino acids and glycyl dipeptides by chiral ligand-exchange capillary electrophoresis comparing Cu(II), Co(II), Ni(II) and Zn(II) complexes of three different sugar acids. *J. Chromatogr. A* 1204:210–8.
28. Horimai, T., Ohara M., and Ichinose, M. 1997. Optical resolution of new quinolone drugs by capillary electrophoresis with ligand-exchange and host-guest interactions. *J. Chromatogr. A* 760:235–44.
29. Davankov, V. A. 2003. Enantioselective ligand exchange in modern separation techniques. *J. Chromatogr. A* 1000:891–915.
30. Lepri, L., Cincinelli, A., and Del Bubba, M. 2007. Planar chromatography enantioseparations on non-commercial CCSPs. In: *Thin layer chromatography in chiral separations and analysis*, ed. T. Kowalska, and J. Sherma, 111–146. Boca Raton, FL: Taylor & Francis.
31. Chai, Z. 2004. Resin microspheres as stationary phase for liquid ligand exchange chromatography. In: *Encyclopedia of Chromatography*, ed. J. Cazes, 243–249. New York: Marcel Dekker.
32. Natalini, B., Sardella, R., Macchiarulo, A., and Pellicciari, R. 2008. S-Trityl-(*R*)-cysteine, a powerful chiral selector for the analytical and preparative ligand-exchange chromatography of amino acids. *J. Sep. Sci.* 31:696–704.
33. Natalini, B., Sardella, R., Carbone, G., Macchiarulo, A., and Pellicciari R. 2009. The effect of the copper(II) salt anion in the chiral ligand-exchange chromatography of amino acids. *Anal. Chim. Acta* 638:225–33.

34. Sanaie, N. and Haynes, C. A. 2006. A multiple chemical equilibria approach to modelling and interpreting the separation of amino acid enantiomers by chiral ligand-exchange chromatography. *J. Chromatogr. A* 1132:39–50.

35. Sanaie, N. and Haynes, C. A. 2006. Interpreting the effects of temperature and solvent composition on separation of amino-acid racemates by chiral ligand-exchange chromatography. *J. Chromatogr. A* 1104:164–72.

36. Pirkle, W. H. and Pochapsky T. C. 1989. Considerations of chiral recognition relevant to the liquid chromatographic separation of enantiomers. *Chem. Rev.* 89:347–62.

37. Ma, G. -J., Gong, B. -L., and Chao, Y. 2008. Preparation of polymer-bonded chiral ligand exchange chromatographic stationary phase and resolution of racemates. *Chin. J. Anal. Chem.* 36:275–79.

38. Davankov, V. A. and Kurganov, A. A. 1983. The role of achiral sorbent matrix in chiral recognition of amino acid enantiomers in ligand-exchange chromatography. *Chromatographia* 17:686–90.

39. Davankov, V. A. 1997. The nature of chiral recognition: it is a three-point interaction? *Chirality* 9:99–102.

40. Lefebvre, B., Audebert, R., and Quivoran, C. 1978. Use of new chiral hydrophilic gels for the direct resolution of alpha-amino acids by high pressure liquid chromatography. *J. Liq. Chromatogr.* 1:761–70.

41. Hyun, M. H., Han, S. C., and Whangbo, S. H. 2003. New ligand exchange chiral stationary phase for the liquid chromatographic resolution of α- and β-amino acids. *J. Chromatogr. A* 992:47–56.

42. Chilmonczyk, Z., Ksycińska, H., Cybulski, J., Rydzewski, M., and Leś, A. 1998. Direct separation of α-hydroxy acid enantiomers by ligand exchange chromatography. *Chirality* 10:821–30.

43. Liu, Y., Zou, H., and Haginaka, J. 2006. Preparation and evaluation of a novel chiral stationary phase based on covalently bonded chitosan for ligand-exchange chromatography. *J. Sep. Sci.* 29:1440–6.

44. Deschamps, P., Kulkarnia, P. P., Gautam-Basakh, M., and Sarkar, B. 2005. The saga of copper(II)–L-histidine. *Coord. Chem. Rev.* 249:895–909.

45. Hyun, M. H., Lim, N. -E., and Choi, S. -N. 1991. Resolution of racemic α-amino acids on a dynamic chiral stationary phase by ligand-exchange chromatography. *Bull. Korean Chem. Soc.* 12:564–6.

46. Hyun, M. H., Ryoo, J. -J., and Lim, N. -E. 1993. Optical resolution of racemic α-amino acids on a dynamic chiral stationary phase by ligand exchange chromatography. *J. Liq. Chromatogr.* 16:3249–61.

47. Hyun, M. H., Yang, D. H., Kim, H. J., and Ryoo, J.-J. 1994. Mechanistic evaluation of the resolution of α-amino acids on dynamic chiral stationary phases derived from amino alcohols by ligand-exchange chromatography. *J. Chromatogr. A* 684:189–200.

48. Gil-Av, E., Tishbee, A., and Hare, P. E. 1980. Resolution of underivatized amino acids by reversed-phase chromatography. *J. Am. Chem. Soc.* 102:5115–7.

49. Galaverna, G., Corradini, R., Dossena, A., Marchelli, R., and Dallavalle, F. 1996. Copper(II) complexes of N^2-alkyl-(S)-amino acid amides as chiral selectors for dynamically coated chiral stationary phases in RP-HPLC. *Chirality* 8:189–196.

50. Ôi, N., Kitahara, H., and Kira, R. 1992. Direct separation of enantiomers by high performance liquid chromatography on a new chiral ligand-exchange phase. *J. Chromatogr.* 592:291–6.

51. Schlauch, M. and Frahm, A. W. 2001. A thermodynamic study of the temperature-dependent elution order of cyclic α-amino acid enantiomers on a copper(II)-D-penicillamine chiral stationary phase. *Anal. Chem.* 73:262–6.

52. Wan, Q. -H., Shaw, P. N., Davies, M. C., and Barrett, D. A. 1997. Chiral chromatography of amino acids on porous graphitic carbon coated with a series of N-substituted L-phenylalanine selectors. Effect of the anchor molecule on enantioselectivity. *J. Chromatogr. A* 765:187–200.

53. Wan, Q. -H., Shaw, P. N., Davies, M. C., and Barrett, D. A. 1997. Role of alkyl and aryl substituents in chiral ligand exchange chromatography of amino acids. Study using porous graphitic carbon coated with N-substituted-L-proline selectors. *J. Chromatogr. A* 786:249–57.

54. Aleksandrov, G. G., Struchkov, Yu. T., Kurganov, A. A., Rogozhin, S. V., and Davankov, V. A. 1972. Crystal and molecular structures of bis-(N-benzylprolinato)copper(2+) complexes. *J. Chem. Soc. Comm.* 1328–9.

55. Natalini, B., Sardella, R., Macchiarulo, A., and Pellicciari, R. 2008. Cysteine-based chiral selectors for the ligand-exchange separation of amino acids. *J. Chromatogr. B* 875:108–117.

56. Diniz, V. and Volesky, B. 2005. Effect of counterions on lanthanum biosorption by *Sargassum polycystum*. *Water Res.* 39:2229–36.

57. Weinstein, S. 1982. Resolution of D- and L-amino acids by HPLC with copper complexes of N,N-dialkyl-α-amino acids as novel chiral additives. A structure-selectivity study. *Angew. Chem. Suppl.* 425–33.

58. Yamauchi, O., Takaba, T., and Sakurai, T. 1980. Solution equilibria of histidine-containing ternary amino acid-copper(II) complexes in 20 v/v% dioxane–water. *Bull. Chem. Soc. Jpn.* 53:106–11.

59. Kostova, A. and Bart, H.-J., 2007. Preparative chromatographic separation of amino acid racemic mixtures. I. Measurement of adsorption isotherms, *Sep. Purif. Technol.* 54:340–8.

60. Kostova, A. and Bart, H.-J. 2007. Preparative chromatographic separation of amino acid racemic mixtures: II. Modelling of the separation process. *Sep. Purif. Technol.* 54:315–21.

61. Kurganov, A. A., Davankov, V. A., Unger, K., Eisenbeiss, F., and Kinkel, J. 1994. Unusual peak distorsion in ligand-exchange chromatography of enantiomers under overloaded conditions. *J. Chromatogr. A* 666:99–110.

62. Pellicciari, R., Filosa, R., Fulco, M. C., et al. 2006. Synthesis and preliminary biological evaluation of 2-substituted 2-(3-carboxybicyclo[1.1.1]pentyl)glycine derivatives as group I selective metabotropic glutamate receptor ligands. *ChemMedChem* 1:358–65.

63. Wagner, J., Gaget, C., Heintzelmann, B., and Wolf, E. 1987. Resolution of the enantiomers of various alpha-substituted ornithine and lysine analogs by high-performance liquid chromatography with chiral eluant and by gas chromatography. *Anal. Biochem.* 164:102–16.

64. Brueckner, H. 1987. Enantiomer resolution of N-methyl-α-amino acids and α-alkyl-α-amino acids by ligand-exchange chromatography. *Chromatographia* 24:725–38.

65. Davankov, V. A., and Rogozhin, S. V. 1971. Ligand chromatography as a novel method for the investigation of mixed complexes: stereoselective effects in α-amino acid copper(II) complexes. *J. Chromatogr.* 60:280–3.

66. Jozefonvicz, J., Petit, M. A., and Szubarga, A. 1978. Preparative resolution of DL-proline by liquid chromatography on a polystyrene resin containing the L-proline copper(II) complex. *J. Chromatogr.* 147:177–83.

67. Arnell, R., Forssén, P., Fornstedt, T., Sardella, R., Lämmerhofer, M., and Lindner, W. 2009. Adsorption behaviour of a quinidine carbamate-based chiral stationary phase: Role of the additive. *J. Chromatogr. A* 1216:3480–7.

68. Sanaie, N. and Haynes, C. A. 2007. Modeling L-DOPA purification by chiral ligand-exchange chromatography. *AIChE* 53:617–26.

69. Martell, A. E. and Smith, R. M. 2001. *The National Institute of Standards and Technology (NIST). Standard Reference Database 46, Version 6.0. Critically Selected Stability Constants of Metal Complexes.* Gaithersburg, MD: NIST.

70. Pettit, L. D. and Powell, K. J. 2003. *The IUPAC Stability Constants Database.* SCDatabase, Academic Software and IUPAC http://www.acadsoft.co.uk/scdbase/scdbase.htm.

71. Koska, J., Mui, C., and Haynes, C. A. 2001. Solvent effects in chiral ligand exchange systems. *Chem. Eng. Sci.* 56:29–41.

72. Natalini, B., Marinozzi, M., Sardella, R., Macchiarulo, A., and Pellicciari, R. 2004. Evaluation of the enantiomeric selectivity in the chiral ligand-exchange chromatography of amino acids by a computational model. *J. Chromatogr. A* 1033:363–7.

73. Rappe, A. K., Casewit, C. J., Colwell, K. S., Goddard, W. A., and Skiff, W. M. 1992. UFF, a rule-based full periodic table force field for molecular mechanics and molecular dynamics simulations. *J. Am. Chem. Soc.* 114:10024–35.

74. Becke, A. D. 1988. Density-functional exchange-energy approximation with correct asymptotic behaviour. *Phys. Rev.* A38:3098–100.

75. Jorgensen, W. L. 2004. The many roles of computation in drug discovery. *Science* 303:1813–8.

76. Fujita, T., Iwasa, J., and Hansch, C. 1964. A new substituent, *"pi"* derived from partition coefficients. *J. Am. Chem. Soc.* 86:5175–80.

77. Hansch, C., Muir, R. M., Fujita, T., Maloney, P. P., Geiger, E., and Streich, M. 1964. The correlation of biological activity of plant growth regulators and chloromycetin derivatives with Hammet constants and partition coefficients. *J. Am. Chem. Soc.* 85:2817–24.

78. Hansch, C. and Leo, A. 1995. QSAR of nonspecific toxicity. In *Exploring QSAR, fundamentals and applications in chemistry and biology,* ed. S.R. Heller, 169–221. Washington, DC: American Chemical Society.

79. Zhou, P. and Shang, Z. 2009. 2D molecular graphics: a flattened world of chemistry and biology. *Brief Bioinform.* 10:247–58.

80. Burkert, U. and Allinger, N. L. 1982. *Molecular Mechanics,* vol. 177, ed. M.C. Cascrio. Washington, DC: American Chemical Society Monograph 339.

81. Loew, G. H. and Burt, S. K. 1990. Quantum mechanics and the modelling of drug properties. In *Quantitative Drug Design,* ed. C. A. Ramsden, 105–124. Oxford: Pergammon Press.

82. Kier, L. B. and Hall, L. H. 1999. *Molecular structure description. The electrotopological state.* San Diego: Academic Press.

83. Roxburgh, R. H. and Jurs, P. C. 1987. Descriptions of molecular shape applied in studies of structure/activity and structure/property relationships. *Anal. Chim. Acta* 199:99–109.

84. Stanton, D. T. and Jurs, P. C. 1990. Development and use of charged partial surface area structural descriptors in computer assissted quantitative structure property relationship studies. *Anal. Chem.* 62:2323–9.

85. Natalini, B., Macchiarulo, A., Sardella, R., Massarotti, A., and Pellicciari, R. 2008. Descriptive structure-separation relationship studies in chiral ligand-exchange chromatography. *J. Sep. Sci.* 2008. 31:2395–403.

86. Natalini, B., Sardella, R., Macchiarulo, A., Natalini, S., and Pellicciari, R. 2007. (*S*)-(-)-alpha,alpha-di(2-naphthyl)-2-pyrrolidinemethanol, a useful tool to study the recognition mechanism in chiral ligand-exchange chromatography. *J. Sep. Sci.* 30:21–7.

87. Draper, N. R. and Smith, H. 1981. *Applied regression analysis,* 2nd ed. New York: John Wiley & Sons.

88. Topliss, J. G. and Edwards, R. P. 1979. Chance factors in studies of quantitative structure-activity relationships. *J. Med. Chem.* 22:1238–44.

89. Breiman, L., Friedman, J., Olshen, R. and Stone, C. 1984. *Classification and Regression Trees*. Belmont: Wadsworth International Group.
90. Quinland, J. R. 1993. *C4.5: Programs for Machine Learning*. San Matteo, CA: Morgan Kaufmann Publishers.
91. MacKay, D. J. C. 2003. *Information theory, inference and learning algorithms*. Cambridge: Cambridge University Press.
92. Quinland, J. R. and Rivest, R. L. 1989. Inferring decision trees using the minimum description length principle. *Inf. Comput.* 80:227–48.

4 Glycosylation Analysis of Proteins, Proteoglycans, and Glycolipids Using Capillary Electrophoresis and Mass Spectrometry

Alina D. Zamfir, Corina Flangea, Friedrich Altmann, and Andreas M. Rizzi

CONTENTS

4.1   INTRODUCTION

4.1.1   THE NEED FOR DETAILED CARBOHYDRATE ANALYSIS

Carbohydrates attached to proteins, peptides, and lipids have attracted particular and major interest during the last decades [1–5]. Many aspects of carbohydrate interaction with other biopolymers influencing enzyme activity, immunogenicity, cell–cell interactions, and other biologically important processes deserve better understanding, and the elucidation of the biological importance of carbohydrates attached to proteins and lipids is still progressing rapidly. Understandably, an impressive number of papers and review articles have been dedicated to carbohydrate and glycoconjugate analysis in all types of biological samples focusing on various methodological and technical approaches [6–17]. Carbohydrates are linked to proteins as rather small (M_r up to 3 kDa) linear or branched glycans or as very long (M_r tens of kDa), mostly linear chains of glycosaminoglycans (GAGs) bound to large proteins giving proteoglycans. Short oligosaccharides are also part of the glycosylphosphatidylinositol (GPI) anchor by which proteins are attached to the plasma membrane, and rather small, branched glycans are linked to glycerolipids and shingolipids embedded into cell membranes. Carbohydrates are further parts of peptidoglycans in prokaryotes, of lipopolysaccharides in bacteria, and of glycoconjugates in plants, and they are present as neat oligo- and polysaccharides.

There is a growing body of knowledge exploring ways protein- and lipid-linked glycans are involved in controlling and influencing biological processes like cell adhesion, immunological response, inflammation [18], cancer metastasis [1,19], morphogenesis, and embryogenesis [20,21], just to name a few. For more detailed readings about correlation between diseases and altered oligosaccharide pattern, see [22].

Glycan structures present in cells are the end product of biosynthesis pathways involving many steps in the endoplasmic reticulum (ER) and the Golgi apparatus. The attachment of the initial oligosaccharides onto certain amino acids (Asn, Thr, Ser) is controlled by an enzymatic machinery positioned in the ER membranes. After moving on the glycoprotein from the ER to the Golgi apparatus, a further enzyme complex consisting of various glycosyl-transferases is active in trimming the glycans into some kind of canonical end structure. Glycan structures predominantly found in glycoproteins are therefore not as manifold but follow more or less some canonical patterns with certain variations. Some of these variations can be due to congenital disorders (CDG) affecting expression and activity of glycosyltransferases, such as by mutations in the respective genes. Otherwise, the variation can be considered a "disorder" because of acquired mutations in the genes, cell transformation to malignancy, or other cell processes like inflammation. Presently, quite a number of congenital and acquired disorders in glycosylation can be understood [19,22].

Complete and detailed characterization of oligosaccharide structures and glycoform distributions is needed at least for the following three major purposes. First, it is still necessary to further improve and consolidate the understanding of the biological role and functions of glycosylation as well as the impact of subtle variations in the glycan structures on biological processes. Second, as variations in glycostructures often are associated with certain diseases, aberrant glycosylation pattern might serve as a kind of biomarker. Third, a detailed glycoanalysis is needed in the context of drug safety when dealing with recombinant therapeutic glycoproteins. In this context it was shown frequently that structural differences in glycoforms are correlated with differences in thermodynamic as well as kinetic constants regarding binding to and dissociation from their physiological partner molecules [23,24]. Most importantly, some of the glycan structures represent immunogenic epitopes giving rise to undesired immunological reactions. Glycostructures attached to the Fc region of IgG–antibodies are known to influence the activation of the complement system. The extent of capping N-linked glycan antennas by sialic acids is relevant, as the presence of sialic acids (or of an unprotected galactose) might affect the clearance rate of a therapeutic protein drug. The effect of differences in the extent of core-fucosylation regarding therapeutic activity seems to be dependent on the respective glycoprotein and is not fully clear yet in many cases. The question of how fucosylation influences protein activity, however, might be important as, depending on the cell type from which a recombinant glycoprotein is gained (CHO and BHK cell cultures or milk-producing cells of transgenic animals), the fucosylation patterns might considerably differ from the normal human signature [25–28].

In this chapter we highlight some major achievements and advances in the analysis of oligosaccharide structures in glycoproteins (GPs) and proteoglycans (PGs) as well as glycolipids (GL). The canonical structures and their variability and observed variance within these groups of glycoconjugates are different and will be discussed in more detail in upcoming sections. The discussion on methodology and applications will primarily focus on CE-MS/MS techniques; however, achievements in the shape-selective discrimination of isobaric glycoforms by means of high-performance liquid chromatography (HPLC) coupled with sensitive tandem mass spectrometry (MS/MS) will be addressed as well. In addition, a brief discussion of the aspects of sample pretreatment is included covering selective enrichment of glycoconjugates and released oligosaccharides, releasing reactions of glycans, selective degradation, as well as labeling techniques. Appropriate sample pretreatment might often be critical when determining glyco structures from biological samples.

Multistage MS is a method suited for the detailed analysis of oligosaccharide structures, and the potential of this technique in this respect has repeatedly been reviewed [cf. 12,13,15,29,30]. Two aspects are decisive—why a high-performance preseparation step prior to MS is often indispensable and significantly enhances the reliability of MS data. First, ion suppression effects in electrospray ionization (ESI) and matrix-assisted laser desorption ionization (MALDI) can be minimized in this way that otherwise they would otherwise become very significant when dealing with glycosylated peptides in the presence of nonglycosylated ones or with differently charged oligosaccharides. When targeting low abundant glycoforms and when interested in quantitative patterns, this problem might become troublesome. Second, it is sometimes of interest to distinguish between isobaric glycans that differ in the branching, linkage, and anomeric pattern. CZE or HPLC is effective in distinguishing between such isobaric isomers, which hardly can be reliably differentiated by MS/MS techniques alone.

This chapter, therefore, will present general analysis strategies for and recent trends in carbohydrate analysis related to the aforementioned groups of glycoconjugates. Because of the broad field covered by this survey, it will not be possible to discuss the large body of related articles in technical detail. In this regard, we fortunately can refer to a number of high-quality technical reviews in this field published in the last years and recently [9–11].

4.1.2 GLYCOPROFILING VERSUS GLYCOPROTEOMICS

Glycosylation analysis regarding glycoproteins can be focused on two different aims. On one hand, it is for attaining a comprehensive and detailed and often quantitative characterization of all glycoforms present in a protein—a procedure referred to as *glycoprofiling* or *glycotyping*. On the other hand, it can be aimed at the identification of (specifically) glycosylated proteins in an entire proteome. The latter procedure, referred to as *glycoproteomics* (or, when dealing with the released glycan pool, as *glycomics*) is often aimed at the detection of glycosylation differences between samples from healthy and diseased individuals.

Glycoprofiling usually includes (1) the determination of the total number of protein glycoforms present, (2) the characterization of the occupancy of glycosylation sites and the site-specific structures of oligosaccharides (as well as the structural variations), and (3) in many cases also the determination of the relative abundances of these glycoforms. These three aspects imply different challenges to the analytic methodology. While the number of protein glycoforms and their relative abundances are investigated best on the level of the intact glycoproteins, the detailed structural analysis of the carbohydrate moieties is usually done on the level of glycopeptides or released glycans. Such an analysis might cover the monosaccharide sequence as well as branching, linkage positions, and anomeric configurations of each monosaccharide unit. However, not in all instances is such a detailed (and costly) analysis required. Often, such as in several cases of clinical glycomics and for controlling the glycosylation in recombinant glycoproteins, simply the knowledge of the monosaccharide sequence allows one to evaluate structural features relevant for the biological activity of the glycan. Such features are, for instance, (1) the number of sialic acids (SAs) present and the completeness in capping the antennas in *N*-linked glycans by SAs, (2) the general type of glycan (complex-, high-mannose-, hybride) and the number of antennas, (3) the presence of fucose (FUC), either in core-linked position or in the antennas, (4) the presence of an intersecting N-acetyl-glucosamine (GlcNAc).

Glycoproteomics, on the other hand, is a different field. It has gained growing interest recently [31–37] as it is assumed that up to 60% of the proteins are glycosylated, particularly those present in the serum. Glycoproteomic strategies beyond the common aspects of glycan analysis, however, are not the focus of this review. Glycoproteomics shares many analytical concepts and protocols with general proteomics. It can proceed along the 2-DE approach dealing with intact proteins or along the 2-D-HPLC approach dealing with peptides after enzymatic cleavage, in both cases using multistage MS for protein identification and glycan characterization. Within both these approaches, preselection and enrichment of glycoproteins is frequently done by use of lectin-affinity chromatography. By using one particular lectin a subpopulation of glycoproteins can be selected, in most cases a set of different lectins is used when targeting for a comprehensive analysis of the glycoproteome. Glycoproteomics faces some particular challenges as glycoproteins are often membrane associated, and it is difficult to ensure that their detachment from the membrane and their solubilization is carried out without significant losses.

4.2 PRETREATMENT STRATEGIES FOR GLYCOCONJUGATE ANALYSIS

4.2.1 CARBOHYDRATE-SPECIFIC PRESEPARATION AND ENRICHMENT

Preseparation and particularly enrichment of glycoconjugates become necessary in most cases when dealing with samples in which the glycoconjugates are present in physiological concentrations like in serum, urine, cerebrospinal fluid (CSF), and tissue specimen, and it is commonly applied in glycoproteomics. Major tools for "glyco-catching" are various formats of AFC employing various carbohydrate-binding ligands. An important and widely used group of such ligands comprises various lectins [34,38–42], many of them being commercially available immobilized on chromatographic support materials or on magnetic beads. Almost all lectins exhibit a preference for glycan structures in a sugar-, branching-, linkage-, and configuration-specific way. ConA lectin for instance binds strongly to a rather broad spectrum of bi-antennary N-linked glycans [34] but only tightly to tri-antennary and tetra-antennary complex type glycans, which, as a consequence, are commonly missed by this approach. Though no lectin to catch all types of glycans is available, by combining various lectins in a "multi-lectin" AFC column a more or less "complete" spectrum of glycan affinity can be covered. The following lectins have widely been employed: *Concanavalin A* (ConA) binding to high-mannose-, hybrid- and bi-antenary complex-type glycans, *Galectin LEC-6* binding terminal galactose, *Peanut agglutinin* binding (Galβ1-3GalNAc) motifs often found as "core" in O-linked glycans, and lectin from *Lotus Tetragonolobus* for binding fucosylated glycans.

Alternatively, selective enrichment of oligosaccharides and glycoconjugates can be obtained by reaction with bead-immobilized phenylboronate groups that are able to catch *cis*-diolic structures under alkaline conditions (pH about 8.5). Other "catching columns" are based on immobilized hydrazine and oxime [43–45] targeting the aldehyde- or keto-group at the reducing end of released free glycans or the aldehyde/keto group at a nonreducing end generated via periodate oxidation when dealing with glycoproteins and peptides [46]. In all instances when working under alkaline or acidic conditions, one has to take care not to lose labile subunits of glycans by hydrolysis.

4.2.2 CARBOHYDRATE RELEASE

Mammalian *N*-linked glycans are commonly released enzymatically with peptide *N*-glycosidase F (PNGase F) [47]. PNGase F cleaves the GlcNAc-1-Asn bond of all types of *N*-glycans except those with Fuc linked to the three-position of the first (Asn-linked) core-GlcNAc [48]. Mammalian cells exhibit 1-6 linked core Fuc only, whereas 1-3 linked Fuc is found on glycoproteins from nonvertebrates and plants. When in doubt, PNGase A should be used [48]. Both PNGases act as amidases and convert the Asn to an Asp residue inducing a mass shift of +1 unit on the peptide moiety. A 1-amino-oligosaccharide is at first generated, which decomposes spontaneously at slightly acidic pH. Glycan release might severely influence protein solubility and does not always pass to completeness, depending on the glycoprotein under consideration.

O-linked glycans can be liberated in some limited cases by treatment with *O*-glycosidase [49]. Thus, *O*-glycans are usually released by chemical treatment. Mild hydrazinolysis yields reducing oligosaccharides amenable to labeling [50]. Reductive β-elimination with NaOH solution in the presence of $NaBH_4$ leads to alditols suitable only for MS detection [51]. Both techniques include the risk of inferring modifications in the glycan structure, like hydrolysis of alkali labile substituents, or fragmentation of the peptide backbone (which might be less problematic when targeting the glyco structures). When carrying out the deglycosylation in the presence of $H_2^{18}O$, the ^{18}O isotope marks the glycan attachment site on the peptide [52,53].

GAGs (most of them *O*-linked) are usually released from proteoglycans as free chains by a β-elimination reaction carried out in NaOH and $NaBH_4$. Glycans from glycosphingolipids (GSLs) are commonly released by treatment with endoglyceroceramidases.

Cleaved glycans can be separated from the deglycosylated proteins by protein precipitation, by reversed-phase (RP)- or size-exclusion (SEC)-type chromatographic preseparations, by lectin-AFC-based trapping, or other "glyco-catching" techniques. Carbohydrate clean-up cartridges based on a Hydrophilic interaction (HILIC) mechanism or on graphitic carbon become increasingly popular [54,55].

4.2.3 OTHER ENZYMATIC PRETREATMENT

A number of further exo- and endo-glycosidases that cleave glycosidic bonds at the end of or within the glycan quite specifically are successfully used for sample pretreatment or for assisting in structure elucidation [56]. The high specificity of these enzymes is used as a major strategy in determining the epimeric form, branching, linkage position, and configuration of glycan constituents. In this approach, the changes in glycan mass upon treatment with the specific glycosidases are monitored by MS. Often unspecific sialidases are employed to enhance the ionization yield in the positive ion mode.

For the purpose of depolymerizing the long polymeric GAG chains into shorter chains or even up to disaccharide units, lyases with known recognition specificity are used. Chondroitin B lyase, for instance, cleaves the glycosidic bond between GalNAc and iduronic acid (IdoA), whereas chondroitin AC lyase cleaves the bond between GalNAc and glucuronic acid (GlcA). This specificity is important when dealing with hybrid chondroitin/dermatan sulfate (CS/DS) GAGs. Both lyases cause the uronic acid at the cleavage site to be unsaturated, leaving a glycan with an 4,5 Δ-IdoA, and 4,5 Δ-GlcA, respectively, at the nonreducing end. The presence of such unsaturated uronic acids after lyase digestion opens the possibility of tagging these oligosaccharides at the nonreducing ends in a 1:1 stoichiometry.

4.2.4 LABELING AND DERIVATIZATION

Labeling reactions on glycans are usually carried out at the reducing hemiacetal carbon (C1, in SAs C2) via the aldehyde/keto function present in the open form of the sugar. Reactions at this side guarantee a 1:1 stoichiometry of the reaction. Labeling may serve various purposes: (1) facilitating or enabling the electrophoretic separation of the glycans by inferring charge; (2) enhancing the detection sensitivity by inferring fluorescent moieties; (3) enhancing the ionization yield in MS by inferring basic or acidic moieties; and (4) attaching isotopically coded labels for relative quantitation by MS. Derivatization with fluorescent and charged labels is usually carried out by reductive amination in the presence of $NaBH_4$ via a primary amino group present in the label. 8-aminopyrene-1,3,6-trisulfonate (APTS), 8-aminonaphthalene-1,3,6-trisulfonate (ANTS), or analogous negatively charged labels were abundantly used for CZE separations of glycans in connection with (laser-induced) fluorescence detection. 2-aminopyridine (2-AP) and 2-aminobenzamide (2-AB) are net positively charged labels widely used for shape-discriminative HPLC separations.

Another derivatization method particularly applicable to neutral glycans is based on the oxidation of the hemiacetal to a lactone, followed by ring opening by an amine (e.g., 1,6-hexamethylenediamine [HMD]) yielding the lactonamide [57]. The introduction of an amino and amido group is expected to significantly increase the ionization efficiency in positive ion mode ESI MS because of the increased proton affinity. This protocol is recommended for large glycan chains lacking easily ionizable groups, which, in their native form, cannot be analyzed by ESI or MALDI-MS. A comprehensive review on derivatization of carbohydrates for chromatographic, electrophoretic, and mass spectrometric structure analysis is given in [58].

Permethylation of all free hydroxyl groups in a glycan is frequently done to enhance the ionization yield, to increase the extent of cross-ring fragmentation, and to stabilize the SA linkages [59–64].

4.3   GENERAL TECHNICAL CONSIDERATIONS REGARDING CE-MS HYPHENATION AND MS/MS OF OLIGOSACCHARIDES

The major advantages of CZE-MS in the context of glycoconjugate analysis lies in the speed of separation attainable and in the fact that protein adsorption onto particle surfaces, walls, or frits can widely be reduced or avoided. A major limitation is the low concentration sensitivity associated with CZE separations. However, this problem can partially be overcome by subtle on-capillary enrichment techniques (e.g., by creating transient isotachophoretic conditions leading to sample stacking) or by inserting a small enrichment zone packed with AFC-support material at the very front end of the CE capillary [65–68]. Finally, ongoing improvements in sensitivity of modern mass spectrometers decisively contribute to reaching lower limits of detection.

4.3.1   INTERFACING CE-MS

General aspects of the hyphenation of CZE to ESI-MS using various interfaces have recently been reviewed by several authors [69–73]. When hyphenated to a standard microspray ESI, most commonly interfaces with a makeup sheath flow are used working with flow velocities between 2 and 4 µL/min. For this purpose a coaxial triple-tube sprayer is commercially available. The sheath liquid is essential for constituting a current bridge and for delivering a sufficient amount of liquid for generating a stable spray. In addition, it allows decoupling optimum CE separation conditions (tuned by the choice of the BGE) from optimum ionization conditions (tuned by the selection of the sheath-liquid composition). Commonly, small alcohols (up to 2-propanol) mixed in aqueous solutions are chosen as sheath-liquid components enhancing volatility. Small amounts of acidic components like formic or acetic acid (typically 0.5 to 1%) or volatile buffers (ammonium acetate) are used for assisting positive mode ionization and NH_4OH for assisting negative mode ionization. Sheath-flow and spray conditions need to be carefully adjusted, as high nitrogen flow often facilitates the spraying, enhancing ionization yield and sensitivity, but it may also create a sucking flow in the CZE capillary that deteriorates separation. ESI under standard microspray conditions works in a concentration-dependent detection regime; the application of a sheath–flow reduces the sensitivity due to dilution. Usually, the minimum amount of glycoprotein needed for a detailed glycosylation analysis lies between 1 and 5 pmol injected amount.

Such a dilution is avoided when working under nanospray conditions; in this instance no makeup flow is needed as the electroosmotic flow (EOF) itself provides the amount of liquid sufficient for generating and maintaining the nanospray (100–200 nL/min). With this type of interface the analyte is infused directly from the CE nanospray tip into the mass spectrometer. Such sheathless interfaces were used also in the context of glycoconjugate analysis. LODs attained in this way were by a factor of about 5 lower than reported with the sheath-flow interface, and other benefits of nanospray could be used as well, such as improved ionization, desolvation, and ion transfer into the MS. To allow an electrical contact needed for CE as well as ESI, most often the sprayer tip is coated by a conductive layer of metal [74,75] or carbon

[76]. This tip can be the CE capillary itself, etched as a microsprayer, or a commercial nanospraying needle is connected with the CE column [77]. Using sheathless interfaces with nanospray, the LOD for glycan analysis was about 0.5 pmols. Other sheathless CE-MS interfaces were reported using a porous junction for maintaining the current [78,81]. This interface has been used for protein and peptide analyses [82,83]. A laboratory-made low-flow interface, employed for the analysis of glycolipids [84], is discussed subsequently.

MALDI interfacing to CE has been done with different technical and methodological approaches [85,90] as described in recent reviews on this topic [73,91]. Particular advantages of this technique lie in the high sensitivity of MALDI and in the higher ruggedness against the presence of salts and surfactants. Detection limits down to the attomol range were reported. Droplet deposition volumes of 50–1000 nL onto the MALDI target were possible, depending on the peak separation aimed for. This technique allows different methods of matrix preparation (an aspect relevant when using multistage MS) and, of course, the use of all types of multistage MALDI-MS instruments (TOF/TOF, IT-TOF, FT-MS). However, only a minority of the interfaces proposed are commercially available, and most often the sheath-flow assisted spotting via a T-piece is used. There are some applications for general peptide [86,92], glycopeptide [93,94], and glycan analyses [95,98], but this hyphenation is still less popular than CZE-ESI-MS.

4.3.2   MS AND MS^N OF GLYCOCONJUGATES

The field of MS/MS of oligosaccharides released or present in glycoconjugates is in fast and highly dynamic development [13,30,61,99]. This chapter addresses particularly those aspects relevant when coupling MS/MS instruments to CE.

Regarding fragmentation spectra, different types of multistage instruments exhibit different features with respect to (1) the molecular mass limit for the precursor ions, (2) the energy transferred in collision induced dissociation (CID), (3) the amount of metastable ions formed during the ionization process, (4) the possibility of using electron transfer dissociation (ETD), electron detachment dissociation (EDD), and electron capture dissociation (ECD) mechanisms, and (5) the possibility of charge reducing of highly charged precursor ions by proton-transfer reactions (PTR). CID fragmentation pattern differs considerably depending on the collision energies involved. Collision energies in ion traps are often lower than in QqTOF instruments and much lower than in TOF/TOF instruments. The mode of ionization (positive or negative), the type of cationization (H$^+$ or Na$^+$, K$^+$), and the choice of the collision gas (He, Ar) might significantly influence the CID pattern as well. When using MALDI sources the extent of postsource decay (PSD) depends not only on the laser fluence and the used matrix but also on the type of analyzer, the residence time of the ion in the analyzer, and the ion-cooling procedure employed. Thus, depending on the used instrument, slightly and sometimes significantly different fragmentation patterns might be obtained.

Whereas ETD is not effective with carbohydrate structures, electron detachment dissociation (EDD) has recently been demonstrated as a powerful fragmentation technique for examination of glycan structural features [100]. EDD produces a radical anion that induces a more extensive fragmentation than that achievable by activation of even electron ions using low energy or threshold dissociation methods. Unlike CID, the characteristics of glycan fragmentation by EDD include abundant cross-ring fragmentation, cleavage of all glycosidic bonds, and the formation of even- and odd-electron product ions [101]. Moreover, whereas CID or electron-induced dissociation methods of epimers produce identical fragmentation patterns, EDD generates spectra from which the epimers can be distinguished by their specific product ions. Therefore, EDD is the ideal MS tool for distinguishing glucuronic and iduronic acids [102]. EDD of heparan sulfate (HS) tetrasaccharide dianions was also shown to produce a radical species that fragments into information-rich glycosidic and cross-ring product ions useful for determination of acetylation and sulfation sites. Moreover, when, in an oligosaccharide, the degree of ionization is greater than the number of sulfate groups, a significant reduction in SO_3 loss is observed in the EDD mass spectra. This observation suggests that SO_3 loss is reduced when an electron is detached from carboxylate instead of sulfate group [103].

4.4 GLYCOPROTEINS

Protein glycosylation, the covalent attachment of a carbohydrate moiety, is a common posttranslational modification of proteins in eukaryotic cells. N-glycans are attached to the amido group of an Asn side chain in a particular consensus sequence (Asn-Xxx-Ser/Thr with Xxx being any amino acid except Pro). Such sequences may be glycosylated only partially or not at all, and this aspect is called site occupancy. In some instances, Asn was found being N-glycosylated when embedded in a Asn-Xxx-Cys sequence [104,105]. Typical monosaccharides present in N-glycans are N-acetylglucosamine (GlcNAc), mannose (Man) and galactose (Gal), fucose (Fuc), and N-acetylneuraminic acid (Neu5Ac), but N-acetylgalactosamine (GalNAc) and N-glycolylneuraminic acid (Neu5Gc) may also be encountered. A variety of O-glycans exists, of which the mucin-type is most often found. In lieu of a simple consensus sequence, a probability for the occurrence of mucin-type O-glycosylation can be predicted by a more complex empirical approach (cf. http://www.cbs.dtu.dk/services/NetOGlyc). In mucin-type O-glycans, GalNAc is linked to a Ser or Thr residue. In other types of O-glycans, Gal, GlcNAc, Fuc, Man, or even glucose (Glc) and xylose (Xyl)—as in certain GAGs—are the linking sugars. In plant glycoproteins, we find arabinose linked to hydroxyproline instead. In addition to these carbohydrate constituents, the presence of noncarbohydrate substituents must be considered, such as sulfate, phosphate, acetate, methyl, or other groups.

The assembly of all these glycans is not template controlled like that of polypeptides. Therefore, an otherwise homogenous protein usually occurs in a large number of glycoforms differing by site occupancy as well as size and structure of the attached oligosaccharides. This "micro heterogeneity" of glycoproteins demands high-resolution and at the same time high-sensitivity methods for its characterization. In the following chapters, the most elaborated approaches for this demanding and important task—that is, the combination of CZE or LC with MS—will be presented, starting with intact glycoproteins and ending with free glycans.

4.4.1 Analysis of Intact Glycoproteins

MS analysis of intact proteins is aimed at determining their total molecular masses including all modifications. Thus, analysis of intact glycoproteins can distinguish glyco- and isoforms of a protein if they differ in the cumulative number of monosaccharide units and other PTMs present. Such other PTMs are commonly acetylation, sulfatation, phosphorylation, γ-carboxylation, and deamidation. On the level of intact proteins one is also able to assess a fairly reliable quantitative pattern of an iso-/glyco-form distribution. Such a quantitative pattern is important particularly for therapeutic protein production, as glycosylation can influence protein stability, ligand binding, immunogenicity, and serum half-life [6,106]. The quantitative glycoform profile might be influenced by production and cell culture conditions, and thus regulatory agencies require a proof for oligosaccharide consistency. Analysis of the detailed glycan structures and their site-specific variability is typically not carried out by analyzing intact proteins; they are analyzed more readily on the level of glycopeptides and particularly released glycans.

Very commonly, protein glycoforms differ in the total number of SAs present in the glycan leading to differences in the total charge of the glycoforms. Due to this charge difference, CZE-MS is particularly well suited for the purpose of glycoform analysis of intact proteins. Typical amounts needed by this method are in the range of a 10–50 ng of intact glycoprotein (corresponding to about 1 pmol) injected into the CZE capillary. The corresponding concentration is about some mg/mL when injecting not more than 10 nL into the CZE system. Sample enrichment prior to CZE-MS is thus often inevitable when dealing with samples from biological sources as discussed above.

Whenever dealing with intact proteins/glycoproteins, wall adsorption is an important issue, and capillary surface coating is crucial in this instance [71,107]. By choosing an appropriate coating, EOF might be prevented, if desired, or might be generated either aimed for accelerating the separation (using analyte/EOF comigration) or for enhancing separation selectivity (by analyte/EOF countermigration). And, importantly, this coating-based EOF is usually found to support the spray and to stabilize the current during electrophoresis. The advantages of the various coatings (neutral, cationic, anionic, double-layer, and multilayer [SMIL]) in the context of glycoprotein analysis was discussed in a recent review of these authors [9], and general aspects of single- and multilayer coatings were recently reviewed by Stutz [108]. A very low EOF, attained by neutral coatings, might be chosen if the transfer of too much BGE into the MS instrument should be avoided. Such a case might be given, if urea has to be added to the BGE for keeping the protein solubilized [109].

Online interfacing of CZE to ESI-MS is in most instances done by using sheath-liquid interfaces built as a triple-tube sprayer [109,115]. Although the sheath flow causes sample dilution, this setup is more robust than the sheathless interfaces and allows altering the composition of the BGE after separation and prior to ESI by mixing with the sheath liquid. Small alcohols (methanol or 2-propanol) mixed with aqueous solutions of formic or acetic acid or volatile buffers frequently serve as sheath liquids. (These conditions are quite commonly used for the CZE-MS analysis of proteins and peptides irrespective of glycosylation.) Low pH supports MS ionization in the positive ion mode that is predominantly used with intact proteins and peptides. Direct (automated off-line) hyphenation of CZE to MALDI-TOF-MS is much less common, probably because of the limited mass accuracy attainable by linear TOF analyzers for intact proteins. Advantages of MALDI, such as being less affected by the presence of salts and surfactants, might be of interest in certain cases.

In the MS analysis of intact proteins, the features of the MS instruments regarding high resolution and mass accuracy become particularly important. High mass spectrometric resolution is needed for distinguishing the ion peaks in very narrow positioned charge distributions of highly charged (charge numbers 30–60) glycoforms present simultaneously. With the high mass resolution of up to 60,000 attainable by recent generations of mass spectrometers like ESI-oaTOF instruments and Fourier-Transform (FT)-instruments like Orbitrap and Ion-Cyclotron-Resonance (ICR) mass spectrometers, the deconvolution of narrow positioned interwoven charge states becomes possible even for large glycoproteins. Mass accuracy attainable with high-resolution instruments such as ESI-oaTOF can be expected below 5 ppm (i.e., better than $\pm$ 1 u) even with proteins up to M_r approximately up to molecular masses (M_r) of 30,000 [116]. With this high accuracy it is possible to distinguish among most of the common PTMs and amino acid exchanges. With ion trap (IT) analyzers the mass accuracy is typically lower (about 100 ppm), giving an uncertainty of about 10 u, and this value still allows one to distinguishing among Hex, desoxy-Hex, HexNAc, Pent, Neu5Ac, and Neu5Gc and also to account for major modifications in the peptide backbone, like oxidation of methionine.

Most of the CZE-MS applications reported in recent years deal with recombinant glycoproteins aimed for therapeutic use [109,112–114,116–119] and available for analysis in concentrations significantly above their physiological values. Nevertheless, there are some papers in which glycoprotein isoforms in human serum or urine were monitored with respect to disease marker function [110] and for controlling drug application [113].

The spectrum of glycoproteins investigated as intact proteins covers the following:

1. Plasma-derived human antithrombin, a therapeutic drug used for controlling blood coagulation. It was characterized by use of a quadrupole ion trap (QIT) analyzer applying 20 to 30 ng of the glycoprotein [109], in which seven glycoforms could be distinguished, and quantifies as being present in relative abundances between 4 and 70%. The molecular masses of the individual glycoforms could be given with an accuracy of approximately 100 ppm. Electrophoretic migration was primarily determined by the sialic acid occupation in the glycans.

2. Various commercially available pharmaceutical formulations of recombinant human erythropoietin (rhu EPO) [112–114,116]. Again, separation could be attained for glycoforms differing in the number of SAs. By use of a high resolution ESI-oaTOF instrument [116], up to nine glycoforms differing in the number of Hex and HexNAc units could be assigned to each of the seven sialoforms separated, and a considerable extent of glycan-assigned acetylation was observed in addition (Figure 4.1). In this way a total number of 44 different glycoforms could be distinguished, giving at

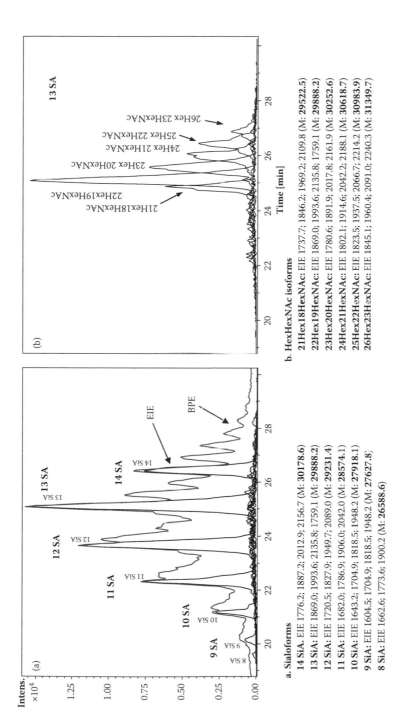

FIGURE 4.1 Electropherograms of intact rHu EOP by CZE-ESI-caTOF-MS. Base peak electropherogram (BPE) and extracted ion electropherograms (EIEs) of (a) glycoforms with different numbers of sialic acids and (b) glycoforms with different content of Hex and HexNAc but identical nubers of sialic acids. Sample concentration: 2.5 μg/μL; CZE: polyacrylamide coated capillary, voltage + 30 kV, BGE: 2 M acetic acid, sheath flow 4 μL/min. (Adapted from E. Balaguer et al., *Electrophoresis*, 27, 2638, 2006. With permission.)

a. Sialoforms

14 SiA. EIE 1776.2; 1887.2; 2012.9; 2156.7 (M: **30178.6**)
13 SiA. EIE 1869.0; 1993.6; 2135.8; 1759.1 (M: **29888.2**)
12 SiA. EIE 1720.5; 1827.9; 1949.7; 2089.0 (M: **29231.4**)
11 SiA. EIE 1682.0; 1786.9; 1906.0; 2042.0 (M: **28574.1**)
10 SiA. EIE 1643.2; 1704.9; 1818.5; 1948.2 (M: **27918.1**)
9 SiA. EIE 1604.5; 1704.9; 1818.5; 1948.2 (M: **27627.8**)
8 SiA. EIE 1662.6; 1773.6; 1900.2 (M: **26588.6**)

b. HexHexNAc isoforms

21Hex18HexNAc: EIE 1737.7; 1846.2; 1969.2; 2109.8 (M: **29522.5**)
22Hex19HexNAc: EIE 1869.0; 1993.6; 2135.8; 1759.1 (M: **29888.2**)
23Hex20HexNAc: EIE 1780.6; 1891.9; 2017.8; 2161.9 (M: **30252.6**)
24Hex21HexNAc: EIE 1802.1; 1914.6; 2042.2; 2188.1 (M: **30618.7**)
25Hex22HexNAc: EIE 1823.5; 1937.5; 2066.7; 2214.2 (M: **30983.9**)
26Hex23HexNAc: EIE 1845.1; 1960.4; 2091.0; 2240.3 (M: **31349.7**)

least 135 isoforms when accounting also for differences in the acetylation of the glycans. For this analysis, concentrations in the range of some mg/mL were necessary. As EPO is present in low to middle ng/L concentrations in urine and in 100 ng/L concentrations in serum, effective preconcentration of the protein will be necessary to monitor glyco- and isoforms of EPO in urine and blood. In recent works, this method was applied to analyze in addition a hyperglycosylated analogue of EPO called novel erytrhropoiesis stimulating protein (NESP) that differs from rHuEPO by five amino acids leading to the attachment of two additional *N*-glycans [117]. By use of CZE-ESI-MS, the authors could separate glycoforms differing in the total number of SAs and partially those differing in the total number of monosaccharide units. The significantly higher degree of glycosylation present in NESP made the resolution of these single glycoforms much more difficult.

3. Glycoforms of transferrin [110] not only as protein standard solutions but also as samples gained from sera of healthy individuals and persons suffering from congenital disorders of glycosylation (CDG), a syndrome associated with enhanced concentrations of carbohydrate-deficient variants of transferrin. Despite the good sensitivity of the aoTOF-MS instrument used, when aimed for monitoring plasma level concentrations the method still needs an increase in sensitivity.

4. Some other applications dealing with standard samples of α_1-acid glycoprotein (AGP) and fetuin (FET) [111].

5. Recently, recombinant vascular endothelial growth factor 165 (VEGF) has been investigated using CZE-MS [118]. VEGF concentration was approximately 100–200 μg/mL, and an injection volume of 50 nL was used. Seven peaks could be partially separated by CZE; after deconvolution each selected "peak" could be assigned to several glyco- as well as isoforms. The protein was expressed in insect cells, and the short glycans did not contain SAs and were thus uncharged. The CZE separation mainly distinguished between different isoforms of the protein; glycoforms were distinguished due to accurate deconvolution of the MS spectra (5 ppm accuracy using an ESI-TOF-MS). BGE composition was crucial. After extensive study of conditions the use of 60 mM formic acid was found best. For enhancing detection sensitivity, formic acid concentration was increased to 90 mM in the sheath liquid.

6. Very recently, recombinant human chorionic gonadotrophin (r-hCG) was analyzed [119] using a linear ion trap (LTQ)-FT-MS instrument. The high resolution of this mass spectrometer allowed the authors to identify over 60 different glycoforms, differing not only in the number of SAs but also in the number of monosaccharide units. Protein samples expressed in CHO cell lines were found different in their glycoform pattern than those obtained from a murine cell line.

It seems that the application of CZE-MS for distinguishing among glycoforms of (mainly recombinant) proteins becomes increasingly attractive with the availability of high-resolution MS instruments. In many cases, the high performance of the mass analyzers allow one to get good spectral information on glycoforms even if they were only partially separated by the CZE. The CZE conditions that have to be chosen for gaining sufficient separation are very dependent on the individual protein/glycoprotein under investigation. The attainable CZE separation depends on the extent of glycosylation, on the protein's total mass, and on whether additives have to be added to the BGE to keep the protein in solution. Under optimized conditions, CZE was shown in many instances to be able to resolve protein glycoforms with different net charges resulting from different extent of sialylation, phosphorylation, or sulfatation. Resolution of equally charged iso- and glycoforms was shown to be partially attained in many cases, the extent of separation depending on the glycoform distribution. The selection of an appropriate coating, particularly its stability, is crucial [117], as any bleeding of the coating polymer impedes the performance of the MS. The appropriate set of tuning parameters of the MS influencing ionization/ desolvation/ion acceleration might be decisive for sensitivity, and the optimum settings might be dependent on the glycoprotein structure.

4.4.2 ANALYSIS OF GLYCOPEPTIDES

Oligosaccharide analysis at the glycopeptide level is chosen for determining the site specificity of glycosylation, and for this purpose multistage MS is required. Commonly, glycopeptides are obtained by enzymatic cleavage of glycoproteins, and with biological samples some kind of preseparation/prefractionation/enrichment is needed as discussed in Section 4.2. It is difficult to directly analyze glycopeptides present in a complex protein digest by MS/MS alone because of ion suppression. Ionization of glycosylated peptides, particularly when SAs are present, is suppressed in the presence of nonglycosylated ones. In a peptide–glycopeptide digestion mixture, minor abundant glycopeptide variants and structures not easily ionized may be completely missed when working without separation.

Reverse-phased (RP)-HPLC peptide mapping, including glycopeptides as well, is a routine approach in all bottom-up proteomic investigations. This extensively used technique is not the subject of this chapter. Rather, it is to underline the particular advantages of CZE-MS of glycopeptides/peptides compared with the corresponding RP-HPLC methods.

First, in CZE we have the possibility of tuning the EOF either to speed up the run (using a high EOF) or to increase selectivity (choosing conditions for electrophoretic and electrokinetic countermigration). Total run times for peptide/glycopeptide mapping with CZE can thus be significantly lower (15–25 minutes) compared with HPLC (usually 60–90 minutes). Second, a more rigorous cleaning and purging procedure is possible, reducing memory effects. Third, CZE supports monitoring small and highly hydrophilic glycopeptides (with many SAs) that are in the case of RP-HPLC eluted near the chromatographic void volume and might become insufficiently ionized (because of short peptide backbone, negative charge, little separation, and occasionally remaining salt burden). And finally, the pH conditions in CZE can be chosen in a way to avoid acid or base catalyzed hydrolysis of SAs, thus reducing the risk of analytical artifacts. The amount of glycopeptides needed for CZE-ESI-MS analyses lies in the range of some pmol (ng) injected amount when working without online sample concentration [93].

With glycopeptides, positively or negatively coated capillaries are used as well as noncoated ones, as adsorption phenomena are not as problematic in this instance as with intact glycoproteins. A detailed discussion on coatings, BGE, sheath liquid, and hyphenation conditions is given in a recent review of the authors [9]. Generally, BGE and sheath-flow conditions are less crucial as with intact proteins. A combination of chip-based CZE with a modified sprayer was reported by [120].

MS analysis of glycopeptides is essentially based on the MS^n and is in the majority of cases carried out in the positive ionization mode. Under low-energy CID conditions as typical for ITs, predominantly the oligosaccharide is fragmented primarily at the glycosidic bonds. Thus, the corresponding spectra exhibit primarily Y and B ions (following the nomenclature of Domon and Costello [121]) with a prevalence of Y ions. (They include the peptide backbone, which contributes in stabilizing of protonation.) The B ions (oxonium ion fragments) are diagnostic for sugars and can be used for an automatic software-based detection of glycopeptides in a peptide mixture. Cross-ring fragmentation (A and X ions) is usually not obtained under these conditions. The spectra are thus comparatively easy to interpret. From such MS^2 spectra without cross-ring fragmentation the total number and the sequence of Hexs, desoxy-Hexs, HexNAcs, and Neu5Acs, Neu5Gcs can be assessed. These data allow one to make conclusions about the SA coverage, the presence or absence of an intersecting GlcNAc, and the extent of fucosylation as well as on the position of Fuc (core-linked or in the antennas—constituting a Lewis-type antigen). The most intense positive ion peak in the MS^2 spectrum often originates from the peptide with one single core-HexNAc remaining, the peak with the lowest m/z value from the peptide having lost all sugars. Choosing these ions as precursor ions for MS^3 CID-fragmentation, ample fragmentation occurs along the peptide backbone giving b and y ions usable for peptide sequence information [93].

Whether glycan- and peptide-backbone fragmentation can be carried out in different stages of MS^n depends on the collision energy and thus on the CID conditions used. CID with higher energy and carried out with negative ions allows cross-ring fragmentation providing information on branching and linkage positions as described by [122]. In contrast to CID, electron-transfer-dissociation does not affect glycan fragmentation but cleaves glycopeptides only at the peptide backbone, yielding predominantly c and z ions. ETD is thus a useful tool for determining the glycan attachment sites (if not unique). Combination of ETD and CID in an alternating way allows one to determine occupancy and site-specificity of glycosylation by ETD, and carbohydrate analysis by CID in one and the same experiment [31,123].

When dealing with glycopeptides, CZE hyphenated to MALDI-MS is a well-suited technique as well. Within the mass range of glycopeptides (up to 5,000 u) the MALDI technique provides good resolution and mass accuracies, and thus an impressive body of experience has been accumulated with MALDI–MS analysis of peptides and glycans [8,61].

CZE-MALDI as well as CIEF-MALDI MS was used for glycopeptide analyses in the sub-pmole range employing sheath-flow mediated automated sample deposition onto MALDI targets [81,82]. MS^2 spectra were obtained in the positive and negative ion mode, allowing sugar sequence analysis. In the CIEF mode, the sheath liquid resembled the catholyte solution (containing ammonium hydroxide) during deposition of the analyte/ampholyte zones. Importantly, residual amounts of neighboring ampholytes present in the spotted analyte fractions did not impede glycopeptide ionization: LODs were comparable to the CZE mode. Using MALDI with MS^2 for the analysis of glycopeptides, one faces the (mostly complete) loss of SAs. The presence or absence of such labile groups has to be ascertained by measuring in a linear TOF mode.

The majority of CZE-ESI- or MALDI-MS applications were aimed for confirming occupancy and glycoform distribution in therapeutic glycoproteins, covering samples from human antithrombin [93,94], rHu EPO [113], and monoclonal antibodies [124]. Other samples were lectins from various seeds [125,126] and from venoms [127], and some standard glycoproteins like cellobiohydrolase I [128], phospholipase A$_2$ [129], and AGP [98].

There were as well samples taken from urine which were prefractionated and concentrated prior to CZE-MS. A series of investigations was dedicated to analyze O-glycosylated peptides and amino acids from human urine being candidates for potential marker compounds for Schindler's disease [29,75,122,130–132]. Nine different compounds corresponding to O-glycosylated sialylated amino acids were fully or partially resolved by CZE and characterized by MS using a sheathless CZE-microsprayer to assure high sensitivity and high-quality spectra down to a small pmol injected amount [132]. This sensitive method resembles earlier methods of this group using different interfacing procedures.

4.4.3 ANALYSIS OF GLYCANS RELEASED FROM PROTEINS

Glycans (i.e., free oligosaccharides) are found in urine where they have diagnostic value in mammals' milk and many other sources. More often, free glycans are obtained by releasing them from glycoproteins or glycolipids. While the release covers up the original linkage sites at the protein, this approach provides a global view of the glycan population of a given sample. This approach allows relative quantitation of glycan structures from different attachment sites without the need of extensive standardization as it eludes the detrimental influence of differences in ionization yields inferred by the peptide moiety. Importantly, the separation of individual glycan structures (e.g., isomeric structures) is facilitated in the absence of the peptide, which strongly dominates, for example, reversed-phase retention. Finally, the fragmentation patterns of glycans are not mingled with peptide backbone fragments that could complicate interpretation especially under high-energy collision conditions applied to obtain cross-ring cleavages.

Regarding MS analysis, negatively charged (derivatized and underivatized) glycans were often ionized in the negative mode and neutral ones in the positive for reasons of sensitivity. Glycans exhibit significantly different fragmentation patterns, depending on the ionization mode used and the status and type of derivatization. And of course, CID-based MS/MS spectra are very dependent on the fragmentation energy, which in turn depends on the type of analyzer employed. Very often, CID-based MS/MS spectra of glycans are dominated by fragments resulting from a loss of SAs; cleavages of the glycosidic bonds are predominant in the positive ionization mode under low-energy CID conditions, and in the negative mode cross-ring cleavage might occur as well.

4.4.3.1 CZE-MS/MS

CZE is particularly successful in separating glycans with respect to charge (resulting, e.g., from different numbers of SAs or other charged groups, like sulfate) and size (resulting, e.g., from different numbers of monosaccharide unit constituents). The branching and linkage pattern of isobaric glycans influence their mobility to a much lesser degree, at least under standard CZE conditions. However, when including a complexation equilibrium by which negatively charged boronate complexes are formed, selective migration with respect to branching and to linkage position and anomery (subsuming these aspects as "shape" selective) can be induced. Isomers exhibiting differences in their boronate complexation constants experience different contributions to their effective net charge and thus to their electrophoretic mobility. This method is widely used in CZE separations with optical detection (CZE-LIF), and a wealth of literature and many excellent surveys about this topic are available [4,133,134]. The use of nonvolatile borate buffers, however, is prohibitive for a robust and long-term CZE-ESI coupling, and it is avoided in most laboratories (despite the dilution effect attainable by a sheath-flow interface).

Another technique is size separation of labeled glycans by a CGE mechanism involving a sieving matrix [135–138]. This technique has been carried out also in a microchip format using for this purpose a well-established DNA sequencer equipment [136–139]. This interesting technique has so far not been hyphenated to MS, as hyphenation of electrophoretically operated microchips to MS is not yet an easy task.

Employing volatile buffers only (without borate and phosphate), as required for robust CZE-ESI-MS/MS coupling, labeled glycans can still be separated in a size-selective way; shape selectivity, however, is limited. Separation selectivity (and efficiency) is influenced by the type of label [106]. Very widely used for this purpose are the 8-aminopyrene-1,3,6-trisulfonate (APTS) and 8-aminonaphthalene-1,3,6-trisulfonate (ANTS) labels inferring multiple negative charge, or 2-aminobenzoic acid (2-AA) [47], less frequently used, and 2-aminopyridine (2-AP) and 2-aminobenzamide (2-AB) labels inferring positive charge. Recently, a new and effective label was introduced: 5-amino-2-naphthalenesulfonic acid (ANSA) [140]. Besides their potential for LIF detection, these labels effect high ionization yield in the ESI process (when the ionization mode corresponds with the label charge). This aspect might be important when sensitive fragment ion detection in multistage MS experiments is crucial.

In most of the CZE-ESI-MS/MS applications, the BGE consisted of a standard buffer solution (commonly ammonium acetate, formiate). In some instances, also the use of 6-aminocaproic acid in the BGE was reported as well as addition of methanol or a low amount of polyethylene glycol to the BGE [9,11]. Glycans were either ANTS- [141], APTS- [142], or 5-amino-2-naphthalenesulfonic acid (ANSA) [140] derivatized, or separated without derivatization.

Recently, Nakano et al. [143] presented CZE-ESI-MS^n-based glycan analyses using 9-fluorenylmethyl oxycarbonyl- (FMOC-) as a neutral label. (In Nakano's derivatization procedure the N-glycan, released by PNGase F in a first step as N-glycosylamine was reacted at pH conditions above eight directly in this form with FMOC-Cl, giving the FMOC-labeled N-glycan.) The analysis of these derivatives was performed under standard conditions (bare fused silica capillary, 40 mM ammonium acetate, pH 6.8 as BGE, direct CE polarity) and positive ion mode ESI–MS. With this system, the FMOC-labeled glycans could be separated according to their charge (number of SAs) and size (total number of monosaccharide units); the discrimination of isobaric structures has still to be demonstrated (cf. Figure 4.2). As the FMOC label does not infer additional charge to the glycan, CZE separation between neutral glycans is not possible. Due to the high sensitivity achieved, good CID-based MS/MS spectra could be gained. The spectrum of the trisialo-triantennary glycan shown in Figure 4.3 exhibits single- or double-protonated Y-type fragment ions generated by stepwise stripping of Neu5Ac, Gal, GlcNAc, and Man from the nonlabeled end. Their B-type counterparts were detected in the low mass region as [M+H]$^+$ ions.

At present, a general assessment of available data seem to indicate that, among the tested CE-MS techniques, online CZE-ESI–MS based on sheath-liquid interfacing and using CID-based MS^n (for instance in ion trap analyzers) emerge as most versatile. Method refinements are oriented toward improving sensitivity and selectivity as well as the effectiveness of the follow-up MS/MS fragmentation so that N-glycan linkage patterns can be rapidly and unambiguously deduced in the picomolar range.

CZE combination with MALDI is not as frequently reported in the context of oligosaccharide analysis of glycoconjugates as CZE-ESI-MS [95–98]. Nevertheless, it has good potential. The easiest way to do CE-MALDI-MS seems to be via automated sheath-flow assisted sample deposition. Multistage MALDI-MS instruments (TOF/TOF, IT-TOF, FT-MS) can be used for gaining subtle fragmentation spectra; in addition, derivatization procedures frequently encountered in the MALDI-based analyses of glycans can be carried out on the target and independently of the CZE separation. This includes particularly permethylation, which was described to significantly enhance ionization yield and strongly influences fragmentation pathways [64]. Much experience is available regarding detailed linkage analyses of permethylated glycans by MALDI-based MS/MS via cross-ring fragmentation [60,61,144,145].

4.4.3.2 HPLC-MS/MS

HPLC has been a key method for complex glycan analysis for many years. Four modes of separation have proven useful and all have—to varying degrees—been hyphenated to ESI-MS [146]. Derivatized oligosaccharides have been analyzed by HPLC on reversed-phase [147], on normal alias HILIC phases [148,149], and also on porous graphitic carbon (PGC), which binds nonderivatized glycans [150]. A plethora of labels has been introduced, a topic that has been dealt with elsewhere [151,152]. Underivatized glycans can be separated by high-pH anion exchange chromatography (HPAEC) with amperometric detection [153]. These methods can of course also be combined with MALDI-MS [154,155].

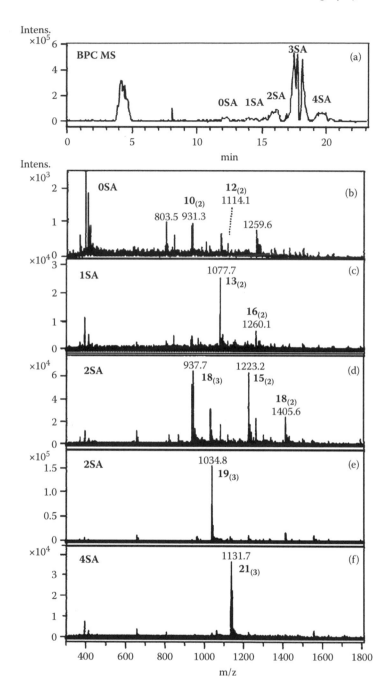

FIGURE 4.2 CZE-ESI–MS analysis of FMOC-labeled *N*-glycans released from bovine fetuin. (a) Base peak chromatogram (BPC). (b–f) MS¹ spectra of asialo-, monosialo-, disialo-, trisialo-, and tetrasialo-glycans (0SA, 1SA, 2SA, 3SA, and 4SA) gained from the corresponding BPC peaks over the indicated sampling interval. Bold numbers *(continued on next page)*

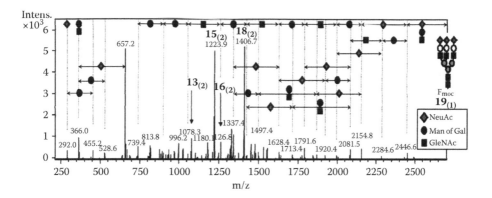

FIGURE 4.3 MS/MS spectrum of the [M+3H]$^{3+}$ion at *m/z* 1034.8 corresponding to FMOC-labeled trisialo-triantennary glycan derived from bovine fetuin. (From M. Nakano et al., *Glycobiology, 19,* 135, 2009. With permission.)

Due to their ability to separate according to subtle structural differences of the glycans including overall topology, branch, and linkage isomers, LC-MS methods perfectly complement CZE-MS, where—in the absence of adsorption processes— molecular shape plays an inferior role.

4.4.3.2.1   High-pH Anion Exchange Mode

When working at pH 13, the high concentrations of NaOH and sodium acetate impede ESI detection. However, the principal manufacturer of HPAEC-PAD systems has developed a desalting cell that removes these involatile solvent ingredients to a point where ESI-MS becomes possible, as was demonstrated by Bruggink and coworkers [156]. A capillary version of this system provided enhanced sensitivity, as was demonstrated with glycans present in urine of gangliosidosis patients [157]. There are only a few further reports on HPAEC-MS hyphenations, though they do not belong to the glycoprotein area [158,159].

4.4.3.2.2   Reversed-Phase Mode

Being notoriously hydrophilic, oligosaccharides as such are essentially not retained by reversed-phase matrices unless they are derivatized [151,152,160]. Given the great selectivity of RP matrices regarding 2-aminopyridine labeled sugars [160,161], this combination appears as being ideally suited for coupling with ESI-MS: Takegawa combined the classical RP separation of pyridylaminated sugars with MS detection [162] and found new features in the glycan profile of women with rheumatoid arthritis [163]. The ionization device used in these studies was a "sonic spray ionization interface." We surmise that the authors tried to evade spray instability arising from

---

FIGURE 4.2 (continued)   refer to glycan structures; numbers in parentheses indicate the charge state of the ion. CZE conditions: BGE: 40 mM ammonium acetate, pH 6.8, 30 kV, 20°C; sheath liquid: 50/49.9/0.1 (v/v/v) MeOH/water/formic acid; sheath flow rate 2 µL/min. ESI conditions: voltage 4 kV; N$_2$ flow rate 4 L/min. (From M. Nakano et al., *Glycobiology, 19,* 135, 2009. With permission.)

the very low content of organic solvent required to elute 2-aminopyridine-labeled sugars [151]. Others have applied a makeup solvent to ensure a proper spray [164].

Another strategy was to use a hydrophobic label such as ANTS [165]. 2-Amino-benzamide-labeled N-glycans were shown to become well separated into groups of similar structures by using RP-HPLC, albeit a rather long gradient was needed [166]. This group showed that by using polarity switching in MS ionization, simple fragmentation data allowing sequencing as well as detail-rich cross-ring fragmentation spectra could be achieved in one LC run [167]. Ion-pairing reversed-phase found use in GAG analysis [168].

4.4.3.2.3 Porous Graphitized Carbon

PGC was at first used as an online desalting tool [164] but very soon as a true separation device [170,171]. Since then, these three groups have produced a large number of papers proving the usefulness of PGC-LC-MS for a wide variety of analytical aims, from neutral to charged (sialylated, sulfated) N-glycans, O-glycans and GAGs [51,172–185]. The use of small column diameters enhanced overall sensitivity [178,185]. This approach was used by some other laboratories as well [186], such as for analyzing GAGs and glycans from glycoproteins [187,188], various types of plant oligosaccharides [189–190], and carragenans [191].

Regarding the sensitivity of this approach, recent examples of PGC-LC-MS applications gave good results for N-glycans and O-glycans from about 10^6 cells [181,192]. For cytokine-treated human cells, subtle quantitative difference in the distribution of isobaric glycoforms could be measured with statistical significance attesting PGC-LC-MS to constitute a real "glycomic" method [192]. The introduction of microfluidic devices and HPLC chips further enhances the overall sensitivity and simplifies the use of PGC as demonstrated with N-glycans released from serum proteins and other samples in high-throughput applications [193–195]. Recently, it was shown that by means of an integrated on-chip PNGase F reactor the release and analysis of glycans can be achieved in only 10 minutes [196]. Ultrasensitive analysis can be achieved using 10 μm I.D. porous layer open tubular columns [197].

PCG-based separations use mobile phase gradients with increasing amounts of acetonitrile added to aqueous buffer solution. The composition of the aqueous solvent in PGC-ESI-MS is limited in the sense that it must be able to promote elution of all components of the sample and should allow sensitive detection of analytes as well. Pabst et al. investigated the influence of solvent composition on the retention of neutral and sialylated glycans and concluded that a certain ionic strength is obligatory when dealing with multiply charged glycans [198]. Both ammonium carbonate (pH around 8, used by Packer's [172] and Hansson's [175] groups) and ammonium acetate (pH 9.6, used by Kawasaki's [173] group) buffer solutions can elute tetra-sialylated N-glycans, similar to formate buffer at various pH values. Generally, ammonium carbonate confers the higher elution strength compared with ammonium acetate and formate. At pH 3.0 formate buffer provided a stronger retention of sialylated glycans and thus a higher selectivity of the whole separation system [198]. Unbuffered formic acid (0.1%, pH about 2.3) must be rated as inappropriate for the analysis of charged

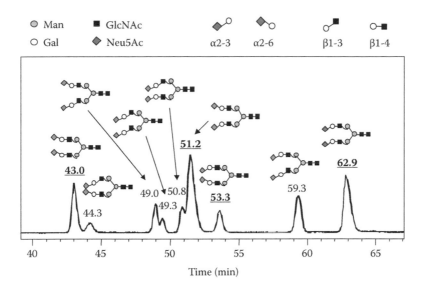

FIGURE 4.4 HPLC-ESI-MS analysis of isobaric *N*-glycans differing in the linkages of sialic acid and galactose using a PGC column. The retention times of the four isomers with only β1,4-Gal residues are underlined. A similar separation is shown in [187], where experimental details are given.

sugars because of its low ionic strength [198]. (The use of this solvent may be the reason for the apparent lack of di- or more highly sialylated glycans in many samples. Researchers may be well advised to countercheck the ability of their separation system to elute and detect all relevant analytes by an independent method or by the use of reference samples.)

Carbon columns are extremely robust and sensitive at the same time. Their ability to strongly adsorb a variety of substances may lead to rapid "fouling" of the column and thus to unstable elution times [199]. This can be prevented by prudent sample preparation and by occasional, rigorous regeneration of the PGC column [198]. Another pitfall arises from the electrode character of carbon that is awakened by the electrospray voltage. Reproducible elution of charged glycans requires a thorough electrical grounding of the column [198]. All these above mentioned drawbacks can be overcome. PGC appears thus as a most promising approach for glycomic analyses—especially if one takes into account the high shape selectivity.

The high shape-discrimination capability of PGC-LC toward isomeric and isobaric glycan structures was used to establish a structure-retention time correlation system based on reference glycans [192,200,201]. Nine disialylated *N*-glycans with different linkages of Neu5Ac and Gal could be separated on PGC and assigned using a reference glycan approach [200], as shown in Figure 4.4. To discriminate those four structures differing only in sialic acid linkages by MS alone, two MS³ transitions were required [202]. Remarkably, these researchers applied MS³ "spectral matching" rather than *de novo* interpretation for identification of the four disialylated glycans with α, 2,3- or α,2-6 linkages [202]. Tandem MS spectra of isomers are often very similar. Moreover, the possibility of sugar rearrangements during MS/MS severely endangers spectrum interpretation [203–206].

4.4.3.2.4 *Hydrophilic Interaction Chromatography*

HILIC chromatography predominantly separates according to the number of polar groups, and hence the size of the glycans. To a lesser extent, it displays shape discrimination of oligosaccharides [149]. Notably, HILIC tolerates all kinds of natural and artificial modifications of the glycans such as sulfate and phosphocholine groups or fluorescent labels [207].

Amide phases have attained high acceptance for the analysis of fluorescently labeled glycans [207,208]. Wuhrer et al. employed a capillary amide column with ESI-MS for detection of underivatized as well as labeled sugars [209,210]. Glycans released from the serum proteins of pancreatic cancer patients were investigated by this method [211]. In another study on serum glycans, Bereman et al. compared results gained by LC-ESI-MS using HILIC and PGC phases [199]. Chip-based amide-HILIC LC/MS was very recently employed for GAG glycomics profiling [212].

Zwitterionic (ZIC)-HILIC columns with sulfobetaine ligands were used for online desalting prior to MS of PA-labeled glycans [202]. Glycopeptides with either *N*- or *O*-glycans from erythropoietin were separated by ZIC-HILIC, and considerable separation according to glycan structural features was obtained [213].

We may conclude with some general considerations regarding the potential and advantages of the various LC-mechanisms if employed prior to an MS/MS-based structure analysis. Using an HILIC mechanism, separation is often predominantly attained according to the size of the oligosaccharides—a subject in which MS is superior. In contrast, RP and particularly PGC provide more orthogonal selectivity that primarily depends on the shape of the analytes. The ability of PGC columns to retain and separate underivatized oligosaccharides makes them the first choice for attaining branching- and linkage-selective analyses of isobaric glycans, as the large number of publications indicates.

4.5   GLYCOSAMINOGLYCANS

Glycosaminoglycans (GAGs) represent a special class of unbranched polysaccharides, which contain repeating units of modified disaccharides: one part of the disaccharide being GlcNAc or GalNAc, and the other one a uronic acid. The majority of GAGs in the body are linked to core proteins, forming proteoglycans (PGs), also called *mucopolysaccharides*. The GAGs extend perpendicularly from the core in a brush-like structure. The linkage of GAGs to the protein core involves a specific tetrasaccharide composed of one D-glucuronic acid, two galactose residues, and a xylose (GAG-GlcAGalGalXyl-O-protein). The linker is coupled to the serine in the protein core through an *O*-glycosidic bond (Figure 4.5). Some forms of keratan sulfate are linked to the protein core through an *N*-asparaginyl bond. The protein cores of PGs are rich in serine and threonine residues, which allow multiple GAG attachments [214].

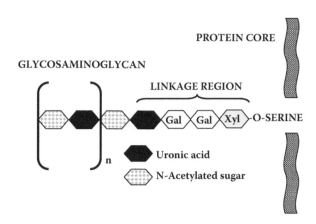

FIGURE 4.5   Schematic of proteoglycan structure.

GAGs are located primarily on the surface of cells or in the extracellular matrix (ECM). The high viscosity of GAG solutions and their associated low compressibility make these molecules ideal components in the lubricating fluid of joints. At the same time, the rigidity of these polymeric structures provides structural integrity to cells and passageways between cells, allowing for cell migration. The specific GAGs of physiological significance are hyaluronic acid (HA), dermatan sulfate (DS), chondroitin sulfate (CS), heparin, heparan sulfate (HS), and keratan sulfate (KS). Although each of these GAGs has a predominant disaccharide component, heterogeneity does exist in the sugars present in the makeup of any given class of GAGs [215,216] (Figure 4.6).

Hyaluronic acid is unique among the GAGs in that it does not contain any sulfate and is not found covalently attached to proteins as a proteoglycan. It is, however, a component of noncovalently formed complexes with proteoglycans in the ECM. Hyaluronan is a polymer composed by repetitive disaccharides, one unit having GlcA and GlcNAc, linked together via alternating β1-4 and β 1-3 glycosidic bonds. Hyaluronic acid has many important functional roles [217], which include signaling during embryonic morphogenesis [218], pulmonary and vascular diseases [219], and wound healing [220]. HA also acts in the lubrication of synovial joints and joint movement, and its function has been described as space filler, wetting agent, flow barrier within the synovium, and protector of cartilage surfaces [221]. The influence of HA on cancer progression has been described repeatedly [219,222,223].

Chondroitin sulfate and dermatan sulfate GAGs are covalently linked to a wide range of core proteins, forming PGs with a widespread distribution in mammalian tissue. CS chain backbone consists of disaccharide repeats containing GlcA and GalNAc residues, whereas DS is the stereoisomeric variant of CS with varying proportions of IdoA. The repeating units of CS/DS involve a β1-3 glycosidic linkage between the uronic acid and GalNAc and a β1-4 glycosidic linkage between GalNAc and the next uronic acid. CS/DS hybrid chains have a variable IdoA content; the highest percentages have been observed in skin PGs [224]. A CS/DS chain

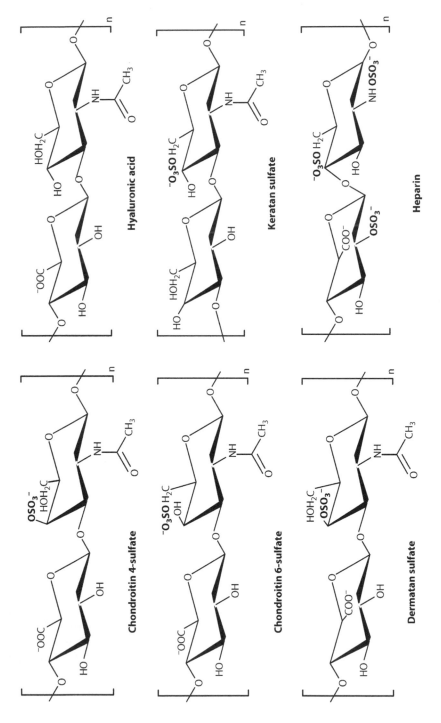

FIGURE 4.6   Detailed structures of repeating disaccharide units in various GAGs.

can encompass over 100 monosaccharides, each of which might be sulfated in variable positions and quantities. CS/DS chains are further modified by the differential sulfation pattern of specific sulfotransferases at C-2 of GlcA/IdoA or C-4 or C-6 of GalNAc to yield enormous structural diversity [225]. Highly sulfated disaccharide units like the E-unit, GlcAβ1-3GalNAc(4S,6S) [226], where 4S and 6S stand for 4-*O*- and 6-*O*-sulfate, respectively, are rare, and E-unit-rich CS preparations show remarkable biological activities like the promotion of neurite outgrowth and high affinity binding to growth factors [227]. Increasing evidence suggests that the expression pattern of cell surface CS/DS-PGs is related to metastatic potential [228]. However, the involvement of highly sulfated structures of CS/DS chains in the metastatic process remains obscure [229].

In contrast to other GAGs, keratan sulfate does not contain uronic acids, and its repeating disaccharide unit is composed of alternating residues of D-Gal and GlcNAc linked β1-4 and β1-3, respectively. In KS from most tissues, the hydroxyl groups at the C-6 positions of both Gal and GlcNAc residues are sulfated. Different degrees of sulfation have been observed in the cornea, various types of cartilage, and the brain. Keratan sulfate plays important roles in corneal transparency, nerve growth cone guidance, and cell adhesion [230].

Heparin and heparan sulfate are the most highly modified GAGs, consisting initially of GlcA β1-4 GlcNAc α1-4 units, which undergo variable processing by GlcNAc *N*-deacetylation and *N*-sulfation, C5 epimerization of some GlcA units to iduronic acid, and *O*-sulfate addition to C2 of the uronic acids and C6 and more rarely at C3 of the glucosamine units [231]. The arrangement of the modified residues along the chain creates binding sites for numerous growth factors, enzymes, and extracellular matrix proteins [232]. Heparin is widely used as an injectable anticoagulant, but the true physiological role in the body remains unclear, because blood anticoagulation is achieved mostly by endothelial cell-derived heparan sulfate proteoglycans [233].

The structural variation that can occur makes sulfated GAG chains one of the most complex classes of macromolecules found in nature, for which massive efforts toward development of specific analytical strategies, including those based on MS and CE/MS, are nowadays invested.

4.5.1 MS OF GAGS

MS of GAGs usually begins with a general identification of glycoforms obtained after detachment from the core protein by chemical or enzymatic methods. Next, exhaustive characterization of single glycoforms ought to be pursued by collecting and assembling detailed data upon epimerization, sulfate content, and its distribution. Last, the presence of regular or irregular domains must be directly recognized by sequencing single components in tandem MS experiments. Because of the high molecular weight, the structural analysis of GAGs by MS and hyphenated MS methods almost always requires a detachment of the GAG chains from the proteoglycan core protein followed by their depolymerization (cf. Section 4.2).

Numerous difficulties encountered in the MS analysis of GAGs [30,234] limit the benefits of this technique when applied to complex mixtures, if no separation is employed prior to MS. When the depolymerization of GAG chains is performed by lyases, complex GAG mixtures containing intact chains of variable length and degree of sulfation are formed. In ESI-MS, because of chain constitution, the monoisotopic peaks corresponding to charge states that equal ½ the number of the repeating units are observed at the same m/z value. This peak overlapping results in another specific challenge in ESI-MS of GAGs, since it makes the mass spectra difficult to interpret and jeopardizes the isolation of the precursor ion for tandem MS, in particular when low-resolution instruments are used.

Other challenges of MS-based GAG analysis are (1) impossibility of obtaining adequate ionization yields for long chains; (2) technical difficulties in hampering the unspecific *in-source* cleavage of the excessively labile sulfate ester group; (3) problems in detecting oversulfated and undersulfated ions in the presence of the regularly sulfated ones of higher abundance; and (4) mutually exclusive sequencing principles that are required for reliable determination of sulfation sites (i.e., cleavage of the glycosidic bond while keeping the sulfate group attached).

4.5.2   CZE of GAGs

Problems associated with the screening and sequencing of single GAG components in complex mixtures may be overcome by combining MS with efficient separation techniques such as SEC [235], HPLC [236], ion-pair liquid chromatography (IP-LC) [237], and CZE [77,238,239] properly optimized for this type of analysis.

For structural investigation [240–242] of GAG oligosaccharides, a further development of specific methods was required, among which CZE [243] and MS [244–246] brought an essential progress. From the separation point of view, a fundamental and general concern is that the best-suited CZE buffers are usually incompatible with GAG ionization by the electrospray process. Therefore, the compatibility of the CZE BGE as spraying solvent for GAGs is one of the major difficulties encountered when interfacing CZE to ESI-MS, either off-line or online.

4.5.3   Off-Line CZE-MS

An accessible approach for ESI-MS characterization of GAG components in complex mixtures is the off-line collection of CZE separated fractions [11]. CZE in off-line conjunction with nanoESI using a hybrid quadrupole time-of-flight (QTOF) MS/MS instrument was reported [238] for the analysis of CS/DS-derived oligosaccharides obtained from bovine aorta by digestion with chondroitin B lyase. The mixture of partially depolymerized oligosaccharides was dissolved in CZE buffer containing 50 mM aq. ammonium acetate/ammonia pH 12.0. For CZE-ultraviolet (UV) screening, the sample was injected into a bare-fused silica capillary by pressure and separated at 25 kV with direct polarity, monitoring UV absorption at 214 nm. Ten different fractions were collected in vials, were transferred to nanoESI capillaries by the aid of a micropipette loader, and were submitted to nanoESI-QTOF–MS screening using negative ion mode detection. Regularly sulfated hexa-, octa-, and decasaccharides, one undersulfated octasaccharide and one oversulfated hexasaccharide were identified [238]. Several new aspects offering the real dimension of the

CZE-MS applicability in GAG characterization were revealed for the first time by this study:

1. CZE-UV could provide signals of quite high intensity for 10 GAG components, although the detection by UV absorption is not the best-suited method for carbohydrate profiling.
2. By CZE, species with high molar sulfate content could be clearly separated from the nonsulfated ones, present in the GAG mixture released by β-elimination.
3. For a strict determination of the degree of sulfation in single GAG species the CZE separation is crucial. In addition it allows for distinguishing between the real under- and nonsulfated species and possible artifacts induced by the in-source decay of the sulfate groups in the MS mode.
4. Besides enhancing the MS signals of minor components, CZE separation eliminated the widely known possibility of misinterpreting the GAG composition because of the overlapping of isobaric peaks.
5. Under optimized nanoESI-MS conditions, the in-source desulfation of the molecules could be avoided, and the formation of multiply charged ions was favored. This feature provided a significant contribution to the successful detection of regularly sulfated and oversulfated saccharides in the CZE fractions.

The last stage of the methodology development included the use of tandem mass spectrometry to provide the elucidation of monosaccharide building block sequences, information on the repeating GlcA-GalNAc, GlcA-GalNAc(S) units, as well as data on the glycosidic linkages. The fully sulfated octasaccharide detected in the first CZE fraction and the sulfated disaccharide from the second CZE fractions were the precursor ions in the MS/MS experiments. The most important outcome of the fragmentation process of both species was the clear indication of the sulfate group substitution pattern along the GAG chain.

In a subsequent application [239] of this protocol, the biologically active sequence of CS/DS chains from decorin secreted by human skin fibroblasts able to interact with fibroblast growth factor-2 (FGF-2) was determined. In this case, however, for assessing the sulfation pattern along the carbohydrate chain of 4,5-Δ-IdoΔGalNAc[GlcAGalNAc]$_2$(5S) species a novel tandem MS protocol with CID at variable energy (CID-VE) was developed, in which the collision energy was readjusted during the ongoing MS experiment within the range of 10–30 eV. The MS/MS obtained under this condition provided evidence for the presence of two structural variants: (1) one containing IdoA sulfated GalNAc from the nonreducing end disulfated and the other two GalNAc moieties monosulfated; and (2) one with IdoA, the first GlcA from the nonreducing end and all GalNAc residues monosulfated.

According to these results, off-line CZE-nanoESI-MS with tandem MS methodology appears as a practical alternative, which overcomes part of the limitations experienced in structural analysis of GAGs. The data suggested that for instance the domain structure of biologically active CS/DS chains may be elucidated by such an off-line approach.

4.5.4 ONLINE CZE-MS

The low GAG yield produced after the application of specific sample extraction and purification protocols, the mild ESI source parameters, and the high ionization efficiency necessary for proper ionization required the developments of an online CZE/nanoESI–MS configuration able to address these issues. In such a setup [77], the CZE column was butted to a commercial nanosprayer needle (New Objectives Company) by using a homemade joint. The resulting two-piece column was incorporated into a stainless steel clenching device [130] to allow the application of the ESI voltage onto the needle, and finally the whole interface was mounted directly on the ESI high-voltage plate of a QTOF–MS as depicted in Figure 4.7. To prevent the misalignment of the CZE capillary and the nanoneedle into the joint and to reduce dead volume formation to a minimum, the CZE capillary and the needle end surfaces were adjusted by filing. Prior to a new coupling or between runs, the joint was dried by air flow to hinder any air bubble formation. The online CZE-(-) nanoESI-QTOF–MS analysis of the separation buffers only, carried out after each washing procedure, indicated a complete removal of the sample or impurities and the absence of any adsorption of sample onto the CZE capillary walls, the joint, and the needle. The spray could be initiated at values of 600–900 V applied to the nanoESI needle and 12–30 V for the sampling cone potential without the need of nebulizer gas. Leakage in the butted region or broadening of the TIC peaks were not observed, proving that no misalignment of the connection or gap between the CZE column and the nanosprayer needle were created. For all online CZE-MS experiments, complete desolvation could be attained at a temperature of 40^0C under moderate drying gas flow, and the CZE emitter tip unit could be left connected to the mass spectrometer during buffer rinsing of the capillary. This sheathless CZE-MS protocol was applied to extended CS/DS GAG chains of recombinantly expressed decorin from HEK293 cells [77] transfected with human decorin cDNA. In this experiment, the GAG chain was released by a β-elimination reaction and depolymerized with chondroitin

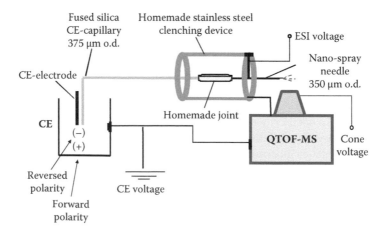

FIGURE 4.7 Schematic drawing of the sheathless online CE-nanoESI–QTOF–MS setup used for GAG analysis. (From A. D. Zamfir et al., *Electrophoresis, 25,* 2010, 2004. With permission.)

B lyase from *Flavobacterium heparinum* following the standard procedure mentioned in Section 4.2. The resulting CS/DS mixture was dissolved in BGE (50 mM ammonium acetate/ammonia pH 12.0) and hydrodynamically injected to the CZE by applying a pressure of 3.45 kPa for 6 s. Separation was carried out under 30 kV CZE voltage in direct polarity (Figure 4.8a) and for MS detection (Figure 4.8b) under the sampling cone potential was set to 15 V, nanoESI voltage to 700 V. For structural elucidation, online CZE-ESI–MS/MS was performed by automatic CID MS to MS/MS switching in data-dependent analysis mode (Figures 4.8c and 4.8d). This protocol applied on an oversulfated eicosasaccharide species provided data that enabled the localization of the additional sulfate group along the GAG chain as depicted in Figure 4.8e, where all sequence ions are assigned according to Domon and Costello nomenclature [121]. While in this case the bisulfation of GalNAc was unambiguously demonstrated, sometimes CID fragmentation in the MS² stage cannot provide sufficient data for a clear-cut discrimination between structural motifs containing either disulfated D-GalNAc or monosulfated D-GlcA. For this reason in many cases, the issue of elucidating proteoglycan glycosylation can be better addressed by the CZE separation of GAG segments having dissimilar sulfation contents combined with multistage MS (MS^n). Multistage MS, involving more than one selection or fragmentation event, $n > 2$, provides a better control of the fragmentation process, and the possibility of resequencing small, irregularly sulfated fragment ions until an unequivocal assignment of the sulfation sites along the chain and within the UroA and/or GalNAc ring can be attained [9].

Online CZE-ESI-MS of HA oligosaccharides was investigated by Kühn et al. [247] as a method for the separation and mass characterization of GAGs obtained by enzymatic digestion with bacterial hyaluronate lyase from *Streptococcus agalactiae*. In Kühn's configuration, a fused-silica capillary coated with polyacrylamide was used in combination with a sheath liquid of methanol/1% triethanolamine (8:2) composition delivered at 10 μL/min flow rate. CZE was run in normal polarity using polyacrylamide as neutral polymer coating and 40 mM ammonium acetate pH 9.0 as BGE solution. Under these conditions and for the separation of oligosaccharides containing 4–16 monomers, it was noticed that the migration behavior follows the chain length of the oligomers, regardless of charge state, although no linear relationship was found between mobility and chain length.

Heparin oligosaccharides were also the subject of online CZE-MS protocol development. One of the successful strategies included sheath-flow CZE coupled online with ESI-QIT–MS [248] tested in positive and negative ionization modes, and normal and reversed polarity using a combination of acrylonitrile/2 mM triethanolamine and 3mM ammonium formate (1:1) as a sheath liquid at 5 μL/min flow rate. It was reported that, as expected, normal polarity CZE allowed shorter analysis times, but at lower resolution, which made the discrimination of isomeric heparin disaccharides difficult to achieve.

In another protocol [249], heparin-derived disaccharides were analyzed by pressure-assisted CZE-MS and MS/MS, which allowed qualitative and quantitative identification of the co-eluting disaccharides. Highly sulfated HS bearing multiple negative charges could be successfully analyzed also in reversed CZE polarity followed by online sheath-flow ESI in negative ion mode; however, a pressure-generated

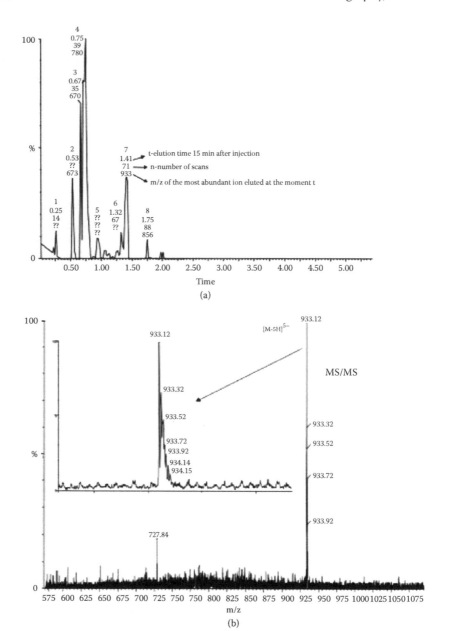

FIGURE 4.8 Sheathless online CZE-(-)nanoESI-QTOF–MS of a CS/DS mixture from human kidney fibroblast decorin. (a) CZE-TIC profile. (b) MS of the seventh TIC-MS peak eluted at min 15.41 after injection. (c–d) auto MS/MS spectra of the precursor ion [M-5H]$^{5-}$ at m/z 933.12 recorded over the m/z ranges (1240–1530) in (c) and (1400–1700) in (d). Fragment ion assignments resulted in a structural proposal for the oversulfated oligosaccharide 4,5-Δ-IdoA-GalNAc[GlcA-GalNAc]$_9$(11S) shown in (e). ESI potential 700V; sampling cone potential 15V; collision energy 35eV. (Adapted from A. D. Zamfir et al., *Electrophoresis, 25*, 2010, 2004. With permission.)

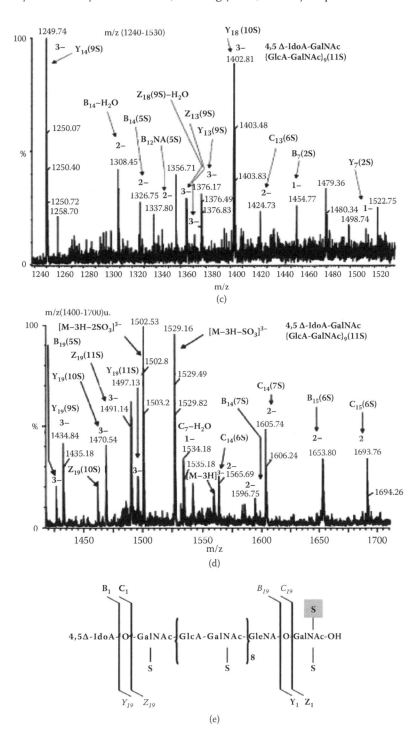

FIGURE 4.8   (continued).

stream to sustain the low EOF and strongly acidic electrolyte conditions (30 mM formic acid, pH 3.2) was required.

4.5.5   FACE-ESI-MS

Unlike the situation in CZE, in frontal analysis capillary electrophoresis (FACE) the sample is continuously injected into the electrophoresis capillary. The FACE mode offers, therefore, a higher sensitivity than the zone electrophoretic mode and reduces the sample handling steps. FACE is also well suited for the online coupling to ESI-MS [250].

By using the online FACE-ESI-MS approach, it was possible to detect a noncovalent complex formed between antithrombin and a sulfated pentasaccharide mimicking the antithrombin-binding sequence in heparin sulfate and to structurally characterize the HS oligosaccharide ligand without the need of previous isolation and purification (Figure 4.9) [250]. For the analysis of this protein–carbohydrate

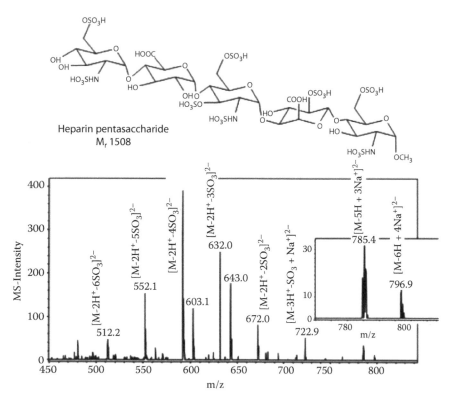

FIGURE 4.9   FACE-ESI–MS of a sulfated oligosaccharide bound to antithrombin. Mass spectrum of a heparin derived pentasacharide involved in complexation with antithrombin and the corresponding structure proposal. FACE conditions: PVA-coated capillary, BGE: 20 mM 6-aminocaproic acid, pH 4.12, CE voltage -10 kV; electrokinetic sample injection, 20 kV for 2 min; Sheath liquid: ACN at 3 mL/min flow rate; ESI voltage 4 kV; target mass 2000 *m/z*. (From S. Fermas et al., *Anal. Chem., 79,* 4987, 2007. With permission.)

complex, antithrombin was incubated with a synthetic heparin pentasaccharide for 1 h at room temperature in 30 mM ammonium hydrogencarbonate, pH 8.5. The protein and the oligosaccharide interacted freely in solution, and the reaction mixture was injected into the capillary at the anodic inlet without prior purification. The injection protocol for the FACE encompassed (1) hydrodynamic injection of the neutral marker benzyl alcohol and the separation electrolyte (30 mbar for 2 s each), and (2) continuous electrokinetic sample injection under a positive voltage of 15 kV. For FACE-ESI-MS the CE capillary was connected to the mass spectrometer using a coaxial sheath-flow interface consisting of organic, aqueous-organic, or aqueous sheath liquids. MS analysis under nondenaturing conditions, favorable to complex preservation, was performed with a sheath liquid consisting of 10 mM ammonium acetate (pH 6.5). Under these conditions, FACE exhibited two migration peaks identified by MS: (1) free protein; and (2) the complex and the free protein coexisting in equilibrium. MS analysis under denaturing conditions, inducing noncovalent complex dissociation, was performed in both positive and negative ionization modes. For the MS analysis in the positive mode, the sheath liquid consisted of 60/40 (v/v) water/acetonitrile with 1% formic acid, whereas for the negative ionization mode acetonitrile was chosen as sheath liquid. Under denaturing conditions FACE-MS characterization of the antithrombin-pentasaccharide complex led to a mass spectrum in positive ion mode, indicating that the detected protein was devoid of bound oligosaccharide, which evidenced that the noncovalent complex was dissociated upon its mixing with the solvent sheath liquid and its subsequent desolvation in the ESI source. As an anionic compound, dissociated HS species migrated gradually toward the ESI-connected cathode driven by the EOF, and it could be detected by switching to negative ion mode ESI. Thus, it was possible to carry on the structural characterization of the heparin bound to a specific protein via complex dissociation in the ESI source and negative ion mode detection. Thus, FACE was demonstrated to allow an online MS analysis not only of the intact noncovalent complex but also of the complex partners.

4.6   GLYCOLIPIDS

Glycolipids represent a class of glycosylated glycerolipids, found primarily in bacteria and plants, as well as glycosylated shingo-lipids (GSLs) found primarily in animals and humans. In the latter case, the sugars are linked to ceramide (Cer). Among all glycolipid species, gangliosides represent the most complex type found in mammalians. They are GSLs that consist of a sialylated (mono- to poly-) oligosaccharide chain of variable length attached to a Cer portion of different compositions with respect to types of sphingoid base and fatty acid residues. This variability of molecular constitution gives rise to a high number of species classified into oligosaccharide series according to the major oligosaccharide core structure (Figure 4.10, Table 4.1) [251,252]. In Svennerholm's nomenclature system [251], G stands for ganglioside; the number of sialic acid residues is identified by M for mono-, D for di-, T for tri-, and Q for tetrasialoglycosphingolipids; and the following number specifies the individual compound. (It initially referred to its migration order in a certain chromatographic system). SA positional isomers are further distinguished by lowercase letters

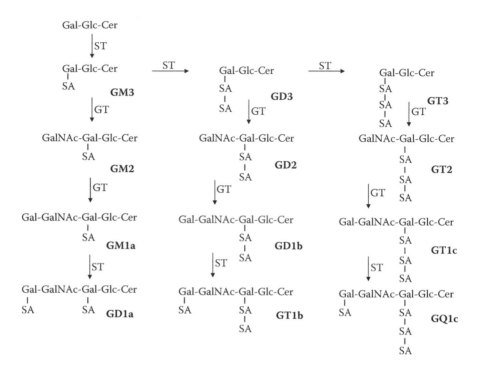

FIGURE 4.10 Biosynthesis pathway of gangliosides. Nomenclature following Svennerholm [251]; SA: sialic acid, ST: sialyltransferase, GT: glycosyltransferase.

(e.g., a, b). In the IUPAC-IUB nomenclature SA positional isomers are distinguished by roman numbers indicating the monosaccharide unit (counted from the Cer-linkage) onto which the Neu5Ac moieties are linked (commonly by α 2-3 linkage).

GSLs are embedded into the plasma membrane with the ceramide portion rooted in the outer leaflet of the membrane, where the hydrophilic oligosaccharide chain protruding into the extracellular environment is part of the glycocalix around a cell. In this way the oligosaccharides are involved in the cell–cell and cell–matrix interaction and adhesion. Gangliosides are enriched in microdomains, functional membrane units, where they participate in cell-to-cell recognition and communication and cell signaling, modulating or triggering various biological events affecting cell proliferation, differentiation, development, apoptosis, and so forth. The central nervous system contains the highest content of gangliosides: neuronal membranes hold at least several times higher concentrations of gangliosides than the extraneural cell types, highlighting their special role at this level [253].

GSLs, however, are found in all vertebrate tissues, and their glycans are part of the serologically active blood group determinants A, B, 0, and Lewis. Ganglioside composition is species and cell type specific and changes specifically during brain development, maturation, aging, and disease or neurodegeneration. GSLs can thus serve as developmental and differentiation markers [254,255], and aberrant GSL glycosylation was observed in all types of experimental human cancers [256,257]. For this reason, GSLs are considered as valuable tissue stage or diagnostic markers and even as potential therapeutic agents [258,259]. Detailed and unambiguous compositional mapping and structural elucidation of individual ganglioside components are therefore of crucial necessity for systematic characterization of ganglioside composition in different tissues in general and different brain regions in particular, or in body fluids in health and disease. Such a study is of major importance for correlating the composition and structure specificity with the functional specialization of the particular region and pathological state, respectively.

In the past few years, determination of glycolipid/ganglioside composition and quantity as well as their distribution and cell surface expression have been achieved primarily by thin-layer chromatographic, immunochemical, and immunohistochemical methods [256]. Such data could, however, be assessed for the major abundant species only due to detection limits of the used methods.

Efficient separation and detailed MS-based structural characterization [261,262] of glycolipids from biological sources are basic prerequisites for further developing research strategies aimed to elucidate the specific functions of each particular structure and to use it, accordingly, as therapeutic agents in treatment of diseases or as specific diagnostic markers [263,266]. For highly sensitive detection and structural analysis of glycolipid species in complex mixtures from biological matrices, CZE-ESI–MS and MS/MS in either off-line or online conjunction and very recently fully automated chip-nanoESI–MS^n combined with computer software platforms for rapid and reliable interpretation of mass spectra represent the newest chosen alternatives.

4.6.1 OFF-LINE CZE-ESI MS

In the analysis of glycolipid mixtures, off-line CE-MS might be regarded as a method of choice (1) for simple mixtures or mixtures for which a partial separation of components is enough to reduce the ion suppression and to enhance single component identification by MS, (2) when flexibility toward system optimization is deemed essential since in off-line method CZE and MS instruments can be adjusted independently and optimized separately, and (3) for cases where rather large amounts of analytes are available since lack of sensitivity is the specific drawback of this approach that limits the extension of its applicability toward minute quantities.

For gangliosides, off-line collection of fractions was found suitable when performed using either a CZE-UV or CZE-LIF prerun to determine the migration times and set the fraction collection interval. Collected fractions may be subsequently analyzed by either ESI or MALDI-MS to identify the partially separated components. When doing sample collection in vials for subsequent nanoESI, a certain limitation lies in the dilution the fractions experience (to about 5 μL) due to simple technical requirements like capillary and electrode immersion and sample handling; in addition, the CZE current becomes interrupted during change of vials. Dealing with GLSs, the analyte concentration must always stay below the limit of micelle formation.

The feasibility of combining high-performance CZE/UV operated in direct polarity (with EOF in a noncoated capillary) in an off-line mode with negative nano(-)ESI-QTOF–MS and CID MS/MS has been tested [257] for partial separation, screening, and sequencing of ganglioside components in a commercially available ganglioside mixture from bovine brain. The chosen BGE for CZE consisted of

TABLE 4.1

Ganglioside Structures Frequently Found in Various Tissues Specified in the Nomenclature of Svennerholm [251] and in the IUPAC-IUB Nomenclature [252]

Svennerholm Nomenclature	IUPAC-IUB Nomenclature
LacCer	Galβ4Glcβ1Cer
GA2, Gg₃Cer	GalNAcβ4Galβ4Glcβ1Cer
GA, Gg₄Cer	Galβ3GalNAcβ4Galβ4Glcβ1Cer
nLc₄Cer	Galβ4GlcNAcβ3Galβ4Glcβ1Cer
Lc₄Cer	Galβ3GlcNAcβ3Galβ4Glcβ1Cer
GM4	I³-α-Neu5Ac-GlcCer
GM3	II³-α-Neu5Ac-LacCer
GD3	II³-α-(Neu5Ac)₂-LacCer
GT3	II³-α-(Neu5Ac)₃-LacCer
GM2	II³-α-Neu5Ac-Gg₃Cer
GD2	II³-α-(Neu5Ac)₂-Gg₃Cer
GM1a	II³-α-Neu5Ac-Gg₄Cer
GM1b	IV³-α-Neu5Ac-Gg₄Cer
LM1	IV³-α-Neu5Ac-Lc₄Cer
GD1a	IV³-α-Neu5Ac,II³-α-Neu5Ac-Gg₄Cer
GD1b	II³-α-(Neu5Ac)₂-Gg₄Cer
LD1	IV³-α-(Neu5Ac)₂-Lc₄Cer
GT1a	IV³-α-(Neu5Ac)₂,II³-α-Neu5Ac-Gg₄Cer
GT1b	IV³-α-Neu5Ac,II³-α-(Neu5Ac)₂-Gg₄Cer
GT1c	II³-α-(Neu5Ac)₃-Gg₄Cer
GQ1b	IV³-α-(Neu5Ac)₂,II³-α-(Neu5Ac)₂-Gg₄Cer
GQ1c	IV³-α-Neu5Ac,II³-α-(Neu5Ac)₃-Gg₄Cer

50 mM aqueous ammonium acetate pH 10.0 and 12.0 (adjusted by 32% ammonia). At pH 10.0, only four major peaks were detected by UV monitoring, which indicated that under these conditions an optimal electrophoretic separation of the components could not be achieved. Obviously, for ganglioside analysis, the separation power of the ESI-compatible buffer systems (lacking borate) is not as high as those of phosphate-borate-based buffers. At this pH value only a rough separation was possible to

attain for gangliosides identical in carbohydrate but differing in the ceramide moiety. By setting the pH value to 12.0, at least 11 different species could be separated as distinct peaks, without jeopardizing structural features of polysialylated species [269]. Besides, at this pH value, except for the ganglioside structures specified by the producer, three minor components, GD3, GQ1b, and GD2, with different electrophoretic migration properties were additionally detected in Cronassial mixture.

Another positive aspect was that ganglioside molecular ions from 10 different fractions, collected after CZE separation at this high pH value, did not show sodium adducts like those collected at pH 10.0. For MS/MS fragmentation experiments, low-energy CID conditions were carefully optimized to generate specific fragmentation patterns indispensable for structural identification. The efficiency of fragmentation plays a crucial role in the assignment of structures from the CZE collected fraction, in which only low amounts of material were present. CID MS/MS experiments revealed that the decomposition of analyte/ammonia clusters and fragmentation could be obtained only under harsh experimental conditions employing 200 V cone potential and 75 eV collision energy. When decreasing the ammonium acetate concentration by adding 50% methanol to the CZE collected fractions, the spray stability was improved, the electrospray and sampling cone potentials could be decreased significantly, and the in-source fragmentation was reduced. Using these conditions elaborated for ganglioside sequencing, it was possible to carry out valuable fragmentation analysis by low-energy CID MS^2 of the precursor ions corresponding to GM1 (d18:1/18:0), GM1 (d18:1/20:0), and GD1 (d18:1/18:0) molecular species.

4.6.2 Online CZE-ESI MS

The utility of CZE in glycolipid analysis can be greatly enhanced by directly online interfacing to ESI-MS [268,269]. In this way, not only can molecular masses of the separated ganglioside components be measured with good accuracy, but also specific fragment ions may be generated by MS/MS from individual components in a complex mixture. This allows deducing the structure of the oligosaccharide core and confirming the composition of the ceramide through its specific Y_0 or Z_0 fragment ion.

Online CZE-ESI–MS was first introduced in ganglioside analysis by employing a direct triaxial sheath flow interface for coupling to a single quadrupole MS [270] tuned in the negative ion mode electrospray. Though for ganglioside CZE-UV separation the best result was obtained using 50 mM borate and 50 mM phosphate buffer containing 20mM α-cyclodextrin at pH 9.9, this system was, certainly, found unsuitable for CZE-ESI–MS because of the incompatibility of the borate/phosphate buffer to electrospray process. Therefore, volatile buffers, such as ammonium acetate or 2-[N-cyclohexylamino]-ethanesulfonic acid, were tested and found to provide an excellent resolution for ganglioside species with different numbers of monosaccharide units.

More recently, a superior variant of the low-makeup CZE-ESI–MS interface was reported [84] and applied to ganglioside separation followed by direct mass analysis. In this configuration, the separation capillary and the sheath-liquid capillary are both tapered and terminated in a tip with a beveled edge (Figure 4.11). The BGE

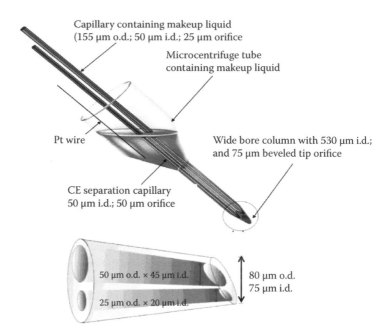

FIGURE 4.11 Schematic representation of the low-makeup flow/beveled tip used as CZE–ESI–MS interface for ganglioside analysis. (From M. C. Tseng, *Anal. Chem., 76,* 6306, 2004. With permission.)

is driven by the electroosmotic flow of the capillary, whereas the flow rate of the makeup liquid is controlled by a syringe pump. This interface allowed the usage of a tip with a larger orifice, which is less prone to clogging than the regular small-sized ones. The main advantage of this configuration resides in the introduction of the makeup flow capillary into the interface, a procedure that enhanced the control of the final spray solution. This system was applied to a commercially available type III ganglioside mixture containing GM1, GD1a, GD1b, and GT1b species purified from bovine brain, which was separated in 40 mM borate and 20mM α-CD in water buffered with ammonia to pH 10.6, under direct CZE polarity conditions (strong EOF) and detected online by a negative ion mode ESI-linear IT-MS, with data collection in SIM mode (Figure 4.12). Interestingly, the interface functioned properly even under the chosen nonvolatile buffer conditions and the four major gangliosides (GM1, GD1a, GD1b, GT1b) were successfully separated. Moreover, as visible in this figure, each ganglioside yielded two peaks, demonstrating that species having the same oligosaccharide core could be distinguished according to their ceramide moieties of (d18:1/18:0) and (d18:1/20:0) constitutions, respectively. In a variation of the beveled-edge theme, reported by the same group [271], the interface contains two columns (a separation and a makeup column). Both columns were tapered, beveled, and combined to yield a dual beveled-edge emitter exhibiting smaller tip, reduced sample dilution, and less clogging probability in comparison with a flat tip (Figure 4.13). This configuration was successfully applied to the same commercial

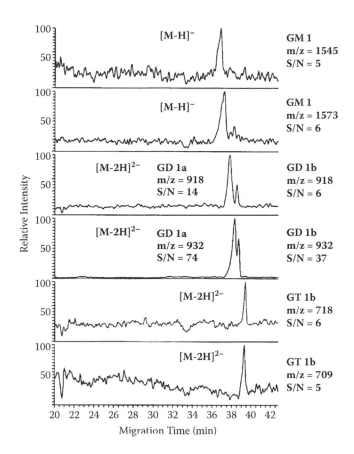

FIGURE 4.12 Extracted ion electropherograms at various mass traces obtained for a type III ganglioside mixture containing GM1, GD1a, GD1b, and GT1b from bovine brain using the low-makeup beveled tip CE-ESI–MS interface in negative ion mode. Mixture concentration: 4 mg/mL; Makeup liquid: 80% 2-propanol with 1% ammonium hydroxide at 400 nL/min flow rate; CE voltage: +20 kV; ESI voltage: –2 kV. (From M. C. Tseng, *Anal. Chem.*, 76, 6306, 2004. With permission.)

type III ganglioside mixture, which was analyzed by CZE-ESI–MS in the same ESI nonfriendly aqueous solution mentioned already including 40 mM borate.

4.6.3 MICROFLUIDICS-BASED PLATFORMS

Currently, a strong trend in analytical sciences is toward high-throughput measurements based on the modern technology achievements in microfluidics, robotics, and computer software for automatic data interpretation. In biomolecule analysis, integrated microfluidic devices for ESI have been demonstrated to provide one of the most rapid, sensitive, and accurate analyses, sometimes without the need of mixture separation prior to MS [272–275]. Some of the recent protocols for glycolipid

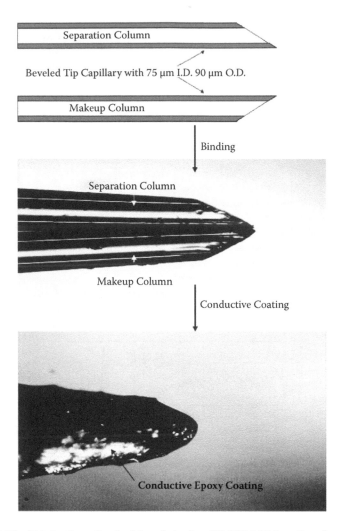

FIGURE 4.13 Fabrication of a dual beveled-edge CE-ESI–MS interface for ganglioside analysis. (From Y. R. Chen et al., *Electrophoresis, 26,* 1376, 2005. With permission.)

analysis included robotized sample delivery by fully automated chip-based nanoESI [276,277] in combination with MS^2 by low-energy CID. This option was guided by several advantages of the system such as (1) high throughput; (2) increase of the sensitivity by drastic reduction of the sample and reagent consumption, sample handling, and potential sample loss; (3) high reproducibility of the experiments; (4) increased ionization efficiency; (5) high signal-to-noise ratio; (6) reduced in-source fragmentation; (7) elimination of possible cross-contamination and carryovers from sample to sample.

However, in this combined strategy a number of limitations were associated with CID-based MS^2: (1) the MS/MS experiment was found unable to provide sufficient diagnostic ions for the assignment of all Neu5Ac positions in polysialylated species;

(2) in the case of GSLs containing modifications such as *O*-Fuc and *O*-Ac, assignment of the attachment sites becomes a challenging if not unfeasible task; and (3) CID MS/MS produces mostly the ions corresponding to intact lipid residue. Ceramide-derived fragment ions are observed only at higher CID energies when ring cleavage ions of the oligosaccharides that complicate the spectrum occur as well. These limitations were recently overcome [274,275,278,279] by combining the robotic chip-based ESI systems with ultrafast CID multistep fragmentations (MS^n) and automatic interpretation of mass spectra by computer software to form all together an integrated high-throughput platform (Figure 4.14). Such a platform encompassing a NanoMate robot coupled to a high-capacity ion trap (HCT) MS was tested on a GT1b polysialylated ganglioside fraction [279]. The fraction dissolved in pure MeOH was infused into HCT MS by chip-nanoESI and subjected to MS^1 screening followed by top-down

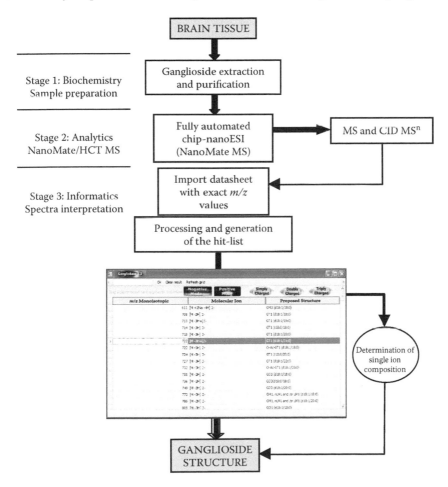

FIGURE 4.14 Schematic of the platform for high-throughput ganglioside analysis based on fully automated chip-nanoESI–MS, CID based MS^n with computer software for mass spectra interpretation.

MS^2-MS^6 analysis of the oligosaccharide sequence and ceramide residue. Within MS^2–MS^5 dissociation events, a complete characterization of the oligosaccharide core including discrimination of sialylation sites was achieved by stepwise sequencing of tri-, di-, mono-, and finally asialo fragment ions. Further, the Cer residue was structurally characterized by CID MS^6 using as a precursor the Cer-associated Y_0 ion produced in the previous fragmentation stage. MS^1–MS^6 analysis was performed in a single experiment, in a high-throughput mode (less than 3 minutes) at a sensitivity situated in the subpicomolar range.

REFERENCES

1. M. M. Fuster and J. D. Esko, The sweet and sour of cancer: glycans as novel therapeutic targets, *Nat. Rev. Cancer 5*: 526 (2005).
2. A. Varki, R. D. Cummings, J. D. Esko, H. Freeze, G. Hart and J. Marth, Essentials of Glycobiology, Cold Spring Harbor Laboratory Press, New York, pp. 537 (1999).
3. M. R. Bond and J. J. Kohler, Chemical methods for glycoprotein discovery, *Curr. Opin. Chem. Biol.* : 52 (2007).
4. Z. El Rassi, Carbohydrate Analysis by Modern Chromatography and Electrophoresis, in *J.Chromatogr. Libr.*, Vol. 66, Elsevier, Amsterdam (2002).
5. H.-J. Gabius, The Sugar Code—Fundamentals in Glycosciences, Wiley-Blackwell, Weinheim (2009).
6. C. Huhn, M. H. J. Selman, L. R. Ruhaak, A. M. Deelder and M. Wuhrer, IgG glycosylation analysis, *Proteomics, 9*: 882 (2009).
7. P. J. Domann, A. C. Pardos-Pardos, D. L. Fernandes, D. I. R. Spencer, C. M. Radcliff, L. Royle, R. A. Dwek and P. M. Rudd, Separation-based glycoprofiling approaches using fluorescent labels, *Practical Proteomics, 2*: 70 (2007).
8. Y. Mechref and M. V. Novotny, Structural investigations of glycoconjugates at high sensitivity, *Chem. Rev., 102*: 321 (2002).
9. S. Amon, A. D. Zamfir and A. Rizzi, Glycosylation analysis of glycoproteins and proteoglycans using capillary electrophoresis-mass spectrometry strategies, *Electrophoresis, 29*: 2485 (2008).
10. C. Campa, A. Coslovi, A. Flamigni and M. Rossi, Overview on advances in capillary electrophoresis-mass spectrometry of carbohydrates: a tabulated review, *Electrophoresis, 27*: 2027 (2006).
11. A. Zamfir and J. Peter-Katalinic, Capillary electrophoresis-mass spectrometry for glycoscreening in biomedical research, *Electrophoresis, 25*: 1949 (2004).
12. W. Morelle, K. Canis, F. Chirat, V. Faid and J.-C. Michalski, Characterization of N-glycans of recombinant human thyrotropin using mass spectrometry, *Proteomics, 6*: 3993 (2006).
13. W. Morelle and J.-C. Michalski, The mass spectrometric analysis of glycoproteins and their glycan structures., *Curr. Anal. Chem., 1*: 29 (2005).
14. W. Morelle and J.-C. Michalski, Sequencing of oligosaccharides derivatized with benzylamine using electrospray ionization-quadrupole time of flight-tandem mass spectrometry, *Electrophoresis 25*: 2144 (2004).
15. Y. Mechref and M. V. Novotny, *Chem. Rev., 102*: 321 (2002).
16. M. Stahl, A. Von Brocke and E. Bayer, in *Carbohydrate Analysis by Modern Chromatography and Electrophoresis*, Vol. 66, Z. El Rassi, Ed., Elsevier, Amsterdam, pp. 961 (2002).

17. D. J. Harvey, Matrix-assisted laser desorption/ionization mass spectrometry of carbohydrates, *Mass Spectrom. Rev., 18*: 349 (1999).

18. R. Schwartz-Albiez, in *The Sugar Code—Fundamentals of Glycosciences*, H.-J. Gabius, Ed., Wiley-Blackwell, Weinheim, 2009, p. 447.

19. T. Hennet, in *The Sugar Code—Fundamentals of Glycosciences*, H.-J. Gabius, Ed., Wiley-Blackwell, Weinheim, 2009, p. 365.

20. F. A. Habermann and F. Sinowatz, in *The Sugar Code—Fundamentals of Glycosciences*, H.-J. Gabius, Ed., Wiley-Blackwell, Weinheim, 2009, p. 403.

21. R. W. Ledeen and G. Wu, in *The Sugar Code—Fundamentals of Glycosciences*, H.-J. Gabius, Ed., Wiley-Blackwell, Weinheim, 2009, p. 495.

22. K. Honke and N. Taniguchi, in *The Sugar Code—Fundamentals of Glycosciences*, H.-J. Gabius, Ed., Wiley-Blackwell, Weinheim, 2009, p. 385.

23. A. J. McCoy, X. Y. Pei, R. Skinner, J.-P. Abrahams and R. W. Carrell, Structure of beta-antithrombin and the effect of glycosylation on antithrombin's heparin affinity and activity, *J. Mol. Biol., 326*: 823 (2003).

24. B. Turk, I. Brieditis, S. C. Bock, S. T. Olson and I. Björk, The oligosaccharide side chain on Asn-135 of alpha-antithrombin, absent in beta-antithrombin, decreases the heparin affinity of the inhibitor by affecting the heparin-induced conformational change, *Biochemistry, 36*: 6682 (1997).

25. A. Plematl, U. M. Demelbauer, D. Josić and A. Rizzi, Determination of the site-specific and isoform-specific glycosylation in human plasma-derived antithrombin by IEF and capillary HPLC-ESI-MS/MS, *Proteomics, 5*: 4026 (2005).

26. T. Edmunds, S. M. Van Patten, J. Pollock, E. Hanson, R. Bernasconi, E. Higgins, P. Manavalan, C. Ziomek, H. Meade, J. M. McPherson and E. S. Cole, Transgenically produced human antithrombin: structural and functional comparison to human plasma-derived antithrombin, *Blood, 91*: 4561 (1998).

27. L. Garone, T. Edmunds, E. Hanson, R. Bernasconi, J. A. Huntington, J. L. Meagher, B. Fan and P. G. W. Gettins, *Biochemistry, 35*: 8881 (1996).

28. R. Jefferis, Glycosylation as a strategy to improve antibody-based therapeutics, *Nat Rev Drug Discov., 8*: 226 (2009).

29. J. Peter-Katalinic, O-glycosylation of proteins, *Methods in Enzymology, 405*: 139 (2005).

30. J. Zaia, Mass spectrometry of oligosaccharides, *Mass. Spectrom. Rev., 23*: 161 (2004).

31. M. Wuhrer, M. I. Catalina, A. M. Deelder and C. H. Hokke, Glycoproteomics based on tandem mass spectrometry of glycopeptides, *J. Chromatogr. B ñ Anal.Techn. Biomed. Life Sci., 849*: 115 (2007).

32. Z. Yang, W. S. Hancock, T. R. Chew and L. Bonilla, *Proteomics, 5*: 3353 (2005).

33. R. Qiu and F. E. Regnier, *Anal. Chem., 77*: 7225 (2005).

34. J. Hirabayasi and K.-I. Kasai, Separation technologies for glycomics, *J. Chromatogr. B, 771*: 67 (2002).

35. M. Geng, X. Zhang, M. Bina and F. Regnier, Proteomics of glycoproteins based on affinity selection of glycopeptides from tryptic digests, *Journal of Chromatography B: Biomedical Sciences and Applications, 752*: 293 (2001).

36. Z. Xu, X. Zhou, H. Lu, N. Wu, H. Zhao, L. Zhang, W. Zhang, Y. L. Liang, L. Wang, Y. Liu, P. Yang and X. Zha, *Proteomics, 7*: 2358 (2007).

37. T. Schwientek, U. Mandel, U. Roth, S. Mueller and F.-G. Hanisch, *Proteomics 7*: 3264 (2007).

38. T. Endo, in *Carbohydrate Analysis by Modern Chromatography and Electrophoresis*, Elsevier Science B.V., 2002, p. 251.

39. A. Monzo, G. K. Bonn and A. Guttman, Lectin-immobilization strategies for affinity purification and separation of glycoconjugates, *TrAC, 26*: 423 (2007).

40. A. Monzo, M. Olajos, L. De Benedictis, Z. Rivera, G. K. Bonn and A. Guttman, Boronic acid lectin affinity chromatography (BLAC). 2. Affinity micropartitioning-mediated comparative glycosylation profiling, *Anal. Bioanal. Chem., 392*: 195 (2008).

41. N. L. Wilson, N. G. Karlsson and N. H. Packer, Enrichment and analysis of glyco-proteins in the proteome, in *Separation Methods in Proteomics*, G. B. Smejkal, and A. Lazareu, Eds., CRC Press LLC, Boca Raton, Fla (2006).

42. K. Yamamoto, T. Tsuji and T. Osawa, in *Protein Protocols Handbook*, J. M. Walker, Ed., Humana Press Inc., Totowa, N. J 2002, p. 917.

43. K. Niikura, R. Kamitani, M. Kurogochi, R. Uematsu, Y. Shinohara, H. Nakagawa, K. Deguchi, K. Monde, H. Kondo and S.-I. Nishimura, Versatile glycoblotting nanopar-ticles for high-throughput protein glycomics, *Chem. Eur. J., 11*: 3825 (2005).

44. H. Zhang, X.-J. Li, D. B. Martin and R. Aebersold, Identification and quantification of N-linked glycoproteins using hydrazide chemistry, stable isotope labeling and mass spectrometry, *Nature Biotechnology, 21*: 660 (2003).

45. H. Zhang and R. Aebersold, in *Methods in Molecular Biology—New and Emerging Proteomic Techniques*, Vol. 328, Humana Press Inc., Totowa, NJ, USA 2006, p. 177.

46. B. Wollscheid, D. Bausch-Fluck, C. Henderson, R. O'Brien, M. Bibel, R. Schiess, R. Aebersold and J. D. Watts, Mass-spectrometric identification and relative quantifica-tion of N-linked cell surface glycoproteins, *Nat Biotechnol., 27*: 378 (2009).

47. R. Nuck, M. Zimmermann, D. Sauvageot, D. Josi and W. Reutter, Optimized deglyco-sylation of glycoproteins by peptide-N4-(N-acetyl-beta-glucosaminyl)-asparagine ami-dase from Flavobacterium meningosepticum, *Glycoconj. J., 7*: 279 (1990).

48. V. Tretter, F. Altmann and L. Marz, Peptide-N4-(N-acetyl-beta-glucosaminyl)asparagine amidase F cannot release glycans with fucose attached alpha 1----3 to the asparagine-linked N-acetylglucosamine residue, *Eur. J. Biochem., 199*: 647 (1991).

49. M. M. Brooks and A. V. Savage, The substrate specificity of the enzyme endo-alpha-N-acetyl-D-galactosaminidase from Diplococcus pneumonia, *Glycoconj. J., 14*: 183 (1997).

50. L. Royle, T. S. Mattu, E. Hart, J. I. Langridge, A. H. Merry, N. Murphy, D. J. Harvey, R. A. Dwek and P. M. Rudd, An analytical and structural database provides a strategy for sequencing O-glycans from microgram quantities of glycoproteins, *Anal. Biochem., 304*: 70 (2002).

51. B. L. Schulz, A. J. Sloane, L. J. Robinson, S. S. Prasad, R. A. Lindner, M. Robinson, P. T. Bye, D. W. Nielson, J. L. Harry, N. H. Packer and N. G. Karlsson, Glycosylation of sputum mucins is altered in cystic fibrosis patients, *Glycobiology, 17*: 698 (2007).

52. J. Gonzales, T. Takao, H. Hori, V. Besada, R. Rodriguez, G. Padron and Y. Shimonishi, A method for determination of N-glycosylation sites in glycoproteins by collision-induced dissociation analysis in fast atom bombardment mass spectrometry: identification of the positions of carbohydrate-linked asparagine in recombinant alpha-amylase by treatment with peptide-N-glycosidase F in 18O-labeled water, *Anal.Biochem., 205*: 151 (1992).

53. L. Xiong and F. E. Regnier, Use of a lectin affinity selector in the search for unusual glycosylation in proteomics, *J. Chromatogr. B ñ Anal. Technol. Biomed. Life Sci. 782*: 405 (2002).

54. L. Ruhaak, C. Huhn, W. J. Waterreus, A. R. de Boer, C. Neusüß, C. Hokke, A. M. Deelder and M. Wuhrer, Hydrophilic interaction chromatography-based high-throughput sample preparation method for N-glycan analysis from total human plasma glycoproteins, *Anal. Chem., 80*: 6119 (2008).

55. N. G. Karlsson, N. L. Wilson, H.-J. Wirth, P. Dawes, J. Hiren and N. H. Packer, Negative ion graphitised carbon nano-liquid chromatography/mass spectrometry increases sen-sitivity for glycoprotein oligosaccharide analysis *Rapid Commun. Mass Spectrom, 18*: 2282 (2004).

56. P. J. Domann, A. C. Pardos-Pardos, D. L. Fernandes, D. I. R. Spencer, C. M. Radcliffe, L. Royle, R. A. Dwek and P. M. Rudd, Separation-based glycoprofiling approaches using fluorescent labels, *Proteomics, 7*: 70 (2007).
57. E. Sisu, W. T. Bosker, W. Norde, T. M. Slaghek, J. W. Timmermans, J. Peter-Katalinić, M. A. Cohen-Stuart and A. D. Zamfir, Electrospray ionization quadrupole time-of-flight tandem mass spectrometric analysis of hexamethylenediamine-modified maltodextrin and dextran,, *Rapid Commun. Mass Spectrom., 20*: 209 (2006).
58. F. N. Lamari, R. Kuhn and N. K. Karamanos, Derivatization of carbohydrates for chromatographic, electrophoretic and mass spectrometric structure analysis, *J. Chromatogr. B 793*: 15 (2003).
59. P. Kang, Y. Mechref, I. Klouckova and M. V. Novotny, Solid-phase permethylation of glycans for mass spectrometric analysis, *Rapid Commun. Mass Spectrom., 19*: 3421 (2005).
60. Y. Mechref, P. Kang and M. V. Novotny, Differentiating structural isomers of sialylated glycans by matrix-assisted laser desorption/ionization time-of-flight/time-of-flight tandem mass spectrometry, *Rapid Commun. Mass Spectrom., 20*: 1381 (2006).
61. Y. Mechref and M. V. Novotny, Glycomic analysis by capillary electrophoresis-mass spectrometry, *Mass Spectrom. Rev., 28*: 207 (2009).
62. S. J. North, P. G. Hitchen, S. M. Haslam and A. Dell, Mass spectrometry in the analysis of N-linked and O-linked glycans, *Curr. Opin. Struct. Biol., 19*: 498 (2009).
63. V. N. Reinhold, B. B. Reinhold and C. E. Costello, Carbohydrate molecular weight profiling, sequence, linkage, and branching data: ES-MS and CID, *Anal. Chem., 67*: 1772 (1995).
64. C. E. Costello, J. M. Contado-Miller and J. F. Cipollo, A glycomic platform for the analysis of permethylated oligosaccharide alditols, *J. Am. Soc. Mass Spectrom., 18*: 1799 (2007).
65. F. Benavente, M. C. Vescina, E. Hernandez, V. Sanz-Nebot, J. Barbosa and N. A. Guzman, Lowering the concentration limits of detection by on-line solid-phase extraction-capillary electrophoresis-electrospray mass spectrometry, *J. Chromatogr. A 1140*: 205 (2007).
66. F. Benavente, E. Hernandez, N. A. Guzman, V. Sanz-Nebot and J. Barbosa, Determination of human erythropoietin by on-line immunoaffinity capillary electrophoresis: a preliminary report, *Anal. Bioanal. Chem., 387*: 2633 (2007).
67. N. A. Guzman and T. M. Phillips, Immunoaffinity capillary electrophoresis for proteomics studies *Anal. Chem., 77*: 60A (2005).
68. N. A. Guzman, T. Blanc and T. M. Phillips, Immunoaffinity capillary electrophoresis as a powerful strategy for the quantification of low-abundance biomarkers, drugs, and metabolites in biological matrices, *Electrophoresis 29*: 3259 (2008).
69. A. von Brocke, G. Nicholson and E. Bayer, Recent advances in capillary electrophoresis/ electrospray-mass spectrometry, *Electrophoresis, 22*: 1251 (2001).
70. G. M. Janini, M. Zhou, L. R. Yu, J. Blonder, M. Gignac, T. P. Conrads, H. J. Issaq and T. D. Veenstra, On-column sample enrichment for capillary electrophoresis sheathless electrospray ionization mass spectrometry: evaluation for peptide analysis and protein identification, *Anal. Chem., 75*: 5984 (2003).
71. J. Hernandez-Borges, C. Neusüss, A. Cifuentes and M. Pelzing, On-line capillary electrophoresis-mass spectrometry for the analysis of biomolecules, *Electrophoresis, 25*: 2257 (2004).
72. P. Schmitt-Kopplin and M. Englmann, Capillary electrophoresis—mass spectrometry: survey on developments and applications 2003-2004, *Electrophoresis, 26*: 1209 (2005).
73. H. Stutz, Advances in the analysis of proteins and peptides by capillary electrophoresis with matrix-assisted laser desorption/ionization and electrospray-mass spectrometry detection, *Electrophoresis, 26*: 1254 (2005).

74. Z. Kele, G. Ferenc, E. Klement, G. K. Toth and T. Janaky, Design and performance of a sheathless capillary electrophoresis/mass spectrometry interface by combining fused-silica capillaries with gold-coated nanoelectrospray tips, *Rapid Commun. Mass Spectrom., 19*: 881 (2005).

75. A. D. Zamfir, N. Dinca, E. Sisu and J. Peter-Katalinic, Copper-coated microsprayer interface for on-line sheathless capillary electrophoresis electrospray mass spectrometry of carbohydrates, *J. Sep. Sci., 29*: 414 (2006).

76. Y. Z. Chang and G. R. Her, Sheathless capillary electrophoresis/electrospray mass spectrometry using a carbon-coated fused-silica capillary, *Anal. Chem., 72*: 626 (2000).

77. A. D. Zamfir, D. G. Seidler, E. Schönherr, H. Kresse and J. Peter-Katalinic, On-line sheathless capillary electrophoresis/nanoelectrospray ionization-tandem mass spectrometry for the analysis of glycosaminoglycan oligosaccharides, *Electrophoresis, 25*: 2010 (2004).

78. J. T. Whitt and M. Moini, Capillary Electrophoresis to Mass Spectrometry Interface Using a Porous Junction, *Anal. Chem., 75*: 2188 (2003).

79. M. Moini, Design and performance of a universal sheathless capillary electrophoresis to mass spectrometry interface using a split-flow technique, *Anal. Chem., 73*: 3497 (2001).

80. M. Moini, Capillary electrophoresis mass spectrometry and its application to the analysis of biological mixtures, *Anal. Bioanal. Chem., 373*: 466 (2002).

81. M. Moini, Simplifying CE-MS operation. 2. Interfacing low-flow separation techniques to mass spectrometry using a porous tip, *Anal. Chem., 79*: 4241 (2007).

82. M. Moini, in *Methods in Molecular Biology* Vol. 276 (Capillary Electrophoresis of Proteins and Peptides), Humana Press Inc., Totowa, NJ, United States, 2004, p. 253.

83. S. Garza and M. Moini, Analysis of Complex Protein Mixtures with Improved Sequence Coverage Using (CE-MS/MS)^n, *Anal. Chem., 78*: 7309 (2006).

84. M. C. Tseng, Y. R. Chen and G. R. Her, A low-makeup beveled tip capillary electrophoresis/electrospray ionization mass spectrometry interface for micellar electrokinetic chromatography and nonvolatile buffer capillary electrophoresis, *Anal. Chem., 76*: 6306 (2004).

85. T. Johnson, J. Bergquist, R. Ekman and E. Nordhoff, A CE-MALDI interface based on the use of prestructured sample supports, *Anal. Chem., 73*: 1670 (2001).

86. A. Zuberovic, M. Wetterhall, J. Hanrieder and J. Bergquist, MALDI-TOF/TOF MS for multiplexed quantification of proteins in human ventricular cerebrospinal fluid, *Electrophoresis 30*: 1836 (2009).

87. J.-M. Busnel, J. Josserand, N. Lion and H. H. Girault, Iontophoretic Fraction Collection for Coupling Capillary Zone Electrophoresis with Matrix-Assisted Laser Desorption/Ionization Mass Spectrometry, *Anal. Chem., 81*: 3867 (2009).

88. O. Müller, F. Foret and B. L. Karger, Design of a high-precision fraction collector for capillary electrophoresis, *Anal. Chem., 67*: 2974 (1995).

89. J. Preisler, P. Hu, T. Rejtar, E. Moskovets and B. L. Karger, Capillary array electrophoresis-MALDI mass spectrometry using a vacuum deposition interface, *Anal. Chem., 74*: 17 (2002).

90. H. Zhang and K. K. C. Yeung, Nanoliter-volume selective sampling of peptides based on isoelectric points for MALDI-MS, *Anal. Chem., 76*: 6814 (2004).

91. C. W. Huck, R. Bakry, L. A. Huber and G. K. Bonn, Progress in capillary electrophoresis coupled to matrix-assisted laser desorption/ionization-time of flight mass spectrometry, *Electrophoresis, 27*: 2063 (2006).

92. Y. Luo, S. Xu, J. W. Schilling, K. H. Lau, J. C. Whitin, T. T. S. Yu and H. J. Cohen, Microfluidic device for coupling capillary electrophoresis and matrix-assisted laser desorption ionization-mass spectrometry, *JALA 14*: 252 (2009).

93. S. Amon, A. Plematl and A. Rizzi, Capillary zone electrophoresis of glycopeptides under controlled electroosmotic flow conditions coupled to electrospray and matrix-assisted laser desorption/ionization mass spectrometry, *Electrophoresis, 27*: 1209 (2006).

94. M. Lechner, A. Seifner and A. Rizzi, Capillary isoelectric focusing hyphenated to single- and multistage MALDI-mass spectrometry using automated sheath-flow assisted sample deposition, *Electrophoresis, 29*: 1974 (2008).

95. G. Choudhary, J. Chakel, W. Hancock, A. Torres-Duarte, G. McMahon and I. Wainer, Investigation of the potential of capillary electrophoresis with off-line matrix-assisted laser desorption/ionization time-of-flight mass spectrometry for clinical analysis: Examination of a glycoprotein factor associated with cancer cachexia, *Anal. Chem., 71*: 855 (1999).

96. H. Suzuki, O. Mueller, A. Guttman and B. L. Karger, Analysis of 1-aminopyrene-3,6,8-trisulfonate-derivatized oligosaccharides by capillary electrophoresis with matrix-assisted laser desorption/ionization time-of-flight mass spectrometry, *Anal. Chem., 69*: 4554 (1997).

97. H. J. An, A. H. Franz and C. B. Lebrilla, Improved capillary electrophoretic separation and mass spectrometric detection of oligosaccharides, *J. Chromatogr. A, 1004*: 121 (2003).

98. S. I. Snovida, V. C. Chen, O. Krokhin and H. Perreault, Isolation and identification of sialylated glycopeptides from bovine a1-acid glycoprotein by off-line capillary electrophoresis MALDI-TOF mass spectrometry, *Anal. Chem., 78*: 6556 (2006).

99. S. J. North, P. G. Hitchen, S. M. Haslam and A. Dell, Mass spectrometry in the analysis of N-linked and O-linked glycans, *Curr. Opin. Struct. Biol., 19*: 498 (2009).

100. J. J. Wolff, I. J. Amster, L. Chi and R. J. Linhardt, Electron detachment dissociation of glycosaminoglycan tetrasaccharides, *J. Am. Soc. Mass Spectrom., 18*: 234 (2007).

101. F. E. Leach, J. J. Wolff, T. N. Laremore, R. J. Linhardt and I. J. Amster, Evaluation of the experimental parameters which control electron detachment dissociation, and their effect on the fragmentation efficiency of glycosaminoglycan carbohydrates, *Int. J. Mass Spectrom., 276*: 110 (2008).

102. J. J. Wolff, T. N. Laremore, A. M. Busch, R. J. Linhardt and I. J. Amster, Electron detachment dissociation of dermatan sulfate oligosaccharides, *J. Am. Soc. Mass Spectrom., 19*: 294 (2008).

103. J. J. Wolff, T. N. Laremore, A. M. Busch, R. J. Linhardt and I. J. Amster, Influence of charge state and sodium cationization on the electron detachment dissociation and infrared multiphoton dissociation of glycosaminoglycan oligosaccharides, *J. Am. Soc. Mass Spectrom., 19*: 790 (2008).

104. G. C. Gil, W. H. Velander and K. E. Van Cott, N-glycosylation microheterogeneity and site occupancy of an Asn-X-Cys sequon in plasma-derived and recombinant protein C, *Proteomics, 9*: 2555 (2009).

105. J. F. Valliere-Douglass, P. Kodama, M. Mujacic, L. Brady, W. Wang, A. Wallace, B. Yan, P. Reddy, M. J. Treuheit and A. Balland, Asparagine-linked oligosaccharides present on a non-consensus amino acid sequence in the CH1 domain of human antibodies, *J. Biol. Chem., 284*: 32493 (2009).

106. S. Kamoda and K. Kakehi, Evaluation of glycosylation for quality assurance of antibody pharmaceuticals by capillary electrophoresis, *Electrophoresis, 29*: 3595 (2008).

107. C. Neusüss and M. Pelzing, in *Methods in Molecular Biology (Mass Spectrometry of Proteins and Peptides)*, Vol. 492, Humana Press Inc., Totowa, NJ, United States, 2009, p. 201.

108. H. Stutz, Protein attachment onto silica surfaces—a survey of molecular fundamentals, resulting effects and novel preventive strategies in CE, *Electrophoresis, 30*: 2032 (2009).

109. U. M. Demelbauer, A. Plematl, L. Kremser, G. Allmaier, D. Josic and A. Rizzi, Characterization of glyco isoforms in plasma-derived human antithrombin by on-line capillary zone electrophoresis-electrospray ionization-quadrupole ion trap-mass spectrometry of the intact glycoproteins, *Electrophoresis, 25*: 2026 (2004).

110. V. Sanz-Nebot, E. Balaguer, F. Benavente, C. Neusüß and J. Barbosa, Characterization of transferrin glycoforms in human serum by CE-UV and CE-ESI-MS, *Electrophoresis, 28*: 1949 (2007).

111. E. Balaguer and C. Neusüß, Glycoprotein characterization combining intact protein and glycan analysis by capillary electrophoresis-electrospray ionization-mass spectrometry, *Anal. Chem., 78*: 5384 (2006).

112. E. Balaguer and C. Neusüß, Intact glycoform characterization of erythropoietin-? and erythropoietin-? by CZE-ESI-TOF-MS, *Chromatographia, 64*: 351 (2006).

113. E. Balaguer, U. Demelbauer, M. Pelzing, V. Sanz-Nebot, J. Barbosa and C. Neusüß, Glycoform characterization of erythropoietin combining glycan and intact protein analysis by capillary electrophoresis—Electrospray—time-of-flight mass spectrometry, *Electrophoresis, 27*: 2638 (2006).

114. B. Yu, H. Cong, H. Liu, Y. Li and F. Liu, Ionene-dynamically coated capillary for analysis of urinary and recombinant human erythropoietin by capillary electrophoresis and online electrospray ionization mass spectrometry, *J. Sep. Sci., 28*: 2390 (2005).

115. C. Neusuess, U. Demelbauer and M. Pelzing, Glycoform characterization of intact erythropoietin by capillary electrophoresis-electrospray-time of flight-mass spectrometry, *Electrophoresis, 26*: 1442 (2005).

116. C. Neusüß, U. Demelbauer and M. Pelzing, Glycoform characterization of intact erythropoietin by capillary electrophoresis-electrospray-time of flight-mass spectrometry, *Electrophoresis, 26*: 1442 (2005).

117. E. Gimenez, F. Benavente, J. Barbosa and V. Sanz-Nebot, Analysis of intact erythropoietin and NESP by CE-electrspray- ion trap MS, *Electrophoresis, 29*: 2161 (2008).

118. A. Puerta and J. Bergquist, Development of a CE-MS method to analyze components of the potential biomarker vascular endothelial growth factor 165, *Electrophoresis, 30*: 2355 (2009).

119. D. Thakur, T. Rejtar, B. L. Karger, N. J. Washburn, C. J. Bosques, N. S. Gunay, Z. Shriver and G. Venkataraman, Profiling of glycoforms of the intact alpha-subunit of recombinant human choronic gonadotropin by high-resolution capillary electrophoresis-mass spectrometry, *Anal. Chem., 81*: 899 (2009).

120. Y. Zheng, H. Li, Z. Guo, J.-M. Lin and Z. Cai, Chip-based CE coupled to a quadrupole TOF mass spectrometer for the analysis of a glycopeptide, *Electrophoresis, 28*: 1305 (2007).

121. B. Domon and C. E. Costello, A systematic nomenclature for carbohydrate fragmentations in FAB-MS/MS spectra of glycoconjugates, *Glycoconj. J., 5*: 397 (1988).

122. L. Bindila, R. Almeida, A. Sterling, M. Allen, J. Peter-Katalinic and A. Zamfir, Off-line capillary electrophoresis/fully automated nanoelectrospray chip quadrupole time-of-flight mass spectrometry and tandem mass spectrometry for glycoconjugate analysis, *J. Mass Spectrom., 39*: 1190 (2004).

123. M. Wuhrer, J. C. Stam, F. E. van de Geijn, C. A. Koeleman, C. T. Verrips, R. J. Dolhain, C. H. Hokke and A. M. Deelder, Glycosylation profiling of immunoglobulin G (IgG) subclasses from human serum, *Proteomics, 7*: 4070 (2007).

124. L. A. Gennaro, O. Salas-Solano and S. Ma, Capillary electrophoresis-mass spectrometry as a characterization tool for therapeutic proteins, *Anal. Biochem., 355*: 249 (2006).

125. E. Bonneil, N. M. Young, H. Lis, N. Sharon and P. Thibault, Probing genetic variation and glycoform distribution in lectins of the Erythrina genus by mass spectrometry, *Arch. Biochem. Biophys., 426*: 241 (2004).

126. J. Li, J. F. Kelly, I. Chernushevich, D. J. Harrison and P. Thibault, Separation and identification of peptides from gel-isolated membrane proteins using a microfabricated device for combined capillary electrophoresis/nanoelectrospray mass spectrometry, *Anal. Chem., 72*: 599 (2000).

127. T. Liu, J.-D. Li, R. Zeng, X.-X. Shao, K.-Y. Wang and Q.-C. Xia, Capillary electrophoresis-electrospray mass spectrometry for the characterization of high-mannose-type N-glycosylation and differential oxidation in glycoproteins by charge reversal and protease/glycosidase digestion, *Anal. Chem., 73*: 5875 (2001).

128. K. Sandra, F. Lynen, B. Devreese, J. Van Beeumen and P. Sandra, On-column sample enrichment for the high-sensitivity sheath-flow CE-MS analysis of peptides, *Anal. Bioanal. Chem., 385*: 671 (2006).

129. C.-C. Lai and G.-R. Her, Analysis of phospholipase A2 glycosylation patterns from venom of individual bees by capillary electrophoresis/electrospray ionization mass spectrometry using an ion trap mass spectrometer, *Rapid Commun. Mass Spectrom., 14*: 2012 (2000).

130. A. Zamfir and J. Peter-Katalinic, Glycoscreening by on-line sheathless capillary electrophoresis/electrospray ionization-quadrupole time of flight-tandem mass spectrometry, *Electrophoresis, 22*: 2448 (2001).

131. A. Zamfir, S. König, J. Althoff and J. Peter-Katalinc, Capillary electrophoresis and off-line capillary electrophoresis-electrospray ionization quadrupole time-of-flight tandem mass spectrometry of carbohydrates, *J. Chromatogr. A, 895*: 291 (2000).

132. L. Bindila, J. Peter-Katalinic and A. Zamfir, Sheathless reverse-polarity capillary electrophoresis-electrospray-mass spectrometry for analysis of underivatized glycoconjugates, *Electrophoresis, 26*: 1488 (2005).

133. Z. El Rassi, in *Adv. in Chromatogr.*, Vol. 34, P. R. Brown, and E. Grushka, Eds., Marcel Dekker, New York, 1994, p. 177.

134. Z. El Rassi, Recent developments in capillary electrophoresis and capillary electrochromatography of carbohydrate species, *Electrophoresis, 20*: 3134 (1999).

135. A. Guttman, F.-T. A. Chen and R. A. Evangelista, High-resolution capillary gel electrophoresis of reducing oligosaccharides labeled with 1-aminopyrene-3,6,8-trisulfonate, *Anal. Biochem., 233*: 234 (1996).

136. N. Callewaert, G. Geysens, F. Molemans and R. Contreras, Ultrasensitive profiling and sequencing of N-linked oligosaccharides using standard DNA-sequencing equipment, *Glycobiology, 11*: 275 (2001).

137. N. Callewaert, R. Contreras, L. Mitnik-Gankin, L. Carey, P. Matsudaira and D. Ehrlich, Total serum protein N-glycome profiling on a CE-microfluidics platform, *Electrophoresis, 25* (2004).

138. W. Laroy, R. Contreras and N. Callewaert, Glycome mapping on DNA sequencing equipment, *Nature protocols, 1*: 397 (2006).

139. J. Schwarzer, E. Rapp and U. Reichl, N-glycan analysis by CGE-LIF: profiling influenza A virus hemagglutinin N-glycosylation during vaccine production, *Electrophoresis, 29*: 4203 (2008).

140. J. B. Briggs, R. G. Keck, S. Ma, W. Lau and A. J. S. Jones, An analytical system for the characterization of highly heterogeneous mixtures of N-linked oligosaccharides, *Anal. Biochem., 389*: 40 (2009).

141. L. A. Gennaro, J. Delaney, P. Vouros, D. J. Harvey and B. Domon, Capillary electrophoresis/electrospray ion trap mass spectrometry for the analysis of negatively charged derivatized and underivatized glycans, *Rapid Commun. Mass Spectrom., 16*: 192 (2002).

142. M. A. Kabel, W. H. Heijnis, E. J. Bakx, R. Kuijpers, A. G. J. Voragen and H. A. Schols, Capillary electrophoresis fingerprinting, quantification and mass-identification of various 9-aminopyrene-1,4,6-trisulfonate-derivatized oligomers derived from plant polysaccharides, *J. Chromatogr. A 1137*: 119 (2006).

143. M. Nakano, D. Higo, E. Arai, T. Nakagawa, K. Kakehi, N. Taniguchi and A. Kondo, Capillary electrophoresis—electrospray ionization mass spectrometry for rapid and sensitive N-glycan analysis of glycoproteins as 9-fluorenylmethyl derivatives, *Glycobiology, 19*: 135 (2009).

144. M. V. Novotny, Glycoconjugate analysis by capillary electrophoresis, *Methods in Enzymology, 271*: 319 (1996).

145. Y. Mechref, J. Muzikar and M. V. Novotny, Comprehensive assessment of N-glycans derived from a murine monoclonal antibody: A case for multimethodological approach, *Electrophoresis 26*: 2034 (2005).

146. M. Wuhrer, A. M. Deelder and C. H. Hokke, Protein glycosylation analysis by liquid chromatography-mass spectrometry, *J. Chromatogr. B Analyt. Technol. Biomed. Life Sci., 825*: 124 (2005).

147. S. Hase, T. Ibuki and T. Ikenaka, Reexamination of the pyridylamination used for fluorescence labeling of oligosaccharides and its application to glycoproteins, *J. Biochem., 95*: 197 (1984).

148. S. Natsuka, S. Hase and T. Ikenaka, Fluorescence method for the structural analysis of oligomannose-type sugar chains by partial acetolysis, *Anal. Biochem., 167*: 154 (1987).

149. N. Tomiya, J. Awaya, M. Kurono, S. Endo, Y. Arata and N. Takahashi, Analyses of N-linked oligosaccharides using a two-dimensional mapping technique, *Anal. Biochem., 171*: 73 (1988).

150. M. J. Davies and E. F. Hounsell, Comparison of separation modes of high-performance liquid chromatography for the analysis of glycoprotein- and proteoglycan-derived oligosaccharides, *J. Chromatogr. A, 720*: 227 (1996).

151. M. Pabst, D. Kolarich, G. Poltl, T. Dalik, G. Lubec, A. Hofinger and F. Altmann, Comparison of fluorescent labels for oligosaccharides and introduction of a new postlabeling purification method, *Anal. Biochem., 384*: 263 (2009).

152. K. R. Anumula, Advances in fluorescence derivatization methods for high-performance liquid chromatographic analysis of glycoprotein carbohydrates, *Anal. Biochem., 350*: 1 (2006).

153. P. Hermentin, R. Witzel, J. F. Vliegenthart, J. P. Kamerling, M. Nimtz and H. S. Conradt, A strategy for the mapping of N-glycans by high-pH anion-exchange chromatography with pulsed amperometric detection, *Anal. Biochem., 203*: 281 (1992).

154. M. P. Gillmeister, N. Tomiya, S. J. Jacobia, Y. C. Lee, S. F. Gorfien and M. J. Betenbaugh, An HPLC-MALDI MS method for N-glycan analyses using smaller size samples: application to monitor glycan modulation by medium conditions, *Glycoconj. J., 26*: 1135 (2009).

155. S. Maslen, P. Sadowski, A. Adam, K. Lilley and E. Stephens, Differentiation of isomeric N-glycan structures by normal-phase liquid chromatography-MALDI-TOF/TOF tandem mass spectrometry, *Anal. Chem., 78*: 8491 (2006).

156. C. Bruggink, R. Maurer, H. Herrmann, S. Cavalli and F. Hoefler, Analysis of carbohydrates by anion exchange chromatography and mass spectrometry, *J. Chromatogr. A, 1085*: 104 (2005).

157. C. Bruggink, M. Wuhrer, C. A. Koeleman, V. Barreto, Y. Liu, C. Pohl, A. Ingendoh, C. H. Hokke and A. M. Deelder, Oligosaccharide analysis by capillary-scale high-pH anion-exchange chromatography with on-line ion-trap mass spectrometry, *J. Chromatogr. B Analyt. Technol. Biomed. Life Sci., 829*: 136 (2005).

158. G. Chataigne, F. Couderc and V. Poinsot, Polysaccharides analysis of sinorhizobial capside by on-line anion exchange chromatography with pulsed amperometric detection and mass spectrometry coupling, *J. Chromatogr. A, 1185*: 241 (2008).

159. T. L. Delattc, M. H. Selman, H. Schluepmann, G. W. Somsen, S. C. Smeekens and G. J. de Jong, Determination of trehalose-6-phosphate in Arabidopsis seedlings by successive extractions followed by anion exchange chromatography-mass spectrometry, *Anal. Biochem.*, *389*: 12 (2009).

160. S. Hase, Precolumn derivatization for chromatographic and electrophoretic analyses of carbohydrates, *J. Chromatogr. A, 720*: 137 (1996).

161. N. Takahashi, H. Nakagawa, K. Fujikawa, Y. Kawamura and N. Tomiya, Three-dimensional elution mapping of pyridylaminated N-linked neutral and sialyl oligosaccharides, *Anal. Biochem., 226*: 139 (1995).

162. Y. Takegawa, K. Deguchi, S. Ito, S. Yoshioka, A. Sano, K. Yoshinari, K. Kobayashi, H. Nakagawa, K. Monde and S. Nishimura, Assignment and quantification of 2-amino-pyridine derivatized oligosaccharide isomers coeluted on reversed-phase HPLC/MS by MSn spectral library, *Anal. Chem., 76*: 7294 (2004).

163. H. Nakagawa, M. Hato, Y. Takegawa, K. Deguchi, H. Ito, M. Takahata, N. Iwasaki, A. Minami and S. Nishimura, Detection of altered N-glycan profiles in whole serum from rheumatoid arthritis patients, *J. Chromatogr. B Analyt. Technol. Biomed. Life Sci., 853*: 133 (2007).

164. Y. Kanie, M. Yamamoto-Hino, Y. Karino, H. Yokozawa, S. Nishihara, R. Ueda, S. Goto and O. Kanie, Insight into the regulation of glycan synthesis in Drosophila chaoptin based on mass spectrometry, *PLoS One, 4*: e5434 (2009).

165. L. A. Gennaro, D. J. Harvey and P. Vouros, Reversed-phase ion-pairing liquid chromatography/ion trap mass spectrometry for the analysis of negatively charged, derivatized glycans, *Rapid Commun. Mass Spectrom., 17*: 1528 (2003).

166. X. Chen and G. C. Flynn, Analysis of N-glycans from recombinant immunoglobulin G by on-line reversed-phase high-performance liquid chromatography/mass spectrometry, *Anal. Biochem., 370*: 147 (2007).

167. X. Chen and G. C. Flynn, Gas-phase oligosaccharide nonreducing end (GONE) sequencing and structural analysis by reversed phase HPLC/mass spectrometry with polarity switching, *J. Am. Soc. Mass Spectrom., 20*: 1821 (2009).

168. J. Zaia, On-line separations combined with MS for analysis of glycosaminoglycans, *Mass Spectrom. Rev., 28*: 254 (2009).

169. N. H. Packer, M. A. Lawson, D. R. Jardine, J. C. Sanchez and A. A. Gooley, Analyzing glycoproteins separated by two-dimensional gel electrophoresis, *Electrophoresis, 19*: 981 (1998).

170. N. Kawasaki, M. Ohta, S. Hyuga, O. Hashimoto and T. Hayakawa, Analysis of carbohydrate heterogeneity in a glycoprotein using liquid chromatography/mass spectrometry and liquid chromatography with tandem mass spectrometry, *Anal. Biochem., 269*: 297 (1999).

171. K. A. Thomsson, N. G. Karlsson and G. C. Hansson, Liquid chromatography-electrospray mass spectrometry as a tool for the analysis of sulfated oligosaccharides from mucin glycoproteins, *J. Chromatogr. A, 854*: 131 (1999).

172. R. P. Estrella, J. M. Whitelock, N. H. Packer and N. G. Karlsson, Graphitized carbon LC-MS characterization of the chondroitin sulfate oligosaccharides of aggrecan, *Anal. Chem., 79*: 3597 (2007).

173. S. Itoh, N. Kawasaki, N. Hashii, A. Harazono, Y. Matsuishi, T. Hayakawa and T. Kawanishi, N-linked oligosaccharide analysis of rat brain Thy-1 by liquid chromatography with graphitized carbon column/ion trap-Fourier transform ion cyclotron resonance mass spectrometry in positive and negative ion modes, *J. Chromatogr. A, 1103*: 296 (2006).

174. S. Itoh, N. Kawasaki, M. Ohta, M. Hyuga, S. Hyuga and T. Hayakawa, Simultaneous microanalysis of N-linked oligosaccharides in a glycoprotein using microbore graphitized carbon column liquid chromatography-mass spectrometry, *J. Chromatogr. A, 968*: 89 (2002).

175. H. Karlsson, J. M. Larsson, K. A. Thomsson, I. Hard, M. Backstrom and G. C. Hansson, High-throughput and high-sensitivity nano-LC/MS and MS/MS for O-glycan profiling, *Methods Mol. Biol., 534*: 117 (2009).

176. N. G. Karlsson, B. L. Schulz and N. H. Packer, Structural determination of neutral O-linked oligosaccharide alditols by negative ion LC-electrospray-MSn, *J. Am. Soc. Mass Spectrom., 15*: 659 (2004).

177. N. G. Karlsson, B. L. Schulz, N. H. Packer and J. M. Whitelock, Use of graphitised carbon negative ion LC-MS to analyse enzymatically digested glycosaminoglycans, *J. Chromatogr. B Analyt. Technol. Biomed. Life Sci., 824*: 139 (2005).

178. N. G. Karlsson, N. L. Wilson, H. J. Wirth, P. Dawes, H. Joshi and N. H. Packer, Negative ion graphitised carbon nano-liquid chromatography/mass spectrometry increases sensitivity for glycoprotein oligosaccharide analysis, *Rapid Commun. Mass Spectrom., 18*: 2282 (2004).

179. K. A. Thomsson, H. Karlsson and G. C. Hansson, Sequencing of sulfated oligosaccharides from mucins by liquid chromatography and electrospray ionization tandem mass spectrometry, *Anal. Chem., 72*: 4543 (2000).

180. N. L. Wilson, L. J. Robinson, A. Donnet, L. Bovetto, N. H. Packer and N. G. Karlsson, Glycoproteomics of milk: differences in sugar epitopes on human and bovine milk fat globule membranes, *J. Proteome Res., 7*: 3687 (2008).

181. M. Backstrom, K. A. Thomsson, H. Karlsson and G. C. Hansson, Sensitive liquid chromatography-electrospray mass spectrometry allows for the analysis of the O-glycosylation of immunoprecipitated proteins from cells or tissues: application to MUC1 glycosylation in cancer, *J. Proteome Res., 8*: 538 (2009).

182. A. Harazono, N. Kawasaki, S. Itoh, N. Hashii, Y. Matsuishi-Nakajima, T. Kawanishi and T. Yamaguchi, Simultaneous glycosylation analysis of human serum glycoproteins by high-performance liquid chromatography/tandem mass spectrometry, *J. Chromatogr. B Analyt. Technol. Biomed. Life Sci., 869*: 20 (2008).

183. N. Hashii, N. Kawasaki, S. Itoh, A. Harazono, Y. Matsuishi, T. Hayakawa and T. Kawanishi, Specific detection of Lewis x-carbohydrates in biological samples using liquid chromatography/multiple-stage tandem mass spectrometry, *Rapid Commun. Mass Spectrom, 19*: 3315 (2005).

184. N. Kawasaki, Y. Haishima, M. Ohta, S. Itoh, M. Hyuga, S. Hyuga and T. Hayakawa, Structural analysis of sulfated N-linked oligosaccharides in erythropoietin, *Glycobiology, 11*: 1043 (2001).

185. N. Kawasaki, S. Itoh, M. Ohta and T. Hayakawa, Microanalysis of N-linked oligosaccharides in a glycoprotein by capillary liquid chromatography/mass spectrometry and liquid chromatography/tandem mass spectrometry, *Anal. Biochem., 316*: 15 (2003).

186. L. R. Ruhaak, A. M. Deelder and M. Wuhrer, Oligosaccharide analysis by graphitized carbon liquid chromatography-mass spectrometry, *Anal. Bioanal. Chem., 394*: 163 (2009).

187. B. Barroso, M. Didraga and R. Bischoff, Analysis of proteoglycans derived sulphated disaccharides by liquid chromatography/mass spectrometry, *J. Chromatogr. A, 1080*: 43 (2005).

188. B. Barroso, R. Dijkstra, M. Geerts, F. Lagerwerf, P. van Veelen and A. de Ru, On-line high-performance liquid chromatography/mass spectrometric characterization of native oligosaccharides from glycoproteins, *Rapid Commun. Mass Spectrom, 16*: 1320 (2002).

189. C. Antonio, C. Pinheiro, M. M. Chaves, C. P. Ricardo, M. F. Ortuno and J. Thomas-Oates, Analysis of carbohydrates in Lupinus albus stems on imposition of water deficit, using porous graphitic carbon liquid chromatography-electrospray ionization mass spectrometry, *J. Chromatogr. A, 1187*: 111 (2008).
190. S. Robinson, E. Bergstrom, M. Seymour and J. Thomas-Oates, Screening of underivatized oligosaccharides extracted from the stems of Triticum aestivum using porous graphitized carbon liquid chromatography-mass spectrometry, *Anal. Chem., 79*: 2437 (2007).
191. A. Antonopoulos, P. Favetta, W. Helbert and M. Lafosse, On-line liquid chromatography-electrospray ionisation mass spectrometry for kappa-carrageenan oligosaccharides with a porous graphitic carbon column, *J. Chromatogr. A, 1147*: 37 (2007).
192. S. Toegel, M. Pabst, S. Q. Wu, J. Grass, M. B. Goldring, C. Chiari, A. Kolb, F. Altmann, H. Viernstein and F. M. Unger, Phenotype-related differential alpha-2,6- or alpha-2,3-sialylation of glycoprotein N-glycans in human chondrocytes, *Osteoarthritis Cartilage* (2009).
193. C. S. Chu, M. R. Ninonuevo, B. H. Clowers, P. D. Perkins, H. J. An, H. Yin, K. Killeen, S. Miyamoto, R. Grimm and C. B. Lebrilla, Profile of native N-linked glycan structures from human serum using high performance liquid chromatography on a microfluidic chip and time-of-flight mass spectrometry, *Proteomics, 9*: 1939 (2009).
194. M. Ninonuevo, H. An, H. Yin, K. Killeen, R. Grimm, R. Ward, B. German and C. Lebrilla, Nanoliquid chromatography-mass spectrometry of oligosaccharides employing graphitized carbon chromatography on microchip with a high-accuracy mass analyzer, *Electrophoresis, 26*: 3641 (2005).
195. M. R. Ninonuevo, P. D. Perkins, J. Francis, L. M. Lamotte, R. G. LoCascio, S. L. Freeman, D. A. Mills, J. B. German, R. Grimm and C. B. Lebrilla, Daily variations in oligosaccharides of human milk determined by microfluidic chips and mass spectrometry, *J. Agric. Food Chem., 56*: 618 (2008).
196. M. A. Bynum, H. Yin, K. Felts, Y. M. Lee, C. R. Moncll and K. Killeen, Characterization of IgG N-glycans employing a microfluidic chip that integrates glycan cleavage, sample purification, LC separation, and MS detection, *Anal. Chem., 81*: 8818 (2009).
197. Q. Luo, T. Rejtar, S. L. Wu and B. L. Karger, Hydrophilic interaction 10 microm I.D. porous layer open tubular columns for ultratrace glycan analysis by liquid chromatography-mass spectrometry, *J. Chromatogr. A, 1216*: 1223 (2009).
198. M. Pabst and F. Altmann, Influence of electrosorption, solvent, temperature, and ion polarity on the performance of LC-ESI-MS using graphitic carbon for acidic oligosaccharides, *Anal. Chem., 80*: 7534 (2008).
199. M. S. Bereman, T. I. Williams and D. C. Muddiman, Development of a nanoLC LTQ orbitrap mass spectrometric method for profiling glycans derived from plasma from healthy, benign tumor control, and epithelial ovarian cancer patients, *Anal. Chem., 81*: 1130 (2009).
200. M. Pabst, J. S. Bondili, J. Stadlmann, L. Mach and F. Altmann, Mass + retention time = structure: a strategy for the analysis of N-glycans by carbon LC-ESI-MS and its application to fibrin N-glycans, *Anal. Chem., 79*: 5051 (2007).
201. J. Stadlmann, M. Pabst, D. Kolarich, R. Kunert and F. Altmann, Analysis of immunoglobulin glycosylation by LC-ESI-MS of glycopeptides and oligosaccharides, *Proteomics, 8*: 2858 (2008).
202. H. Ito, K. Yamada, K. Deguchi, H. Nakagawa and S. Nishimura, Structural assignment of disialylated biantennary N-glycan isomers derivatized with 2-aminopyridine using negative-ion multistage tandem mass spectral matching, *Rapid Commun. Mass Spectrom., 21*: 212 (2007).

203. A. Broberg, High-performance liquid chromatography/electrospray ionization ion-trap mass spectrometry for analysis of oligosaccharides derivatized by reductive amination and N,N-dimethylation, *Carbohydr. Res., 342*: 1462 (2007).

204. L. P. Brull, V. Kovacik, J. E. Thomas-Oates, W. Heerma and J. Haverkamp, Sodium-cationized oligosaccharides do not appear to undergo 'internal residue loss' rearrangement processes on tandem mass spectrometry, *Rapid Commun. Mass Spectrom., 12*: 1520 (1998).

205. M. Wuhrer, C. A. Koeleman and A. M. Deelder, Hexose rearrangements upon fragmentation of N-glycopeptides and reductively aminated N-glycans, *Anal. Chem., 81*: 4422 (2009).

206. D. J. Harvey, T. S. Mattu, M. R. Wormald, L. Royle, R. A. Dwek and P. M. Rudd, "Internal residue loss": rearrangements occurring during the fragmentation of carbohydrates derivatized at the reducing terminus, *Anal. Chem., 74*: 734 (2002).

207. M. Wuhrer, A. R. de Boer and A. M. Deelder, Structural glycomics using hydrophilic interaction chromatography (HILIC) with mass spectrometry, *Mass Spectrom. Rev., 28*: 192 (2009).

208. P. J. Domann, A. C. Pardos-Pardos, D. L. Fernandes, D. I. Spencer, C. M. Radcliffe, L. Royle, R. A. Dwek and P. M. Rudd, Separation-based glycoprofiling approaches using fluorescent labels, *Proteomics, 7 Suppl 1*: 70 (2007).

209. M. Wuhrer, C. A. Koeleman, A. M. Deelder and C. H. Hokke, Normal-phase nanoscale liquid chromatography-mass spectrometry of underivatized oligosaccharides at low-femtomole sensitivity, *Anal. Chem., 76*: 833 (2004).

210. M. Wuhrer, C. A. Koeleman, C. H. Hokke and A. M. Deelder, Nano-scale liquid chromatography-mass spectrometry of 2-aminobenzamide labeled oligosaccharides at low femtomole sensitivity, *Int. J. Mass Spec., 232*: 51 (2004).

211. J. Zhao, W. Qiu, D. M. Simeone and D. M. Lubman, N-linked glycosylation profiling of pancreatic cancer serum using capillary liquid phase separation coupled with mass spectrometric analysis, *J. Proteome Res., 6*: 1126 (2007).

212. G. O. Staples, M. J. Bowman, C. E. Costello, A. M. Hitchcock, J. M. Lau, N. Leymarie, C. Miller, H. Naimy, X. Shi and J. Zaia, A chip-based amide-HILIC LC/MS platform for glycosaminoglycan glycomics profiling, *Proteomics, 9*: 686 (2009).

213. Y. Takegawa, H. Ito, T. Keira, K. Deguchi, H. Nakagawa and S. Nishimura, Profiling of N- and O-glycopeptides of erythropoietin by capillary zwitterionic type of hydrophilic interaction chromatography/electrospray ionization mass spectrometry, *J. Sep. Sci., 31*: 1585 (2008).

214. S. O. Kolset, K. Prydz and G. Pejler, Intracellular proteoglycans, *Biochem. J., 379*: 217 (2004).

215. N. S. Gandhi and R. L. Mancera, The Structure of glycosaminoglycans and their interactions with proteins, *Chem. Biol. Drug. Des., 72*: 455 (2008).

216. D. G. Seidler, J. Peter-Katalinić and A. D. Zamfir, Galactosaminoglycan function and oligosaccharide structure determination, *Sci. World. J., 7*: 233 (2007).

217. N. Itano, Simple primary structure, complex turnover regulation and multiple roles of hyaluronan, *J. Biochem., 144*: 131 (2008).

218. B. P. Toole, Hyaluronan in morphogenesis, *J. Intern. Med., 242*: 35 (1997).

219. B. P. Toole, T. N. Wight and M. I. Tammi, Hyaluronanñcell interactions in cancer and vascular disease, *J. Biol. Chem., 277*: 4593 (2002).

220. W. Y. Chen and G. Abatangelo, Functions of hyaluronan in wound repair, *Wound Repair Regen., 7*: 79 (1999).

221. A. Migliore and M. Granata, Intra-articular use of hyaluronic acid in the treatment of osteoarthritis, *Clin. Interv. Aging 3*: 365 (2008).

222. A. Aruffo, I. Stamenkovic, M. Melnick, C. B. Underhill and B. Seed, CD44 is the principal cell surface receptor for hyaluronate, *Cell 61*: 1303 (1990).
223. P. Heldin, E. Karousou, B. Bernert, H. Porsch, K. Nishitsuka and S. S. Skandalis, Importance of hyaluronan-CD44 interactions in inflammation and tumorigenesis, *Connect. Tissue Res., 49*: 215 (2008).
224. N. K. Karamanos, P. Vanky, A. Syrokou and A. Hjerpe, Identity of dermatan and chondroitin sequences in dermatan sulfate chains determined by using fragmentation with chondroitinases and ion-pair high-performance liquid chromatography, *Anal. Biochem., 225*: 220 (1995).
225. M. Kusche-Gullberg and L. Kjellen, Sulfotransferases in glycosaminoglycan biosynthesis, *Curr. Opin. Struct. Biol., 13*: 605 (2003).
226. K. Sugahara and S. Yamada, Structure and function of oversulfated chondroitin sulfate variants: unique sulfation patterns and neuroregulatory activities, *Trends Glycosci. Glycotechnol., 12*: 321 (2000).
227. K. Sugahara and T. Mikami, Chondroitin/dermatan sulfate in the central nervous system, *Curr. Opin. Struct. Biol., 17*: 536 (2007).
228. J. Timar, K. Lapis, J. Dudas, A. Sebestyen, L. Kopper and I. Kovalszky, Proteoglycans and tumor progression: Janus-faced molecules with contradictory functions in cancer, *Semin. Cancer Biol., 12*: 173 (2002).
229. F. Li, G. B. ten Dam, S. Murugan, S. Yamada, T. Hashiguchi, S. Mizumoto, K. Oguri, M. Okayama, T. H. van Kuppevelt and K. Sugahara, Involvement of Highly Sulfated Chondroitin Sulfate in the Metastasis of the Lewis Lung Carcinoma Cells, *J. Biol. Chem., 283*: 34294 (2008).
230. Y. Zhang, Y. Kariya, A. H. Conrad, E. S. Tasheva and G. W. Conrad, Analysis of Keratan Sulfate Oligosaccharides by Electrospray Ionization Tandem Mass Spectrometry, *Anal. Chem., 77*: 902 (2005).
231. B. Mulloy and M. J. Forster, Conformation and Dynamics of Heparin and Heparan Sulfate, *Glycobiology, 10*: 1147 (2000).
232. R. Lawrence, S. K. Olson, R. E. Steele, L. Wang, R. Warrior, R. D. Cummings and J. D. Esko, Evolutionary Differences in Glycosaminoglycan Fine Structure Detected by Quantitative Glycan Reductive Isotope Labeling, *J. Biol. Chem., 283*: 33674 (2008).
233. U. Lindahl, Heparan sulfate-protein interactions ñ A concept for drug design?, *Thromb. Haemost., 98*: 109 (2007).
234. M. J. Miller, C. E. Costello, A. Malmström and J. A. Zaia, A tandem mass spectrometric approach to determination of chondroitin/dermatan sulfate oligosaccharide glycoforms, *Glycobiology, 16*: 502 (2006).
235. A. Ziegler and J. A. Zaia, Separation of heparin oligosaccharides by size-exclusion chromatography, *J. Chromatogr. B—Analytical Technologies in the Biomedical and Life Sciences, 837*: 76 (2006).
236. A. M. Hitchcock, C. E. Costello and J. A. Zaia, Glycoform quantification of chondroitin/dermatan sulfate using an LC/MS/MS platform, *Biochemistry, 45*: 2350 (2006).
237. A. K. Korir, J. F. Limtiaco, S. M. Gutierrez and C. K. Larive, Ultraperformance ion-pair liquid chromatography coupled to electrospray time-of-flight mass spectrometry for compositional profiling and quantification of heparin and heparan sulfate, *Anal. Chem., 80*: 1297 (2008).
238. A. D. Zamfir, D. G. Seidler, H. Kresse and J. Peter-Katalinic, Structural characterization of chondroitin/dermatan sulfate oligosaccharides from bovine aorta by capillary electrophoresis and electrospray ionization quadrupole time-of-flight tandem mass spectrometry, *Rapid Commun. Mass Spectrom., 16*: 2015 (2002).

239. A. D. Zamfir, D. G. Seidler, H. Kresse and J. Peter-Katalinic, Structural investigation of chondroitin/dermatan sulfate oligosaccharides from human skin fibroblast decorin, *Glycobiology, 13*: 733 (2003).
240. D. J. Mahoney, R. T. Aplin, A. Calabro, V. C. Hascall and A. J. Day, Novel methods for the preparation and characterization of hyaluronan oligosaccharides of defined length, *Glycobiology, 11*: 1025 (2001).
241. J. Zaia, B. Liu, R. Boynton and F. Barry, Structural analysis of cartilage proteoglycans and glycoproteins using matrix-assisted laser desorption/ionization time-of-flight mass spectrometry, *Anal. Biochem., 277*: 94 (2000).
242. N. Keiser, G. Venkataraman, Z. Shriver and R. Sasisekharan, Direct isolation and sequencing of specific protein-binding glycosaminoglycans, *Nature Med., 7*: 123 (2001).
243. T. N. Mitropoulou, F. Lamari, A. Syrokou, A. Hjerpe and N. Karamanos, Identification of oligomeric domains within dermatan sulfate chains using differential enzymic treatments, derivatization with 2-aminoacridone and capillary electrophoresis, *Electrophoresis 22*: 2458 (2001).
244. J. Zaia, J. E. McClellan and C. E. Costello, Tandem mass spectrometric determination of the 4S/6S sulfation sequence in chondroitin sulfate oligosaccharides, *Anal. Chem., 73*: 6030 (2001).
245. A. Tawada, T. Masa, Y. Oonuki, A. Watanabe, Y. Matsuzaki and A. Asari, Large-scale preparation, purification, and characterization of hyaluronan oligosaccharides from 4-mers to 52-mers, *Glycobiology, 12*: 421 (2002).
246. S. Yamada, Y. Okada, M. Ueno, S. Iwata, S. S. Deepa, S. Nishimura, M. Fujita, I. van Die, Y. Hirabayashi and K. Sugahara, Determination of the glycosaminoglycan-protein linkage region oligosaccharide structures of proteoglycans from Drosophila melanogaster and Caenorhabditis elegans, *J. Biol. Chem., 277*: 31877 (2002).
247. A. V. Kühn, H. H. Rüttinger, R. H. Neubert and K. Raith, Identification of hyaluronic acid oligosaccharides by direct coupling of capillary electrophoresis with electrospray ion trap mass spectrometry, *Rapid Commun. Mass Spectrom., 17*: 576 (2003).
248. S. Duteil, P. Gareil, S. Girault, A. Mallet, C. Feve and L. Siret, Identification of heparin oligosaccharides by direct coupling of capillary electrophoresis/ionspray-mass spectrometry, *Rapid Commun. Mass Spectrom., 13*: 1889 (1999).
249. V. Ruiz-Calero, E. Moyano, L. Puignou and M. T. Galceran, Pressure-assisted capillary electrophoresis-electrospray ion trap mass spectrometry for the analysis of heparin depolymerised disaccharides, *J. Chromatogr. A, 914*: 277 (2001).
250. S. Fermas, F. Gonnet, A. Varenne, P. Gareil and R. Daniel, Frontal analysis capillary electrophoresis hyphenated to electrospray ionization mass spectrometry for the characterization of the antithrombin/heparin pentasaccharide complex, *Anal. Chem., 79*: 4987 (2007).
251. L. Svennerholm, Gangliosides and synaptic transmission, *Adv. Exp. Med. Biol., 125*: 11 (1980).
252. UPAC-IUB Joint Commission on Biochemical Nomenclature, Nomenclature of glycolipids, *Eur. J. Biochem.,, 257*: 293 (1998).
253. R. K. Yu, Y. Nakatani and M. Yanagisawa, The role of glycosphingolipid metabolism in the developing brain, *J. Lipid Res., 50*: 440 (2009).
254. J. S. Ryu, K. Ko, J. W. Lee, S. B. Park, S. J. Byun, E. J. Jeong, K. Ko and Y. K. Choo, Gangliosides are involved in neural differentiation of human dental pulp-derived stem cells, *Biochem. Biophys. Res. Commun., 387*: 266 (2009).
255. P. McJarrow, N. Schnell, J. Jumpsen and T. Clandinin, *Nutr. Rev., 67*: 451 (2009).
256. M. Taniguchi, T. Tashiro, N. Dashtsoodol, N. Hongo and H. Watarai, The specialized iNKT cell system recognizes glycolipid antigens and bridges the innate and acquired immune systems with potential applications for cancer therapy, *Int. Immunol, 22*: 1 (2010).

257. G. Zhang, H. Zhang, Q. Wang, P. Lal, A. M. Carroll, M. de la Llera-Moya, X. Xu and M. I. Greene, Suppression of human prostate tumor growth by a unique prostate-specific monoclonal antibody F77 targeting a glycolipid marker, *Proc. Natl. Acad. Sc. U.S.A., 107*: 732 (2010).

258. S. Hakomori, Tumor-associated carbohydrate antigens defining tumor malignancy: basis for development of anti-cancer vaccines, *Adv. Exp. Med. Biol., 491*: 369 (2001).

259. S. Hakomori and K. Handa, Glycosphingolipid-dependent cross-talk between glyco-synapses interfacing tumor cells with their host cells: essential basis to define tumor malignancy, *FEBS Lett., 531*: 88 (2002).

260. S. B. Levery, Glycosphingolipid structural analysis and glycosphingolipidomics, *Methods Enzymol., 405*: 300 (2005).

261. U. Distler, M. Hülsewig, J. Souady, K. Dreiswerd, J. Haier, N. Senninger, A. W. Friedrich, H. Karch, F. Hillenkamp, S. Berkenkamp, J. Peter-Katalinić and J. Müthing, Matching IR-MALDI-o-TOF mass spectrometry with the TLC overlay binding assay and its clinical application for tracing tumor-associated glycosphingolipids in hepatocellular and pancreatic cancer, *Anal. Chem., 80*: 1835 (2008).

262. M. C. Sullards, J. C. Allegood, S. Kelly, E. Wang, C. A. Haynes, H. Park, Y. Chen and A. H. Merrill Jr., Structure-specific, quantitative methods for analysis of sphingolipids by liquid chromatography-tandem mass spectrometry: "inside-out" sphingolipidomics, *Methods Enzymol., 432*: 95 (2007).

263. T. Ariga, M. P. McDonald and R. K. Yu, Role of ganglioside metabolism in the pathogenesis of Alzheimer's disease, *J. Lipid Res., 49*: 1157 (2008).

264. W. R. Jung, H. G. Kim and K. L. Kim, Ganglioside GQ1b improves spatial learning and memory of rats as measured by the Y-maze and the Morris water maze tests, *Neurosci. Lett., 439*: 220 (2008).

265. Y. Fujimoto, S. Izumoto, T. Suzuki, M. Kinoshita, N. Kagawa, K. Wada, N. Hashimoto, M. Maruno, Y. Nakatsuji and T. Yoshimine, Ganglioside GM3 inhibits proliferation and invasion of glioma, *J. Neurooncol., 71*: 99 (2005).

266. Z. Vukelić, S. Kalanj-Bognar, M. Froesch, L. Bindila, B. Radić, M. Allen, J. Peter-Katalinić and A. D. Zamfir, Human gliosarcoma-associated ganglioside composition is complex and distinctive as evidenced by high-performance mass spectrometric determination and structural characterization, *Glycobiology, 17*: 504 (2007).

267. A. D. Zamfir, Ž. Vukelic and J. Peter-Katalinić, A capillary electrophoresis and off-line capillary electrophoresis/electrospray ionization-quadrupole time of flight-tandem mass spectrometry approach for ganglioside analysis, *Electrophoresis, 23*: 2894 (2002).

268. J. Li, A. Martin, A. D. Cox, E. R. Moxon, J. C. Richards and P. Thibault, in *Methods in Enzymology (Mass Spectrometry: Modified Proteins and Glycoconjugates)*, Vol. 405 Elsevier, Amsterdam, 2005, p. 369.

269. C. W. Reid, J. Stupak, M. M. Chen, B. Imperiali, J. Li and C. M. Szymanski, Affinity-Capture Tandem Mass Spectrometric Characterization of Polyprenyl-Linked Oligosaccharides: Tool to Study Protein N-Glycosylation Pathways, *Anal. Chem., 80*: 5468 (2008).

270. D. D. Ju, C. C. Lai and G. R. Her, Analysis of gangliosides by capillary zone electrophoresis and capillary zone electrophoresis-electrospray mass spectrometry, *J. Chromatogr. A, 779*: 195 (1997).

271. Y. R. Chen, M. C. Tseng and G. R. Her, Design and performance of a low-flow capillary electrophoresis-electrospray-mass spectrometry interface using an emitter with dual beveled edge, *Electrophoresis, 26*: 1376 (2005).

272. P. Liuni, T. Rob and D. J. Wilson, A microfluidic reactor for rapid, low-pressure proteolysis with on-chip electrospray ionization, *Rapid Commun. Mass Spectrom., 24*: 315 (2010).

273. J. Lee, S. A. Soper and K. K. Murray, Microfluidic chips for mass spectrometry-based proteomics, *J. Mass Spectrom., 44*: 579 (2009).

274. A. Serb, C. Schiopu, C. Flangea, Ž. Vukelić, E. Sisu, L. Zagrean and A. D. Zamfir, High-throughput analysis of gangliosides in defined regions of fetal brain by fully automated chipbased nanoelectrospray ionization multistage mass spectrometry, *Eur. J. Mass Spectrom., 15*: 541 (2009).

275. R. Almeida, C. Mosoarca, M. Chirita, V. Udrescu, N. Dinca, Ž. Vukelić, M. Allen and A. D. Zamfir, Coupling of fully automated chip-based electrospray ionization to high-capacity ion trap mass spectrometer for ganglioside analysis, *Anal. Biochem., 378*: 52 (2008).

276. A. D. Zamfir, L. Bindila, N. Lion, M. Allen, H. H. Girault and J. Peter-Katalinic, Chip electrospray mass spectrometry for carbohydrate analysis, *Electrophoresis, 26*: 3650 (2005).

277. A. D. Zamfir, Z. Vukelić, L. Bindila, J. Peter-Katalinić, R. Almeida, A. Sterling and M. Allen, Fully-automated chip-based nanoelectrospray tandem mass spectrometry of gangliosides from human cerebellum, *J. Am. Soc. Mass Spectrom.*: 1649 (2004).

278. A. Serb, C. Schiopu, C. Flangea, E. Sisu and A. D. Zamfir, Top-down glycolipidom-ics: fragmentation analysis of ganglioside oligosaccharide core and ceramide moiety by chip-nanoelectrospray collision-induced dissociation MS2-MS6, *J. Mass Spectrom., 44*: 1434 (2009).

279. C. Schiopu, A. Serb, F. Capitan, C. Flangea, E. Sisu, Z. Vukelic, S. Kalanj-Bognar, M. Przybylski and A. D. Zamfir, Determination of ganglioside composition and structure in human brain hemangioma by chip-based nanoelectrospray ionization tandem mass spectrometry, *Anal. Bioanal. Chem., 395*: 2465 (2009).

5 Oligonucleotide Adducts as Biomarkers for DNA Damages
Analysis by Mass Spectrometry Coupled to Separation Methods

Qing Liao and Paul Vouros

CONTENTS

5.1 INTRODUCTION

5.1.1 INITIATION OF CANCER

Cancer is a genetic disease that affects millions of people and is, at least in part, due to DNA damage caused by chemical, viral, or physical carcinogens. Among all diseases, cancer is the second lethal killer after cardiovascular disease. It kills nearly 1,500 Americans a day. In 2004 alone, the deaths caused by cancer were 553,888 (NIH and American Cancer Society). The cancer deaths estimated for 2007 by the American Cancer Society are 559,650, despite the fact that the federal government alone has spent $69 billion on cancer research since 1971. The most common types of cancer are breast cancer, prostate cancer, and colorectal cancer. Cancer can be attributed to a single underlying malady of the genetic program that directs the lives of our cells. When organisms fall prey to uncontrolled proliferation of defective, undifferential cells, cancer occurs. A so-called carcinogenic event is usually set off by some external factors such as exposure to sunlight, x-rays, smoking, and dangerous chemicals. Exposure to genotoxic compounds can alter DNA sequence, lead to mutated genes, and give rise to genetic instability. The myriad alterations within the cell include switching on or off genes, aberrant protein expression, or altering cell cycle control.

There is overwhelming evidence that mutations in relevant target sequences, such as oncogenes or tumor suppressor genes, are associated with the carcinogenic process. A number of cancer-susceptible genes have been identified including *p53*, *ras*, *rb*, *BRCA1*, *BRCA2*, *hprt*, *aprt*, and *dhfr*. The mechanism of mutation is quite complex; it can include frameshift, base deletion, or base substitution that occurs as a result of chemical reactions between genotoxic compounds and DNA. It has been estimated that every gene in our DNA is damaged some 10 billion times in our lifetime, yet the rate of the mutations is far lower. This is mainly due to the ability of the cell to repair its damaged DNA. However, if not repaired, mutation can occur. Therefore, it is of paramount importance to a cell to maintain the integrity of DNA structure to have less chance of mutation and likelihood of cancer occurrence.

5.1.2 DNA ADDUCTS AS BIOMARKERS IN CANCER RESEARCH

Keys to success in the fight against cancer include prevention, early detection, and more effective treatment. In recent years, there were declines in annual cancer deaths owing largely to screening efforts. We are seeing some real dividends as earlier detection was made possible by advances in technology. What are called biomarkers has played a significant role in this.

Biomarkers are distinctive biological indicators of molecular and cellular information that can mark the presence or progression of a disease. They can be acquired from patients from host cells or tissues of tumors and body fluids. Detecting biomarkers specific to a disease can help in identification, diagnosis, and treatment of those with the disease as well as those who might be at risk but are as yet asymptomatic. In cancer research specifically, the study of cancer biomarkers may help to (1) identify who is at risk of the disease,(2) diagnose the disease at an early stage, (3) select

the best treatment, and (4) monitor response to treatment. To date, cancer biomarkers have included indicators such as carcinogen–protein adducts, carcinogen–DNA
adducts, chromosomal aberrations, polymorphisms in drug-metabolizing enzymes,
and host DNA repair capacity.

A major goal of DNA adduct studies is to use the information in predicting human
cancer risk. In our daily life we are constantly exposed to genotoxic compounds that
may interact with cellular DNA to form chemical products called DNA adducts, where
adducting species (carcinogens) are covalently bonded to DNA. DNA adducts formation is generally considered one of the key events in tumor initiation during chemical
carcinogenesis. Considerable epidemiological and experimental evidence has indicated that exposure to most carcinogens results in damage to the structural integrity
of DNA. DNA adducts are the precursor lesions for mutation, as mutations often
occur at (or close to) the carcinogen adduction sites [1–3]. DNA adduct formation
often leads to malfunction of DNA that includes major distortions in DNA and even
to formation of cross-links between two strands. The majority of DNA adducts are
eliminated by DNA repair processes but, if not repaired, may result in misreplication,
inhibition of DNA synthesis, or termination of transcription that will lead to mutation
in important growth-controlling genes or loci, resulting in mutagenesis. Alterations
in DNA sequence may also occur when adducts are subjected to erroneous repair, as
damage to DNA can increase the frequency of miscopying of DNA, which may lead
to apoptosis or frameshift mutation. The consequence of mutagenesis is aberrant cellular growth and ultimately cancer. Since formation of DNA adducts is a potential
first step in the initiation of cancer, their measurement can provide at the very least
an estimate of internal exposure and ideally a measure of biological outcome. For
these reasons, DNA adducts are viewed as biomarkers to assess genetic damage that
may also be used in the molecular epidemiology of cancer for risk assessment.

Evidence of the mutagenicity of DNA adducts came from studies involving incorporation of single adducts into defined DNA sequences using site-specific techniques
as well as analyzing induced mutational spectra in specific genes in cells of treated
animals. Mutational spectra are used to provide information on the frequency and
types of mutations that arise from a particular DNA-damaging compound. Numerous
studies support the hypothesis that DNA adduct levels in target and in surrogate tissues are appropriate biomarkers of DNA damage induced by carcinogen exposure.
Extensive evidence gained through carcinogenicity testing in experimental animals
suggests that, while DNA adduct formation alone is insufficient to cause tumor formation, it is at least one prerequisite [4–8].

Although the data are less definitive in humans, it appears that the relationship
between DNA adduct levels and tumor incidences following lifetime administration
of different doses of chemical carcinogens to experimental animals and the mechanisms derived from such studies can be extrapolated to provide the best basis for
human cancer risk assessment.

The quantitative correlation between DNA adducts formation and cancer risk is
still unclear. There are uncertainties concerning the biological significance of low levels of DNA adduct formation in particular. The scientific community has yet to agree
on whether there is a threshold of DNA adducts below which there is no measurable
biological effect. It is unlikely that a single "universal" value could be given to such

an acceptable level of adducts, as their mutational effectiveness varies according to the nature of the carcinogen and the chemical structure of the DNA adduct. There is no simple algorithm for translating DNA adducts levels into cancer risk in a particular tissue. It is generally accepted by many regulatory authorities that the dose–response relationship for genotoxic carcinogens does not have a threshold, and thus the presence of any amount of exposure presents a carcinogenic risk. The dose–response relationship for genotoxic carcinogens is likely nonlinear as it is not always possible to detect any increase in mutations above background at administered doses of genotoxic agents that produce detectable adducts. On the other hand, the lack of adduct formation might suggest that no mutation could occur, and demonstration of the inability of a chemical to produce adducts may be used to eliminate the possibility that the compound is a genotoxic carcinogen. This may have a major influence on regulatory decisions on such compounds. Although the relative roles of chemical reactivity in the formation of a lesion and the carcinogenic potency of a particular lesion in the establishment of clonal growth advantage remain unclear, nonetheless as a means to cancer prevention it is crucial to decrease the amount of DNA damage. Put in simple terms, any modification to genetic material cannot be overlooked.

The advantage of using DNA adducts as biomarkers and comparison with environmental monitoring data is that it can provide several aspects of assessment. These include information leading to exposure levels, absorption, distribution, metabolic activation (or inactivation), genetic susceptibility, and DNA repair capacity. There is, however, consensus that the use of adduct data in risk extrapolation has the greatest value when adduct structure has been characterized and the roles of adduct removal and biological relevance of specific adducts are understood. The biological potential of a given DNA adduct depends on its mutagenicity, ability to be repaired, location within a target gene, and the nature of the target gene. Total DNA adducts levels, and in some cases specific adducts, has been correlated with in vitro mutation, with chromosomal aberrations, and generally with carcinogenicity. Thus, quantitation of DNA adducts in human cells provides a useful parameter of risk assessment in terms of monitoring human carcinogen exposure, the determination of a biologically effective dose, and individual cell type-specific DNA repair capacity. Detection and quantitation of DNA adducts in human tissues also confirms epidemiological associations of cancer and risk factors and provides information on the identification of carcinogenic hazard and quantitative risk assessments of accumulative genetic damage.

Risk assessment with DNA adducts as biomarkers does have limitations. For example, it does not take into account interindividual variation in absorption, metabolism, excretion, and bioavailability of the carcinogens. The presence of relatively high levels of certain endogenous DNA adducts, such as oxidized bases induced by endogenously formed oxygen radicals, may further complicate the interpretation of data. This makes further characterization necessary, including structure–activity relationship studies, mechanistic studies on species differences in metabolism, and investigation of genetic polymorphisms.

Although a multitude of factors affects the interpretation of the relationship between DNA adducts levels and carcinogenesis, it should be realized that when DNA adducts, including decomposed forms of bases, are detected with remarkable frequency, the following situations have arisen: (1) there are mutagenic chemicals, sources of dangerous electromagnetic waves, or serious stresses to cause DNA mutation around our lives; and (2) the possibility of falling victim to serious diseases such as cancers caused by changes in the DNA sequences is increased drastically. We are thus required to decrease the quantity of mutagenic subjects in our surroundings and, if possible, to remove them completely. On the other hand, we should investigate the mechanisms expressing toxicity for each DNA adduct and also reveal the strength of the toxicity.

In summary, since the formation of specific DNA modifications appears to be a critical event in carcinogenesis, measurement of carcinogen–DNA adducts should provide biologically relevant information on the net result of exposure, absorption, metabolism, DNA adduct formation, and DNA repair. Thus, analysis of DNA adduct levels may be one of the best tools available to characterize exposure to complex mixtures of genotoxic chemicals in different environmental and occupational exposure settings. Consequently, the study of DNA adducts as biomarkers for cancer involves determination of the nature and extent of adduct formation and if possible also the position of these adducts within the DNA structure, which is the primary focus of this presentation. It is anticipated that the characterization of DNA adducts in conjunction with other biological information (e.g., gene expression analysis, mutation analysis) will play a greater role in the assessment of carcinogenic hazard and possibly of risk in the foreseeable future.

5.1.3 CLASSES OF GENOTOXIC COMPOUNDS AND THEIR TARGET SITES ON DNA BASES

There are two types of DNA-damaging agents: the exogenous compounds and the endogenous compounds. The exogenous compounds are what are commonly known as carcinogens that have been demonstrated by epidemiological studies to be genotoxic chemicals of a wide range. They are chemicals that can be divided into two groups: (1) the genotoxic, which cause damage to DNA; and (2) the nongenotoxic, which cause no direct damage but indirectly initiate DNA damage because of their metabolic pathway. They often require activation to reactive metabolites that bind to nucleophilic centers in DNA to form covalent adducts. The exogenous compounds include mainly these classes of compounds: polycyclic aromatic hydrocarbons (PAHs), nitro-PAHs, N-nitrosamines, aromatic amines (AA), heterocyclic aromatic amines (HAA), and aflatoxins. DNA adducts can also be described in terms of their hydrophobicity and divided into small or polar adducts and bulky or nonpolar adducts.

The sources of these carcinogens can be dietary, lifestyle related (e.g., cigarette smoking), medicinal, occupational, and environmental/industrial (air, water) and ultraviolet (UV) radiation. For example, carcinogens found in the diet include mycotoxins and N-nitroso compounds. There are also compounds generated by the cooking process including HAA, PAHs, and acrylamide. Many potential human genotoxic carcinogens are formed following the heating of carbohydrate-rich foods to high temperatures. Environmental air pollution is another major source of carcinogens. In highly industrialized cities, elevated levels of PAHs are found. Lifestyle is another important factor when considering the total scheme of exposure events, of which cigarette smoke is the major culprit. In summary, one may encounter multiple sources of exposures in daily life. Cigarette smoking, environmental tobacco smoke, exhaust by-products, and foods may all contribute to a person's cumulative risk of cancer.

For example, cigarette smoke contains approximately 4000 chemicals, about 50 compounds of which have been identified as either animal or human carcinogens according to the International Agency for Research on Cancer (IARC). The major players are PAHs and the alkylating tobacco-specific nitrosamines that most likely lead to lung cancer. In addition, cigarette smoke can generate reactive oxidizing species (ROS) or free radicals that can hydroxylate DNA bases. They can induce DNA single-strand breaks in vitro as well as cause oxidative DNA damage in cultured human cells. Exogenous modifications often involve aromatic or polycyclic aromatic metabolites to form bulky DNA adducts at concentrations ranging from 1 adduct in 10^6 to 10^9 unmodified nucleotides.

Endogenous compounds are derived from endogenous electrophiles released from normal cellular function such as metabolism of nutrients and other natural dietary components. Many of these agents are highly reactive and thus do not need further metabolic activation. Various types of endogenous DNA damage include those from DNA instability, errors in replication and repair, oxidatively damaged bases, and adducts derived from reaction of bases with aldehydic lipid peroxidation products. Endogenous DNA damage, formed by small modifications such as by oxygen radicals, is present at relatively high levels in mammalian tissues, and such endogenous adducts can be found at concentrations of 1 adduct in 10^4 to 10^7 unmodified nucleotides. The exact significance of endogenous damage in terms of human carcinogenesis is not totally clear.

More than 800 different compounds have been evaluated by IARC. Over 75 specific adducts or closely related groups of adducts have been determined to be human carcinogens [5,6]. Many compounds form the same adducts and also diastereomers or enantiomers of the same structural type. Hemminki and Garner each compiled a table of DNA adducts detected in human tissues and the method of detection [4,6]. The sources from which the DNA adducts can be obtained are usually total white blood cells (WBC) or peripheral blood lymphocytes and to a lesser extent skin, placenta, kidney, liver, lung, breast, and pancreas (Table 5.1).

General classes of these carcinogens include alkylating agents, aromatic amines, epoxides, PAHs, nitro-PAHs, food-derived HAAs, aldehydes, allylic compounds, and chemotherapeutic drugs. Under the PAHs category the most studied is benzo[a] pyrene (BaP) and its metabolite benzo[a]pyrene diol epoxide (BPDE) and to a lesser extent benzo[a]anthracene (BaA) and benzo[b]fluoranthrene (BkF) [1,9–13]. In the class of nitro-PAHs there are 4-nitropyrene (4-NP), 6-nitrochrysene (4-NC), and 3-nitrofluoranthrene (3-NF). In the lifestyle-related carcinogens, the methylating nitrosamines are mostly found in tobacco smoke. Some of the most extensively studied are 4-(methylnitrosamino)-1-(3-pyridyl)-1-butanone (NNK) and its metabolite 4-(methylnitrosamino)-1-(3-pyridyl)-1-butanal (NNAL) and N'-nitrosonornicotine (NNN) [14–23]. Under the category of PAAs there are 2-aminofluorene, benzidine (BZ), and 3',3'-dichlorobenzidine (DCB). The occupational environmental carcinogens often studied are arylamines (AA), nitroarenes, polychlorinated biphenyls (PCB), and compounds such as urethane and N, N-dimethylformamide (DMF).

TABLE 5.1
Genotoxic Compounds Detected in Human Tissues and Methods of Detection

Chemical	Tissue	Method
4-Aminobiphenyl	Bladder, lung	Postlabeling, GC/MS
4,4'-Methyleneebis(2-chloroaniline)	Bladder	Postlabeling
Methylating agents	Lymphocytes, liver, lung, bronchus, WBC	IAC/postlabeling, LC-MS
Aflatoxin B$_1$	Urine, lung, kidney, liver	Immunoassay, Fluorescence, LC-MS
Cisplatin	Lymphocytes, WBC	HPLC/immunoassay, LC-MS
Benzo(a)pyrene	WBC, lung	Fluorescence, LC-MS
Cigarette smoke	Lung, bladder, breast lymphocytes, cervix, placenta, buccal mucosa	Postlabeling
Foundry fumes	Lymphocytes	Postlabeling
Air pollution	Lymphocytes	Postlabeling
MeIQx	Colon, kidney, rectum	Postlabeling, LC-MS
PhIP	Colon	GC/MS, LC-MS
Aristolochic acid	Kidney	Postlabeling
Malondiadehyde	Lymphocytes, liver, pancreas, breast, WBC	GC/MS, Postlabeling
Styrene oxide	Lymphocytes, WBC	Postlabeling, LC-MS
4-Hydroxynonenal	Lymphocytes, liver	Immunoaffinity/ Postlabeling
Etheno A, C, G (vinyl chloride)	Liver, WBC, pancreas	IA/Postlabeling, LC-MS
Acrolein	Liver, WBC, mammary gland	Postlabeling, LC-MS
Crotonaldehyde	Liver, WBC, mammary gland	Postlabeling, LC-MS
Tamoxifen	WBC, endometrium	Postlabeling, LC-MS
8-OxodG	WBC, pancreas	EC, LC-MS
Benzidine	WBC	Postlabeling, LC-MS
Butadiene	Lymphocytes	Postlabeling, LC-MS
UV photoproducts	Skin	Postlabeling

Source: Adapted and modified from Garner, R. C. The Role of DNA Adducts in Chemical Carcinogenesis. *Mutat. Res.* 1998, *402*, 67–75. Hemminki, K.; Koskinen, M.; Rajanierni, H.; Zhao, C. DNA Adducts, Mutations, and Cancer. *Regul. Toxicol. Pharmacol.* 2000, *32*, 264–275. (With permission.)

In the class of arylamines the most studied are 4-aminobiphenyl (4-ABP) and its analogs such as 2ABP and 3ABP. Less studied are 2-chloroaniline (2CA) and 4CA, 4-methylaniline (4MA), 2,4-dimethylaniline (2,4DMA), and 2,6DMA. In the class of nitroarenes there are 2-chloronitrobenzene and 4-chloronitrobenzene, 2-nitrotoluene (2-NT), 4-NT, and nitrofluorenes. There are thioarenes such as 5-nitrobenzo[b] naphtha[2,1-d]thiophene (5-nitro-BNT) and dihaloalkanes such as 1,2-dihaloethane. In diet there are carcinogens such as mycotoxin A, N-nitrosodiethanolamine

(NDELA) and furan. Some cancer chemotherapeutic agents have a secondary carcinogenic effect. One class most often studied is nitrogen mustards, which are bifunctional alkylating agents. Two nitrogen mustards, melphalan and tamoxifen (TAM),
are model carcinogens in many studies. In addition, some sugar adducts and antibiotics have some carcinogenic effects. One radiation-induced damage to DNA results in
formation of 5-hydroxymethyluracil (5-HMU).

There have been many studies on model carcinogens including platinum-DNA
adducts such as cisplatin [24–28], estrogens, including the metabolite 4-hydroxy-
equilenin (4-OHEN), 4-hydroxy-estradiol and 4-hydroxy-estrone [29–33], and styrene
oxide (SO) [34–39]. Other important carcinogens receiving attention are 2-amino-
1-methyl-6-phenylimidazo[4,5-b]pyridine (PhIP), 2-amino-r-methylimidazo[4,5-f]
quinoline (IQ), aflatoxin B (AFB), acrylamide, N-methyl-N-nitrosourea (MNU), 2-amino-
3,8-dimethylimdazo[4,5-f]quinxaline (MeIQx), phenyl glycidyl ether (PGE), bisphenol A digylcocidy ether, choloroambucil, ochratoxin A (OTA), hydroxyethylvaline
(HOEtVal), 7-(2-hydroxyethyl)guanine (7-HEG), S-(1-Acetoxymethyl)glutathione
(GSCH2OAc), bleomycin (BLM), hypochlorous acid (HOCl), 1,2-epoxy-
butene (BDO), redelline, and leukotiene A4 (LTA4). Aacetaldehyde (AA), crotonaldehyde (CR), formaldehyde, acrolein and 1, 3-butadiene (BD), depoxybutane (DEB),
and methyl methanesulfonate (MMS) are also among the carcinogens that appear
often in the literature.

The most extensively studied classes of carcinogens perhaps are lipid hydroper-
oxidation products and ROS generated from them [40–46]. Among these are etheno
DNA adducts such as N6-ethenoadenine (epsilon A) and vepsilonA, MDA, vinyl
halides such as vinyl chloride (VC), pyrimido[1,2-a]-purin-10(3H)-one (M1G) and
to a lesser extent 4-oxy-2-nonenal (4-ONE), 4,5-epoxy-2(E)-decenal (4,5-EDE),
2,3-epoxy-4-hydroxynonenal (EH), 4-hydroxy-2-nonenal (4-HNE), 4-oxo-2-hexane-
nal (4-OHE), and 4-hydroperoxy-2-nonenal (HPNE) [47–56]. Chen and Chang [57],
Chen and Chiu [58], Chen et al. [59–61], of the National Chung Cheng University
in Taiwan conducted an extensive study of lipid peroxide products, particularly
etheno DNA adducts, using gas chromatography/negative ion/chemical ionization
mass spectrometry (GC/NICI/MS). In more recent studies, they explored the use of
LC-MS and CE-MS [57,58]. They also reviewed the formation, analysis, and repair
of exocyclic etheno DNA adducts [62].

Among the various forms of oxidative lesions, the ROS 7,8-dihydro-8-oxo-2′-
deoxyguanisine (8-oxodG) is the oxidation product from the aforementioned species.
8-oxodG is the most extensively studied biomarker for oxidative damage [63–69]
often detected in urine and has been shown to be carcinogenic. Urinary excretion
of 8-oxodG has been used extensively as a noninvasive biomarker of oxidative DNA
damage in humans.

Carcinogens are usually electrophilic species or become electrophilic after
metabolism in vivo and bind covalently to the nucleophilic sites of DNA. In general,
adduction can occur on the DNA base as well as on the phosphate and deoxyribose
moieties in the nucleic acid structure (Figure 5.1). On the bases, the nitrogen and
oxygen atoms are active sites. Of the four DNA bases, the one most susceptible to
carcinogen modification is guanine (G) and the C8 carbon site on the guanine is the
place attacked most often by carcinogens. The N7 position of G is predominantly

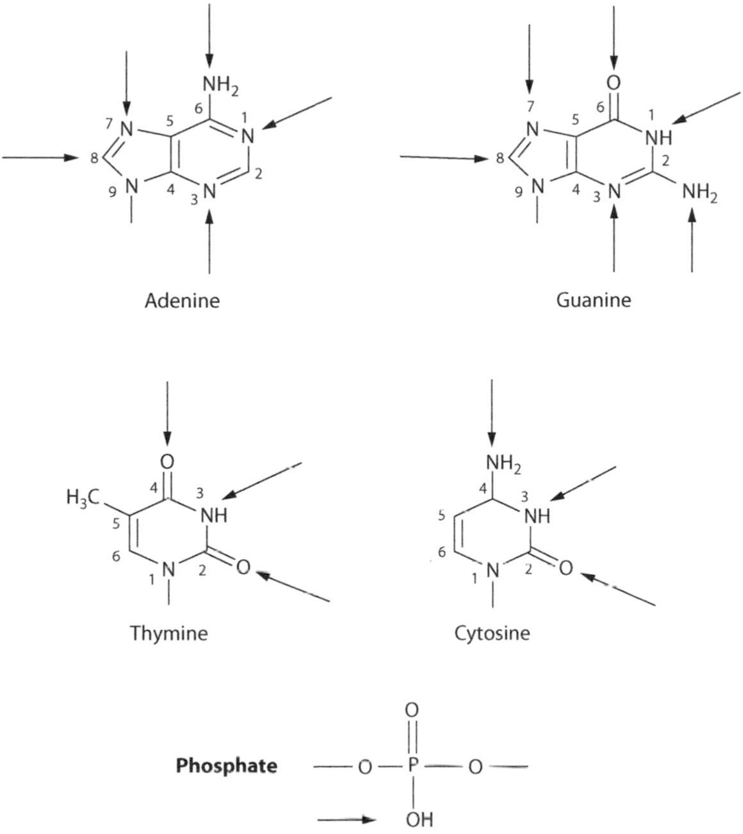

FIGURE 5.1 Potential sites of addition reactions for the DNA bases. (Adapted and modified from Esaka et al., *J. Chromatogr., B: Anal. Technol. Biomed. Life Sci.* 2003, *797*, 321–329. With permission.)

modified by alkylation (methylation or ethylation) agents, whereas aromatic amines and PAHs prefer the C8 and N2 position, respectively. Common carcinogens are shown in Table 5.2.

5.1.4 ANALYTICAL CHALLENGES

The major challenge in analyzing DNA adducts lies in the extremely high sensitivity and selectivity that is required to detect and quantify these compounds in vivo where they are present in only minute amounts. The damaged bases in DNA of living organisms are usually in the range of 1 in 10^6 to 10^8 and can be as low as 1 in 10^{11} normal nucleotides. This means that one may expect ca. 10 pg of adduct in 1 mg of DNA, effectively searching for a needle in a haystack. Assuming an analytical mass detection limit of 10 pg and setting as a goal the detection of 1 adduct in 10^8 nucleotides, this would translate into a requirement of ca. 1 g of tissue sample (considering that 1 g of tissue may contain 1 mg of DNA). In practice, it is desirable to have about a 10-fold excess of analyte. Thus, there appear to be two alternatives: either have a

TABLE 5.2

Some Common Carcinogens and Their Binding Sites on DNA

Adduct Bonding Site	Genotoxic Compounds Bound to DNA
N-1, N^2	Lipid peroxidation products
	Malondialdehyde (MDA)
	4-Hydroxy-2-nonenal
	Crotonaldehyde (CR)
	2-Hexenal (HX)
	Vinyl chloride (VC)
	Ethanol
	Estrogen
	N-Nitrosodiethanolamine
N^2	2-Amino-r-methylimidazo[4,5-f]quinoline (IQ)
	2-Amino-1-methyl-6-phenylimidazo[4,5-b]pyridine (PhIP)
	2-Amino-3,8-dimethylimdazo[4,5-f]quinxaline (MeIQx)
	Tamoxifen (TAM)
	Ethanol
	Phenyl glycidyl ether (PGE)
	Benzo[a]pyrene (BaP)
	2-Aminofluorene
	2-Acetoxy-Acetyl-Aminofluorene (AAAF)
N^2, N-3	Lipid peroxidation products
	4-Hydroxy-2-nonenal
O^- (on phosphate)	Ethylating agents
	PGE
C-8	Hydroxyl radicals
	Ethanol
	4-Aminobiphenyl (ABP)
	PhIP
	MeIQx
	IQ
	Peroxynitrite
	2-Aminofluorene
	AAAF
N-7	Methylating agents
	Ethylating agents
	Ethylene
	1, 3-Butadiene (BD)
	Acrylamide
	BaP
	Benzo[a]pyrene diol epoxide (BPDE)
	7,12-Dimethylbenz[a]anthracene (DMBA)
	Styrene oxide (SO)
	PGE

TABLE 5.2 (continued)
Some Common Carcinogens and Their Binding Sites on DNA

Adduct Bonding Site	Genotoxic Compounds Bound to DNA
	Aflatoxin B (AFB)
	Estrone
O^6	N-Nitrosodiethanolamine (NDEA)
	N-Methyl-N-nitrosourea (MNU)
	N-Ethyl-N-nitrosourea (ENU)
	4-(Methylnitrosamino)-1-(3-pyridyl)-1-butanone

Source: Adapted and modified from Singh, R. et al., *Chem. Res. Toxicol.* 2006, *19*, 868–878. (With permission.)

10 g sample of tissue; or improve the detectability by a factor of 10 or more. The availability of DNA samples from human is actually quite restricted. DNA human samples mainly are obtained from white blood cells, and the yield is low—typically 1 mL of blood will yield 20–40 µg of DNA. Other sources of DNA are accessible tissues such as placenta and buccal swabs, and the quantity of DNA that can be obtained is only some hundred µg maximally. Even if the absolute mass sensitivity of the analysis is adequate, it still must be determined that the adduct can be detected in the presence of significantly greater quantities of unmodified bases. As a consequence, analytical techniques for analyzing DNA must

- Be sensitive enough to detect low levels of adducts
- Require only microgram quantities of DNA
- Produce results that can be quantitatively related to exposure
- Be applicable to unknown adducts that may be formed from complex mixtures
- Be able to resolve, quantify, and identify adducts
- Be inexpensive
- Be rapid
- Be able to analyze large numbers of DNA samples
- Produce low risk to the person carrying out the procedure

5.1.5 ANALYSIS OF ADDUCTS BY METHODS OTHER THAN MASS SPECTROMETRY

Since the early 1960s when it was unequivocally established that DNA was a macromolecular target for chemical carcinogens, various methods have been developed for adduct analysis. These include ^{32}P-postlabeling, immunoassays, fluorescence spectroscopy, and mass spectrometry. It was previously stated that analytical techniques employed in DNA adduct analysis must offer powerful identification ability and excellent limits of detection (LOD). Most techniques currently favored are deficient in at least one of these requirements. As interest has grown to better understand the relationship between DNA adducts formation and cancer risk, advances have been made in both respects. Table 5.3 summarizes and compares different techniques

TABLE 5.3
Comparison of Different Techniques Applied for the Analysis of DNA Adducts

Technique	Chromatography	Sensitivity (adduct/nucleotide)	Amount of DNA Needed (ug)	Quantification	Comments
^{32}P-Postlabeling	TLC	$1/10^{10}$	2–10	Radioactivity detection of postlabeled species Recovery determined with standard adducted nucleotide	• Useful for screening variety of carcinogen adducts • Applicable to analysis of complex mixtures • Danger of underestimation of adduct levels due to inefficient chromatographic recovery and phosphorylation • Interference of "indogeneous" spots
	HPLC	$1/10^{10}$	10		• Less sensitive compared to TLC (~factor 10), but a relatively large amount of DNA can be analyzed to compensate the loss of sensitivity
	CE	$1/10^{9}$	10		• Enables multiple injections of a single sample
Immunoassay	Standard ELISA	$1/10^{8}$	>100	Calibration curve using standard adduct	• Usual high specificity for carcinogen or class of carcinogens • Interference of substances that compete with antibody recognition
	CE	$1/10^{7}$	0.5–2		• Less DNA needed but also less sensitive

Method	Technique			Principle	Comments
Fluorescence	HPLC	$1/10^8$	>100	Calibration curve using standard adduct	• Only applicable to fluorescent compounds • Information on adduct identify needed
Mass Spectrometry	GC/LC	$1/10^9$	>100	Ion intensity. Calibration curve using standard adduct and stable isotope labeled internal standard	• Structural identification of adduct • Derivatization needed for GC • Sample stacking used to lower detection limit
	CE	$1/10^9$	>100		
Accelerated Mass Spectrometry	HPLC	$1/10^{12}$	500	Separation of isotope ions based on mass, followed by quantification in a gas ionization detector to give an isotope ratio	• Not applicable in standard human biomonitoring studies due to use of ^{14}C-labled substances

Source: Appendix, adapted and modified from De Kok et al., *J. Chromatogr., B: Anal. Technol. Biomed. Life Sci., 778*, 345–355, 2002 and Farmer, P. B. et al., *Toxicol. Appl. Pharmacol.* 2005, *207*, (Suppl. 2), S293–S301. With permission.

used for the analysis of DNA adducts. In choosing among the different techniques, the amount of DNA sample available and the chemical nature of the DNA adduct should be considered.

5.2   ANALYSIS OF DNA ADDUCTS BY MASS SPECTROMETRY

5.2.1   GENERAL OVERVIEW

Initial applications of mass spectrometry to the study of carcinogen-related biomarkers focused on protein adducts [71–82]. There are several reasons for this. First, adducted proteins are available in greater quantity (several hundred milligrams) from in vivo studies in contrast to adducted DNA where only microgram quantities are normally available. Second, the earlier MS technology was based on compound separation by gas chromatography (GC) using electron impact ionization (EI) or chemical ionization (CI), which was more applicable to modified amino acids than to modified nucleotides. In addition, the generally favorable and well-defined response of proteins or amino acids to MS ionization methods, especially in the positive ion detection mode, made the analysis of protein adducts simpler.

However, over the past 10 to 15 years, interest in the use of MS for the analysis of DNA adducts has been on the rise. DNA and DNA adduct analysis by MS has two major advantages over that of protein adducts: (1) With its four bases, DNA is quite "unique" in composition, and any sequence can be analyzed with virtually the same setting of the mass spectrometer; and (2) each nucleated cell carries two copies of genomic DNA, and as a consequence there is no problem with dynamic range.

Historically, the role of MS in the determination of DNA adducts had been limited to providing information for the identification of new DNA adducts or for the structural characterization of DNA adduct standards that have been used to determine adduct levels by other detection methods, such as ^{32}P-postlabeling. However, MS has gained quite a momentum in recent years largely owing to the development of improved chromatographic interfaces and ongoing technological advances in ionization methods, in particular ESI as well as in ion transmission and detection. These improvements have led to significant improvements in sensitivity and efficiency compared with a decade ago. In contrast, the sensitivity of ^{32}P-postlabeling has not improved any further while, at least for targeted analysis, the limit of detection of MS can reach 1 adduct in 10^9 nucleotides and the DNA quantity used for such adduct analysis can be as low as ca. 10 μg. Therefore, the power of MS to achieve qualitative and quantitative analyses of human DNA adducts has increased greatly, and it is rapidly becoming the method of choice for DNA adduct analysis.

In fact, the growing use of MS for the analysis of DNA adducts can be gauged by the increasing number of publications between year 1996 and 2006 (SciFinder). Traditional ionization methods such as EI and CI found limited use but the development of fast atom bombardment (FAB) offered new opportunities for direct analysis of DNA adducts, albeit with somewhat disappointing sensitivity. MALDI accounted for a small portion of the DNA adduct analysis as it is an off-line technique. Gas chromatography in combination with negative ion chemical ionization (GC-NICI-MS) has provided significant improvement in sensitivity, but its usage

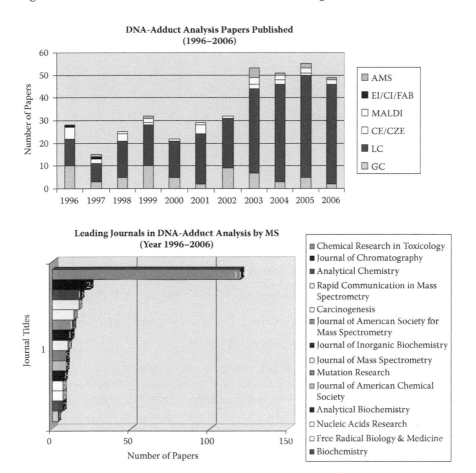

FIGURE 5.2 (*See color insert following page 152.*) Number of papers on DNA adducts categorized by the techniques used and leading journals with most publications on the analysis of DNA adducts between years 1996 and 2006, searched through SciFinder.

has gradually diminished with the steady growth of LC-MS following the development of electrospray ionization techniques. These trends in MS applications to DNA adducts analysis are illustrated in Figure 5.2, and in this regard, it is also important to identify the leading journals in which these publications have been appearing.

The fundamental approach to the analysis of DNA adducts by MS methods is to some extent analogous to that used in ^{32}P-postlabeling. Specifically, using as an illustration Figure 5.3, which shows a short segment of DNA with a PAH carcinogen attached to a nucleobase, the first step in the analysis involves enzymatic hydrolysis of the DNA polymer (or oligomer) into monomeric units, either nucleosides or nucleotides. For purposes of the mass spectrometric analysis, these monomeric units can be viewed as consisting of effectively three "blocks": the base (typically **G** and occasionally **A**), the deoxyribose (or deoxyribose monophosphate), and the carcinogen. Alternatively, if interested in identifying the sequence context of adduct attachment (which is the main objective of this review), enzymes capable of hydrolyzing

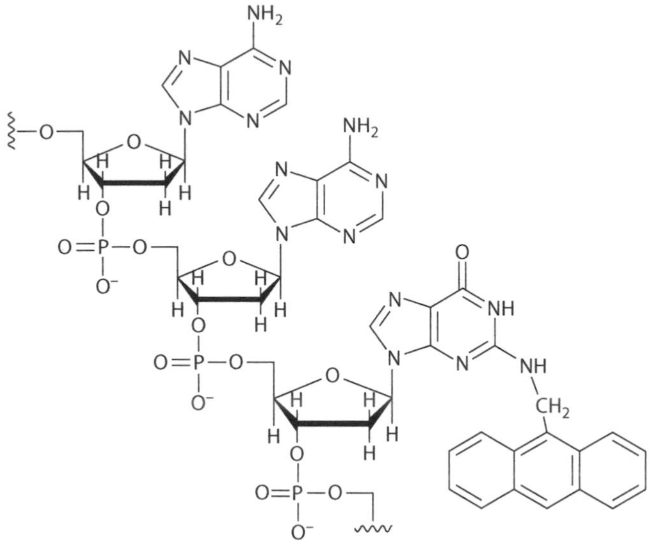

FIGURE 5.3 A segment of DNA showing formation of a PAH adduct on the exocyclic N^2-position of deoxyguanosine.

the DNA into oligonucleotide adducts may be used. In either case, extensive sample pretreatment needs to be conducted to remove the bulk of the unmodified DNA and reconstitute the adducts into a final volume enriched in the analytes of interest. The specific sample cleanup procedures may differ depending on the type of adducts involved and will not be discussed any further in this review.

In line with the previous considerations, the approaches taken for the analysis of DNA adducts, whether monomers or oligomers, by mass spectrometry have varied over the years, and, as might be expected, they followed the technological developments associated with MS ionization and ion separation methods. While this review is intended to focus primarily on developments related to the analysis of oligonucleotide adducts, a brief discussion of representative MS approaches used for the analysis of monomeric adducts have been included to provide the broader perspective of the field.

The clear advantage of high-performance liquid chromatography (HPLC) over GC is the ability to work with thermally labile compounds, and, when viewed in the context of DNA adducts, this translates into the ability to handle intact DNA adducts either in their nucleoside, nucleotide, or oligonucleotide forms. We present here highlights of HPLC-MS applications to the analysis of DNA adducts in that sequence. Straub and Burlingame as early as 1981 wrote an extensive review on the use of mass spectrometry for the analysis of xenobiotic-modified nucleic acids [83]. At that time DNA adducts were often first separated by LC or CE and analyzed by off-line mass spectrometric methods. Subsequent reviews have focused on the analysis of DNA adducts by capillary methods, both CE and LC, coupled to mass spectrometry [84,85]. Turesky and Vouros [86] reviewed the principal methods for synthesis and characterization of DNA adducts and summarized the detection of the heterocyclic aromatic amine adducts in experimental animals and humans. The features of different analytical methods including LC-MS, accelerator mass spectrometry (AMS), and ^{32}P-postlabeling as applied to the measurement of heterocyclic aromatic amine (HAA-DNA) adducts in vitro and in vivo were discussed. Other important reviews include those of Bohr [87], Esaka et al. [70], Esmans et al. [88], Garner [4], Phillips et al. [89], de Kok [90], Doerge et al. [91], Stiborova et al. [92], and Farmer et al. [2]. Perhaps the most detailed and comprehensive coverage of the most recent literature can be found in a review by Singh et al. [1].

5.2.2 ANALYSIS OF DNA ADDUCTS IN THE FORM OF MONONUCLEOSIDES

The introduction of electrospray ionization in combination with tandem mass spectrometry opened numerous new opportunities for the analysis of DNA adducts. In 1995 Chaudhary et al. published the first paper using LC-ESI-MS methods for nucleoside adducts analysis [93]. Referring back to the fundamental structural features of a DNA adduct (Figure 5.3), it is almost fortuitous that the MS/MS behavior makes them ideally suited for low-level detection and quantification. Notably, a favorable collision-induced dissociation (CID) of the [M+H][+] ions of adducts in positive ESI ionization involves the loss of the deoxyribose block (116 amu). A low-intensity fragment at m/z 117 amu is also observed occasionally and corresponds to the protonated 2′-deoxyribose "block." The facile cleavage of the glycosidic bond lends itself for trace level detection and quantification with selective reaction monitoring (SRM) using the transition [M+H][+] → [M+H-116][+] in triple quadrupole mass spectrometers. In addition, since the fragmentation appears to be common to almost all examples of nucleoside adducts reported in the literature, operation of the tandem MS in the constant neutral loss (CNL) mode allows selective screening of complex chromatographic mixtures for the presence of adducts. Although the sensitivity of CNL is much lower than SRM, its selectivity does yield information for the determination of different DNA adducts in mixtures.

Numerous applications of LC-MS/MS are based on this characteristic behavior of nucleoside adducts. In what was perhaps one of the earliest applications of capillary LC-MS/MS to the analysis of DNA adducts, Wolf and Vouros [94] capitalized on this process to confirm the presence of deoxyguanosine adducts of N-acetylaminofluorene (AAF) in calf thymus DNA (ctDNA) reacted with N-acetoxy-2-acetylaminofluorene (AAAF). Moreover, a CNL screen of the same mixture allowed the identification of deoxyadenosine adducts whose occurrence had not been previously verified.

The commonality of this fragmentation is illustrated in the example of Figure 5.4, which shows the CID spectra of two isomeric adducts, dG-C8-IQ and dG-N²-IQ, produced from the heterocyclic aromatic amine 2-amino-3-methylimidazo[4,5-*f*] quinoline (IQ), a carcinogen generated during the cooking of beef or fish. Both spectra exhibit the characteristic loss of 116 amu, whereas an additional major peak at m/z 331 (loss of NH_3 from m/z 348) is distinctly more intense in the spectrum of the dG-N²-isomer [95]. These general trends in the fragmentation patterns of DNA adducts are further illustrated in Figure 5.5, which shows the MS/MS spectra of a series of nucleoside, nucleotide, and depurinated adducts [1].

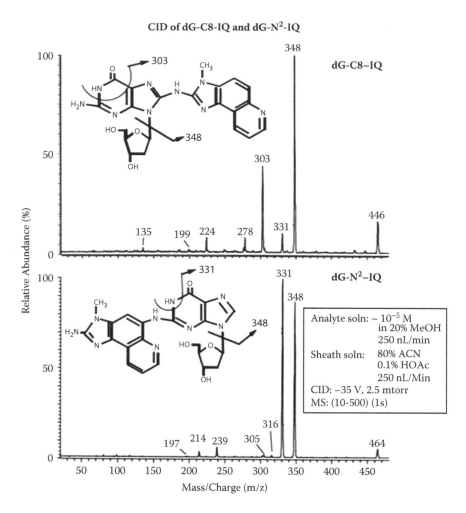

FIGURE 5.4 Product ion spectra of (a) dG-C8-IQ and (b) dG-N²-IQ. (From Soglia, J. R. et al., *Anal. Chem.* 2001, *73*, 2819–2827. With permission.)

5.2.2.1 Capillary LC-MS

The advances in capillary LC (CapLC) technology and the high ionization efficiency of nanoESI have now brought LC-MS within the range of human applications for the analysis of DNA adducts. The improved mass sensitivity was significant when compared with normal bore LC-ESI-MS due to improved ionization efficiency at the lower flow rates. Abian et al. [96] reviewed the practical considerations and technical aspects that need to be taken into account for the implementation of capLC with ESI-MS. The benefits derived from the use of capillary chromatography and nano (or micro) ESI systems, specifically as they pertain to DNA adducts have been discussed by the Esmans group. Applications to the analysis of a wide variety of adducts have been reported by this group including those of estrogen metabolites and other mustards [97–102].

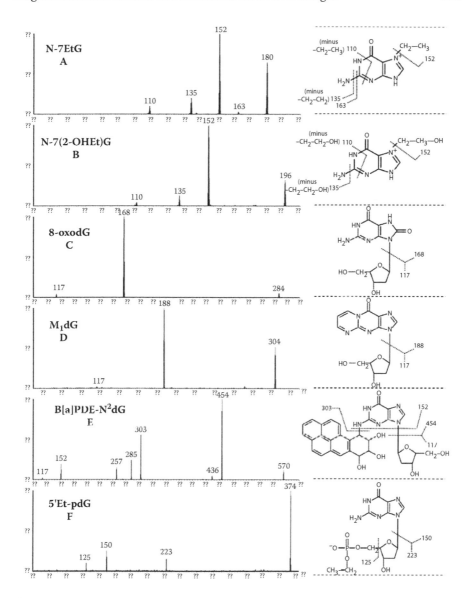

FIGURE 5.5 Typical ESI-MS/MS CID product ion spectra obtained using a Micromass Quattro Ultima Pt. tandem quadrupole mass spectrometer (Micromass, Waters Ltd., Manchester, UK) for different guanine $[M+H]^+$ precursor ions of nucleic acid base, 2′-deoxynucleoside and $[M−H]^−$ precursor ion of 2′-deoxynucleotide DNA adducts. (From Singh, R. et al., *Chem. Res. Toxicol.* 2006, *19*, 868–878. With permission.)

The special features offered by capillary LC have been extensively exploited by various groups, and considerable improvements have been achieved over the past several years by exploring the use of progressively smaller diameter columns, with a 75 μm i.d. column providing a reasonable compromise in terms of sample capacity (loadability) and sensitivity. An early design of a capillary interface was used in

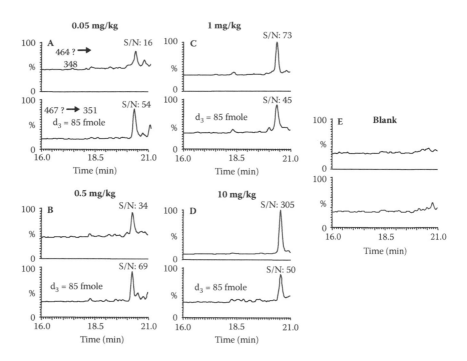

FIGURE 5.6 Selected chromatograms of liver DNA from rats dosed with IQ at different levels. (From Soglia, J. R. et al., *Anal. Chem.* 2001, *73*, 2819–2827. With permission.)

a dose–response study in which rats were administered the carcinogen IQ and the levels of dG-C8-IQ in liver were determined by LC-MS/MS [95]. Each isolated DNA sample was spiked with 85 fmol of isotopically labeled internal standard, and, following its purification, the digest was reconstituted into a final volume of 20 μL, of which 1 μL was injected into the LC-MS for analysis. Chromatograms representing the detection of the adduct at each level of administration are shown in Figure 5.6. The profiles of the isotopically labeled internal standard are included in each panel, where the actual mass of internal standard injected into the LC-MS is 1/20 of the 85 fmol (i.e., *ca.* 4 fmol). Adduct levels ranged from as little as 3.5 modifications per 10^8 normal bases at the lowest dose (0.05 mg/kg) to as high as 38 per 10^8 at the highest dose (10 mg/kg) with relative standard deviation (RSD) of less than 20%.

5.2.3 Analysis of DNA Adducts in the Form of Mononucleotides

As noted already, DNA adducts are analyzed by LC-ESI/MS primarily in the form of nucleosides and in the positive ion mode. Detection of nucleoside adducts in positive ESI is typically at least an order of magnitude more sensitive than detection of the corresponding nucleotide adduct, which has to be analyzed in the negative ion mode due to the presence of the phosphate group. Moreover, nucleotide adducts are often too hydrophilic to be separated by the most popular reversed-phase-HPLC (RP-HPLC) methods, and, as a result, best separations are achieved using ion pairing (IP) methods. The high electrolyte content of the mobile phase, however, introduces ion suppression effects in the electrospray ionization, which impacts negatively on the sensitivity of the assays. There are of course exceptions such as adducts of thymidine glycol and 5-hydroxy-2'-deoxyuridine, which are actually detected with greater sensitivity in negative ESI compared with positive ESI and undergo cleavage of the glycosidic bond following CID. Nucleotides may also be first separated by anion-exchange SPE followed by enzymatic dephosphorylation and LC-MS analysis of the resultant nucleosides. The addition of extra cleanup steps, however, is not encouraged as it may lead to sample losses particularly when dealing with trace levels of adducts. To overcome these problems, attention has been turned to the use of capillary zone electrophoresis (CZE-MS), a brief discussion of which follows later.

Despite the aforementioned limitations, analysis of DNA adducts, whether monomeric or oligomeric, in the nucleotide form offers the additional advantage of being able to detect alkylation to the phosphate group (i.e., damage to the DNA backbone), which would be missed in the analysis of the dephosphorylated analog. An excellent example of particular current interest is the identification of phosphate adducts of phenylglycidyl ether in studies conducted by the Esmans group in the 1990s on bisphenol A (BPA) and related analogs. BPA was first synthesized in 1891, and BPA products have been on the market for over 50 years as more than 2 billion lb are used annually in the United States [103]. BPA is a principal reactant in the preparation of polycarbonate (PC) plastics and is also widely used to make epoxy-based polymer resins or to strengthen plastics as plastic additives, lacquers, or surfactants. BPA and analogs show up in canned food, beverages, and utensils [104–107]. Reusable baby bottles and other housewares are often made of PC. In consumer products, BPA analogs can also be found in compact discs, eyeglasses, bicycle helmets, and automotive parts. Their presence is found in environmental samples as well, namely, wastewaters and sewage [108]. BPA on dental treatment could be released into saliva from dentures, which contain resins [109].

BPA has been studied for 40 years, and there have always been controversial reports regarding its benefits and risks. In recent years much concern has been raised regarding the health risks of low-level exposure to BPA despite two opposing views expressed by its advocates and opponents [110,111]. Environmental exposure to these phenolic compounds that can act like the hormone estrogen has been associated with adverse reproductive and developmental effects in wildlife and humans, which potentially leads to breast cancer and other hormonally-mediated health outcomes [112]. Numerous studies published in the past decades have linked BPA exposure to increased rates of prostate and breast cancer, reproductive abnormalities, decreased sperm count, accelerated puberty in females, neurological effects similar to attention deficit hyperactivity disorder (ADHD), diabetes, and obesity in animal studies [113]. It was also linked to heart disease. 92% of Americans age 6 or older test positive for BPA according to the data from the Center for Disease Control and Prevention (CDC). Bisphenol A can be present in urine, serum, and plasma samples [114–116].

A broad range of DNA adducts of bisphenol A diglicydyl ether (BPADGE) have been identified by LC-MS/MS in in vitro and in vivo studies conducted by the Esmans group [103]. The detection of pairs of isomeric BPADGE adducts of dGMP, dAMP, and dCMP is illustrated in the SRM chromatograms of Figure 5.7. In each case, the presence of the prominent ion at m/z 195 (the deoxyribose phosphate ion)

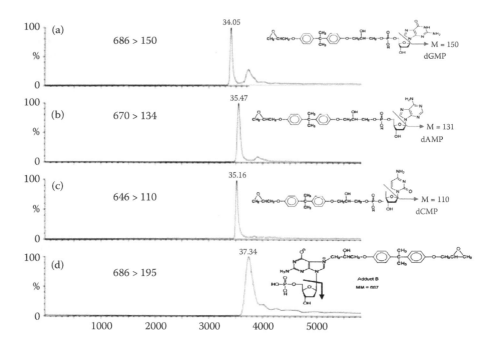

FIGURE 5.7 Injection of 1 µL of the dAMP/dCMP/dGMP/BPADGE in vitro reaction mixture onto the NanoFlow ES LC/MS column-switching system under MRM conditions. (a–c) Phosphate-alkylated adducts (dGMP/BPADGE, dAMP/BPADGE and dCMP/BPADGE, respectively). (d) Base-alkylated adduct (dGMP/BPADGE). (From Vanhoutte, K. et al., *Anal. Chem.* 1997, *69*, 3161–3168. With permission.)

in the MS/MS spectra indicates adduct formation through base modification, and the transition [M-H]⁻ → m/z 195 can be used to recognize those adducts formed by alkylation on the nucleobase (Figure 5.7d). On the other hand, phosphate alkylation results in alternative fragmentation pathways and spectra void of m/z 195 fragments (Figures 5.7a through 5.7c). Instead, the MS/MS spectra of the phosphate alkylated adducts exhibit characteristic fragments corresponding in mass to the respective nucleobases (m/z 150 for G; m/z 134 for A; m/z 110 for C). The structure of a base-alkylated adduct identified in these studies are shown in Figure 5.7d.

5.2.4   OLIGONUCLEOTIDE ADDUCTS

5.2.4.1   Significance of Oligonucleotide Adducts

Analysis of DNA adducts as monomeric species such as nucleosides or nucleotides fails to provide information about the adducts location within the DNA sequence. However, it has been observed that reaction of carcinogens with DNA often exhibits sequence selectivity, the recognition of which may be important to understand their mutagenic activities [117–137].

Following exposure to these genotoxic carcinogens, characteristic mutation spectra then result, which show a degree of site selectivity dependent on base-sequence context in production and potency of mutations. Sequence selectivity of adduct formation is manifested in a number of oncogenes. For example, mutational hotspots in the *p53* gene are particularly frequent in domains that include CpG dinucleotides [6]. This gene has been under extensive scrutiny and shows strong evidence linking adducts, mutation, and cancer. The specific mutations appear to be present in liver and lung cancers in particular. There are at least three separate databases on *p53* mutations, one in IARC with over 10,000 mutations. The predominant hotspots often correlate with major DNA adducts formation [118,120,121]. The mutation efficiencies of different adducts are different, ranging from barely mutagenic to those that frequently cause mutations. It has been determined that several factors affect the ability of an adduct to induce mutation. These include orientation of the modifying group especially with respect to the base-pairing face, size, and conformation of the group, nature of polymerase, sequence context, and rearrangement of the adducted base due to tautomerization, rotation, and wobbling. Many DNA repair deficiency syndromes involve vast increases in risk of cancer relevant to DNA adducts. The sequence selectivity or, in other words, the site specificity found in a few mutants is striking yet unexplained. Still, the prevailing hypothesis is that the occurrence of a DNA lesion via formation of an adduct is dependent on base sequence, as is the potential for its subsequent repair via, for example, nucleotide excision or base excision repair (NER or BER) processes [138,139].

In view of these considerations, the assignment of the sites of modification by different carcinogens within a target sequence is a very important topic in cancer etiology since different adduction sites may induce different mutagenic activities due to different biological responses to the presence of the adducts. Therefore, a critical part in the analysis of DNA adducts is to develop a method to determine the site and frequency of modification on the DNA strands to better understand the relationship between the chemical behavior of different carcinogens and mutation hot spots. To accomplish this goal, it is necessary to have a fast and accurate methodology for structure characterization of oligonucleotide–carcinogen adducts from in vitro and in vivo sources using online separation coupled with tandem MS. Although HPLC is presently the preferred method for oligonucleotide analysis, other separation methods, most notably capillary zone electrophoresis (CZE) and capillary electrochromatography (CEC), have also been used.

After a brief summary of the principles associated with the CID mass spectra of oligonucleotides, the present status and key developments in the characterization of oligonucleotide adducts by electrophoretic and chromatographic separation methods coupled to MS/MS are summarized. Much of the effort in this area has as a major objective the investigation of the effect of neighboring bases on adduction site and the potential relationship between chemoselective and mutation sites. Examples of pertinent applications of these techniques are presented to illustrate the analytical challenges.

5.2.4.2 Sequencing of Oligonucleotides and Oligonucleotide Adducts by MS/MS

The ability of ESI-MS to form multiply charged ions in combination with improved mass range in modern MS instruments has made possible the detection of increasingly larger oligonucleotide fragments. In practice, because oligonucleotides have an acidic phosphodiester backbone, oligonucleotide adducts are usually analyzed by ESI in negative ion more than in the positive ion mode. Negative ESI requires careful adjustment of sample conditions and instrument tuning to obtain sensitivity comparable to positive ESI. MS analysis can provide a basis for detecting both length and sequence variations of oligonucleotides, yet this is complicated by multiply charged ions and metal adduct ions. In ESI a mass spectrum may contain an envelope of peaks that corresponds to ions with various charge states with a Gaussian-like distribution. To further complicate the picture, metal adduct ions are formed due to the affinity of the polyanionic backbone for ubiquitous cations, such as sodium and potassium. These multiply charged ions and the $Na+$ and $K+$ adduct ions (often referred to as quasi-molecular ions) contribute their own signals adding a further degree of complexity to the spectral pattern. As a result, the ability to characterize mixtures of oligonucleotides may be compromised. To overcome this problem, many different methods are used for sample desalting and purification to exchange the nonvolatile cations (e.g., $Na+$, $K+$) with protons ($H+$) or ammonium ions (NH_4+).

Sequence determination of DNA was made possible by MS/MS based on the gas-phase CID of multiply charged oligonucleotide ions. McLuckey et al. [140–142] developed a system for sequence determination of oligonucleotides based on well-defined fragmentation pathways in their product-ion spectra. Starting from the 5′-terminus (3′-terminus), a-, b-, c-, and d- fragment ions (z-, y-, x-, and w-fragment ions) are generated on fragmentation at the different positions of the phosphodiester group as illustrated in Figure 5.8.

The first fragmentation step is usually the elimination of a nucleobase, followed by cleavage of the 3′-phosphodiester bond of the nucleotide that suffered the base loss. Under low-energy CID conditions, fragments of the a-B and w- ions are the predominant types of species, where -B indicates the loss of the nucleobase. The resulting series of 5′- and 3′- ions (Figure 5.8b) have characteristic mass differences, from which the sequence can be deduced (Figure 5.8c).

Because of the large number of fragment ions present in the MS/MS spectra of oligonucleotides, manual spectral interpretation is difficult, time-consuming, and prone to error. As a result, efforts have been made to simplify the process of relating MS fragmentation pattern with structure. A computer-based algorithm for the sequencing of oligonucleotides of completely unknown sequence was developed by Ni et al. [143] to automatically derive sequence information from the MS/MS spectra. The algorithm works by extending from the 5′ (a-B ions) and 3′ (w- ions) ends ion series that encode the complete DNA sequence. Mass ladders are identified by sequentially adding each of the four possible nucleotide masses and by searching the spectrum for the best match of the expected ions. Rozenski and McCloskey [144] also developed a novel approach for nearest-neighbor determination based on the analysis of fragment ions of the nucleic acid formed in the ionization region of the mass spectrometer along with the fragment ions from MS/MS. They later demonstrated the ab initio determination of unknown oligonucleotide sequences at approximately the 12-mer level and below, and the approach was termed simple oligonucleotide sequencer (SOS) [145]. However, this method cannot be extended to

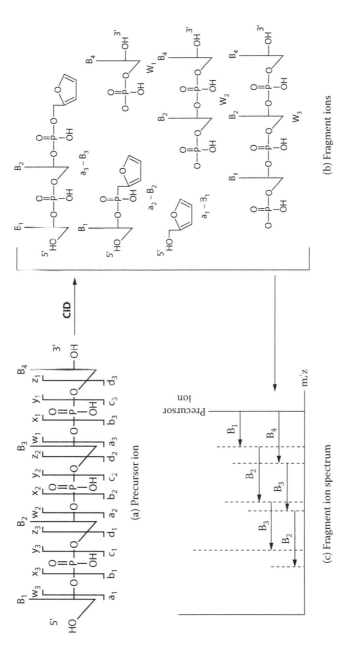

FIGURE 5.8 Scheme for the sequencing of DNA by MS/MS upon CID of a multiply charged precursor ion (charges are omitted in the structures).

longer oligonucleotide sequences because usually several missing fragments in the series of a-B or w ions prevent the successful identification of the whole sequence.

5.2.4.3 Capillary Electrophoresis MS

Because of their negative charge, nucleotide and oligonucleotide adducts are amenable to analysis by electrophoretic methods. Since the phosphate group is negatively charged over a wide pH range, electroosmotic flow conditions in traditional CZE provide an ideal mode for their separation. Compared with HPLC, another advantage of CZE is the ability to work with small sample quantities, and, in cases where one is sample limited, it is possible to perform multiple measurements from a single sample. In addition, sample throughput may be considerably higher because there is no dead time for column reequilibration. Despite these inherent advantages, a clear limitation of CE/CZE is the relatively small size of sampling volume (1–10 nL) that can be introduced into the CE-MS system. Thus, although the mass detection limit in CE can be very low, the concentration detection limit may be as high as 10^{-6} M or greater depending on the detection method.

The concentration detection limits in CZE can be improved by online enrichment using sample stacking techniques [146]. As shown by Wolf and Vouros [147], the method resulted in 1000-fold improvement in the CZE-MS analysis of nucleotide adducts of AAF and injection volumes as high as 2–3 µL were possible. Deforce and Van den Eeckhout [148] described the methodology developed in the analysis of DNA damage by CZE-ESI-MS and a sample stacking procedure that was able to improve the detection. Willems et al. [149] analyzed BPDE DNA adducts by CE-MS in conjunction with sample stacking. Gennaro et al. [150] developed a sample preparation approach that involved a variation on the digestion procedure, in combination with the use of metal affinity ZipTips to achieve more efficient cleanup of the BPDE-DNA adducts formed in vitro and subjected the bulky hydrophobic adducts to CE/MS analysis. The value of sample stacking was further illustrated in a study of calf thymus DNA exposed to BPDE, a major metabolite of benzo[a]pyrene (BaP) that is the prototypic carcinogen in the class of PAHs. The CZE-MS electropherogram of the sample obtained by hydrodynamic injection of approximately 20 nL of the solution on-column shows the presence of two isomeric BPDE-dG adducts generated by the cis/trans-opening of the epoxide ring upon reaction with deoxyguanosine. Subsequent analysis of the sample by introduction of 2 µL (i.e., 100-fold larger volume) of the same solution via sample stacking produced an electropherogram in which the two isomers were no longer resolved due to column overloading. However, despite the loss of resolution, the increased dynamic range revealed the presence of minor components, in this case identified as oligonucleotide adducts that were present because of incomplete digestion of the DNA [151].

An alternative online enrichment technique using a constant pressure-assisted electrokinetic injection (PAEKI) was developed by Feng et al. of Health Canada [152]. The technique uses external pressure to counterbalance the electroosmotic flow (EOF) during sample introduction and was shown to achieve a 300- to 800-fold improvement in sample concentration for ss and ds oligonucleotides during a 90 second injection. PAEKI was used for the identification and measurement of DNA oligonucleotides and their BPDE adducts by CZE-MS with sensitivities at the low

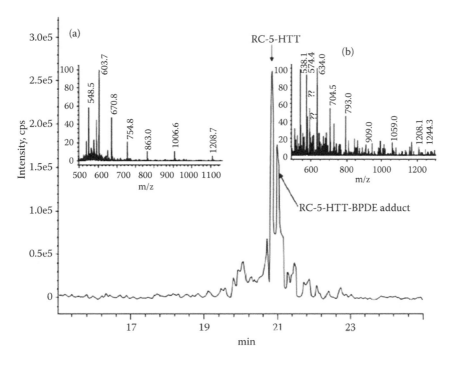

FIGURE 5.9 CE/MS electropherogram and negative ion mass spectra of BPDE and RC-5-HTT reaction mixture (TIC MS signal). PAEKI injection: −16 kV with counter-balance pressure of 50 mbar for 90 s. (a) Mass spectrum of RC-5-HTT. (b) Mass spectrum of BPDE-RC-5-HTT adduct. Sample: RC-5-HTT and RC-5-HTT-BPDE adducts mixture in water from the procedure in synthesis of BPDE-oligonucleotide adducts. (From Feng, Y. L. et al., *J. Chromatogr., A*. 2007, *1148*, 244–249. With permission.)

micromolar level as shown in Figure 5.9. Figures 5.9a and 5.9b show the mass spectra of the oligonucleotide RC-5-HTT (5′-CCAGACATTGCCCAGGTCCA-3′, a serotonin transporter, MW 6047 Da) and of its BPDE-RC-5-HTT adduct (MW 6349 Da), respectively. The mass difference of 302 Da corresponds to the incorporation of the BPDE moiety and provides for their good chromatographic resolution. The study demonstrated that the PAEKI technique can be used to deliver a considerably greater amount of oligonucleotides and their adducts onto a capillary column for the determination of their structures by CZE-MS.

Separation of isomeric oligonucleotide adducts by capillary electrophoretic methods is also common. For example, Barry et al. [153] described the CE-MS analysis of a series of modified oligonucleotides four, five, and six bases in length using a coated fused-silica column filled with an aqueous solution of polyvinylpyrrolidone (PVP). Oligonucleotides of the same length and sequence but differing only by the presence or absence of a small modification (e.g., a methyl group) on a single base were readily resolved. In a subsequent publication, Harsch and Vouros [154] further improved on this methodology to identify isomeric AAF-modified oligonucleotides as shown in Figure 5.10.

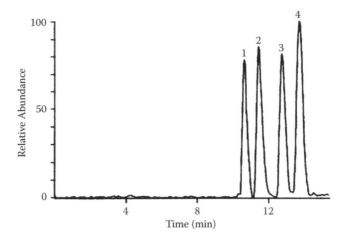

FIGURE 5.10 CE/MS in a PVP matrix. Analysis of four isomeric, AAF-modified oligo-nucleotides: ATG*CTA, ATTCAG*, TG*TAAC, CATG*AT. Field strength, ~450 V/cm. On-line MS detection of the doubly charged molecular ion at m/z 1005. (From Harsch, A. and Vouros, P., *Anal. Chem.* 1998, *70*, 3021–3027. With permission.)

5.2.4.4 Analysis by Liquid Chromatography–Mass Spectrometry

As noted earlier, nucleotide analysis has proven to be cumbersome for RP-LC owing to the polar nature of the phosphate groups. Due to the acidity of the phosphodiester bridges, DNA dissociates into polyanions in aqueous or hydro-organic solution as well as in the gas phase. Although adducts are quite stable in neutral or slightly basic solution, loss of the nucleobases followed by backbone cleavage at acidic pH or hydrolysis of the phosphodiester groups at basic pH results in rapid degradation of DNA. IP-RP-HPLC can overcome many of these problems by virtue of its high-resolving capability for DNA although often at the expense of sensitivity due to ion suppression effects in ESI.

The magnitude of electrostatic interaction, and thus retention, is determined by several factors, including (1) hydrophobicity of the column packing; (2) charge; (3) hydrophobicity and concentration of the pairing ion; (4) ionic strength, temperature, and dielectric constant of the mobile phase; (5) concentration of organic modifier; and (6) charge and size of the DNA molecule. Elution of the adsorbed DNA is effected by a decrease in the surface potential due to desorption of the amphiphilic ions from the stationary phase with a gradient of increasing organic modifier concentration. Because the number of charges uniformly increases with size, double-stranded (ds) DNA molecules are separated according to chain length in IP-RP-HPLC. The separations of DNA are usually more efficient at elevated temperature such as between 25°C and 80°C to reduce the probability of formation of double-stranded oligomers.

Challenges associated with the chromatographic analysis of oligonucleotides have been reviewed by Huber and Oberacher [155]. For example, Belicher and Bayer separated oligonucleotides up to 24-mers by RP-HPLC in a 100 × 2 mm i.d.

Nucleosil C18 column with gradients of acetonitrile in 10 mM ammonium acetate [156,157]. However, more efficient separations can be accomplished via the use of various trialkyl- and tetraalkylammonium salts. The positively charged, hydrophobic triethylammonium ions are adsorbed onto the nonpolar surface of the stationary phase, resulting in the formation of an electric double layer that has an excess of positive charges near the surface.

Huber and Krajete [158] used 100 mM triethylammonium acetate buffer for the separation of single-stranded (ss) oligonucleotides with single nucleotide resolution ranging in size from 3-mers to longer than 80-mers. Other reagents include diisopropylammonium acetate used to separate oligonucleotides up to 20-mers in a 150 × 0.5 mm i.d. column packed with a polymeric PLRP-S 5 μm 100 A stationary phase in acetonitrile-water [159]. Apffel et al. [160,161] analyzed oligonucleotides up to 75-mers by using 400 mM 1,1,1,3,3,3-hexafluoro-2-propanol (HFIP)/2.2 mM TEA as ion-pairing reagents and methanol as organic modifier. Triethylammonium (TEAB) and butyldimethylammonium bicarbonate (BDAB) have provided attractive alternatives in several applications [158,162,163]. In general, the criteria for HPLC analysis of oligonucleotide adducts have been based on those developed for separation of unmodified oligonucleotides.

5.2.4.4.1 Monolithic Columns

The technique's history and developments have been reviewed by Svec and Huber [164]. Monolithic columns are noted for their extremely fast chromatographic separations at high flow rates and at reasonably low back pressure. The high-throughput feature is very advantageous toward biomolecule analysis, as the mass transfer of large molecules such as polymers, peptides, proteins, and nucleic acids is considerably slower than that of small molecules due to their much lower (by several orders of magnitude) diffusion coefficients. Monolithic columns have unique properties such as tolerance to high flow rates and rapid speed of separation and combined with nano LC can achieve very high sensitivity. Higher separation efficiency goes hand in hand with higher permeability and higher flow rates. It has been demonstrated that low-attomole (10^{-18} *mol*) and even zeptomole (10^{-21} *mol*) detection sensitivity can be achieved for peptide/protein analysis using a monolithic column in LC-ESI-MS [165].

A great deal of activity is currently focused on the optimization of monolithic columns for separation of small and large molecules. In the preparation of monolithic columns, two sets of parameters can be simultaneously optimized by using chemical methods to polymerize liquid precursors into a continuous porous mass of coalesced particles. One set are the parameters relating to the nature of material, porosity, and other properties that affect the separation, and another set are the size of the channels and open spaces that dictate the materials' permeability and give them their sponge-like structures. Effective monolithic columns can be prepared from styrene, divinylbenzene, and other monomers. In a direct comparison, monolithic columns made of polystyrenedivinyl benzene (PSDVB) phase performed 30–40% better than 2 μm, nonporous PSDVB counterparts packed with microparticulate stationary phases. Single- and double-stranded DNA fragments ranging in size from 51 to 587 base pairs were analyzed by IP-RP-HPLC-ESI-MS [166]. An example of their applicability to the analysis of oligonucleotide adducts are discussed later in this review.

5.2.4.4.2   Hydrophobic Interaction Liquid Chromatography (HILIC)

As noted already, analysis of oligonucleotide adducts by conventional reversed-phased LC requires the use of ion pairing techniques that, however, introduces ion suppression effects in their detection by electrospray ionization in LC-MS. Polar columns in combination with mobile phases containing a large percentage of organic solvents can overcome or at least reduce many of these problems [167,168]. Interesting recent examples include the use of an aminopropyl column for the analysis 2'-C-methylcytidine triphosphate in rat liver [169] and a silica hydride column for the analysis of nucleotide mixtures [170]. The use of mobile phases containing volatile components such as acetonitrile/water solvents and ammonium acetate buffer facilitated their use in combination with electrospray ionization MS. While the application HILIC chromatography in combination with MS to the analysis of oligonucleotide adducts has not been actively pursued as yet, this may well be a promising approach for future research in areas of trace level detection of such markers.

5.2.4.5   Mapping of Adduction Sites in Oligonucleotide Adducts by LC-MS: Selected Examples

In analyzing oligonucleotide adducts, two fundamentally different approaches are being pursued. The first, introduced by Tretyakova et al. [171], may be viewed as an indirect approach and relies on the use of stable isotope labeling of individual bases to analyze the distribution of DNA adducts by detecting the nucleoside adducts level produced by stepwise hydrolysis of the parent modified oligonucleotide [172,173]. The second involves the direct analysis of modified oligonucleotides using HPLC in combination with CID in tandem mass spectrometry. We will discuss the two methods and provide selected examples with primary emphasis on the second approach.

5.2.4.5.1   Recognition of Adduction Sites by Indirect LC-MS Approaches

Much of the work of Tretyakova and coworkers at the University of Minnesota Cancer Center has focused on adducts of tobacco specific carcinogens [13,17–20, 22,134,174,175]. The roles of DNA adducts of NNK, NNAL, BaP, BPDE, and POB in carcinogenesis with relation to sequence context at specific guanines within *K-ras* and *p53* gene sequences that are the mutation hotspots in lung tumors of smokers were investigated.

In the stable isotope approach described by Tretyakova et al. [176], ^{15}N-labeled guanine nucleobases are first placed at specific positions within DNA oligodeoxynucleotides representing a gene sequence of interest. The oligonucleotides and DNA are then subjected to enzymatic digestion to break down to mononucleoside adducts that are subsequently analyzed by HPLC-ESI/MS/MS in the positive ionization mode. The adducts formed at the ^{15}N-labeled guanine can be distinguished from the other adduction sites by their molecular weight, which is increased due to the presence of the ^{15}N atom. The nucleoside adducts are quantified by MRM, based on the facile loss of deoxyribose (M = 116) under the CID conditions. As demonstrated in

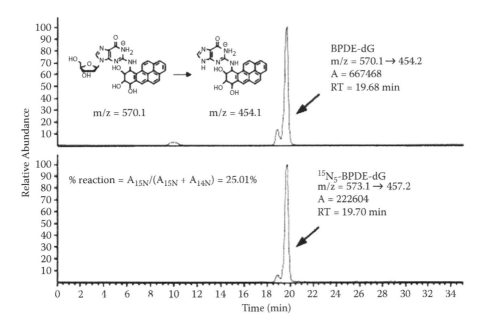

FIGURE 5.11 HPLC-ESI-MS/MS analysis of N^2-BPDE-dG in [$^{15}N_3$]-labeled *K-ras* derived DNA fragment, CCC GGC ACC ^{Me}CXC GTC CGC G (X = [$^{15}N_3$-dG]). Selected reaction monitoring was performed using the transitions: BPDE-dG, *m/z* 570.1 *m/z* 454.0; [$^{15}N_3$]-BPDE-dG, *m/z* 573.1 *m/z* 457.0. HPLC: Agilent 1100 series capillary liquid chromatograph (Agilent technologies). A Zorbax SB-C18 column (150 × 0.5 mm, 5 μm, Agilent Technologies) was eluted at a flow rate of 15 μL/min. HPLC solvents: A = 33% methanol in 15 mM ammonium acetate, B = acetonitrile, gradient 0–30% B in 22.5 min. MS: Finnigan MAT TSQ 7000 (ThermoQuest, San Jose, CA) operated in ESI+ mode. Spray voltage, 5 kV; collision gas pressure, 2 mT; heated capillary, 200C; electron multiplier, 2200 V. (From Tretyakova, N. et al., *Biochemistry* 2002, *41*, 9535–9544. With permission.)

Figure 5.11, the extent of adduct formation at the isotopically labeled nucleobase is calculated directly from the areas under the SRM peaks corresponding to ^{15}N and unlabeled adducts, respectively.

In the specific example shown in Figure 5.11, the authors compared the reactivity of different guanines within *p53* exon 5 toward BPDE by preparing synthetic oligodeoxynucleotides $CCG_1G_2CACCCG_3CG_4TCCG_5CG_6$ containing [^{15}N] label at one of the highlighted positions, G_1, G_3, G_4, or G_5 (codon 157 = G_4TC, codon 158 = CG_5C). They then replaced cytosine nucleobases with 5-Me-C in accordance with the patterns of endogenous cytosine methylation within CpG dinucleotides of the *p53* gene. They observed higher-than-average reactivity toward BPDE for all four methylated positions G_1, G_3, G_4, and G_1. The relative abundances of N^2-BPDE-dG adducts were in the following order: **G3** (codon 156) >> **G5** (codon 158) > G_1 (codon 154) ≈ G_4 (codon 157) (Figure 5.12). It was ascertained that N^2-BPDE-dG adduct formation is strongly affected by local sequence environment in the *K-ras* gene and there are important differences between the distributions of lung-cancer-associated mutation and the N^2-BPDE-dG adduct within a *p53* exon 5-derived gene.

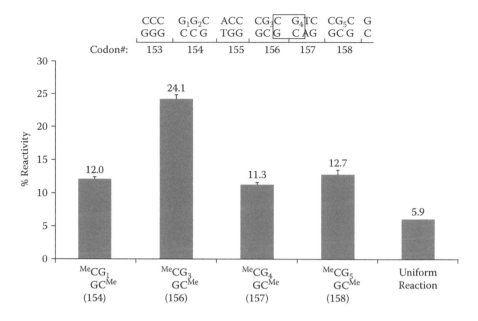

FIGURE 5.12 Relative formation of N^2-BPDE-dG at guanine nucleobases within a double-stranded oligodeoxynucleotide representing a region *p53* exon 5 containing frequently mutated codons 157 and 158: CCG$_1$G$_2$CACCCG$_3$CG$_4$TCCG$_5$CG). The data were compiled from two separate experiments (N = 3–6). The relative adduct formation at each guanine was calculated from the area ratio of HPLC-ESI-MS/MS peak corresponding to [^{15}N]-N^2-BPDE-dG to the sum of unlabeled and [^{15}N]-adduct peak areas. The random reaction value was determined from the total number of guanine nucleobases in both DNA strands. (From Tretyakova, N. et al., *Biochemistry* 2002, *41*, 9535–9544. With permission.)

This paper was their first report to fully describe a mass-spectrometry-based approach for determination and of DNA adducts at specific positions within DNA sequences to investigate the sequence selectivity of the adducts. The study also revealed a complex relationship between DNA sequence and reactivity toward adducts that cannot be simply explained by the influence of neighboring nucleobases. The same general approach was also used effectively by Guza et al. [14] to study the kinetics of O6-methyl-dG repair by O6-alkylguanine DNA alkyltransferase within *K-ras* gene-derived DNA sequences.

BaP exerts genotoxic effects through metabolic activation prior to reaction with DNA, which forms metabolite BPDE (18′-20′). Singh et al. [1] compared postlabeling method and LC-MS/MS on measuring the BaP-DNA formation and obtained a very good correlation between BaP-DNA adduct determination by ^{32}P-postlabeling and LC-MS/MS SRM [1]. Arlt et al. conducted extensive investigations of the role of BPDE in relation with mutation hotspot by various methods including ^{32}P-postlabeling, fluorescence, LC-MS, and LC-MS/MS. In a recent study (2008), they quantitated BPDE-DNA adduct levels by ^{32}P-postlabeling and subsequently confirmed the formation of dG-N2-BPDE by LC-MS/MS analysis using the stable-isotope approach with selected reaction monitoring (SRM) [177]. The major adduct was unequivocally identified as the production of the reaction of BPDE with guanine residues in DNA by ^{32}P-postlabeling and confirmed by MS. They reported the results of experiments on BaP activation by mouse hepatic microsomes and in various organs. Based on the results, they drew contrasting conclusions on the role of cytochrome P450s (CYPs) in the genotoxicity of BaP in vivo and in vitro. They also showed the specific role of hepatic CYP enzymes in the activating pathways of BaP in vitro. The biological consequence of the binding site in the context of their position in the genome is implied in these and other investigations.

5.2.4.5.2 Direct Sequencing of Oligonucleotide Adducts by LC-MS/MS

The method of Tretyakova and coworkers represented a significant advance in both sensitivity and adduct quantitation over traditional bioanalytical methods such as PAGE and fluorescence. However, the extensive synthetic effort, sample preparation, and handling in addition to digestion of oligonucleotides to monomers suggests that a more direct method based on characterization of the intact oligonucleotide adduct by CID would be more efficient for determination of the carcinogen-binding spectrum within a given ologonucleotide sequence. Tandem MS using CID was the obvious route, and, in accordance with this reasoning, Iannitti et al. [178] at the University of Wollongong of Australia used ESI-MS and ESI-MS/MS to examine the sequence selectivity of adducts formation between the antibiotic hedamycin and model 6-mer oligonucleotides. Samples were purified off-line and introduced by direct infusion into the ESI-MS/MS system. In a follow-up paper, Colgrave et al. [179] determined the sequence selectivity and stability of alkylated oligonucleotide adducts of duocarmycin C_2 (pyrindamuycin A), duocarmycin C_1 (pyrindamuycin B) hedamycin, and DC92-B by ESI-MS and ESI-MS/MS offline. Glover et al. [180] used tandem MS to fully sequence a 7-mer para-benzoquinone adducted oligonucleotide. They compared the different fragmentation pathways on different charge states and CID collision energy. Their results demonstrated that the modified oligonucleotides follow the same fragmentation patterns of unmodified oligonucleotides as discussed in a previous section.

In related work, Marzilli et al. [181] used ESI-ITMS (electrospray ionization ion trap mass spectrometry) to unambiguously identify the site of adduction in three isomeric oligonucleotides of the sequence 5'-CCGGAGGCC modified by the carcinogen aflatoxin B_1 (AFB$_1$) using MS, MS/MS and MS^3. Figure 5.13 (left panel) shows the product ions spectrum of the [M–3H]$^{3-}$ (m/z 1016) ion of one of the three isomers in which the site AFB$_1$ was located at the G4 position along the oligonucleotide chain. The structure was further confirmed by CID (MS^3) of the [M-B$_4$*]$^-$ ion at m/z 1286 by which two complementary product ions were generated (Figure 5.13; right panel).

The previously reported results demonstrated the utility of CID-MS/MS for mapping adduct distribution in short oligonucleotide sequences. However, all these investigations were conducted using model oligonucleotide sequences purified off-line and introduced into the MS by direct infusion. A natural follow-up to that was the coupling of CID with chromatographic separation and the challenging separation

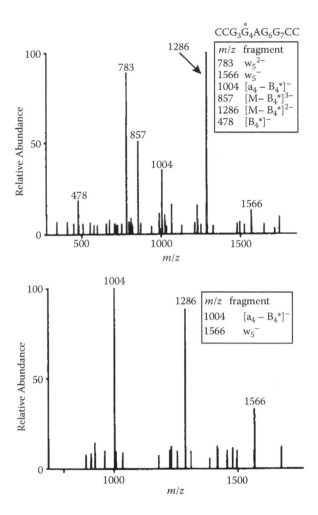

FIGURE 5.13 Left panel: MS/MS spectrum of the $[M-3H]^{3-}$ ion at m/z 1016 of AFB_1– 5'–CCGG*AGGCC adduct peak C. The modified base is designated with an asterisk. *Right Panel*: MS^3 spectrum (m/z 1016 → m/z 1286 → ?) of $[M-B^*_4]^{2-}$ ion. The result is cleavage of the phosphodiester backbone at the site of modification. (From Marzilli, L. A. *J. Am. Soc. Mass Spectrom.* 1998, *9*, 676–682. With permission.)

of the resulting positional isomers. As a first step toward that goal, a number of studies have been carried out in which short oligonucleotides (generally < 15-mers) were reacted either as single- or double-stranded with reactive metabolites of carcinogens and the reaction products analyzed by HPLC-MS/MS to determine the sequence selectivity of DNA alkylating agents.

In work from our laboratory, the ds 12-mer sequence TAGTCA^{579}A^{580}GGGCA from the coding region of the *hprt* gene in Chinese hamster V79 cells was reacted with benzo[a]pyrene diol epoxide (BPDE), the reactive metabolite of the carcinogen benzo[a]pyrene (B[a]P) [182]. In this sequence the two indicated dA nucleotides are known to differ markedly in their extents of mutation after treatment with the epoxide. Analysis by RP-HPLC-ESI-MS/MS allowed assignment and semi quantitative assessment of the sites of modification for positional isomers. Figure 5.14 displays the TIC derived from the RP-HPLC-ESI-MS/MS analysis of the 12-mer modified with BPDE. The sites of adduct formation in the separated compounds were identified by an analysis of the characteristic series of a-B and w fragment ions generated by CID of the triply deprotonated molecule. It is seen that dA adducts predominated over dG adducts in the case of ds DNA, whereas the reverse was true for ss adducts (Figure 5.14). Additionally, the method permitted a facile distinction of the two strands of an oligonucleotide duplex, based on the differing masses of the strands. In consequence, adduct formation in both strands of a DNA duplex with multiple target bases can be successfully detected by RP-HPLC-ESI-MS/MS. This study represented one of the very first examples of the use of LC-MS/MS for the identification of adduct-containing sequences with multiple target bases in a ds oligonucleotide.

2-Amino-1-methyl-6-phenylimidazo[4,5-b]pyridine (PhIP) and 2-amino-3-methyl-imidazo[4,5-f] quinoline (IQ) are two important food-borne carcinogens associated with colon and breast cancer and have been the subject of numerous related studies. The propensity for site selectivity in the adduction of IQ was investigated by Stover et al. [183] who incorporated both IQ adducts into the G_1- and G_3- positions of the NarI recognition sequence 5'-$G_1G_2CG_3CC$-3', which is a well-known mutation hotspot for arylamine modification. The extension products with *E. coli* pol I Klenow fragment exo-, *E coli* pol II exo-, and *Sulfolobus solfataricus* P2 DNA polymerase IV (Dpo4) were sequenced by tandem mass spectrometry. Replication of the C8-adduct at the G_3 position with all three polymerases resulted in two-base deletion and error-free product at the G_1 position. Whereas the N^2-adduct was bypassed and extended at the G_1-position to produce an error-free product, the N^2-adduct at the G_3 position was bypassed and extended only by Dpo4 and the error-free product was observed. The results for the translesion synthesis of IQ-modified oligonucleotides by Dpo4 are summarized in Table 5.4 and indicate that the replication of the IQ-adducts of dGuo is strongly influenced by the local sequence and the environment of the adduction site.

The question of the relationship between adduction and mutation site for both PhIP and the related food-borne carcinogen IQ was recently addressed by Jamin et al. [184] of the National Institute for Agricultural Research of France (INRA). The authors investigated systematically the influence of neighboring base effect along the nucleotide sequence on the formation of adducts of the two carcinogens. Modified oligonucleotides were analyzed by HPLC-ESI-MS/MS to identify the adduct location along the oligonucleotide sequence. Modification yields of the various model sequences were compared semiquantitatively from the ratios of their respective ion currents by assuming similar responses of isomeric oligomers to ESI. MS/MS spectra allowed the unambiguous localization of the modification on the oligonucleotide sequence by the attribution of diagnostic fragment ions. Distinct differences for adduct formation as a function of base sequence were observed for the two carcinogens. Specifically, IQ was found to favor adduct formation at the sequence 5'-GGG-3', whereas the sequence 5'-GGA/G/T-3' was favored by PhIP. Though as also pointed out by the authors it is obviously dangerous to make biological extrapolation from

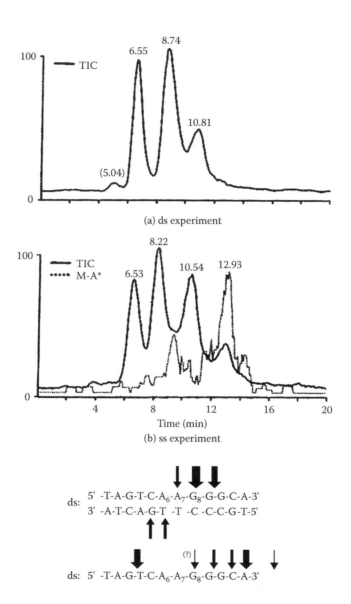

FIGURE 5.14 (a) LC-MS/MS analysis of ds (a) and ss (b) samples. The TIC of the MS/MS fragmentation of BcPh DE-modified TAGTCAAGGGCA under CID conditions was monitored in both experiments. In addition, panel b displays the extracted mass data trace of M-A* corresponding to loss of the modified adenine (A*) from the full-length oligonucleotide by depurination under CID conditions. (From Harsch et al., 2000. With permission.) (b) Identified sites of (-)-BcPh DE-2 adduct formation on the ds and ss oligonucleotides. Sizes of the arrows indicate an approximation of the relative amounts of adducts formed in each individual (ds or ss) sample and do not imply a direct comparison between the two samples. (From Harsch, A. et al., *Chem. Res. Toxicol.* 2000, *13*, 1342–1348. With permission.)

TABLE 5.4
Summary of the Dpo4 Extension of the IQ-dGuo-Modified Oligonucleotides

Template	Extension Products Identified by MS	Comments
3′-CCXCGGCTCA-5′ X = C8-IQ (**1b**)	5′-pGGCCCGCGAG-3′	Two- base deletion and error prone extension
X = N^2-IQ (**1c**)	5′-pGGCGCCGAGTA-3′	Error-free product
3′-CCGCGXCTCA-5′ X = C8-IQ (**2b**)	5′-pGGCGCCGAGT-3′	Error-free product
X = N^2-IQ (**1c**)	5′-pGGCGCCGAGT-3′	Error-free product

Source: Adapted from Stover, J. S. et al. *Chem. Res. Toxicol.* 2006, *19*, 1506–1517.

mere chemical data, it is still interesting to note that the sequence 5′GGGA3′ is also the mutational hotspot for PhIP.

Certain anticancer drugs act by binding covalently to DNA, and one of the most notable is cisplatin whose interaction with oligonucleotides was probed by Egger et al. [185] of the University of Vienna using FT-ICR MS and nanoLC-ESI-MS/MS. FT-ICR is well suited to study the interaction of matallodrugs with biomolecules owing to its unrivaled resolution and accuracy. Their study considered the binding kinetics, the nature of the adducts formed, and the location of the binding site within the specifically designed ds DNA oligonucleotides, ds GTATTGGCACGTA, and ds GTACCGGTGTGTA. By examining the characteristic isotopic patterns of platinum containing adduct ions and the MS/MS data, it was unambiguously revealed that the major binding sites correspond to the known preferred binding sites for cisplatin and demonstrated the preference for binding to guanosine. This FT-ICR MS-based method proved to be capable of locating the Pt-binding site via selective fragmentation using CID and IRMPD and much more rapid than NMR in binding site determination and more generally applicable than x-ray analysis.

In a separate consideration of the binding of cisplatin to DNA, Nyakas et al. [186] of the University of Bern investigated the gas-phase dissociation of cisplatin DNA and RNA adducts by negative ion ESI-MS/MS. The fundamental mechanistic aspects of fragmentation were elucidated to provide the basis for the tandem mass spectrometric detection of binding motifs and binding sites of this important anti-cancer drug. The results showed that the binding of cisplatin to vicinal guanines drastically altered the gas-phase fragmentation behavior of oligonucleotides.

LC-ESI-MS/MS continues to be used extensively to investigate the relationship between sequence-specificity and mutation site for a wide variety of carcinogens. Sherman et al. [187] of the University of Texas at Austin studied the interactions between a novel enediyne [1-methyl-2-(phenylethynyl)-3-(3-phenylprop-2-ynyl)-3H-benzimidazolium] and various cytosine-containing oligonucleotides using ESI-MS in a flow injection (FIA) mode. The structures of these adducts were examined and a sequence dependence of the 2′-dC-specific cleavage was noted. MS/MS was also conducted to confirm the strength of the interactions between the enediyne and DNA. The results indicated that the interaction between the enediyne and DNA is of comparable strength to the bonds in the oligonucleotide and suggested that adduct forms through a covalent linkage. The evaluation of adducts provided strong evidence for attachment of the enediyne directly to the cytosine nucleobase. Chan et al. [188] of Hong Kong Baptist University developed an ESI-MS method for the analysis of unmodified and AA-modified oligonucleotides (5'-TTTATT-3', 5'-TTTGTT-3' and 5'-TACATGTGT-3'). Evison et al. [189] of La Trobe University, Australia, applied mass spectrometry to analyze drug-oligonucleotide complexes that confirmed the formation of covalent pixantrone–DNA adducts is mediated by a single methylene linkage provided by formaldehyde. They found that adduction occurred only with guanine-containing double-stranded oligonucleotide. In addition, they concluded that CpG methylation, an epigenetic modification of the mammalian genome, significantly enhanced the generation of pixantrone–DNA adducts within a methylated DNA substrate, which indicates that the methylated dinucleotide may be a favored target in a cellular environment.

5.2.4.5.3   Oligonucleotide Adducts Analyzed by Monolithic Columns

The previously presented examples have relied on the use of "standard"-type HPLC columns (i.e., packed narrow bore or capillary columns). A significant improvement toward the use of HPLC-MS/MS for the analysis of oligonucleotide adducts has been realized via the use of monolithic columns as shown in a recent publication by Xiong et al. [190], which employed a monolithic PSDVB column coupled to nanoESI-MS for separation and identification of isomeric oligonucleotide adducts derived from the covalent binding of BPDE to the ds 14-mer 5'-PO$_4$⁻-ACCCGCGTCCGCGC-3'/5'-GCGCGGGCGCGGGT-3'. The in-house fabricated monolithic column was able to effectively separate four positional isomeric BPDE oligonucleotide adducts that were identified subsequently by MS/MS. Moreover, based on the MS/MS fragmentation patterns it could be deduced that peaks III-1 and III-2 represent well known diastereomeric adducts in which (±)-BPDE is attached to the same guanine base (G$_7$) of the oligonucleotide. Peak III-3 represents the case in which (±)-BPDE adduct is attached to the guanine G$_{11}$ of the oligonucleotides. Peaks III-4, III-6, and III-7 represent diastereomeric adducts in which BPDE is attached to the same guanine G$_5$ of the oligonucleotide. Peaks III-5, III-8, and III-9 represent diastereomeric adducts in which BPDE is attached to the same adenine A$_1$ of the oligonucleotide. Surprisingly, no adduct was found in which (±)-BPDE is attached to the guanine G$_{13}$ of the oligonucleotides. The chromatogram depicted in Figure 5.15 reveals the partial separation of at least nine peaks, with shoulders on some of the peaks indicating the presence of even more compounds. However, no further information could be obtained on the specific nature of the stereoselectivity of the adducts studied in this work. In that regard, the indirect approach of Tretyakova et al. may be advantageous, at least when working with model compounds.

5.2.4.5.4   HPLC-MS Analysis of Oligonucleotide Adducts Produced
from Enzymatic Digestion of Modified Oligomers

It is evident from the previous discussion that HPLC-MS/MS represents an ideal method for investigating the chemical selectivity in DNA sequences that contain

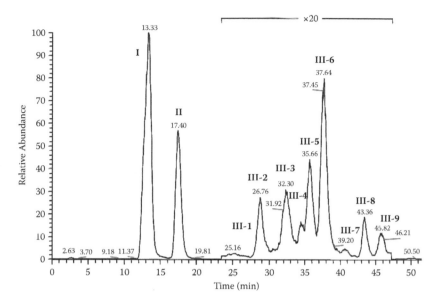

FIGURE 5.15 LC-MS composite EICs for the primary oligonucleotide strand obtained from the injection of a crude reaction mixture of (±)-BPDE with ds-oligonucleotide on an optimized monolithic PS-DVB nanocolumn (I) the unmodified 5'-phosphorylated oligonucleotide (m/z 1420) (II) the unmodified 5'-dephosphorylated oligonucleotide (m/z 1394) and (III) the modified 5'-phosphorylated oligonucleotides (m/z 1521). (From Xiong, W. et al., *Anal. Chem.* 2007, *79*, 5312–5321. With permission.)

multiple target bases competing for adduction. However, as illustrated from the previously shown examples, most current studies have focused on the examination of model systems in which an oligomer of known sequence has been analyzed by HPLC-MS/MS following its reaction with an activated carcinogen. Of more relevance is the application of HPLC-MS to the analysis of in vitro or in vivo systems where the enzymatic digestion of DNA produces a mixture rich in fragments containing the covalently bound carcinogen. In addition, for generation of meaningful data, it is necessary for the enzymatic hydrolysis to produce fragments of sufficient length (at least 15- to 20-mers) to be able to relate adduction site to a specific gene sequence. Hence, the challenges are not only the selection of the proper enzymatic digestion process but also the complexity of the mixtures that require use of high-resolution chromatography and mass spectrometry as well as the ability to handle the voluminous amount of data from their LC-MS analysis. Efforts in these directions are currently in progress in several groups, including the Linscheid group [191–193], the Vouros group [190,194–196], the Chiarelli group [197,198], the Fu group [62,199].

The significance of the direct LC-MS/MS approach for the analysis of oligonucleotide adducts produced from the enzymatic digestion of DNA, or large oligomers to map adduction sites, was first described in a 1994 publication by Janning et al. [200]. Modified DNA was digested to short DNA fragments ranging in length from dinucleotides to hexanucleotides using benzonase/alkaline phosphatase, and styrene oxide adducts were characterized by CZE-MS/MS in negative ion mode. The dinucleotides were present as singly charged species, whereas trinucleotides were detected as doubly and triply charged species. In subsequent publications, the Linscheid group of the University of Berlin determined styrene oxide (SO) adducts in DNA in the form of oligonucleotides at lengths of 2–8 bases by CZE-MS and LC-ESI-MS [201,202]. More recently, Mohamed and Linscheid [192] investigated melphalan modified oligonucleotide prepared enzymatically from modified calf thymus DNA. DNA was incubated in vitro with melphalan, and the resulting modifications were cleaved by benzonase and nuclease S1. Enzymatic digestion produced trinucleotide adducts that were sample preconcentrated by SPE and subsequently separated and detected by RP-HPLC-IT-ESI-MS. Mono-alkylated adducts were in much higher abundance than bi-alkylated adducts, and the alkylation site was located on the nucleobases. In a follow-up study, Mawaka et al. [191] digested calf thymus DNA-oxaliplatin adducts into dinucleotides with benzonase, alkaline phosphatase, and nuclease S1. They separated and identified these adducts by HPLC-MS and subsequently used MS^n to obtain the detailed structure of the oxaliplatin adducts of dinucleotides and dinucleoside monophosphatates. In a most recent study, Mohamed et al. [193] presented the results for the different chloroambucil-trinucleotide and tetranucleotide adducts with respect to their fragmentation behavior, abundance, and biological relevance (Figure 5.16). They demonstrated that the combination of Benzonase, Alkaline Phosphatase, and nuclease S1 are capable of digesting the chlorambucil treated DNA to trimers and tetramers, and the control of the nuclease S1 digestion time is the crucial factor in obtaining adducts of increased chain lengths. The data complement those obtained in their study on melphalan, which is another nitrogen mustard. From these data it is evident that examination of the oligomers is necessary to determine more of the details of adduct formation.

A variety of enzymes or enzyme cocktails have been used to digest DNA into shorter oligonucleotide adducts to establish the dependence of adduct formation in the context of base sequence context. Wang et al. [203] used Nuclease P1 to digest oligonucleotide photoproducts to small photoproduct-containing trinucleotides whose structures were assigned on the bases of molecular weight and MS/MS data. The partial sequence information of these oligo fragments was preserved by such a method, and this approach paved the way to do nearest-neighbor analysis and identify hotspots for photodamage in cellular DNA. The authors suggested the method to be generally useful for studying all types of DNA modifications. Chou et al. identified a series of 6,7-Dihydro-7-hydroxy-1-hydroxymethyl-5H-pyrrolizine (DHP) modified dinucleotide adducts of riddelline by negative ion LC-ESI-MS/MS. Adducted calf thymus DNA was enzymatically digested by micrococcal nuclease (MN) and spleen phosphodiesterase (SPD) and separated by HPLC [204]. In Figure 5.17, the left panel shows an example of the MRM chromatograms for the set of isomeric adducts derived from DHP modification of either adenine or thymine base in ApTp and TpAp in intact ctDNA. The right panel shows the corresponding mass chromatograms for product ion scans from the reaction of DHP with either ApTp and TpAp.

Yang et al. of the University of Rhode Island [205] described the preparation of enantiomeric 2-fluoro-BPDEs as models for probing the BPDE-induced

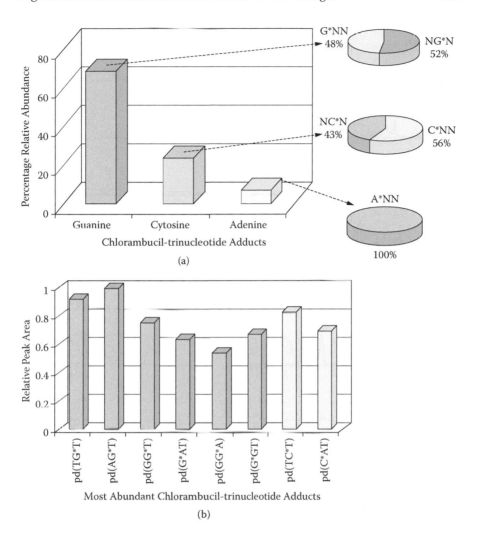

FIGURE 5.16 Comparison between the percentage relative abundance of the different Chlorambucil-trinucleotide adducts formed at guanine, cytosine, and adenine bases (a) and the relative peak areas of the most abundant monoalkylated Chlorambucil-trinucleotide adducts. All adducts possess the hydroxyethyl arm of Chlorambucil (b). The relative peak area is calculated as an average of three experiments with coefficient of variation (as percentage) (The position of alkylation is assigned with an asterisk). (From Mohamed, D. et al. *Chem. Res. in Toxi.* 2009, *22*(8), 1435–1446. With permission.)

conformational heterogeneity. Subsequently, the structures of FBP-modified oligodeoxynucleotides were characterized by enzyme digestion and HPLC ESI-MS. Andrews et al. [194] analyzed the in vitro digestion products of oligonucleotides and DNA modified by AAAF using IP-RP-LC-MS and LC-MS/MS. Benzonase/alkaline phosphatase was used to digest the oligonucleotides and DNA into fragments of different size and allowed examination of the sequence context of adducted segments

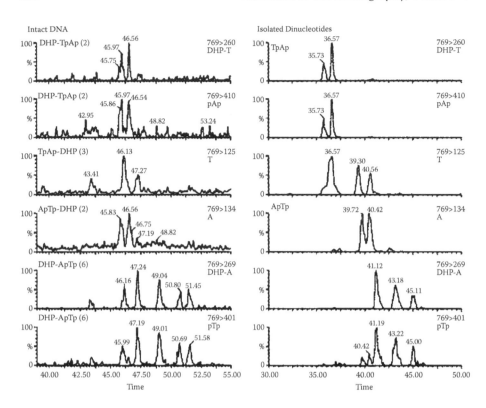

FIGURE 5.17 Characterization of representative DHP-modified dinucleotides from either isolated dinucleotides or intact ctDNA. Product ion chromatograms of selected transitions that characterize DHP adducts of TpAp and ApTp produced by reaction of DHP with the isolated dinucleotides (either TpAp or ApTp, right panel) or MRM chromatograms of the same adducts derived from DHP modification of intact ctDNA (left panel). The transitions monitored and the associated structural assignments are listed at the right of each chromatogram. (From Chou, M. W. et al., *Chem. Res. Toxicol.* 2003, *16*, 1130–1137. With permission.)

to determine binding preference of adduct to DNA by presenting the entire molecule as a target. The methodology produced primarily trimer fragments. Comparisons of MS/MS spectra with reference standards were made to confirm sequence determination and peak assignment from characteristic fragmentation patterns. The results pointed to the possibilities of an uneven base distribution in the DNA sequence. In line with some of the aforementioned requirements for the characterization of oligomeric adducts in DNA digests, Liao et al. (2007) developed an ion-pairing HPLC-MS method that has sufficient separation power, selectivity, and sensitivity to investigate the enzymatic behavior of benzonase/alkaline phosphatase on digestion of oligonucleotides and DNA. Analysis of the digest mixture revealed that this enzyme pair can nonspecifically digest oligonucleotides and DNA into fragments ranging from 2 to 10 nucleotides (i.e., sizes suitable for routine mass spectrometric measurements). Adducts of trimers, tetramers, and pentamers were the most prominent digestion products. When the digests were analyzed by nanoLC-MS (75 μm i.d.

columns; flow rates of ca. 200–300 nL/min) detection limits approaching 1 fmole (or 1.36 pg) were observed for the shorter oligomers in the mixtures. This represented a nearly 200-fold improvement in LOD when compared with analysis conditions using 1 mm i.d. columns. Based on these results it was concluded that benzonase/alkaline phosphatase is a promising choice for DNA and DNA adduct-related studies that require a nonspecific enzyme.

A novel and promising approach for determining the structure of oligonucleotide adducts was recently reported by Gao et al. using HPLC-ESI-MS [198]. The method permits determination of adduction sites in short carcinogen modified oligonucleotides by examining the fragments produced upon enzymatic digestion by the combined action of 3'- and 5'-exonucleases. Snake venom phosphodiesterase (3'-PDE) and bovine spleen phosphodiesterase (5'-PDE) were used to cleave sequentially the ODN from their respective ends and enzymatically generate fragments of different length as a function of digestion time. The presence of a lesion stops the enzymatic cleavage, and the site of modification can be ascertained by effectively overlapping the digestion fragments containing AF-modified guanines. The method can be particularly effective when mapping adduction sites in a known gene sequence, as demonstrated by using as a model the AF-modified dodecamer 5'-CTCGGCGCCATC-3' containing the NarI sequence, a hotspot for deletion mutations in *Escherichia coli*.

Described in more detail subsequently is the sequence of steps involved in the determination of the adduction site in the "unknown" modified dodecamer 5'-CTCGGCG(AF)CCATC-3' (P1). In the first step, the MW of the 12-mer was determined by exact mass measurement (e.g., ESI-TOF) to fully establish the oligonucleotide composition of the dodecamer (Figure 5.18). Subsequently, analyses of the digest fragments, many of which consisted of oligonucleotide adducts of different lengths, allowed determination of the AF modification site(s) (Figure 5.19). The mass spectra of these fragments were identified as the corresponding B-type ions ad Y-type ions using the system of nomenclature proposed by McLuckey and Habibigoudarzi [141]. By analysis of both 3'- and 5'-digests, the position of the AF modified G was located (Figure 5.20). As in their studies the modified oligomers all have the same base sequence; the key to differentiating the positional isomers is simply to identify the base compositional oligomer digest ions that are specific for a particular sequential isomer. Their work indicated that this methodology is useful for sequencing modified oligonucleotides with single or multiple adduction sites to serve as templates for site-specific mutagenesis studies. The previous approach was also used to assign the sequence of di-AF modified oligonucleotides of the digestion products, although the signals for the signature ions were low. It was suggested that the sequence information content for these di-AF modified oligonucleotide products could be improved by extending digestion times.

5.2.4.5.5 Mass Spectrometric Analysis with Advanced Computational Screening of DNA Adducts

As expected, digestion of the DNA macromolecule produces a large and diverse number of oligonucleotides or oligonucleotide adducts of varying length, the recognition of which poses a formidable task. In fact, identification of nucleotide compositions

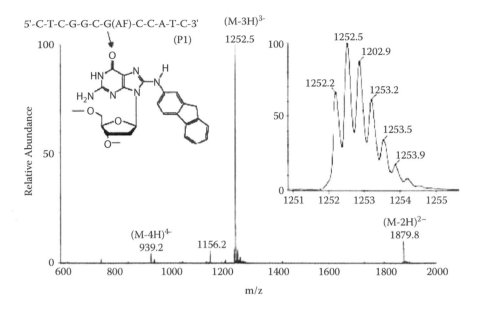

FIGURE 5.18 ESI TOF mass spectrum of the ODN product 1. The structure of AF-adduct [N-(2′-deoxyguanosin-8-yl)-2-aminofluorene in the NarI sequence is shown in the inset. (From Gao, L., J. Am. Soc. Mass Spectrom. 2008, 19, 1147–1155. With permission.)

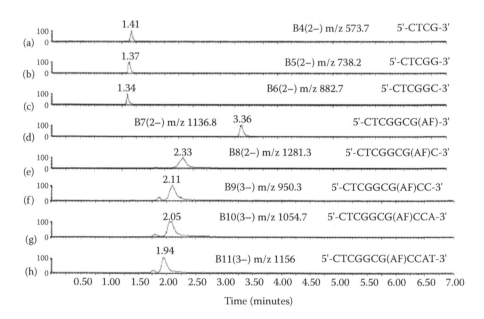

FIGURE 5.19 LC/MS single-ion chromatograms of the predominant ions formed by ODN digest fragments generated during the 3″-exonuclease digestion of product 1. Chromatograms (a–e) were acquired after 40 minutes' digestion time and (f–h) after 20 minutes. (From Gao, L., J. Am. Soc. Mass Spectrom. 2008, 19, 1147–1155. With permission.)

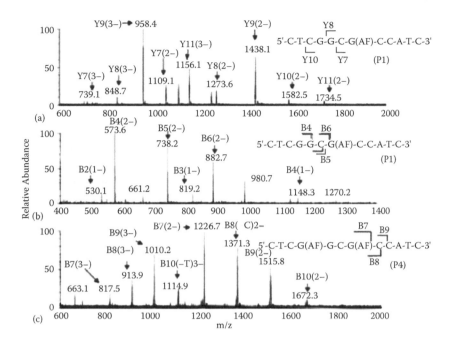

FIGURE 5.20 ESI-TOF mass spectrum of the hydrolysis fragments formed from (a) the 5′-exonuclease digestion of product 1 acquired after 30 minutes, (b) the hydrolysis fragments formed from the 3′-exonuclease digestion of product 1 after 40 minutes, (c) the hydrolysis fragments formed from the 3′-exonuclease digestion of product 4 after 40 minutes. (From Gao, L. et al., *Soc. Mass Spectrom.* 2008, *19*, 1145–1155. With permission.)

becomes increasingly difficult with increasing oligomer length as the number, **N**, of compositional isomers increases exponentially according to

$$N = 4^n$$

where **n** is the number of nucleotides or chain length.

The possible unique masses for oligmers of different size (chain length) are

$$M = (4 + n - 1)!/n!/3!$$

Therefore, for 2-mers, there are $N = 4^2 = 16$ different compositions and $M = 10$ unique masses; for 3-mers, there are $N = 4^3 = 64$ different compositions and $M = 20$ unique masses; for 4-mers, there are $N = 4^4 = 256$ different compositions and $M = 35$ unique masses; for 5-mers, there are $N = 4^5 = 1024$ different compositions and $M = 56$ unique masses. The numbers increase exponentially and make it a daunting task to analyze the data.

In light of these considerations, Liao et al. [196] developed a computer software program named GenoMass to automate the processing of mass spectral data generated from the LC-MS analysis of enzymatic digestion of DNA, with specific focus on the characterization of oligonucleotide adducts. In this program, isomer libraries are generated in silico in a "reversed pseudo-combinatorial" way to represent the digestion products of oligonucleotides, DNA or DNA adducts of various sizes.

The software automatically calculates ion masses of each isomeric segment of the library, searches for them in complicated LC-MS data, lists their intensities, and plots extracted ion chromatograms (EIC) (Figure 5.21). This customized new data analysis tool has enabled a study of the enzymatic behavior of a nuclease system in the digestion of normal and adducted DNA and in the recognition of oligomers containing a carcinogen bound to a nucleobase. The combined enzymatic and software methodology was applied to the analysis of enzymatic digests of oligonucleotide adducts produced from the exposure of progressively larger DNA segments and finally to calf thymus DNA following their in vitro reaction with AAAF. The methodology described provides a systematic approach for evaluating the behavior of DNA-cleaving enzymes by mass spectrometry. In the early version of this methodology, CID was used independently to determine base sequence and identification of adduction site; however, the software program potentially can be further expanded to postulate unknown DNA sequences and recognize the adduction sites.

5.3   CONCLUSIONS

There is strong evidence that, following exposure of DNA to genotoxic carcinogens, characteristic mutation spectra may result that show a degree of site selectivity. It is also claimed that the potency of an adduct is related to its base sequence context and that the latter also influences both the types and the frequencies of ensuing mutation. Therefore, the ability to analyze oligonucleotide adducts will add a very important complement to the field of DNA adduct analysis, which thus far has been largely focused on the detection and characterization of monomeric nucleosides or nucleotides. Assignment of the sequence selectivity of adduct formation and the mutational consequences is of paramount importance and can help to

- Develop biomarkers indicative of the risk associated with the carcinogen exposures
- Identify the nature of the most mutagenic constituents of the complex mixtures of carcinogens to which humans are exposed

As shown from the previously presented examples, structural characterization of oligonucleotide adducts of the order of 10–20 mer in length by HPLC-ESI-MS/MS can be conducted relatively routinely using commonly available mass spectrometers of moderate resolving power (M/ΔM ca. 1000–2000). Based on studies conducted with model oligomers of this order of magnitude it appears that, at least for some of the common bulky adducts, adduction is generally restricted to a single base; formation of multiple adducts in such short sequences is very rarely observed, and, if it does, it is usually in much lower relative abundance presumably due to steric effects. In addition, further improvements in both chromatography and MS detection methods are expected to improve the efficiency and accuracy of these analyses.

Studies conducted to this date using model oligomers have confirmed the general applicability of LC-MS/MS approaches to the analysis of oligonucleotide adducts. These studies have also enabled an investigation of the potential relationships between adduction and mutation sites. However, the practical utility of these analytical

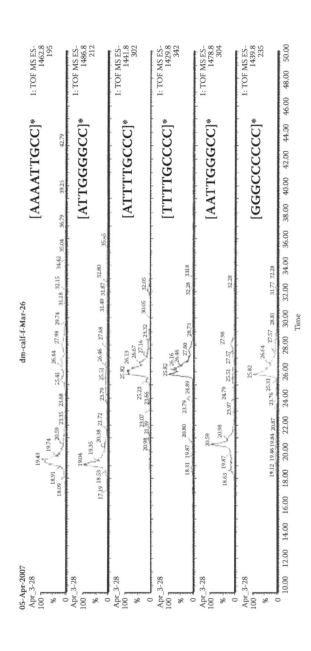

FIGURE 5.21 EICs of identified six most abundant AAF-modified 9-mers in the digest of calf thymus DNA following exposure to AAAF. Chromatograms were automatically generated and plotted by GenoMass. (From Liao, Q., Shen, C., Vouros, P. GenoMass *J. Mass Spectrom.* 2009, 44(4), 549–560. With permission from John Wiley & Sons.)

methodologies will ultimately be realized only by their ability to view DNA adduct formation in the context of base sequences in in vitro and in vivo applications. Thus, a more comprehensive approach to the characterization of oligonucleotide adducts produced from the enzymatic digestion of DNA will require as a minimum:

1. The development of DNA enzymatic digestion procedures that would cut DNA into longer fragments, providing more complete information about neighboring base effects on adduction.
2. The development of data handling software to handle myriad data produced from the LC-MS analysis of DNA digests, bearing in mind that the generation of progressively longer oligonucleotides is likely to yield a multitude of isomeric species.
3. The ability to analyze the oligonucleotide markers in the digests using minimal quantities of DNA with sufficient sensitivity to detect a small number of adducts per genome.

Although there have been a lot of efforts, a sound approach capable of detecting unknown DNA adducts, preserving the base sequence information, and even the supramolecular requirements of the modification reaction still remain to be developed. The prospects, however daunting, for achieving each one of these analytical goals is discussed here in the context of the material previously reviewed.

To be able to draw reasonable deductions about the relationship between chemoselectivity and mutation within a given gene sequence, examination of oligomeric adducts of at least 15–20 base pairs in length would be necessary. Restriction enzymes can cleave genomic DNA into oligonucleotides of varying lengths that could be isolated off-line (e.g., gels) and further subjected to enzymatic digestion using nonspecific enzymes. Unclear at this point is how the presence of an adduct site may influence the action of a restriction enzyme. Moreover, most of the investigations carried out to date using nonspecific enzymes are reported to yield predominantly adducts of three to four base pairs in length. Further research using new enzyme systems and better sample workup by incorporation of antibodies for isolation of oligomers containing specific carcinogen adducts is likely to produce samples more friendly to LC-MS analysis. In this regard, some of the key problems may be overcome in the foreseeable future.

Compared with the analysis of monomeric DNA adducts, less well defined is the current state-of-the-art in terms of the quantitative analysis of oligonucleotide adducts. However, some projections regarding sample requirements can still be made based on the wealth of quantitative information gathered in numerous laboratories over the past several years for the analysis of DNA adducts, especially in the form of mononucleosides. For the latter, a typical target level of physiological relevance is the ability to quantify 2–5 modifications in $ca.$ 10^8 normal bases. Using as reference the LOQ of 75 attomol typically achieved for dG-C8-4-ABP the quantification of 1: 10^8 modifications would require a sample of ca. 2.5 μg of DNA corresponding to detection of 30 adducts per genome. Results from our laboratory for an AAF-modified 14-mer indicate a comfortable LOQ of ca. 1 femtomol of the oligomer injected on column. On that basis, it can be extrapolated that it will be

TABLE 5.5
Calculated Moles of Adduct Expected for Different Quantities of DNA

Total µg DNA	Number of Adducts in 10^x Nucleotides	Femtomoles of Adduct	Number of Adducts per Genome
Experimentally Determined			
2.5	5 in 10^9	1 (LOD)	16
2.5	10 in 10^9	75 (LOQ)	32
Adduct Quantity Estimates			
2.5	1 in 10^9	7.5	3.2
5.0	3.3 in 10^8	500	10^6
10	3.3 in 10^8	1000	10^6

necessary to analyze a sample of 30–40 µg of DNA for the detection of 400–500 adducts per genome. The ability to detect lower amounts of modification or damage is obviously desirable but would in turn require larger amounts of DNA or better mass analytical sensitivity. While these numbers are at best approximations based on incomplete data, they are presented to place the problem of the analysis of oligonucleotide adducts in the context of this review (Table 5.5).

In summary, the characterization of oligonucleotide adducts may provide important information regarding the nature of genomic damage. Advances in this area very much depend on further improvements in LC-MS methodologies as outlined here. It is envisioned that with fully developed analytical methods based on LC-MS/MS identification of an adducted nucleotide sequences could be automatically used to interrogate the human genome database to identify specific genes susceptible to adduction or to develop a genomic wide "fingerprint" of DNA adduction. Ultimately, it should thus be possible to address in a comprehensive fashion the susceptibility of a gene for adduction and its predilection for mutation.

ACKNOWLEDGMENT

This work was supported in part by grants 1RO1CA69390 and 1RO1CA112231 from the National Cancer Institute.

REFERENCES

1. Singh, R.; Gaskell, M.; Le Pla, R. C.; Kaur, B.; Azim-Araghi, A.; Roach, J.; Koukouves, G.; Souliotis, V. L.; Kyrtopoulos, S. A.; Farmer, P. B. Detection and Quantitation of Benzo[a]pyrene-Derived DNA Adducts in Mouse Liver by Liquid Chromatography-Tandem Mass Spectrometry: Comparison with 32P-Postlabeling. *Chem. Res. Toxicol.* 2006, *19*, 868–878.
2. Farmer, P. B.; Brown, K.; Tompkins, E.; Emms, V. L.; Jones, D. J. L.; Singh, R.; Phillips, D. H. DNA Adducts: Mass Spectrometry Methods and Future Prospects. *Toxicol. Appl. Pharmacol.* 2005, *207*, (Suppl. 2), S293–S301.

3. Turesky, R. J.; Vouros, P. Formation and Analysis of Heterocyclic Aromatic Amine-DNA Adducts in Vitro and in Vivo. *J. Chromatogr., B: Anal. Technol. Biomed. Life Sci.* 2004, *802*, 155–166.

4. Garner, R. C. The Role of DNA Adducts in Chemical Carcinogenesis. *Mutat. Res.* 1998, *402*, 67–75.

5. Hemminki, K. DNA Adducts, Mutations and Cancer. *Carcinogenesis* 1993, *14*, 2007–2012.

6. Hemminki, K.; Koskinen, M.; Rajaniemi, H.; Zhao, C. DNA Adducts, Mutations, and Cancer. *Regul. Toxicol. Pharmacol.* 2000, *32*, 264–275.

7. Santella, R. M.; Gammon, M.; Terry, M.; Senie, R.; Shen, J.; Kennedy, D.; Agrawal, M.; Faraglia, B.; Zhang, F. DNA Adducts, DNA Repair Genotype/Phenotype and Cancer Risk. *Mutat. Res.* 2005, *592*, 29–35.

8. Vulimiri, S. V.; DiGiovanni, J. DNA Adducts as Biomarkers of DNA Damage in Lung Cancer. In *Biomarkers of Environmentally Associated Disease: Technologies, Concepts, and Perspectives;* Wilson, S. H., Suk, W. A., Eds.; Lewis Publishers: Boca Raton, FL, 2002; pp. 109–126.

9. Beland, F. A.; Churchwell, M. I.; Von Tungeln, L. S.; Chen, S.; Fu, P. P.; Culp, S. J.; Schoket, B.; Gyorffy, E.; Minarovits, J.; Poirier, M. C.; Bowman, E. D.; Weston, A.; Doerge, D. R. High-Performance Liquid Chromatography Electrospray Ionization Tandem Mass Spectrometry for the Detection and Quantization of Benzo[a]pyrene-DNA Adducts. *Chem. Res. Toxicol.* 2005, *18*, 1306–1315.

10. Chen, W.-H.; Qin, Y.; Cai, Z.; Chan, C.-L.; Jiang, Z.-H. Spectrometric Studies of Cytotoxic Protoberberine Alkaloids Binding to Double-Stranded DNA. *Bioorg. Med. Chem.* 2005, *13*, 1859–1866.

11. Churchwell, M. I.; Yan, J.; Xia, Q.; Fu, P. P.; Beland, F. A.; Doerge, D. R. LC/MS/MS Analysis of Benzo[a]pyrene DNA Adducts Derived from Diol Epoxide and Cation Radical Pathways. *Abstracts of Papers,* 232nd National Meeting of the American Chemical Society, San Francisco, CA, Sept 10–14, 2006; American Chemical Society: Washington, DC, 2006; TOXI-052.

12. Marzilli, L. A.; Koertje, C.; Vouros, P. Capillary Electrophoresis-Mass Spectrometric Analysis of DNA Adducts. *Methods Mol. Biol. (Totowa, NJ, U.S.)* 2001, *162*, 395–406.

13. Wang, M.; Cheng, G.; Sturla, S. J.; Shi, Y.; McIntee, E. J.; Villalta, P. W.; Upadhyaya, P.; Hecht, S. S. Identification of Adducts Formed by Pyridyloxobutylation of Deoxyguanosine and DNA by 4-(Acetoxymethylnitrosamino)-1-(3-Pyridyl)-1-Butanone, a Chemically Activated Form of Tobacco Specific Carcinogens. *Chem. Res. Toxicol.* 2003, *16*, 616–626.

14. Guza, R.; Rajesh, M.; Fang, Q.; Pegg, A. E.; Tretyakova, N. Kinetics of O6-Methyl-2′-Deoxyguanosine Repair by O6-Alkylguanine DNA Alkyltransferase within K-Ras Gene-Derived DNA Sequences. *Chem. Res. Toxicol.* 2006, *19*, 531–538.

15. Haglund, J.; Henderson, A. P.; Golding, B. T.; Tornqvist, M. Evidence for Phosphate Adducts in DNA from Mice Treated with 4-(N-Methyl-N-Nitrosamino)-1-(3-Pyridyl)-1-Butanone (NNK). *Chem. Res. Toxicol.* 2002, *15*, 773–779.

16. Hecht, S. S.; Villalta, P. W.; Sturla, S. J.; Cheng, G.; Yu, N.; Upadhyaya, P.; Wang, M. Identification of O_2-Substituted Pyrimidine Adducts Formed in Reactions of 4-(Acetoxymethylnitrosamino)-1-(3-Pyridyl)-1-Butanone and 4-(Acetoxymethylnitros-Amino)-1-(3-Pyridyl)-1-Butanol with DNA. *Chem. Res. Toxicol.* 2004, *17*, 588–597.

17. Lao, Y.; Villalta, P. W.; Wang, M.; Cheng, G.; Sturla, S. J.; Hecht, S. S. Development of a Sensitive Method for Quantitation of DNA Pyridyloxobutylation Adducts Derived from the Tobacco Specific Nitrosamine, 4-(Methylnitrosamino)-1-(3-Pyridyl)-1-Butanone (NNK). *Abstracts of Papers,* 230th National Meeting of the American Chemical Society, Washington, DC, Aug 28–Sept 1, 2005; American Chemical Society: Washington, DC, 2005; TOXI-039.

18. Lao, Y.; Villalta, P. W.; Sturla, S. J.; Wang, M.; Hecht, S. S. Quantitation of Pyridyloxobutyl DNA Adducts of Tobacco-Specific Nitrosamines in Rat Tissue DNA by High-Performance Liquid Chromatography-Electrospray Ionization-Tandem Mass Spectrometry. *Chem. Res. Toxicol.* 2006, *19*, 674–682.

19. Lao, Y.; Yu, N.; Kassie, F.; Hecht, S. S. Formation and Accumulation of Pyridyloxobutyl (POB)-DNA Adducts in F344 Rats Treated with Tobacco-Specific Carcinogens. *Abstracts of Papers,* 232nd National Meeting of the American Chemical Society, San Francisco, CA, Sept 10–14, 2006; American Chemical Society: Washington, DC, 2006; TOXI-031.

20. Rajesh, M.; Wang, G.; Jones, R.; Tretyakova, N. Stable Isotope Labeling-Mass Spectrometry Analysis of Methyl- and Pyridyloxobutyl-Guanine Adducts of 4-(Methylnitrosamino)-1-(3-Pyridyl)-1-Butanone in p53-Derived DNA Sequences. *Biochemistry* 2005, *44*, 2197–2207.

21. Tretyakova, N. Stable Isotope Labeling-Mass Spectrometry to Map the Reactivity of DNA Bases towards Carcinogens and Antitumor Agents. *Abstracts of Papers,* 226th National Meeting of the American Chemical Society, New York, NY, Sept 7–11, 2003; American Chemical Society: Washington, DC, 2003; TOXI-003.

22. Upadhyaya, P.; Sturla, S. J.; Tretyakova, N.; Ziegel, R.; Villalta, P. W.; Wang, M.; Hecht, S. S. Identification of Adducts Produced by the Reaction of 4-(Acetoxymethylnitrosamino)-1-(3-Pyridyl)-1-Butanol with Deoxyguanosine and DNA. *Chem. Res. Toxicol.* 2003, *16*, 180–190.

23. Villalta, P. W.; Hecht, S. S.; Sturla, S. J.; Wang, M.; Scott, J.; Lao, Y.; Cheng, G. Investigating DNA Adduct Formation from Tobacco Smoke Exposure using Mass Spectrometry. *Abstracts of Papers,* 61st Northwest Regional Meeting of the American Chemical Society, Reno, NV, June 25–28, 2006; American Chemical Society: Washington, DC, 2006; RE06–221.

24. McDonald, E. S.; Randon, K. R.; Knight, A.; Windebank, A. J. Cisplatin Preferentially Binds to DNA in Dorsal Root Ganglion Neurons in Vitro and in Vivo: a Potential Mechanism for Neurotoxicity. *Neurobiol. Dis.* 2005, *18*, 305–313.

25. Qu, Y.; Scarsdale, N. J.; Tran, M.-C.; Farrell, N. P. Cooperative Effects in Long-Range 1,4 DNA-DNA Interstrand Cross-Links Formed by Polynuclear Platinum Complexes: an Unexpected Syn Orientation of Adenine Bases Outside the Binding Sites. *J. Biol. Inorg. Chem.* 2003, *8*, 19–28.

26. Qu, Y.; Scarsdale, N. J.; Tran, M.-C.; Farrell, N. Comparison of Structural Effects in 1,4 DNA-DNA Interstrand Cross-links Formed by Dinuclear and Trinuclear Platinum Complexes. *J. Inorg. Biochem.* 2004, *98*, 1585–1590.

27. Tu, C.; Wu, X.; Liu, Q.; Wang, X.; Xu, Q.; Guo, Z. Crystal Structure, DNA-Binding Ability and Cytotoxic Activity of Platinum(II) 2,2'-Dipyridylamine Complexes. *Inorg. Chim. Acta* 2004, *357*, 95–102.

28. Warnke, U.; Rappel, C.; Meier, H.; Kloft, C.; Galanski, M.; Hartinger, C. G.; Keppler, B. K.; Jaehde, U. Analysis of Platinum Adducts with DNA Nucleotides and Nucleosides by Capillary Electrophoresis Coupled to ESI-MS: Indications of Guanosine 5'-Monophosphate O6-N7 Chelation. *ChemBioChem* 2004, *5*, 1543–1549.

29. Borges, C.; Lemiere, F.; Embrechts, J.; Van Dongen, W.; Esmans, E. L. Characterisation of Estrone-Nucleic Acid Adducts Formed by Reaction of 3,4-Estrone-O-Quinone with 2'-Deoxynucleosides/Deoxynucleotides Using Capillary Liquid Chromatography/Electrospray Ionization Mass Spectrometry. *Rapid Commun. Mass Spectrom.* 2004, *18*, 2191–2200.

30. Chen, Y.; Liu, X.; Pisha, E.; Constantinou, A. I.; Hua, Y.; Shen, L.; Van Breemen, R. B.; Elguindi, E. C.; Blond, S. Y.; Zhang, F.; Bolton, J. L. A Metabolite of Equine Estrogens, 4-Hydroxyequilenin, Induces DNA Damage and Apoptosis in Breast Cancer Cell Lines. *Chem. Res. Toxicol.* 2000, *13*, 342–350.

31. Debrauwer, L.; Rathahao, E.; Jouanin, I.; Paris, A.; Clodic, G.; Molines, H.; Convert, O.; Fournier, F.; Tabet, J. C. Investigation of the Regio- and Stereo-Selectivity of Deoxyguanosine Linkage to Deuterated 2-Hydroxyestradiol by Using Liquid Chromatography/ESI-Ion Trap Mass Spectrometry. *J. Am. Soc. Mass Spectrom.* 2003, *14*, 364–372.

32. Embrechts, J.; Lemiere, F.; Van Dongen, W.; Esmans, E. L.; Buytaert, P.; Van Marck, E.; Kockx, M.; Makar, A. Detection of Estrogen DNA-Adducts in Human Breast Tumor Tissue and Healthy Tissue by Combined Nano LC-Nano ES Tandem Mass Spectrometry. *J. Am. Soc. Mass Spectrom.* 2003, *14*, 482–491.

33. Li, K.-M.; Todorovic, R.; Devanesan, P.; Higginbotham, S.; Kofeler, H.; Ramanathan, R.; Gross, M. L.; Rogan, E. G.; Cavalieri, E. L. Metabolism and DNA Binding Studies of 4-Hydroxyestradiol and Estradiol-3,4-Quinone in Vitro and in Female ACI Rat Mammary Gland in Vivo. *Carcinogenesis* 2004, *25*, 289–297.

34. Edler, M.; Jakubowski, N.; Linscheid, M. Styrene Oxide DNA Adducts: Quantitative Determination Using 31P Monitoring. *Anal. Bioanal. Chem.* 2005, *381*, 205–211.

35. Shao, H.; Li, J.; Shi, Y. A Study on the Adduct Characteristics of Styrene and DNA. *Zhonghua Laodong Weisheng Zhiyebing Zazhi* 2002, *20*, 347–349.

36. Tarun, M.; Rusling, J. F. Measuring DNA Nucleobase Adducts Using Neutral Hydrolysis and Liquid Chromatography-Mass Spectrometry. *Crit. Rev. Eukaryotic Gene Expression* 2005, *15*, 295–316.

37. Tarun, M.; Rusling, J. F. Quantitative Measurement of DNA Adducts Using Neutral Hydrolysis and LC-MS. Validation of Genotoxicity Sensors. *Anal. Chem.* 2005, *77*, 2056–2062.

38. Tarun, M.; Bajrami, B.; Rusling, J. F. Genotoxicity Screening Using Biocatalyst/DNA Films and Capillary LC-MS/MS. *Anal. Chem.* 2006, *78*, 624–627.

39. Yang, J.; Wang, B.; Rusling, J. F. Genotoxicity Sensor Response Correlated with DNA Nucleobase Damage Rates Measured by LC-MS. *Mol. Biosystems* 2005, *1*, 251–259.

40. Churchwell, M. I.; Beland, F. A.; Doerge, D. R. Quantification of Multiple DNA Adducts Formed through Oxidative Stress Using Liquid Chromatography and Electrospray Tandem Mass Spectrometry. *Chem. Res. Toxicol.* 2002, *15*, 1295–1301.

41. Churchwell, M. I.; Beland, F. A.; Doerge, D. R. Quantification of O6-Methyl and O6-Ethyl Deoxyguanosine Adducts in C57BL/6N/Tk +/- Mice Using LC/MS/MS. *J. Chromatogr., B: Anal. Technol. Biomed. Life Sci.* 2006, *844*, 60–66.

42. Churchwell, M. I.; Yan, J.; Xia, Q.; Fu, P. P.; Beland, F. A.; Doerge, D. R. LC/MS/MS Analysis of Benzo[a]pyrene DNA Adducts Derived from Diol Epoxide and Cation Radical Pathways. *Abstracts of Papers,* 232nd National Meeting of the American Chemical Society, San Francisco, CA, Sept 10–14, 2006; American Chemical Society: Washington, DC, 2006; TOXI-052.

43. Doerge, D. R.; Yi, P.; Churchwell, M. I.; Preece, S. W.; Langridge, J.; Fu, P. P. Mass Spectrometric Analysis of 2'-Deoxyribonucleoside and 2'-Deoxyribonucleotide Adducts with Aldehydes derived from Lipid Peroxidation. *Rapid Commun. Mass Spectrom.* 1998, *12*, 1665–1672.

44. Doerge, D. R.; Churchwell, M. I.; Fang, J.-L.; Beland, F. A. Quantification of Etheno-DNA Adducts Using Liquid Chromatography, On-Line Sample Processing, and Electrospray Tandem Mass Spectrometry. *Chem. Res. Toxicol.* 2000, *13*, 1259–1264.

45. Jian, W.; Lee, S. H.; Blair, I. A. Lipoxygenase Mediated Endogenous DNA Damage. *Abstracts of Papers,* 228th National Meeting of the American Chemical Society, Philadelphia, PA, Aug 22–26, 2004; American Chemical Society: Washington DC, 2004; TOXI-010.

46. Xu, J.; Blair, I. A. Quantitative Analysis of DNA Adducts Generated through Nonenzymatic Pathways Using LC/MS. *Abstracts of Papers,* 228th National Meeting of the American Chemical Society, Philadelphia, PA, Aug 22–26, 2004; American Chemical Society: Washington, DC, 2004; TOXI-093.

47. Gonzalez-Reche, L. M.; Koch, H. M.; Weiss, T.; Muller, J.; Drexler, H.; Angerer, J. Analysis of Ethenoguanine Adducts in Human Urine Using High Performance Liquid Chromatography-Tandem Mass Spectrometry. *Toxicol. Lett.* 2002, *134*, 71–77.

48. Ham, A.-J. L.; Engelward, B. P.; Koc, H.; Sangaiah, R.; Meira, L. B.; Samson, L. D.; Swenberg, J. A. New Immunoaffinity-LC-MS/MS Methodology Reveals that Aag Null Mice Are Deficient in Their Ability to Clear 1,N6-Etheno-Deoxyadenosine DNA Lesions from Lung and Liver in Vivo. *DNA Repair* 2004, *3*, 257–265.

49. Hillestrom, P. R.; Hoberg, A.-M.; Weimann, A.; Poulsen, H. E. Quantification of 1,N6-Etheno-2'-Deoxyadenosine in Human Urine by Column-Switching LC/APCI-MS/MS. *Free Radical Biol. Med.* 2004, *36*, 1383–1392.

50. Hillestrom, P. R.; Nyssonen, K.; Tuomainen, T.-P.; Pukkala, E.; Salonen Jukka, T.; Poulsen Henrik, E. Urinary Excretion of Epsilond A Is Not Predictive of Cancer Development: A Prospective Nested Case-Control Study. *Free Radical Res.* 2005, *39*, 51–53.

51. Hillestrom, P. R.; Covas, M.-I.; Poulsen, H. E. Effect of Dietary Virgin Olive Oil on Urinary Excretion of Etheno-DNA Adducts. *Free Radical Biol. Med.* 2006, *41*, 1133–1138.

52. Hillestrom, P. R.; Weimann, A.; Jensen, C. B.; Storgaard, H.; Vaag, A. A.; Poulsen, H. E. Consequences of Low Birth Weight On Urinary Excretion of DNA Markers of Oxidative Stress in Young Men. *Scand. J. Clin. Lab. Invest.* 2006, *66*, 363–370.

53. Hillestrom, P. R.; Weimann, A.; Poulsen, H. E. Quantification of Urinary Etheno-DNA Adducts by Column-Switching LC/APCI-MS/MS. *J. Am. Soc. Mass Spectrom.* 2006, *17*, 605–610.

54. Kaddiska, M. B.; Gladen, B. C.; Baird, D. D.; Germolec, D.; Graham, L. B.; Parker, C. E.; Nyska, A.; Wachsman, J. T.; Ames, B. N.; Basu, S.; Brot, N.; FitzGerald, G. A.; Floyd, R. A.; George, M.; Heinecke, J. W.; Hatch, G. E.; Hensley, K.; Lawson, J. A.; Marnett, L. J.; Morrow, J. D.; Murray, D. M.; Plastaras, J.; Roberts, L. J.; Rokach, J.; Shigenaga, M. K.; Sohal, R. S.; Sun, J.; Tice, R. R.; Van Thiel, D. H.; Wellner, D.; Walter, P. B.; Tomer, K. B.; Mason, R. P.; Barrett, J. C. Biomarkers of Oxidative Stress Study II: Are Oxidation Products of Lipids, Proteins, and DNA Markers of CCl4 Poisoning? *Free Radical Biol. Med.* 2005, *38*, 698–710.

55. Lee, S. H.; Blair, I. A. Analysis of Etheno 2'-Deoxyguanosine Adducts as Dosimeters of Aldo-keto Reductase Mediated-Oxidative Stress. *ACS Symp. Ser.* 2004, *865*, 139–152.

56. Wu, K.-Y.; Scheller, N.; Ranasinghe, A.; Yen, T.-Y.; Sangaiah, R.; Giese, R.; Swenberg, J. A. A Gas Chromatography/Electron Capture/Negative Chemical Ionization High-Resolution Mass Spectrometry Method for Analysis of Endogenous and Exogenous N7-(2-Hydroxyethyl) Guanine in Rodents and Its Potential for Human Biological Monitoring. *Chem. Res. Toxicol.* 1999, *12*, 722–729.

57. Chen, H.-J. C.; Chang, C.-M. Quantification of Urinary Excretion of 1,N6-Ethenoadenine, a Potential Biomarker of Lipid Peroxidation, in Humans by Stable Isotope Dilution Liquid Chromatography-Electrospray Ionization-Tandem Mass Spectrometry: Comparison with Gas Chromatography-Mass Spectrometry. *Chem. Res. Toxicol.* 2004, *17*, 963–971.

58. Chen, H.-J. C.; Chiu, W.-L. Association Between Cigarette Smoking and Urinary Excretion of 1,N2-Ethenoguanine Measured by Isotope Dilution Liquid Chromatography-Electrospray Ionization/Tandem Mass Spectrometry. *Chem. Res. Toxicol.* 2005, *18*, 1593–1599.

59. Chen, H.-J. C.; Hong, C.-L.; Wu, C.-F.; Chiu, W.-L. Effect of Cigarette Smoking on Urinary 3, N4-Ethenocytosine Levels Measured by Gas Chromatography/Mass Spectrometry. *Toxicol. Sci.* 2003, *76*, 321–327.

60. Chen, H.-J. C.; Zhang, W.; Song, R.; Ma, H.; Dong, G. S.; Zhou, Z.; Fu, J. Analysis of DNA Methylation by Tandem Ion-Pair Reversed-Phase High-Performance Liquid Chromatography/Electrospray Ionization Mass Spectrometry. *Rapid Commun. Mass Spectrom.* 2004, *19*, 1120–1124.

61. Chen, H.-J. C.; Wu, C.-F.; Hong, C.-L.; Chang, C.-M. Urinary Excretion of 3,N4-Etheno-2′-deoxycytidine in Humans as a Biomarker of Oxidative Stress: Association with Cigarette Smoking. *Chem. Res. Toxicol.* 2004, *17*, 896–903.57.

62. Chen, H.-J. C. Formation, Analysis and Repair of Exocyclic Etheno DNA Adducts. *Chin. Pharm. J. (Taipei, Taiwan)* 2004, *56*, 1–16.

63. Hofer, T.; Badouard, C.; Bajak, E.; Ravanat, J.-L.; Mattsson, A.; Cotgreave, I. A. Hydrogen Peroxide Causes Greater Oxidation in Cellular RNA Than in DNA. *Biol. Chem.* 2005, *386*, 333–337.

64. Li, C.-S.; Wu, K.-Y.; Chang-Chien, G.-P.; Chou, C.-C. Analysis of Oxidative DNA Damage 8-Hydroxy-2′-Deoxyguanosine as a Biomarker of Exposures to Persistent Pollutants for Marine Mammals. *Environ. Sci. Technol.* 2005, *39*, 2455–2460.

65. Matter, B.; Malejka-Giganti, D.; Csallany, A. S.; Tretyakova, N. Quantitative Analysis of the Oxidative DNA Lesion, 2,2-Diamino-4-(2-Deoxy-B-D-Erythro-Pentofuranosyl) Amino]-5(2H)-Oxazolone (Oxazolone), in Vitro and in Vivo by Isotope Dilution-Capillary HPLC-ESI-MS/MS. *Nucleic Acids Res.* 2006, *34*, 5449–5460.

66. Rajesh, M.; Ramesh, A.; Ravi, P. E.; Balakrishnamurthy, P.; Coral, K.; Punitham, R.; Sulochana, K. N.; Biswas, J.; Ramakrishnan, S. Accumulation of 8-Hydroxy-deoxyguanosine and Its Relationship with Antioxidant Parameters in Patients with Eales' Disease: Implications for Antioxidant Therapy. *Curr. Eye Res.* 2003, *27*, 103–110.

67. Shiota, G.; Maeta, Y.; Mukoyama, T.; Yanagidani, A.; Udagawa, A.; Oyama, K.; Yashima, K.; Kishimoto, Y.; Nakai, Y.; Miura, T.; Ito, H.; Murawaki, Y.; Kawasaki, H. Effects of Sho-Saiko-to on Hepatocarcinogenesis and 8-Hydroxy-2′-Deoxyguanosine Formation. *Hepatology (Philadelphia, PA, U.S.)* 2002, *35*, 1125–1133.

68. Wang, J.; Xiong, S.; Xie, C.; Markesbery, W. R.; Lovell, M. A. Increased Oxidative Damage in Nuclear and Mitochondrial DNA in Alzheimer's Disease. *J. Neurochem.* 2005, *93*, 953–962.

69. Xi, Z.; Chao, F.; Sun, Y.; Li, G.; Liu, Z.; Li, Y.; Yang, D.; Zhang, H.; Zhang, W. Analysis on Diesel Vehicle Exhaust and Its Oxidative DNA Damage. *Zhongguo Gonggong Weisheng* 2004, *20*, 419–421.

70. Esaka, Y.; Inagaki, S.; Goto, M. Separation Procedures Capable of Revealing DNA Adducts. *J. Chromatogr., B: Anal. Technol.Biomed. Life Sci.* 2003, *797*, 321–329.

71. Bailey, E.; Farmer, P. B.; Shuker, D. E. Estimation of Exposure to Alkylating Carcinogens by the GC-MS Determination of Adducts to Hemoglobin and Nucleic Acid Bases in Urine. *Arch. Toxicol.* 1987, *60*, 187–191.

72. Bailey, E.; Brooks, A. G.; Bird, I.; Farmer, P. B.; Street, B. Monitoring Exposure to 4,4′-Methylenedianiline by the Gas Chromatography-Mass Spectrometry Determination of Adducts to Hemoglobin. *Anal. Biochem.* 1990, *190*, 175–181.

73. Bryant, M. S.; Skipper, P. L.; Tannenbaum, S. R.; Maclure, M. Hemoglobin Adducts of 4-Aminobiphenyl in Smokers and Nonsmokers. *Cancer Res.* 1987, *47*, 602–608.

74. Bryant, M. S.; Vineis, P.; Skipper, P. L.; Tannenbaum, S. R. Hemoglobin Adducts of Aromatic Amines: Associations with Smoking Status and Type of Tobacco. *Proc. Natl. Acad. Sci. U.S.A.* 1988, *85*, 9788–9791.

75. Carmella, S. G.; Kagan, S. S.; Kagan, M.; Foiles, P. G.; Palladino, G.; Quart, A, M.; Quart, E.; Hecht, S. S. Mass Spectrometric Analysis of Tobacco-Specific Nitrosamine Hemoglobin Adducts in Snuff Dippers, Smokers, and Nonsmokers. *Cancer Res.* 1990, *50*, 5438–5445.

76. Day, B. W.; Naylor, S.; Gan, L. S.; Sahali, Y.; Nguyen, T. T.; Skipper, P. L.; Wishnok, J. S.; Tannenbaum, S. R. Molecular Dosimetry of Polycyclic Aromatic Hydrocarbon Epoxides and Diol Epoxides via Hemoglobin Adducts. *Cancer Res.* 1990, *50*, 4611–4618.

77. Farmer, P. B.; Bailey, E. Protein-Carcinogen Adducts in Human Dosimetry. *Arch. Toxicol. Suppl.* 1989, *13*, 83–90.

78. Farmer, P. B.; Bailey, E.; Gorf, S. M.; Törnqvist, M.; Osterman-Golkar, S.; Kautiainen, A.; Lewis-Enright, D. P. Monitoring Human Exposure to Ethylene Oxide by the Determination of Haemoglobin Adducts Using Gas Chromatography-Mass Spectrometry. *Carcinogenesis* 1986, *7*, 637–640.

79. Farmer, P. B.; Bailey, E. Protein-Carcinogen Adducts in Human Dosimetry. *Arch. Toxicol. Suppl.* 1989, *13*, 83–90.

80. Hecht, S. S.; Kagan, S. S.; Kagan, M.; Carmella, S. G. Quantification of 4-Hydroxy-1-(3-Pyridyl)-1-Butanone Released from Human Haemoglobin as a Dosimeter for Exposure to Tobacco-Specific Nitrosamines. *IARC Sci. Publ.* 1991, *105*, 113–118.

81. Lynch, A. M.; Murray, S.; Boobis, A. R.; Davies, D. S.; Gooderham, N. J. The Measurement of MeIQx Adducts with Mouse Haemoglobin in Vitro and in Vivo: Implications for Human Dosimetry. *Carcinogenesis* 1991, *12*, 1067–1072.

82. Sabbioni, G.; Neumann, H. G. Biomonitoring of Arylamines: Hemoglobin Adducts of Urea and Carbamate Pesticides. *Carcinogenesis* 1990, *11*, 111–115.

83. Straub, K. M.; Burlingame, A. L. Mass Spectrometry as a Tool for the Analysis of Xenobiotic-Modified Nucleic Acids. *Soft Ionization Biological Mass Spectrometry: Proceedings of the Chemical Society Symposium on Advances in Mass Spectrometry Soft Ionization Methods, London, July 1980*; Morris. H. R., Ed.; Heyden: London, 1981; pp. 39–52.

84. Andrews, C. L.; Vouros, P.; Harsch, A. Analysis of DNA Adducts Using High-Performance Separation Techniques Coupled to Electrospray Ionization Mass Spectrometry. *J. Chromatogr., A* 1999, *856*, 515–526.

85. Apruzzese, W. A.; Vouros, P. Analysis of DNA Adducts by Capillary Methods Coupled to Mass Spectrometry: A Perspective. *J. Chromatogr., A* 1998, *794*, 97–108.

86. Turesky, R. J.; Vouros, P. Formation and Analysis of Heterocyclic Aromatic Amine-DNA Adducts in Vitro and in Vivo. *J. Chromatogr., B: Anal. Technol.Biomed. Life Sci.* 2004, *802*, 155–166.

87. Bohr, V. A. Gene Specific DNA Repair. *Carcinogenesis* 1991, *12*, 1983–1992.

88. Esmans, E. L.; Broes, D.; Hoes, I.; Lemiere, F.; Vanhoutte, K. Liquid Chromatography-Mass Spectrometry in Nucleoside, Nucleotide and Modified Nucleotide Characterization. *J. Chromatogr., A* 1998, *794*, 109–127.

89. Phillips, D. H.; Farmer, P. B.; Beland, F. A.; Nath, R. G.; Poirier, M. C.; Reddy, M. V.; Turteltaub, K. W. Methods of DNA Adduct Determination and Their Application to Testing Compounds for Genotoxicity. *Environ. Mol. Mutagen.* 2000, *35*, 222–233.

90. de Kok, T. M. C. M.; Moonen, H. J. J.; Van Delft, J.; Van Schooten, F. J. Methodologies for Bulky DNA Adduct Analysis and Biomonitoring of Environmental and Occupational Exposures. *J. Chromatogr., B: Anal. Technol. Biomed. Life Sci.* 2002, *778*, 345–355.

91. Doerge, D. R.; Churchwell, M. I.; Beland, F. A. Analysis of DNA Adducts from Chemical Carcinogens and Lipid Peroxidation Using Liquid Chromatography and Electrospray Mass Spectrometry. *J. Environ. Sci. Health, Part C, Environ. Carcinog. Ecotoxicol. Rev.* 2002, *20*, 1–20.

92. Stiborova, M. R., M.; Hodek, P.; Frei, E.; Schmeiser, H. H. Monitoring of DNA Adducts in Humans and ^{32}P-Postlabelling Methods: A Review. *Collect. Czech. Chem. Commun.* 2004, *69*, 476–498.

93. Chaudhary, A. K.; Nokubo, M.; Oglesby, T. D.; Marnett, L. J.; Blair, I. A. Characterization of Endogenous DNA Adducts by Liquid Chromatography Tandem Mass Spectrometry *J. Mass Spectrom.* 1995, *30*, 1157–1166.

94. Wolf, S. M.; Vouros, P. Application of Capillary Liquid Chromatography Coupled with Tandem Mass Spectrometric Methods to the Rapid Screening of Adducts Formed by the Reaction of N-Acetoxy-N-Acetyl-2-Aminofulorene with Calf Thymus DNA. *Chem. Res. Toxicol.* 1994, *7*, 82–88.

95. Soglia, J. R.; Turesky, R. J.; Paehler, A.; Vouros, P. Quantification of the Heterocyclic Aromatic Amine DNA Adduct N-(Deoxyguanosin-8-yl)-2-Amino-3-Methylimidazo[4,5-f]Quinoline in Livers of Rats Using Capillary Liquid Chromatography/Microelectrospray Mass Spectrometry: A Dose-Response Study. *Anal. Chem.* 2001, *73*, 2819–2827.

96. Abian, J.; Oosterkamp, A. J.; Gelpi, E. Comparison of Conventional, Narrow-Bore and Capillary Liquid Chromatography/Mass Spectrometry for Electrospray Ionization Mass Spectrometry: Practical Consideration. *J. Mass Spectrom.* 1999, *34*, 244–254.

97. Embrechts, J.; Lemiere, F.; Van Dongen, W.; Esmans, E. L. Equilenin-2'-Deoxynucleoside Adducts: Analysis with Nano-Liquid Chromatography Coupled to Nano-Electrospray Tandem Mass Spectrometry. *J. Mass Spectrom.* 2001, *36*, 317–328.

98. Embrechts, J.; Lemiere, F.; Van Dongen, W.; Esmans, E. L.; Buytaert, P.; Van Marck, E.; Kockx, M.; Makar, A. Detection of Estrogen DNA-Adducts in Human Breast Tumor Tissue and Healthy Tissue by Combined Nano LC-Nano ES Tandem Mass Spectrometry. *J. Am. Soc. Mass Spectrom.* 2003, *14*, 482–491.

99. Hoes, I.; Van Dongen, W.; Lemiere, F.; Esmans, E. L.; Van Bockstaele, D.; Berneman, Z. N. Comparison Between Capillary and Nano Liquid Chromatography-Electrospray Mass Spectrometry for the Analysis of Minor DNA-Melphalan Adducts. *J. Chromatogr., B: Biomed. Sci. Appl.* 2000, *748*, 197–212.

100. Van den Driessche, B.; Lemiere, F.; Van Dongen, W.; Esmans, E. L. Structural Characterization of Melphalan Modified 2'-Oligodeoxynucleotides by Miniaturized LC-ES MS/MS. *J. Am. Soc. Mass Spectrom.* 2004, *15*, 568–579.

101. Van den Driessche, B.; Lemiere, F.; Van Dongen, W.; Van der Linden, A.; Esmans, E. L. Qualitative Study of in Vivo Melphalan Adduct Formation in the Rat by Miniaturized Column-Switching Liquid Chromatography Coupled with Electrospray Mass Spectrometry. *J. Mass Spectrom.* 2004, *39*, 29–37.

102. Vanhoutte, K.; Van Dongen, W.; Hoes, I.; Lemiere, F.; Esmans, E. L.; Van Onckelen, H.; Van den Eeckhout, E.; van Soest, R. E. J.; Hudson, A. J. Development of a Nanoscale Liquid Chromatography/Electrospray Mass Spectrometry Methodology for the Detection and Identification of DNA Adducts. *Anal. Chem.* 1997, *69*, 3161–3168.

103. Erickson, B. E. Bisphenol A Under Scrutiny. *Chem. Eng. News* 2008, *86* (22), 36–39.

104. Kuo, H.-W.; Ding, W.-H. Trace Determination of Bisphenol A and Phytoestrogens in Infant Formula Powders by Gas Chromatography-Mass Spectrometry. *J. Chromatogr., A* 2004, *1027*, 67–74.

105. Maragou, N. C.; Lampi, E. N.; Thomaidis, N. S.; Koupparis, M. A. Determination of Bisphenol A in Milk by Solid Phase Extraction and Liquid Chromatography-Mass Spectrometry. *J. Chromatogr., A* 2006, *1129*, 165–173.

106. Shao, B.; Han, H.; Hu, J.; Zhao, J.; Wu, G.; Xue, Y.; Ma, Y.; Zhang, S. Determination of Alkylphenol and Bisphenol A in Beverages Using Liquid Chromatography/Electrospray Ionization Tandem Mass Spectrometry. *Anal. Chim. Acta* 2005, *530*, 245–252.

107. Uematsu, Y.; Hirata, K.; Suzuki, K.; Iida, K.; Saito, K. Chlorohydrins of Bisphenol A Diglycidyl Ether (BADGE) and of Bisphenol F Diglycidyl Ether (BFDGE) in Canned Foods and Ready-to-Drink Coffees from the Japanese Market. *Food Addit. Contam.* 2001, *18*, 177–185.

108. Markham, D. A.; McNett, D. A.; Birk, J. H.; Klecka, G. M.; Bartels, M. J.; Staples, C. A. Quantitative Determination of Bisphenol-A in River Water by Cool On-Column Injection-Gas Chromatography-Mass Spectrometry. *Int. J. Environ. Anal. Chem.* 1998, *69*, 83–98.

109. Manabe, A.; Kaneko, S.; Numazawa, S.; Itoh, K.; Inoue, M.; Hisamitsu, H.; Sasa, R.; Yoshida, T. Detection of Bisphenol-A in Dental Materials by Gas Chromatography-Mass Spectrometry. *Dent. Mater. J.* 2000, *19*, 75–86.

110. Hillem, B. Bisphenol A Vexations. *Chem. Eng. News* 2007, *85* (36), 31–33.

111. Schulz, W.; Moore, K. Momentum Builds Against Bisphenol A. *Chem. Eng. News* 2008, *86* (17): 11.

112. Tominaga, T.; Negishi, T.; Hirooka, H.; Miyachi, A.; Inoue, A.; Hayasaka, I.; Yoshikawa, Y. Toxicokinetics of Bisphenol A in Rats, Monkeys and Chimpanzees by the LC-MS/MS Method. *Toxicology* 2006, *226*, 208–217.

113. Hillem, B. More Concerns Over Bisphenol A. *Chem. Eng. News* 2007, *85* (32), 8.

114. Kuklenyik, Z.; Ekong, J.; Cutchins, C. D.; Needham, L. L.; Calafat, A. M. Simultaneous Measurement of Urinary Bisphenol A and Alkylphenols by Automated Solid-Phase Extractive Derivatization Gas Chromatography/Mass Spectrometry. *Anal. Chem.* 2003, *75*, 6820–6825.

115. Sambe, H.; Hoshina, K.; Hosoya, K.; Haginaka, J. Direct Injection Analysis of Bisphenol A in Serum by Combination of Isotope Imprinting with Liquid Chromatography-Mass Spectrometry. *Analyst (Cambridge, U.K.)* 2005, *130*, 38–40.

116. Tsukioka, T.; Brock, J.; Graiser, S.; Nguyen, J.; Nakazawa, H.; Makino, T. Determination of Trace Amounts of Bisphenol A in Urine by Negative-Ion Chemical-Ionization-Gas Chromatography/Mass Spectrometry. *Anal. Sci.* 2003, *19*, 151–153.

117. Belguise-Valladier, P.; Fuchs, R. P. P. Strong Sequence-Dependent Polymorphism in Adduct-Induced DNA Structure Analysis of Single N-2-Acetylaminofluorene Residues Bound within the NarI Mutation Hot Spot. *Biochemistry* 1991, *30*, 10091–10100.

118. Hartley, J. A.; Bingham, J. P.; Souhami, R. L. DNA Sequence Selectivity of Guanine-N7 Alkylation by Nitrogen Mustards Is Preserved in Intact Cells. *Nucleic Acids Res.* 1992, *20*, 3175–3178.

119. Koehl, P.; Valladier, P.; Lefevere, J.-F.; Fuchs, R. P. P. Strong Structure Effect of the Position of a Single Acetyleaminofluorene Adduct Within a Mutation Hot Spot. *Nucleic Acids Res.* 1989, *17*, 9531–9541.

120. Kohn, K. W.; Hartley, J. A.; Mattes, W. B. Mechanism of DNA Sequence Selective Alkylation of Guanine-N7 Positions by Nitrogen Mustard. *Nucleic Acids Res.* 1987, *15*, 10531–10549.

121. Mattes, W. B.; Hartley, J. A.; Kohn, K. W. DNA Sequence Selectivity of Guanine-N7 Alkylation by Nitrogen Mustards. *Nucleic Acids Res.* 1986, *14*, 2971–2987.

122. Shukla, R.; Liu, T.; Geacintov, N. E.; Loechler, E. L. The Major, N2-dG Adduct of (+)-anti-B[a]PDE Shows a Dramatically Different Mutagenic Specificity (Predominantly, G → A) in a 5'-CGT-3' Sequence Context, *Biochemistry* 1997, *36* (33), 10256–10261.

123. Arghavani, M. B.; SantaLucia, Jr. J.; Romano, L. J. Effect of Mismatched Complementary Strands and 5'-Change in Sequence Context on the Thermodynamics and Structure of Benzo[a]pyrene-Modified Oligonucleotides, *Biochemistry* 1998, *37* (23), 8575–8583.

124. Page, J. E.; Pilcher, A. S.; Yagi, H.; Sayer, J. M.; Jerina, D. M.; Anthony, D. Mutational Consequences of Replication of M13mp7L2 Constructs Containing Cis-Opened Benzo[a]pyrene 7,8-Diol 9,10-Epoxide–Deoxyadenosine Adducts, *Chem. Res. Toxicol.* 1999, *12* (3), 258–263.

125. Burnouf, D.; Koehl, P.; Fuchs, R. P. P. Single Adduct Mutagenesis: Strong Effect of the Position of a Single Acetylaminofluorene Adduct within a Mutation Hot Spot. *Proc. Natl. Acad. Sci. U.S.A.* 1989, *86*, 4147–4151.

126. Cloutier, J. F.; Drouin, R.; Castonguay, A. Treatment of Human Cells with N-Nitroso(acetoxymethyl)methylamine: Distribution Patterns of Piperidine-Sensitive DNA Damage at the Nucleotide Level of Resolution Are Related to the Sequence Context, *Chem. Res. Toxicol.* 1999, *12* (9), 840–849.

127. Bacolod, M. D.; Krishnasamy, R.; Basu, A. K. Mutagenicity of the 1-Nitropyrene–DNA Adduct N-(Deoxyguanosin-8-yl)-1-aminopyrene in *Escherichia coli* Located in a Nonrepetitive CGC Sequence, *Chem. Res. Toxicol.* 2000, *13* (6), 523–528.

128. Pontén, I.; Kroth, H.; Sayer, J. M; Dipple, A.; Jerina, D. M. Differences between the Mutational Consequences of Replication of Cis- and Trans-Opened Benzo[a]pyrene 7,8-Diol 9,10-Epoxide-deoxyguanosine Adducts in M13mp7L2 Constructs. *Chem. Res. Toxicol.* 2001, *14* (6), 720–726.

129. Shibutani, S.; Suzuki, N.; Tan, X.; Johnson, F.; Grollman, A. P. Influence of Flanking Sequence Context on the Mutagenicity of Acetylaminofluorene-Derived DNA Adducts in Mammalian Cells, *Biochemistry* 2001, *40* (12), 3717–3722.

130. Zhuang, P.; Kolbanovskiy, A.; Amin, S.; Geacintov, N. E. Base Sequence Dependence of in Vitro Translesional DNA Replication past a Bulky Lesion Catalyzed by the Exo-Klenow Fragment of Pol, *Biochemistry* 2001, *40* (22), 6660–6669.

131. Huang, X.; Colgate, K.C.; Kolbanovskiy, A.; Amin, S.; Geacintov, N. E. Conformational Changes of a Benzo[a]pyrene Diol Epoxide-N2-dG Adduct Induced by a 5′-Flanking 5-Methyl-Substituted Cytosine in a MeCG Double-Stranded Oligonucleotide Sequence Context, *Chem. Res. Toxicol.* 2002, *15* (3), 438–444.

132. Feng, Z.; Hu, W.; Rom, W. N.; Beland, F. A.; Tang, M-S. N-Hydroxy-4-aminobiphenyl-DNA Binding in Human p53 Gene: Sequence Preference and the Effect of C5 Cytosine Methylation, *Biochemistry* 2002, *41* (20), 6414–6421.

133. Tan, X.; Suzuki, N.; Grollman, A. P.; Shibutani, S. Mutagenic Events in Escherichia coli and Mammalian Cells Generated in Response to Acetylaminofluorene-Derived DNA Adducts Positioned in the Nar I Restriction Enzyme Site, *Biochemistry* 2002, *41* (48), 14255–14262.

134. Ziegel, R.; Shallop, A.; Jones, R.; Tretyakova, N. K-Ras Gene Sequence Effects on the Formation of 4-(Methylnitrosamino)-1-(3-Pyridyl)-1-Butanone (NNK)-DNA Adducts. *Chem. Res. Toxicol.* 2003, *16*, 541–550.

135. Watt, D. L.; Utzat, C. D.; Hilario, P.; Basu, A. K. Mutagenicity of the 1-Nitropyrene-DNA Adduct N-(Deoxyguanosin-8-yl)-1-aminopyrene in Mammalian Cells, *Chem. Res. Toxicol.* 2007, *20* (11), 1658–1664.

136. Matter, B.; Guza, R.; Zhao, J.; Li, Z-z.; Jones, R.; Tretyakova, N. Sequence Distribution of Acetaldehyde-Derived N2-Ethyl-dG Adducts along Duplex DNA, *Chem. Res. Toxicol.* 2007, *20* (10), 1379–1387.

137. Vooradi, V.; Romano, L. J. Effect of N-2-Acetylaminofluorene and 2-Aminofluorene Adducts on DNA Binding and Synthesis by Yeast DNA Polymerase η, *Biochemistry* 2009, *48* (19), 4209–4216.

138. Donigan, K. A.; Sweasy, J. B. Sequence context-specific mutagenesis and base excision repair, *Mol. Carcinog.* 2009, *48*, 362–368.

139. Kropachev, K.; Kolbanovskii, M.; Cai, Y.; Rodriguez, F.; Kolbanovskii, A.; Liu, Y.; Zhang, L.; Amin, S.; Patel, D.; Broyde, S.; Geacintov, N.E. The Sequence Dependence of Human Nucleotide Excision Repair Efficiencies of Benzo[a]pyrene-derived Lesions: Insights into the Structural Factors that Favor Dual Incisions." *J. Mol. Biol.* 2009, *386*, 1193–1203.

140. McLuckey, S.; Berkel, G.; Glish, G. Tandem Mass Spectrometry of Small, Multiply Charged Oligonucleotides. *J. Am. Soc. Mass Spectrom.* 1992, *3*, 60–70.

141. McLuckey, S.; Habibi-Goudarzi, S. Decompositions of Multiply Charged Oligonucleotide Anions. *J. Am. Chem. Soc.* 1993, *115*, 12085–12095.
142. McLuckey, S.; Habibi-Goudarzi, S. Ion Trap Tandem-Mass Spectrometry Applied to Small Multiply-Charged Oligonucleotides with a Modified Base. *J. Am. Soc. Mass Spectrom.* 1994, *5*, 740–747.
143. Ni, J.; Pomerantz, S.; Rozenski, J.; Zhang, Y.; McCloskey, J. Interpretation of Oligonucleotide Mass Spectra for Determination of Sequence Using Electrospray Ionization and Tandem Mass Spectrometry. *Anal. Chem.* 1996, *68*, 1988–1999.
144. Rozenski, J.; McCloskey, J. A. Determination of Nearest Neighbors in Nucleic Acids by Mass Spectrometry. *Anal. Chem.* 1999, *71*, 1454–1459.
145. Rozenski, J.; McCloskey, J. A. SOS: A Simple Interactive Program for Ab Initio Oligonucleotide Sequencing by Mass Spectrometry. *J. Am. Soc. Mass Spectrom.* 2002, *13*, 200–203.
146. Chien, R.-L.; Burgi, D. S. Sample Stacking of an Extremely Large Injection Volume in High-Performance Capillary Electrophoresis. *Anal. Chem.* 1992, *64*, 1046–1050.
147. Wolf, S. M.; Vouros, P. Incorporation of Sample Stacking Techniques into the Capillary electrophoresis CF-FAB Mass Spectrometric Analysis of DNA Adducts. *Anal. Chem.* 1995, *67*, 891–900.
148. Deforce, D. L.; Van den Eeckhout, E. G. Analysis of DNA Damage Using Capillary Zone Electrophoresis and Electrospray Mass Spectrometry. *Methods in Mol. Biol. (Totowa, NJ, U.S.)* 2001, *162*, 429–441.
149. Willems, A. V.; Deforce, D. L.; Van den Eeckhout, E. G.; Lambert, W. E.; Van Peteghem, C. H.; De Leenheer, A. P.; Van Bocxlaer, J. F. Analysis of Benzo[a]pyrene Diol Epoxide-DNA Adducts by Capillary Zone Electrophoresis-Electrospray Ionization-Mass Spectrometry in Conjunction with Sample Stacking. *Electrophoresis* 2002, *23*, 4092–4103.
150. Gennaro, L. A.; Vadhanam, M.; Gupta, R. C.; Vouros, P. Selective Digestion and Novel Cleanup Techniques for Detection of Benzo[a]pyrene Diol Epoxide-DNA Adducts by Capillary Electrophoresis/Mass Spectrometry. *Rapid Commun. Mass Spectrom.* 2004, *18*, 1541–1547.
151. Barry, J. P.; Norwood, C.; Vouros, P. Detection and Identification of Benzo[a]pyrene Diol Epoxide Adducts to DNA Utilizing Capillary Electrophoresis-Electrospray Mass Spectrometry. *Anal. Chem.* 1996, *68*, 1432–1438.
152. Feng, Y.-L.; Lian, H.; Zhu, J. Application of Pressure Assisted Electrokinetic Injection Technique in the Measurements of DNA Oligonucleotides and their Adducts Using Capillary Electrophoresis-Mass Spectrometry. *J. Chromatogr., A.* 2007, *1148*, 244–249.
153. Barry, J. P.; Muth, J.; Law, S. J.; Karger, B. L.; Vouros, P. Analysis of Modified Oligonucleotides by Capillary Electrophoresis in a Polyvinylpyrrolidone Matrix Coupled with Electrospray Mass Spectrometry. *J. Chromatogr., A.* 1996, *732*, 159–166.
154. Harsch, A.; Vouros, P. Interfacing of CE in a PVP Matrix to Ion Trap Mass Spectrometry: Analysis of Isomeric and Structurally Related (N-Acetylamino) Fluorene-Modified Oligonucleotides. *Anal. Chem.* 1998, *70*, 3021–3027.
155. Huber, C. G.; Oberacher, H. Analysis of Nucleic Acids by On-line Liquid Chromatography-Mass Spectrometry. *Mass Spectrom. Rev.* 2002, *20*, 310–343.
156. Bleicher, K.; Bayer, E. Various Factors Influencing the Signal Intensity of Oligonucleotides in Electrospray Mass Spectrometry. *Biol. Mass Spectrom.* 1994, *23*, 320–322.
157. Bleicher, K.; Bayer, E. Analysis of Oligonucleotides Using Coupled High Performance Liquid Chromatography-Electrospray Mass Spectrometry. *Chromatographia* 1994, *39*, 405–408.
158. Huber, C. G.; Krajete, A. Analysis of Nucleic Acids by Capillary Ion-Pair Reverse-Phase HPLC Coupled to Negative Ion-Electrospray Ionization Mass Spectrometry. *Anal. Chem.* 1999, *71*, 3730–3739.

159. Bothner, B.; Chatman, K.; Sarkisian, M.; Siuzdak, G. Liquid Chromatography Mass Spectrometry of Antisense Oligonucleotides *Bioorg. Med. Chem. Lett.* 1995, *5*, 2863–2868.

160. Apffel, A.; Chakel, J. A.; Fischer, S.; Lichtenwalter, K.; Hancock, W. S. Analysis of Oligonucleotides by HPLC-Electrospray Ionizaiton Mass Spectrometry. *Anal. Chem.* 1997, *69*, 1320–1325.

161. Apffel, A.; Chakel, J. A.; Fischer, S.; Lichtenwalter, K.; Hancock, W. S. New Procedure for the Use of High Performance Liquid Chromatography-Electrospray Ionizaiton Mass Spectrometry for the Analysis of Nucelotides and Oligonucleotides. *J. Chromatogr., A* 1997, *777*, 3–21.

162. Oberacher, H.; Oefner, P. J.; Parson, W.; Huber, C. G. On-Line Liquid Chromatography-Mass Spectrometry: A Useful Tool for the Detection of DNA Sequence Variation. *Angew. Chem., Int. Ed.* 2001, *40*, 3828–3830.

163. Oberacher, H.; Parson, W.; Muhlmann, R.; Huber, C. G. Analysis of Polymerase Chain Reaction Products by On-line Liquid Chromatography-Mass Spectrometry for Genotyping of Polymorphic Short Tandem Repeat Loci. *Anal. Chem.* 2001, *73*, 5109–5115.

164. Svec, F.; Huber, C. G. Monolithic Materials: Promise, Challenge, Achievements. *Anal. Chem.* 2006, *78*, 2100–2107.

165. Yue, G.; Luo, Q.; Zhang, J.; Wu, S.-L.; Karger, B. L. Ultratrace LC/MS Proteomic Analysis Using 10-mm-i.d. Porous Layer Open Tubular Poly (Styrene-Divinylbenzene) Capillary Columns. *Anal. Chem.* 2007, *79*, 938–946.

166. Premstaller, A.; Oberacher, H.; Huber, C. G. High-Performance Liquid Chromatography-Electrospray Ionization Mass Spectrometry of Single- and Double-Stranded Nucleic Acids Using Monolithic Capillary Columns. *Anal. Chem.* 2000, *72*, 4386–4393.

167. Bajad, S. U.; Lu, W.; Kimball, E. H.; Yuan, J.; Peterson, C.; Rabinowitz, J. D. Separation and Quantitation of Water Soluble Cellular Metabolites by Hydrophilic Interaction Chromatography-Tandem Mass Spectrometry. *J. Chromatogr., A.* 2006, *1125*, 76–88.

168. Hsieh, Y. Potential of HILIV-MS in Quantitative Bioanalysis of Drugs and Drug Metabolites. *J. Sep. Science* 2008, *31*, 1481–1491.

169. Pucci, V.; Giuliano, C.; Zhang, R.; Koeplinger, K. A.; Leone, J. F.; Monteagudo, E.; Bonelli, F. HILIC LC-MS for the Determination of 2′-C-methyl-cytidine-triphosphate in Rat Liver. *J. Sep. Science*, 2009, *32*, 1275–1283.

170. Pesek, J. J.; Matyska, M. T.; Milton, T.W.; Hearn, W.; Boysen, R. I. Aqueous Normal-phase Retention of Nucleotides on Silica Hydride Columns. *J. Chromatogr., A.* 2009, *1216*, 1140–1146.

171. Tretyakova, N. Y.; Ling, Y.-P.; Upton, P. B.; Sangaiah, R.; Swenberg, J. A. Macromolecular Adducts of Butadiene. *Toxicology* 1996, *113*, 70–76.

172. Koc, H.; Tretyakova, N. Y.; Walker, V. E.; Henderson, R. F.; Swenberg, J. A. Molecular Dosimetry of N-7 Guanine Adduct Formation in Mice and Rats Exposed to 1,3-Butadiene. *Chem. Res. Toxicol.* 1999, *12*, 566–574.

173. Wang, M.; McIntee, E. J.; Cheng, G.; Shi, Y.; Villalta, P. W.; Hecht, S. S. A Schiff Base Is a Major DNA Adduct of Crotonaldehyde. *Chem. Res. Toxicol.* 2001, *14*, 423–430.

174. Sturla, S. J.; Scott, J.; Lao, Y.; Hecht, S. S.; Villalta, P. W. Mass Spectrometric Analysis of Relative Levels of Pyridyloxobutylation Adducts Formed in the Reaction of DNA with a Chemically Activated Form of the Tobacco-Specific Carcinogen 4-(Methylnitrosamino)-1-(3-Pyridyl)-1-Butanone. *Chem. Res. Toxicol.* 2005, *18*, 1048–1055.

175. Ziegel, R.; Shallop, A.; Upadhyaya, P.; Jones, R.; Tretyakova, N. Endogenous 5-Methylcytosine Protects Neighboring Guanines from N7 and O6-Methylation and O6-Pyridyloxobutylation by the Tobacco Carcinogen 4-(Methylnitrosamino)-1-(3-Pyridyl)-1-Butanone. *Biochemistry* 2004, *43*, 540–549.

176. Tretyakova, N.; Matter, B.; Jones, R.; Shallop, A. Formation of Benzo[a]pyrene Diol Epoxide-DNA Adducts at Specific Guanines within K-Ras and *p53* Gene Sequences: Stable Isotope-Labeling Mass Spectrometry Approach. *Biochemistry* 2002, *41*, 9535–9544.

177. Arlt, V. M.; Stiborova, M.; Henderson, C. J.; Thiemann, M.; Frei, E.; Aimova, D.; Singh, R.; da Costa, G. G.; Schmitz, O. J.; Farmer, P. B.; Wolf, C. R.; Phillips, D. H. Metabolic Activation of Benzo[a]pyrene in vitro by Hepatic Cytochrome P450 Contrasts with Detoxification in vivo: Experiments with Hepatic Cytochrome P450 Reductase Null Mice. *Carcinogenesis* 2008, *29* (3), 656–665.

178. Iannitti, P.; Sheil, M. M.; Wickham, G. High Sensitivity and Fragmentation Specificity in the Analysis of Drug-DNA Adducts by Electrospray Tandem Mass Spectrometry. *J. Am. Chem. Soc.* 1997, *119*, 1490–1491.

179. Colgrave, M. L.; Iannitti-Tito, P.; Wickham, G.; Sheil, M. M. Rapid Determination of Sequence Selectivity and Stability of Alkylated Oligonucleotide Adducts by Electrospray Tandem Mass Spectrometry. *Aust. J. Chem.* 2003, *56*, 401–413.

180. Glover, R. P.; Lamb, J. H.; Farmer P. B. Tandem Mass Spectrometry Studies of a Carcinogen Modified Oligonucleotide. *Rapid Commun. Mass Spectrom.* 1998, *12*, 368–372.

181. Marzilli, L. A.; Wang, D.; Kobertz, W. R.; Essigmann, J. M.; Vouros, P. Mass Spectral Identification and Positional Mapping of Aflatoxin B1-Guanine Adducts in Oligonucleotides. *J. Am. Soc. Mass Spectrom.* 1998, *9*, 676–682.

182. Harsch, A.; Sayer, J. M.; Jerina, D. M.; Vouros, P. HPLC-MS/MS Identification of Positionally Isomeric Benzo[c]phenanthrene Diol Epoxide Adducts in Duplex DNA. *Chem. Res. Toxicol.* 2000, *13*, 1342–1348.

183. Stover, J. S.; Chowdhury, G.; Zang, H.; Guengerich, F. Peter; R., Carmelo, J. Translesion Synthesis Past the C8- and N2-Deoxyguanosine Adducts of the Dietary Mutagen 2-Amino-3-methylimidazo[4,5-f]quinoline in the NarI Recognition Sequence by Prokaryotic DNA Polymerases. *Chem. Res. Toxicol.* 2006, *19*, 1506–1517.

184. Jamin, E. L.; Arquier, D.; Tulliez, J.; Debrauwer, L. Mass Spectrometric Investigation of the Sequence Selectivity for Adduction of Heterocyclic Aromatic Amines on Single-Strand Oligonucleotides. *Rapid Commun. Mass Spectrom.* 2008, *22*, 3100–3110.

185. Egger, A. E.; Hartinger, C. G.; Ben Hamidane, H.; Tsybin, Y. O.; Keppler, B. K.; Dyson, P. J. High Resolution Mass Spectrometry for Studying the Interactions of Cisplatin with Oligonucleotides. *Inorg. Chem.* 2008, *47*, 10626–10633.

186. Nyakas, A.; Eymann, M.; Schuerch, S. The Influence of Cisplatin on the Gas-Phase Dissociation of Oligonucleotides Studied by Electrospray Ionization Tandem Mass Spectrometry. *J. Am. Soc. Mass Spectrom.* 2009, *20*, 792–804.

187. Sherman, C. L.; Pierce, S. E.; Brodbelt, J. S.; Tuesuwan, B.; Kerwin, S. M. Identification of the Adduct Between a 4-Aza-3-ene-1,6-diyne and DNA Using Electrospray Ionization Mass Spectrometry. *J. Am. Soc. Mass Spectrom.* 2006, *17*, 1342–1352.

188. Chan, W.; Yue, H.; Wong, R. N. S.; Cai, Z. Characterization of the DNA adducts Induced by Aristolochic Acids in Oligonucleotides by Electrospray Ionization Tandem Mass Spectrometry. *Rapid Commun. Mass Spectrom.* 2008, *22*, 3735–3742.

189. Evison, B. J.; Chiu, F.; Pezzoni, G.; Phillips, D. R.; Cutts, S. M. Formaldehyde-activated Pixantrone is a Monofunctional DNA Alkylator that Binds Selectively to CpG and CpA Doublets. *Mol. Pharmacol.* 2008, *74*, 184–194.

190. Xiong, W.; Glick, J.; Lin, Y.; Vouros, P. Separation and Characterization of Diastereo-isomeric Oligonucleotide Adducts Using Monolithic Columns by Ion-Pair Reversed-Phase Nano-HPLC Coupled to Ion Trap Mass Spectrometry. *Anal. Chem.* 2007, *79*, 5312–5321.

191. Mowaka, S. Mohamed, D.; Linscheid, M. Separation and Characterization of Oxaliplatin Dinucleotides from DNA Using HPLC-ESI Ion Trap Mass Spectrometry. *Anal. and Bioanal. Chem.* 2008, *392* (5), 819–830.

192. Mohamed, D.; Linscheid, M. Separation and Identification of Trinucleotide-melphalan Adducts from Enzymatically Digested DNA Using HPLC-ESI-MS. *Anal. and Bioanal. Chem.* 2008, *392* (5), 805–817.

193. Mohamed, D.; Mowaka, S. Thomale, J.; Linscheid, M. Analysis of Chloramucil-Oligocucleotide Adducts of Enzymatically Digested DNA using HPLC-ESI-MS. *Chem. Res. in Toxi.* 2009, *22* (8), 1435–1446.

194. Andrews, C. L.; Harsch, A.; Vouros, P. Analysis of the in Vitro Digestion of Modified DNA to Oligonucleotides by LC-MS and LC-MS/MS. *Int. J. Mass Spectrom.* 2004, *231*, 169–177.

195. Liao, Q.; Chiu, N. H. L.; Shen, C.; Chen, Y.; Vouros, P. Investigation of Enzymatic Behavior of Benzonase/Alkaline Phosphatase in the Digestion of Oligonucleotides and DNA by ESI-LC/MS. *Anal. Chem.* 2007, *79*, 1907–1917.

196. Liao, Q.; Shen, C.; Vouros, P. GenoMass—A Computer Software for Automated Identification of Oligonucleotide DNA Adducts from LC-MS Analysis of DNA Digests. *J. Mass Spectrom.* 2009, *44* (4), 549–560.

197. Li, L.; Chiarelli, M. P.; Branco, P. S.; Antunes, A. M.; Marques, M. M.; Gonçalves, L. L.; Beland, F. A. Differentiation of Isomeric C8-substituted Alkylaniline Adducts of Guanine by Electrospray Ionization and Tandem Quadrupole Ion Trap Mass Spectrometry. *J. Am. Soc. Mass Spectrom.* 2003, *14*, 1488–1492.

198. Gao, L.; Zhang, L.; Cho, B. P.; Chiarelli, M. P. Sequence Verification of Oligonucleotides Containing Multiple Arylamine Modifications by Enzymatic Digestion and Liquid Chromatography Mass Spectrometry (LC/MS). *J. Am. Soc. Mass Spectrom.* 2008, *19*, 1147–1155.

199. Song, R.; Zhang, W.; Chen, H.; Ma, H.; Dong, Y.; Sheng, G.; Zhou, Z.; Fu, J. Site Determination of Phenyl Glycidyl Ether-DNA Adducts Using High-Performance Liquid Chromatography with Electrospray Ionization Tandem Mass Spectrometry. *Rapid Commun. Mass Spectrom.* 2005, *19*, 1120–1124.

200. Janning, P.; Schrader, W.; Linscheid, M. A New Mass Spectrometric Approach to Detect Modifications in DNA. *Rapid Commun. Mass Spectrom.* 1994, *8*, 1035–1040.

201. Schrader, W.; Linscheid, M. Determination of Styrene Oxide Adducts in DNA and DNA Components. *J. Chromatogr., A* 1995, *717*, 117–125.

202. Schrader, W.; Linscheid, M. Styrene Oxide DNA Adducts: in Vitro Reaction and Sensitive Detection of Modified Oligonucleotides Using Capillary Zone Electrophoresis Interfaced to Electrospray Mass Spectrometry. *Arch. Toxicol.* 1997, *71*, 588–595.

203. Wang, Y.; Taylor, J.-S.; Gross, M. L. Nuclease P1 Digestion Combined with Tandem Mass Spectrometry for the Structure Determination of DNA Photoproducts. *Chem. Res. Toxicol.* 1999, *12*, 1077–1082.

204. Chou, M. W.; Jian, Y.; Williams, L. D.; Xia, Q.; Churchwell, M.; Doerge, D. R.; Fu, P. P. Identification of DNA Adducts Derived from Riddelliine, a Carcinogenic Pyrrolizidine Alkaloid. *Chem. Res. Toxicol.* 2003, *16*, 1130–1137.

205. Yang, T.; Huang, Y.; Cho, B. P. Synthesis and Characterization of Enantiomeric anti-2-Fluoro-benzo[a]pyrene-7,8-dihydro-diol-9,10-epoxides and Their 2′-Deoxyguanosine and Oligodeoxynucleotide Adducts. *Chem. Res. Toxicol.* 2006, *19*, 242–254.

6 Quantitative *in Silico* Analysis of the Specificity of Graphitized (Graphitic) Carbons

Toshihiko Hanai

CONTENTS

6.1   INTRODUCTION

Since publication of the previous review regarding the use of graphitized carbon (GC) in chromatography [1], few studies have reported on its use not in chromatography columns but rather as a strong adsorbent. New graphitized carbon columns and surface-modified graphitized carbons have now been developed to expand their utility. For example, the Carbopack, coated with 5% carbowax, has been used for a dual-phase headspace sorptive extraction of roasted Arabica coffee and dried sage leaves [2], and primary and secondary amine-coated graphitized carbon has been used for pesticide residue analysis in food [3,4].

Single-walled [5] and multiwalled [6] carbon nanotubes have been developed for gas chromatography columns, and a graphitized carbon monolithic column was synthesized [7]. A graphitized carbon column was used as an ion exchanger after coating with fluorine-containing surfactants and for dynamic ion-exchange liquid chromatography (LC) of anions [8]. Graphitized carbon was also used in supercritical fluid chromatography [9,10]. The redox activity has been used as a reaction column [11,12].

The surface properties of graphitized carbon were studied by inverse gas chromatography. Milling graphitized carbon induces the formation of high-energetic sites and increases the nanoroughness of the surface [13], and the surface energy has been determined by inverse gas chromatography at zero surface coverage [14]. The anion-exchange-like and hydrophobic-like retention mechanism was studied based on the retention of different types of compounds. According to the chromatographic data on benzene, indene, and some polar compounds, retention is determined by the interaction of polarized or polarizable functional groups in the samples with the graphitized carbon surface [15]. The retention of glycosaminoglycan disaccharides was analyzed on a flat polycyclic aromatic carbon surface [16]. A charged-induced dipole interaction is assumed to be the dominating factor, and an expression relating the logarithm of retention as a function of the dielectric constant of the medium and the number of sulfate groups on an analyte has been suggested [16]. Molecular mechanics were used for the optimization of hypothetical molecular models of the gas chromatography of perhydroxanthene and perhydro-4-thia-s-indacene isomers to improve the accuracy of predictions of retention parameters, and the Henry constant was described [17].

Quantitative *in silico* analysis of the retention (adsorption) mechanisms has been proposed [18]. The previous results were recalculated using upgraded computer software, and new discoveries about the retention mechanisms are summarized in the review in this chapter. The recent applications of graphitized (graphitic) carbons as a solid-phase adsorbent and chromatography columns are discussed in a later chapter.

6.2   SPECIFICITY OF GRAPHITIZED CARBONS

Graphite is a polycyclic aromatic hydrocarbon (PAH) comprising more than 10^5 carbons. The number of atoms in graphitized carbon is estimated from electric conductive studies. The surface acts as a Lewis base toward polar solutes and has a Lewis acid base and dispersive interactions with aromatic solutes. Based on theoretical calculations, Hosoya [19] reported that 100 carbon atoms are adequate to demonstrate the graphite properties. Based on theoretical calculations of the model, the electrons localize at the edge of the graphitized material. The calculation capacity of a computer, however, is limited. Several polycyclic aromatic hydrocarbons were constructed; the structures were optimized using the CAChe molecular mechanics program (MM2), and the calculated energy values and the center and outer average atomic partial charge were calculated using MOPAC PM5, as summarized in Table 6.1. The electron density and atomic partial charge of coronene (PAH7) are shown in Figure 6.1.

TABLE 6.1
Molecular Properties of Model PAHs

PAH	Cn	fs	vw	hof	apc_c	apc_o
PAH7	24	−51.644	16.168	780.078	−0.013	−0.104
PAH7x2	48	−88.300	17.003	807.212	−0.025	−0.047
PAH14	42	−92.665	29.590	161.242	−0.006	−0.096
PAH14x2	84	−217.600	27.043	336.914	−0.004	−0.068
PAH19	54	−119.034	38.732	200.243	−0.004	−0.093
PAH22	62	−77.482	42.418	1832.489	−0.004	−0.089
PAH22x2	124	−194.608	45.505	2938.829	0.014	0.007
PAH37	96	−134.433	67.729	2224.114	0.002	−0.006
PAH61	150	−225.094	109.174	3006.100	0.002	−0.014
PAH67	160	−86.465	164.359	3268.000	0.002	−0.016

Notes: Cn, carbon number. fs, MM2/final structure energy (kcal/mol). vw,
MM2/van der Waals energy (kcal/mol). hof, PM5/heat of formation
(kcal/mol). apc_c, average apc of center carbon (au). apc_o, average apc
of outer carbon (au).

FIGURE 6.1 (*See color insert following page 152.*) Electron density and atomic partial
charge (unit: au) of PAH7.

The final structure energy of PAH7 decreases about 37 kcal, and van der Waals values increase about 1 kcal after pair formation (PAH7x2). The hydrogen bonding and electrostatic energy values are zero. The final heat of formation is approximately 30 kcal/mol higher. Based on the MOPAC PM5 calculation, the average atomic partial charge of the center carbons is −0.013 au, and that of the outer carbons is −0.104 au. These values change after pair formation in graphite; the atomic partial charges of the center and outer carbons are −0.025 and −0.047, respectively.

In general, graphite comprises multilayers of large PAH. An increase in the molecular size decreases the final structure energy values; the chemical stability increases with an increase in the size of the structure. The final heat of formation of a pair of PAHs is increased compared with that of the single layer. How does it affect the size of a single layer? A pair formation of the molecule (PAH14) reduces the average atomic partial charge of the center atoms from −0.006 to −0.004 au and that of the outer atoms from −0.096 to −0.068 au. The average atomic partial charge values of the center and outer atoms of PAH19 are −0.004 and −0.093, respectively; those of PAH37 are 0.002 and −0.006, respectively. Increasing the number of rings decreases the atomic partial charge value of the center atoms further. The value is affected by the symmetry of the molecule. The value is −0.004 au for PAH22 and 0.002 au for PAH37 and depends on the length and width ratio of the PAH. The electrostatic potentials of PAH37 are shown in Figure 6.2, where the center and outer atomic partial charge indicates the electron concentration difference.

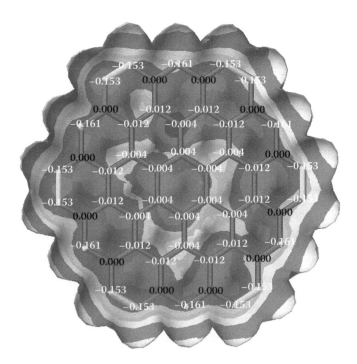

FIGURE 6.2   (*See color insert following page 152.*) Electrostatic potential and atomic partial charge (unit: au) of PAH37.

The structure allows for two types of molecular interactions: a hydrophobic interaction at the center of the large molecules and an electrostatic interaction at the edge of the graphitized carbon. Electron density is low at the center and high at the edge of the molecule. The electron charge of the center atoms of larger molecules is lower than that of small molecules. These observations suggest that the electron charge of the center of graphitized carbon is close to zero and neutral. Similar phenomena are also observed in a saturated carbon-layer of 22 rings, which has a structure similar to that of PAH22 with a more homogeneous net atomic charge distribution. The atomic partial charge of the center carbon is -0.104, and that of the outer carbon is -0.148 au. Therefore, a PAH may be used as a model of graphitized carbon phase.

6.3 CALCULATION OF MOLECULAR INTERACTION ENERGY VALUES

When two molecules form a complex, the energy value decreases from the sum of two molecules' energy values due to the formation of a stable structure. When the energy value of a pair of molecules is the same as the sum of the two molecules' energy value, then a complex has not formed. The molecular interaction (MI) energy value is calculated by subtracting the energy value of the complex from the sum of the individual energy values and the energy value of the model phase. MI values of the final (optimized) structure, hydrogen bonding, electrostatic interaction, and van der Waals interaction are given in the following equations.

MIFS = fs(analyte) + fs(model phase) – FS(analyte-model phase complex)

MIHB = hb(analyte) + hb(model phase) – HB(analyte-model phase complex)

MIES = es(analyte) + es(model phase) – ES(analyte-model phase complex)

MIVW = vw(analyte) + vw(model phase) – VW(analyte-model phase complex)

where fs, hb, es, and vw are energy values of the final (optimized) structure, hydrogen bonding, electrostatic force, and van der Waals force, respectively. FS, HB, ES, and VW are energy values of the final structure, hydrogen bonding, electrostatic force, and van der Waals force of the complexes, respectively. The unit is kcal/mol (1 kj = 4.18 kcal/mol).

The correlation coefficient between individual MI values and logarithmic capacity ratios indicates the contribution of the main molecular interaction factor to retention in chromatography. MIHB, MIES, and MIVW indicate the main interaction factors in normal-phase, ion-exchange, and reversed-phase chromatography, respectively. Steric hindrance affects the molecular interaction in enantiomer separation.

6.3.1 RETENTION OF IONS

The possibility of computational chemical analysis of molecular interactions can be understood from the analysis of simple model compounds. The ion–ion interaction was studied for the combination of an ammonium ion (cation) and an acetic acid ion (anion) by the molecular mechanics calculation CAChe MM2 program. The ion-pair formation and their calculated energy values are as follows.

	NH_4^+ +	CH_3COO^- =	CH_3COONH_4	Balance
Stretch	0.000	0.010	0.012	0.002
Stretch bend	0.000	0.003	0.003	0.000
Improper torsion	0.000	0.000	0.001	0.001
Electrostatic	0.000	0.000	−5.671	−5.671
Angle	1.526	0.038	1.729	0.165
Dihedral angle	0.000	−0.802	−0.803	−0.001
Van der Waals	0.000	0.346	0.224	−0.122
Hydrogen-bonding	0.000	0.000	0.000	0.000
Final structure (optimized)	1.626	−0.404	−5.505	6.726

Note: unit, kcal/mol.

The balance (MI energy values) after a complex (ion-pair) formation indicates the contribution of the ion-pair formation. The electrostatic energy mainly contributes to the ion-pair formation between ammonia and acetic acid ions. The structure of ammonium and acetic acid ions and the ion pair are shown in Figure 6.3, where atomic distance of ammonium hydrogen expanded 1.3Å, and their atomic partial charge using the MOPAC PM5 is shown in Figure 6.4, where atomic partial charge of ammonium hydrogen was doubled. The retention mechanisms of these cations and anions were then studied using a model graphitized carbon phase (PAH14). The optimized complex form is shown in Figure 6.5, where the anions were retained by ion–ion interaction at the model graphitized carbon edge, but cations did not show ion–ion interaction. HOMO and LUMO also indicate the interaction.

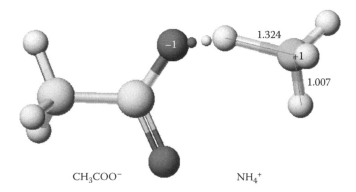

CH_3COO^- NH_4^+

FIGURE 6.3 *(See color insert following page 152.)* Ion pair of ammonium and acetic acid ions. Black balls oxygen, dark gray balls nitrogen, large white balls carbon, small white balls hydrogen. Atomic distance unit Å.

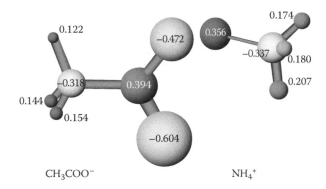

CH₃COO⁻   NH₄⁺

FIGURE 6.4 (*See color insert following page 152.*) Atomic partial charge (unit: au) of ammonium and acetic acid ion pair.

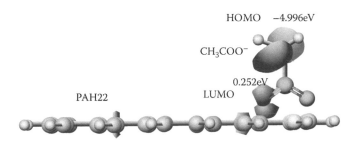

FIGURE 6.5 (*See color insert following page 152.*) Interaction between acetic acid ion and a model graphitized carbon, PAH22, with HOMO and LUMO.

The combination of a tetrabutyl ammonium (cation) and a methylphosphate (anion) was examined by the molecular mechanics calculation CAChe MM2 program. The ion-pair formation and their calculated energy values are as follows [1].

	$(CH_3)_4N^+ + CH_3HPO_4^- = (CH_3)_4N\ CH_3HPO_4$			Balance
Stretch	0.458	0.968	1.420	−0.006
Stretch bend	0.149	−2.018	−1.873	−0.004
Improper torsion	0	0	0	0
Electrostatic	0	2.193	−0.327	−2.520
Angle	0.634	6.596	7.215	−0.015
Dihedral angle	0	0.055	0.017	−0.038
Van der Waals	3.114	0.962	2.038	−2.038
Hydrogen-bonding	0	−0.622	−0.609	0.013
Final structure (optimized)	4.356	8.134	7.880	−4.710

Note: unit, kcal/mol.

These results indicate that van der Waals and electrostatic energies contribute to their interaction. The addition of alkyl groups indicates a van der Waals force contribution.

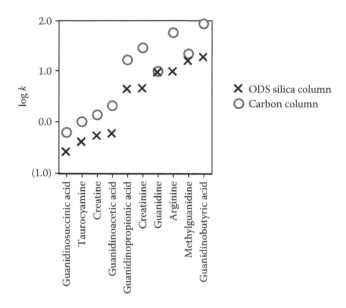

FIGURE 6.6   Selective retention of guanidino compounds on a carbon column.

6.3.2   Retention of Guanidino Compounds [20]

Guanidino compounds are very polar and usually exist in ionic form due to the high pKa value of the guanidine group. Although separation has been performed by ion-exchange LC or reversed-phase ion-pair LC, the separation efficiency has not been satisfactory to date. The retention capability of a graphitized carbon column for guanidino compounds is shown in Figure 6.6, where retention was measured in 50 mM sodium citrate buffer at ambient temperature on an octadecyl-bonded silica gel column and a graphitized carbon column. Guanidino compounds are not retained on the octadecyl-bonded silica gel column even under low or high pH solutions. Several guanidino compounds were separated on the graphitized carbon column. Total separation was performed by the addition of an ion-pair reagent in pH 4.5 sodium citrate buffer. The difference in the hydrophobic retention capacity between an octadecyl-bonded silica gel column and a graphitized carbon column was examined in the separation of 10 guanidino compounds using an isocratic elution with a column switching system. The parallel separation of guanidine compounds with different polarities using an isocratic elution shortens the analysis time. The number of samples was doubled from the gradient elution. This type of column selection will be useful to shorten the time required for the separation of complex mixtures.

6.3.3   Retention of Saccharides

The retention mechanism of polar compounds was studied by analyzing the retention of saccharides in LC. The model phase was a 22-ring PAH. The molecular interaction energy value was calculated using MM2. The retention times were measured

TABLE 6.2
Molecular Interaction Energy Calculated Using MM2

Chemicals	MIFS	MIHB	MIES	MIVW	k
D-(−)-Arabinose	26.833	21.086	−0.036	6.261	0.387
D-(−)-Fructose	33.297	35.526	−0.733	3.017	0.505
D-(+)-Galactose	43.235	45.001	−1.313	3.700	0.822
D-(+)-Glucose	36.566	31.776	−0.202	6.368	0.709
D-(+)-Mannose	40.364	40.161	−1.510	4.527	0.804
D-(−)-Ribose	31.639	30.277	−1.912	4.842	0.694
D-(+)-Xylose	25.591	21.667	1.085	4.200	0.506

Notes: Unit, kcal/mol. Flow rate, 0.5 mL/min at ambient.

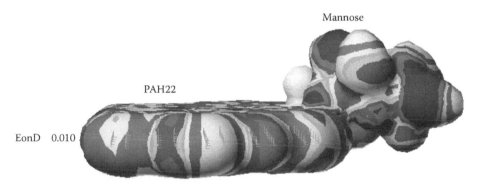

FIGURE 6.7   (*See color insert following page 152.*) Electron density isosurface of mannose and a model graphitized carbon (PAH22) complex.

in water using a graphitized carbon column (TCAS-070510CS), 100 × 4.6 mm i.d. column and are shown in Table 6.2.

The correlation coefficient between these capacity ratios and final structure energy was 0.879, $n = 7$. The correlation coefficients between the capacity ratios and the hydrogen bonding and electrostatic energies were 0.815 and 0.655, respectively, but correlation coefficient between the capacity ratios and van der Waals energy was 0.152. These results indicate that hydrogen-bonding energy mainly contributes to the retention of saccharides on a graphitized carbon in LC, as shown in Figure 6.7, where the electron density isosurface of the complex demonstrates the bonding form. The van der Waals energy values are nearly equal, and their contribution to the retention is weak. The contribution of electrostatic energy was negligible.

6.3.4 RETENTION OF VOLATILE COMPOUNDS IN GC

Little retention data of similar organic volatile compounds are found in the literature of gas chromatographic analyses in which a graphitized carbon column was used for gas analysis. In general, unsaturated compounds elute faster than saturated

TABLE 6.3
Molecular Interaction Energy Calculated Using MM2

Chemicals	MIFS	MIVW	MIHB	MIES
Isobutane	6.689	6.696	0	0
n-Butane	7.923	7.965	0	0
1-Butene	7.695	6.673	0	0
t-2-Butene	11.382	7.022	0	0.099
Cyclohexane	8.235	8.268	0	0
Benzene	8.591	8.601	0	0
1,3,5-Hexene	9.868	9.883	0	0
1,3-Hexene	10.390	9.835	0	0
1-Hexene	11.069	9.777	0	0
Hexane	11.288	11.341	0	0

Note: Unit, kcal/mol.

compounds. The retention time of ethylene is shorter than that of ethane [21], and benzene elutes before n-hexane [22]. The retention order of four carbon hydrocarbons is isobutane, 1-butene, n-butane, and trans-2-butene [23]. The retention mechanism was studied using model standard compounds. The model phase was a 22-ring PAH. The molecular interaction energy value was calculated using MM2. No retention time was measured under isocratic conditions; therefore, the correlation coefficient was not calculated. The final structure energy appears to be related to the retention time, and van der Waals energy contributes to the retention, whereas other energy values do not contribute to the retention according to the MM2 calculation, as summarized in Table 6.3.

The MOPAC calculation does not clearly indicate the contribution force of molecular interaction. It demonstrates the molecular interaction center from the constructed electron density map by the tabulator of the CAChe program and the change in the atomic partial charge before and after optimization of the complex form. These hydrocarbons interact at the center of the PAH molecule as seen on the electron density map, as shown in Figure 6.8, where trans-2-butene located at the center of PAH22.

The atomic partial charge of the model phase and these hydrocarbons do not change significantly after optimization of the complex form of these hydrocarbons, while the electron potential is slightly shifted toward the molecular interaction side. These results clearly indicate the existence of different retention mechanisms on a graphitized carbon phase, a hydrophobic interaction, and hydrogen bonding.

6.4 SELECTIVITY OF GRAPHITIZED CARBON IN NONAQUEOUS SYSTEM [24]

Two types of graphitized carbon packing materials are available. One is made from a high-porosity, high-performance liquid chromatography (HPLC) silica gel

PAH22

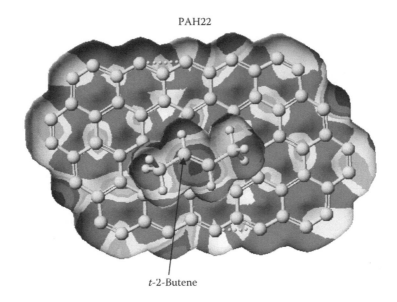

t-2-Butene

FIGURE 6.8   (*See color insert following page 152.*) Retention of *trans*-2-butene on a model graphitized carbon (PAH22).

impregnated with a phenol-formaldehyde resin. The resin is carbonized at 2000 to 2800°C in nitrogen or argon, and the silica particles are dissolved out with alkali (C1 column). Another is made from 100% organic compounds (C2 column). See the synthesis details in Section 6.7. These packing materials (C1 and C2) demonstrate different selectivities.

The analytes, similar to those used in [25], and their properties are summarized in Table 6.4. The eluent was n-heptane. N-Hexane was used first, but the retention time of these compounds is very short, and handling is not easy due to the volatility of n-hexane. The retention times are summarized in Table 6.4. The molecular size of a model graphitized carbon phase is determined based on the analyte's molecular size. The model phase was a 22 aromatic-ring PAH, as shown in Figure 6.9, with *p*-cresol adsorbed on the PAH22. The calculated energy values are summarized in Table 6.4. The energy values of complexes are summarized in Table 6.5.

The correlation coefficient between the calculated individual MI and logarithmic capacity ratio indicates the contribution of individual factors to the retention. MIVW is the main energy value for the interaction in reversed-phase LC [26], and MIES is the main energy value for the retention in ion-exchange LC [27]. Steric hindrance affects the molecular interaction in enantiomer separation [28].

MI energy values were correlated with log k on C1 column (log k_{C1}) and on C2 column (log k_{C2}), and the results are given in the following:

$$\log k_{C1} = 0.053 \text{ (MIFS)} - 0.913, r = 0.362, n = 18$$

$$\log k_{C1} = 0.056 \text{ (MIHB)} - 0.411, r = 0.482, n = 18$$

$$\log k_{C1} = -0.193 \text{ (MIES)} - 0.162, r = 0.021, n = 18$$

$$\log k_{C1} = -0.077 \text{ (MIVW)} + 0.619, \, r = 0.296, \, n = 18$$

$$\log k_{C2} = 0.239 \text{ (MIFS)} - 4.000, \, r = 0.908, \, n = 14$$

$$\log k_{C2} = 0.128 \text{ (MIHB)} - 1.076, \, r = 0.504, \, n = 14$$

$$\log k_{C2} = -92.767 \text{ (MIES)} - 0.653, \, r = 0.384, \, n = 14$$

$$\log k_{C2} = 0.168 \text{ (MIVW)} - 2.296, \, r = 0.563, \, n = 14$$

TABLE 6.4
Properties of Analytes

Chemicals	fs	hb	es	vw	logk$_{C1}$	logk$_{C2}$
Acetophenone	−4.511	−1.459	0	5.846	0.283	—
Anisole	−6.495	0	0	3.301	−0.317	—
Benzene	−8.077	0	0	3.006	−1.060	−1.846
Benzylalcohol	−9.040	−0.606	−0.708	3.481	−0.211	—
Carvacrol	−9.542	−1.485	−0.500	4.146	−0.194	—
Chlorobenzene	−7.802	0	0	3.213	−0.606	—
p-Cresol	−10.755	−1.459	0.017	2.974	0.456	−0.620
Ethylbenzene	−6.784	0	0	4.051	—	−1.557
Nitrobenzene	−8.498	0	0	4.161	0.799	−0.526
Phenol	−10.209	−1.462	0	2.957	0.431	−1.076
2-tButylphenol	−5.156	−1.537	−0.540	6.475	−0.419	—
4-tButylphenol	−6.377	−1.452	0.017	5.426	−0.022	—
4-Ethylphenol	−8.967	−1.456	0.017	3.983	0.277	—
3,4-Dimethylphenol	−10.550	−1.460	0.171	3.414	—	−0.024
2,4,6-Trimethylphenol	−11.719	−1.364	−1.379	4.006	—	−0.374
4-Propylphenol	−8.277	−1.456	0.017	4.470	0.312	—
4-Chlorophenol	−10.024	−1.463	−0.087	3.153	—	−0.642
2,4-Dichlorophenol	−9.706	−1.516	−0.177	3.567	—	−0.283
2,4,6-Trichlorophenol	−12.681	−1.961	−3.207	4.095	—	−0.102
2,3,4,6-Tetrachlorophenol	−4.512	−1.980	2.393	6.099	—	0.456
4-Chloro-2-methylphenol	−10.250	−1.485	−0.069	3.403	—	0.058
4-Phenylphenol	−16.147	−1.459	0	10.989	—	1.246
Thymol	−7.532	−1.302	−0.419	4.961	−0.148	—
Toluene	−8.600	0	0	3.036	−0.791	−1.721
o-Xylene	−8.395	0	0.130	3.498	−0.433	—
m-Xylene	−9.270	0	−0.125	3.047	−0.666	—
p-Xylene	−9.133	0	0.024	3.042	−0.668	—
PAH22 Carbon phase	39.052	0	0	65.823	—	—

Notes: fs, hb, es, and vw, energy values (kcal/mol) of final structure, hydrogen bonding, electrostatic contribution, and van der Waals, respectively, of analytes.

Source: Hanai, T., *Anal. Bioanal. Chem.* 390, 369–375, 2008 (with permission).

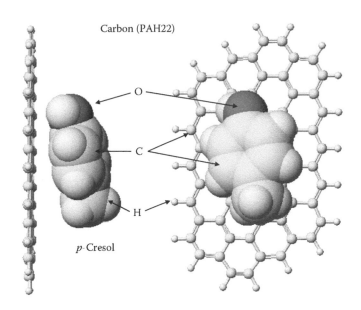

FIGURE 6.9 (*See color insert following page 152.*) Retention of *p*-cresol on a model graphitized carbon (PAH22). Atomic size of the carbon phase is 20%. Black balls oxygen, large white balls carbon, small white balls hydrogen.

The poor correlation coefficient for log k_{C1} on the C1 column seems to relate to the strong retention of phenol and *p*-cresol as shown in Figure 6.9, compared with the result of log k_{C2} on the C2 column. The hydrogen-bonding energy seems to contribute to the retention on the C1 column but not the C2 column. Then, only the alkyl group-substituted compounds were selected, and their log k values were correlated with MI values. The correlation coefficient was 0.718 ($n = 11$) for log k_{C1}. The log k_{C1} values of benzene, toluene, *o*-xylene, *m*-xylene, *p*-xylene, *o-tert*.butylphenol, *p*-ethylphenol, *p*-propylphenol, *p-tert*.butylphenol, phenol, and *p*-cresol were used for the calculation. The correlation coefficient increased to 0.889 ($n = 9$) without phenol and *p*-cresol. The correlation coefficient for log k_{C2} was 0.945. The log k_{C2} values of benzene, toluene, ethylbenzene, phenol, *p*-cresol, 3,4-dimethylphenol, and 2,4,6-trimethylphenol were used for the correlation.

$$\log k_{C1} = 0.134 \ (\text{MIFS}) - 2.128, \ r = 0.718, \ n = 11$$

$$\log k_{C1} = 0.110 \ (\text{MIHB}) - 0.718, \ r = 0.827, \ n = 11$$

$$\log k_{C1} = -4.543 \ (\text{MIES}) - 0.224, \ r = 0.542, \ n = 11$$

$$\log k_{C1} = -0.080 \ (\text{MIVW}) + 0.566, \ r = 0.347, \ n = 11$$

$$\log k_{C2} = 0.945 \ (\text{MIFS}) - 4.089, \ r = 0.945, \ n = 7$$

$$\log k_{C2} = 0.126 \ (\text{MIHB}) - 1.518, \ r = 0.698, \ n = 7$$

$$\log k_{C2} = -255.556 \ (\text{MIES}) - 1.141, \ r = 0.408, \ n = 7$$

$$\log k_{C2} = 0.037 \ (\text{NIVW}) - 1.394, \ r = 0.151, \ n = 7$$

TABLE 6.5
Energy Values of Complexes with PAH22 Carbon Phase

Chemicals	FS	HB	ES	VW
Acetophenone	18.511	−9.147	−0.001	62.219
Anisole	21.324	0	0	57.841
Benzene	22.203	0	0	60.045
Benzylalcohol	6.824	−13.144	−0.941	58.994
Carvacrol	13.435	−9.563	−0.495	61.248
Chlorobenzene	20.883	0	0	58.658
p-Cresol	13.642	−8.612	0.017	60.310
Ethylbenzene	20.137	0	0	57.713
Nitrobenzene	19.730	0	0	59.149
Phenol	15.038	−10.801	0	64.145
2-tButylphenol	17.261	−10.224	−0.539	64.175
4-tButylphenol	15.481	−9.053	0.017	60.855
4-Ethylphenol	13.945	−8.532	0.017	59.832
3,4-Dimethylphenol	12.400	−7.992	0.171	58.725
2,4,6-Trimethylphenol	10.449	−5.338	−1.376	56.221
4-Propylphenol	12.986	−8.328	0.017	58.357
4-Chlorophenol	14.374	−9.076	−0.088	61.057
2,4-Dichlorophenol	13.326	−9.101	−0.178	59.968
2,4,6-Trichlorophenol	10.888	−5.519	−3.198	57.343
2,3,4,6-Tetrachlorophenol	17.299	−5.612	2.403	57.696
4-Chloro-2-methylphenol	12.818	−7.876	−0.066	58.884
4-Phenylphenol	1.384	−8.571	0	61.004
Thymol	17.263	−4.993	−0.419	59.669
Toluene	19.725	0	0	58.153
o-Xylene	18.502	0	0.130	57.130
m-Xylene	17.200	0	−0.125	56.343
p-Xylene	17.299	0	−0.024	56.298

Notes: FS, HB, ES, and VW, energy values (kcal/mol) of final structure,
hydrogen bonding, electrostatic contribution, and van der Waals,
respectively, of complexes.

The correlation coefficients obtained for log k_{C1} remained poor based on the strong retention of phenol and p-cresol. Why were these compounds so strongly retained on the graphitized carbon column? The graphitized carbon was synthesized by washing the silica from graphitized carbon using potassium hydroxide [29]. In this method, a high-porosity HPLC silica gel is impregnated with a phenol-formaldehyde resin. The resin is carbonized at 2000 to 2800°C in nitrogen or argon, and the silica particles dissolved out with alkali. This synthetic process suggests that there are silica and trace amounts of metals, in the silica and potassium hydroxide based on the relatively high correlation coefficient related to the hydrogen-bonding energy values for log k_{C1}.

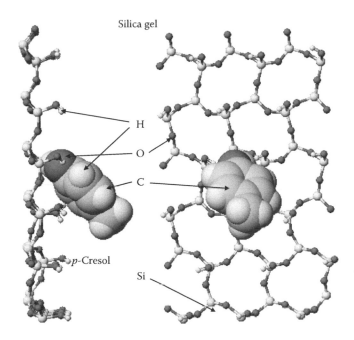

FIGURE 6.10   (*See color insert following page 152.*) Retention of *p*-cresol on a model silica gel. Atomic size of the silica phase is 20%. Black balls oxygen, large white balls carbon, small white balls hydrogen and silicone.

The probability of a silanol effect of the graphitized carbon was studied using a model silanol phase that is used to study retention on silica gels [26]. The molecular interaction energy values were calculated using a model silanol phase shown in Figure 6.10, with *p*-cresol adsorbed on the model silica gel. The correlation coefficients are given in the following equations:

$$\log k_{C1} = 0.098 \text{ (MIFS)} - 2.485, \ r = 0.895, \ n = 11$$

$$\log k_{C1} = 0.094 \text{ (MIHB)} -1.902, \ r = 0.947, \ n = 11$$

$$\log k_{C1} = 0.526 \text{ (MIES)} - 0.048, \ r = 0.713, \ n = 11$$

$$\log k_{C1} = -0.116 \text{ (MIVW)} + 0.517, \ r = 0.351, \ n = 11$$

The correlation coefficient between $\log k_{C1}$ and the hydrogen-bonding energy values was very high. The results were typical of normal-phase LC using silica gels even if the $\log k_{C1}$ values are measured using a graphitized carbon column. On the other hand, the correlation coefficient between $\log k_{C2}$ and MI calculated using the model silica gel was not improved but rather decreased. The results are given in the following equations:

$$\log k_{C2} = 0.150 \text{ (MIFS)} - 4.511, \ r = 0.895, \ n = 7$$

$$\log k_{C2} = 0.093 \text{ (MIHB)} - 2.757, \ r = 0.803, \ n = 7$$

$$\log k_{C2} = 0.378 \text{ (MIES)} - 0.882, \ r = 0.552, \ n = 7$$

$$\log k_{C2} = -0.166 \text{ (MIVW)} - 0.077, \ r = 0.540, \ n = 7$$

The graphitized carbon used in this experiment was 100% organic according to the synthesis method. These results support the speculation that one of the graphitized carbon columns had silanol activity. It seems that the premier matrix silica gel was not completely washed out, and therefore silanol activity remained. An additional calculation was performed using the model silanol phase for the other compounds reported in [25]. The correlations between MI and log k_{C1} values were 0.807 for MIFS, 0.856 for MIHB, 0.373 for MIES, and 0.188 for MIVW ($n = 17$).

The correlation coefficients significantly improved compared with the MI calculated using the graphitized carbon model. The correlation coefficients for MIFS and MIHB were 0.807 and 0,856 with the silanol model compared with those of 0.558 and 0.706 with the graphitized carbon model. No reasonable correlation was obtained for MIES and MIVW because these interactions are not the main factors for retention in normal-phase LC.

6.5 SELECTIVITY OF GRAPHITIZED CARBON COLUMNS IN AQUEOUS PHASE [30]

The analytes used and their properties are listed in Table 6.6. The chromatographic behavior was analyzed to determine whether the retention mechanism was based on reversed-phase LC. These compounds were classified into two groups according to their acidity and basicity. The octanol-water partition coefficients of the analytes, the log P values, were related to log k values measured at pH 2 and 10. There were only 6 compounds. The correlation was better for basic compounds ($r = 0.825$, $n = 6$). Silanol groups exist in their sodium salt form and do not contribute to hydrogen bonding at a high pH but do affect the retention of acidic compounds at a low pH. The correlation coefficient was 0.352 ($n = 6$) for acidic compounds at pH 2.

The MI energy values were calculated using the model carbon phase shown in Figure 6.11, with p-phenitidine on the model phase.

The correlation coefficient between the calculated individual MI and logarithmic capacity ratio indicates the contribution of each individual factor to retention. MIVW is the main energy value for the interaction in reversed-phase LC [26], and MIES is the main energy value for retention in ion-exchange LC [27]. Steric hindrance affects the molecular interaction in enantiomeric separation [28].

MI energy values correlated with reference log k values, and the results are given in the following equations.

At pH 10:

MIFS = 2.854 (log k_{C1}) + 12.303, $r = 0.902$, $n = 6$

MIHB = 0.048 (log k_{C1} + 2.769, $r = 0.061$, $n = 6$

MIES = no correlation was calculated due to identical MIES values

MIVW = 2.796 (log k_{C1}) + 9.703, $r = 0.822$, $n = 6$

TABLE 6.6
Molecular Properties of Analytes and a Model
Graphitized Carbon [30]

Chemicals	fs	hb	es	vw
o-Aminobenzoic acid	−18.021	−5.740	−11.260	6.452
m-Aminobenzoic acid	−15.827	−5.526	−7.258	5.435
p-Aminobenzoic acid	−15.851	−5.514	−7.265	5.384
o-Anisic acid	−15.214	−4.250	−7.200	6.069
m-Anisic acid	−16.296	−4.186	−7.375	5.303
p-Anisic acid	−16.311	−4.182	−7.309	5.207
o Anisidine	−4.376	−1.783	0.580	3.729
m-Anisidine	−4.551	−1.340	−0.010	3.718
p-Anisidine	−4.565	−1.330	−0.008	3.695
o-Phenetidine	−3.809	−1.787	0.580	4.229
m-Phenetidine	−3.915	−1.340	−0.010	4.289
p Phenetidine	−3.927	−1.330	−0.080	4.270
o-Toluidine	−14.610	−4.184	−6.088	6.476
m-Toluidine	−18.415	−4.180	−7.371	5.029
p-Toluidine	−18.433	−4.179	−7.351	4.971
Carbon phase	38.902	0	0	65.672

Chemicals	MIFS	MIHB	MIES	MIVW
o-Aminobenzoic acid	3.976	−16.312	−11.135	64.200
m-Aminobenzoic acid	6.690	−13.697	−7.185	61.433
p-Aminobenzoic acid	7.012	−12.840	−7.190	60.945
o-Anisic acid	6.908	−10.888	−7.119	60.539
m-Anisic acid	6.011	−10.550	−7.297	59.597
p-Anisic acid	5.746	−10.465	−7.230	59.121
o-Anisidine	21.009	−4.823	0.580	58.778
m-Anisidine	20.751	−4.255	−0.010	58.536
p-Anisidine	20.839	−3.914	−0.008	58.289
o-Phenetidine	20.085	−4.432	0.580	57.397
m-Phenetidine	19.838	−4.339	−0.010	57.642
p-Phenetidine	19.986	−3.961	−0.008	57.405
o-Toluidine	8.415	−10.880	−5.990	61.791
m-Toluidine	4.132	−10.814	−7.286	59.673
p-Toluidine	4.223	−10.543	−7.268	59.497

Notes: fs, hb, es, and vw, energy values (kcal/mol) of final structure,
hydrogen bonding, electrostatic contribution, and van der Waals,
respectively, of analytes. FS, HB, ES, and VW, energy values
(kcal/mol) of final structure, hydrogen bonding, electrostatic
contribution, and van der Waals, respectively, of complexes.

Source: Hanai, T. and Homma, H., *J. Liq. Chromatogr. Rel. Technol.*
32, 647–655, 2009 (with permission).

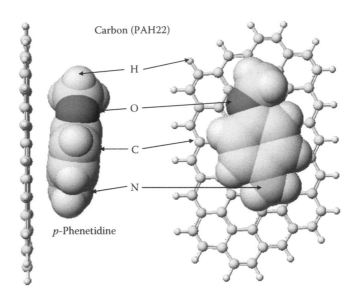

Carbon (PAH22)

p-Phenetidine

FIGURE 6.11   (*See color insert following page 152.*) Retention of *p*-phenetidine on a model graphitized carbon (PAH22). Atomic size of the carbon phase is 20%. Black balls oxygen, large gray balls nitrogen, large white balls carbon, small white balls hydrogen.

At pH 2:

$$\text{MIFS} = 0.328 \ (\log k_{Cl}) + 16.078, \ r = 0.140, \ n = 6$$

$$\text{MIHB} = -0.969 \ (\log k_{Cl}) + 7.595, \ r = 0.844, \ n = 6$$

$$\text{MIES} = 0.018 \ (\log k_{Cl}) - 0.104, \ r = 0.372, \ n = 6$$

$$\text{MIVW} = 1.588 \ (\log k_{Cl}) + 9.346, \ r = 0.534, \ n = 6$$

Why are these compounds retained so strongly on the graphitized carbon column based on the relatively high correlation coefficient of the hydrogen-bonding energy values with log k? The probability of silanol affecting the graphitized carbon was again studied using the model silanol phase that was used to study the retention of silica gels [24]. The MI energy values were calculated using the model silanol phase shown in Figure 6.12, with *p*-phenetidine adsorbed on the model silanol phase. The correlation coefficients in pH 10 eluent were 0.922 for MIFS, 0.706 for MIHB, 0.408 for MIES, and 0.595 for MIVW ($n = 6$). Those in pH 2 eluent were 0.929 for MIFS, 0.969 for MIHB, 0.768 for MIES, and 0.846 for MIVW ($n = 6$). There was a strong correlation between log k and the hydrogen-bonding energy values. The result was typical of normal-phase LC using silica gels even when the log k values were measured using a graphitized carbon column. These results support the speculation regarding the silanol activity of one of the graphitized carbon columns, suggesting that the premier matrix silica gel is not completely washed out and silanol activity persists.

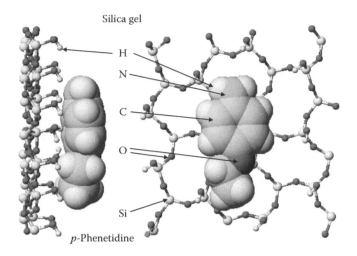

FIGURE 6.12 (*See color insert following page 152.*) Retention of *p*-phenetidine on a model silica gel. Atomic size of the carbon phase is 20%. Black balls oxygen, large gray balls nitrogen, large white balls carbon, small white balls hydrogen and silicone.

Log *P* (octanol–water partition coefficient) values do not relate to MI energy values. They relate weakly with van der Waals energy values, because molecular size is a very important factor for estimating log *P* values.

A semiempirical molecular statistical theory of adsorption based on an atom–atom approximation of the potential function of the intermolecular adsorbate–adsorbent interaction was studied to obtain Henry's constant based on gas chromatography data using graphitic thermal carbon, even though porous adsorbents are inhomogeneous. A simple quantitative correlation of the thermodynamic characteristics of adsorption is applied in LC [31]. Based on several experimental measurements of the isotherms for one of the test solutes followed by nonlinear model fitting, the graphite surface is considered to be homogeneous and to have only one type of adsorption site [17]. Hence, an MM2 calculation was performed to obtain Henry's constant using a flat model [17]. The latter method is straightforward for analyzing a variety of chromatographic data, but it is not straightforward for synthesizing model phases other than a graphitized carbon phase. Chromatographic phases are synthesized homogeneously, but the steric structure is inhomogeneous. The original computer software was used for the conformational analysis of proteins; therefore, such an approach can be applied to the analysis of chromatographic retention if a prospective model is designed. The MI energy values calculated using MM2 support a set of data measured on a graphitized carbon column synthesized using 100% organic materials but not the retention data measured on a graphitized carbon column synthesized using a silica matrix. The chromatographic behavior of these graphitized carbon columns differs, as clearly demonstrated by the aforementioned computational chemical approach. A strong retention of phenolic compounds was described due to the electron-rich surface studied in super/subcritical fluid chromatography [32], but, if the silica matrix of the packing material contributes to the retention of phenolic compounds, this conclusion may be changed.

6.6   STEREO SELECTIVITY OF GRAPHITIZED CARBON

Quantitative *in silico* analysis of polyunsaturated fatty acid methylester (FAME) retention on porous graphitized carbon based on an MM2 calculation demonstrates a good correlation between the capacity ratios and MI energy values. The van der Waals energy is predominant, and the contribution of electrostatic energy is negligible for the retention. The results indicate that van der Waals force, a hydrophobic interaction, is predominant for the retention of polyunsaturated fatty acid methylester on the graphitized carbon in acetonitrile or methylalcohol and chloroform mixtures [33].

Based on theoretical calculations of the model, electrons are observed at the edge of the graphitized material [19], which may indicate that the center of the graphitized carbon is electron poor and thus neutral except for induced effects [1]. Such a flat model has been considered for the study of the retention in gas chromatography [17,31]. The graphitized carbon surface is considered homogeneous based on the liquid chromatographic data of heparin and chondroitin disaccharides [16]. To confirm this, a 195-ring polycyclic aromatic hydrocarbon is considered as a model of graphitized carbon in this study. The molecular interaction energy between a FAME and the 195-ring PAH is calculated using MM2 to quantitatively analyze the retention of FAMEs. Adsorption of C20:5 *cis*-5,8,11,14,17 on the 195-ring PAH is shown in Figure 6.13.

Retention depends on the contact surface area of the molecules. The alkenes with multiple double bonds have less contact surface area, and the molecular interaction energy is smaller than that expected from the carbon atom numbers. The electrons of the double bond and the hydroxyl group do not affect the retention based on *in silico* experiments to examine the relationship between MI energy values of alkyl alcohols, alkanes, and alkenes, and the carbon atom numbers. Increasing the number of carbon double bonds reduces the MI energy values, especially for multiple *cis*- compounds. The electrostatic energy does not change after complex formation. No π–π interaction influences the retention of a graphitized carbon surface [24]. The behavior is the same for all FAME structures. The structure of all *cis*- polyunsaturated FAMEs is neither flat nor straight, as seen in Figure 6.14.

Increasing the number of carbon double bonds decreases the contact surface area with the model phase as seen in Figure 6.15. The correlation coefficient between ΔFS and ln k is 0.968 ($n = 25$). The contribution of van der Waals energy indicates the van der Waals interaction is predominant ($r = 0.954$, $n = 25$) for the FAMEs retention. The phenomenon is supported by the difference in van der Waals energy values. The van der Waals energy values are decreased by increasing the number of carbon double bonds, and the electrostatic energy values do not contribute to the interaction. The MI values are correlated with the capacity ratios, indicating that such model MM2 calculations using a model phase is feasible and confirming that van der Waals energy is predominant. The correlation coefficients between log k and MIVW are 0.946 to 954 in chloroform and acetonitrile mixtures and 0.921 to 0.930 in chloroform and methanol mixtures ($n = 25$). Those for MIFS are 0.965 to 0.968 in chloroform and acetonitrile mixtures and 0.931 to 0.944 in chloroform and methanol mixtures. The contribution

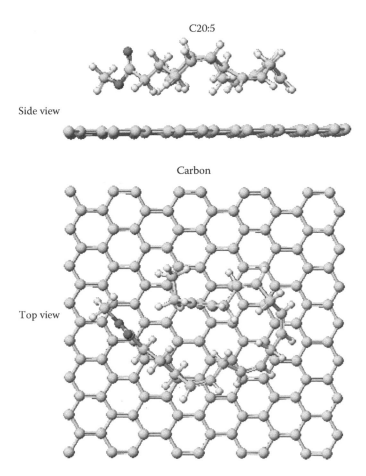

C20:5

Side view

Carbon

Top view

FIGURE 6.13   (*See color insert following page 152.*) Adsorption of C20:5 *cis*-5,8,11,14,17 fatty acid methylester on a model graphitized carbon (PAH195). Black balls oxygen, gray balls carbon, white balls hydrogen.

of hydrogen-bonding energy is zero, and that of electrostatic energy is less than 0.1 kcal/mol. The previous results support Hosoya's observation [19] and indicate that the π–π interaction is not predominant for the retention of FAMEs on the porous glaphitized carbon.

A semiempirical molecular statistical theory of adsorption based on an atom–atom approximation to study the potential function of intermolecular adsorbent–adsorbent interaction was applied to obtain Henry's constant based on gas chromatography data using graphitic thermal carbon, even if porous adsorbents are heterogeneous. A simple quantitative correlation of the thermodynamic characteristics of adsorption is applied in LC [24]. The molecular interaction—hence a molecular mechanics calculation—is performed to obtain Henry's constant using a flat model [31]. The MI energy values can be obtained by an MM2 calculation as demonstrated for the analysis of FAME's retention. The latter method is simple way to analyze a variety of chromatographic data; however, it is not simple to synthesize the model phase,

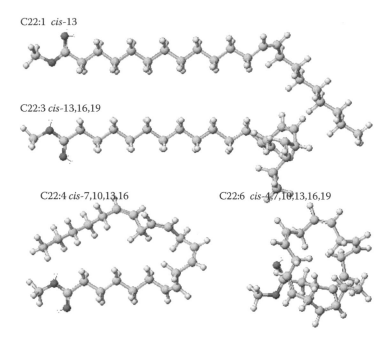

C22:1 *cis*-13

C22:3 *cis*-13,16,19

C22:4 *cis*-7,10,13,16   C22:6 *cis*-4,7,10,13,16,19

FIGURE 6.14 (*See color insert following page 152.*) Stereo structure of 22:1 *cis*-13, 22:3 *cis*-13, 16, 19, 22:4 *cis*-7, 10, 13, 16, and 22:6 *cis*-4, 7, 10, 13, 16, 18 fatty acid methylesters. Black balls oxygen, gray balls carbon, white balls hydrogen.

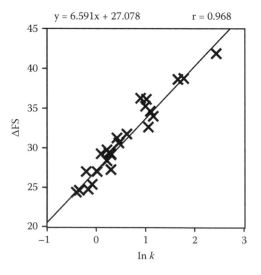

$y = 6.591x + 27.078$ $r = 0.968$

FIGURE 6.15 The correlation between ln k and molecular interaction energy values ΔFS.

except the graphitized carbon phase. Chromatographic phases are synthesized homogeneously, but the steric structure is heterogeneous. The original computer software was used for the conformation analysis of proteins; therefore, such an approach can be applied for the analysis of chromatographic retention only if a prospective model is designed.

6.7   SYNTHESIS OF GRAPHITIZED CARBON FOR CHROMATOGRAPHY [1]

There are three types of porous graphitic carbons. Carbonization of the organic precursors below 1000°C results in an amorphous pyrolytic carbon containing micropores and mesopores. It possesses an oxygenated surface bearing various functional groups such as -OH, -COOH, or -C-O-C-. When pyrolytic carbon is heated to about 1500°C, an amorphous glassy carbon is produced. Carbon heated above 2000°C has the atomic structure of two-dimensional graphite. These materials show different adsorption properties [34].

In 1978, Knox and Gilbert patented a method of making porous carbon [29]. In this method, a high-porosity HPLC silica gel is impregnated with a phenol-formaldehyde resin. The resin is carbonized at 1000°C in nitrogen or argon, and the silica particles dissolve out with alkali. They called this material "porous glassy carbon." The material is microporous, and its LC performance is poor. Upon heating to 2000 ~ 2800°C, the micropores close, and the material is graphitized [35] and marketed under the trade name Hypercarb™. Obayashi et al. of the Tonen Corporation [36] reported an entirely different procedure for making porous graphites. In their method, a roughly 50/50 mixture of low molecular weight pitch (MW ~ 300) and a polymerizable monomer such as styrene or divinylbenzene along with a suitable initiator are suspended in water, and the monomer is polymerized. The beads are then separated and heated in stages to 1100°C and then finally to around 2800°C. The typical material resulting from this method has pores with diameters from 200 to 500 Å. BioTech Research Co. (Kawagoe, Japan) markets the material under the trade name BTR Carbon column. Another method was developed by Ichikawa et al. of the Nippon Carbon Company and the Tosoh Corporation [37]. In their procedure, equal amounts of carbon black colloid particles with approximate diameters of 300 Å and a phenolic resin are dissolved in methanol, made into a slurry, and then spray-dried to give roughly spherical particles ranging in size between 3 to 100 μm. Polymerization of the phenolic resin is complete at about 140°C, and the particles are then heated at a controlled rate in nitrogen to 1000°C and then to 2800°C. The final materials have specific surface areas ranging from 20 to 120 m²/g, and specific pore volumes ranging from 0.3 to 1.0 mL/g. The Tosoh Corporation markets the material under the trade name TSKgel Carbon-500.

Rittenhouse and Olesik developed a low-temperature graphitized porous carbon (LTGC) [38]. Poly-(phenylene diethyl) compounds are first synthesized at low temperature (200–800°C) and then are slowly heated to produce glassy carbon. If the low-temperature glassy carbon is treated at low processing temperatures (i.e., 200°C), then its surface specificity is similar to those of bonded phenyl polysiloxane stationary phases [39]. As the processing temperature is increased, however, conjugation within the LTGC increases, and the dipolarity/polarizability of the solute becomes more important. At processing temperatures close to 550°C, the LTGC shows chemical specificity and retention very similar to those of porous glassy carbon.

Hirayama et al. prepared spherical carbon packing materials from spherical cellulose particles and described the retention mechanism based on the π-electron and the stereo selectivity [40].

6.8 APPLICATIONS

6.8.1 SOLID-PHASE EXTRACTION

Several graphitized carbon related solid-phase extraction (SPE) materials that are suitable for the extraction of volatile organic compounds are on the market [41]. Since publication of the recent review of graphitic carbon columns [1], a number of applications have been reported. The strong adsorption power is applicable for solid-phase extraction of volatile organic compounds in drinking water [42], monoterpenes of western red cedar [43], phenolic acids, and phenolic diterpenes and triterpenes in rosemary [44]. The SPE method is also useful in food analysis [45], such as for the detection of perchlorate in cantaloupe, carrots, lettuce, and spinach [46,47]; spinosad in vegetables and fruits [48]; pesticide residues in agricultural products [4,49–52], green leafy vegetables [53], tea [54], vegetables [55], produce [56], and grains, fruits, and vegetables [57]; terbumeton and its metabolites [58]; pyrifenox in powdered tea [59]; insecticides and fungicides in leafy vegetables [60]; neonicotinoid insecticides in agricultural samples [61]; fungicides in grapes and wines [62] and honeybees [63]; selective separation of 2-acetyl-4-(1,2,3,4-tetrahydroxybutyl)imidazole in Class III caramel color [64]; and polychlorinated biphenyls in children's fast-food containers [63]. The method has also been used in environmental analysis, such as to detect oxygenated volatile organic compounds and nitrile compounds in ambient air [65], benzene, toluene, ethylbenzene, and xylenes in air [66]; cytostatics and metabolites in waste water [67]; poly- and perfluoroalkyl substances in environmental and human matrices [68]; highly leachable pesticides residues in water [69]; pesticides in ground water [70]; polychlorinated naphthalenes, polychlorinated biphenyls, and short-chain chlorinated paraffins [71]; benzo[α]pyrene in soil [72]; polycyclic aromatic hydrocarbons in soil [73]; and steroid hormones in dairy waste disposal systems [74]. The strong adsorption power of graphitized carbon, however, is not suitable for several compounds due to poor desorbtion [66,75–77]. Liquid–liquid extraction is better than SPE for the detection of brominated flame retardants in environmental water and industrial effluent analysis [78].

6.8.2 COLUMN CHROMATOGRAPHY

Graphitized carbon columns are suitable for reversed-phase LC of polar compounds such as acrylamide in food [79], acromelic acid in a poisonous mushroom [80], purine bases in shellfish [81], abscisic acid in grapevine leaf [82], carbohydrate content of Lupinus albus stems [83], oligosaccharides [84–88], underivatized oligosaccharides in the stems of Triticum aestivum [89], Lex and Lewis y[(Fucα1-2)GalB1-4(Fucα1-3)βlcNAc] oligosaccharides [90], oligosaccharides from κ-carrageenan

[91], F2-isoprostanes in human cerebrospinal fluid [92] and urine [93], sulphated disaccharides [94], carbohydrates in a glycoprotein [95], sulfated glycosaminoglycan disaccharides [16], monosubstituted sulfobutyl ether-β-cyclodextrin [96], nitrate ester, nitramine and nitroaromatic explosives [97,98], chemical warfare agent degradation products [99], cyanuric acid in water [100], aniline in freshwater [101], siderophores in natural soil solutions [102], levetiracetam in human serum [103], benazepril, benazeprilat, and hydrochlorothiazide in plasma [104], polar neurotransmitters [105], reduced and oxidized glutathione in human plasma [106], catecholamines and related compounds [107], barbituric acid derivatives [108], haloperidol and its degradation products [109], non-nucleoside reverse transcriptase inhibitor [110], 2′-2′-difluorodeoxycytidine and 2′-2′-difluorodeoxyuridine nucleosides and nucleotides [111], adenosine 5′-monophosphate, adenosine 5′-diphosphate, and adenosine 5′-triphosphate in HepG-2 cells [112], intracellular 1-β-D-arabinofuranosylcytosine triphosphate, cytidine triphosphate, and deoxycytidine triphosphate in a human follicular lymphoma cell line [113], S-adenosylmethionine and S-adenosylhomocysteine in mouse tissues [114], eight nucleoside triphosphates in cell lines [115], zidovudine and stavudine in human serum [116], and organic selenium compounds [117]. The separation of hydrophilic peptide is reviewed [118].

The chemical stability of the column means it can also be used in normal-phase LC for the diastereomeric separation of drug candidates [119], optical and chemical purities of drug with two chiral centers [120], ceramides [121], isomeric tropane alkaloids [122], organic explosives [123], the hydroxamate siderophores ferrichrome and ferricrocin in soil solutions [124], naproxen and its main degradation products [125], and nitroaromatic compounds in air samples [126] and in water [98]. Automated fractionation procedures for polycyclic aromatic hydrocarbons in sediment extracts have also been developed with combination of cyanopropyl- and nitrophenylpropyl-bonded silica gel and porous graphitized carbon columns [127]. Carbon columns can also be used in gas chromatography as a single-walled carbon nanotube [5] and as multwalled carbon nanotubes [6] and in the analysis of ozone precursors [128].

Graphitized carbon columns are also useful in surpercritical fluid chromatography [9]. Stability is maintained under high temperatures [129], and therefore they can be applied in the chromatography of sterols [130]. Peak efficiency is severely lost after 50 h of heating at 185°C [131]. If the phenomenon is due to the silica matrix [30,31], this may change the conclusion.

6.9 CONCLUSIONS

The retention mechanism on graphitized carbons was quantitatively analyzed using a computational chemical method. Different sizes of polycyclic aromatic hydrocarbons have been constructed to study the properties of graphitized carbon using MM2 and PM5 calculations. The atomic partial charge of the center and outer carbons indicates the electron gradient on the surface. The basic interaction is either a hydrophobic interaction, electrostatic interaction, or hydrogen bonding, depending on the properties of the analytes and eluents. Anions are retained, but cations are not. Saccharides are retained by hydrogen bonding. Hydrophobic compounds are retained by hydrophobic interaction in gas chromatography. Aromatic compounds are retained by hydrophobic interaction. The retention mechanisms depend on the method of synthesizing the graphitized carbon. Graphitized carbon synthesized using a silica gel matrix retains silica gel properties. The recovery of trace organic compounds from a variety of samples is excellent, but desorption from the graphitized carbon is poor for hydrophobic compounds. Graphitized carbon is suitable for the collection of volatile organic compounds and relatively polar compounds and for chromatography of polar compounds. The strong adsorption power is useful as a support for different types of polymers and expands the utility of graphitized carbon in a variety of fields as miniaturized sample condensers.

REFERENCES

1. Hanai, T., Separation of polar compounds using carbon columns, *J. Chromatogr. A,* 989, 183–196, 2003.
2. Bicci, C., Cordero, C., Liberto, E., Sgorbini, B., David, F., Sandra, P., and Rubiolo, P., Influence of polymethylsiloxane outer coating and packing material on analyte recovery in dual-phase headspace sorptive extraction, *J. Chromatogr. A,* 1164, 33–39, 2007.
3. He, Y., and Liu, Y.-H., Assessment of primary and secondary amine adsorbents and elution solvents with or without graphitized carbon for the SPE clean-up of food extracts in pesticide residue analysis, *Chromatographia,* 65, 581–590, 2007.
4. Shimelis, O., Yang, Y., Stenerson, K., Kanekoo, T., and Ye, M., Evaluation of a solid-phase extraction dual-layer carbon/primary secondary amine for clean-up of fatty acid matrix components from food extracts in multiresidue pesticide analysis, *J. Chromatogr. A,* 1165, 18–25, 2007.
5. Yuan, L.-M., Ren, C.-X., Li, L., Ai, P., Yan, Z.-H., Min, Z., and Li, Z.-Y., Single-walled carbon nanotubes used as stationaly phase in GC, *Anal. Chem.* 78, 6384–6390, 2006.
6. Li, Q., and Yuan, D., Evaluation of multi-walled carbon nanotubes as gas chromatographic column packing, *J. Chromatogr. A,* 1003, 203–209, 2003.
7. Liang, C., Dai, S., Guiochon, G., A graphitized carbon monolithic column, *Anal. Chem.,* 75. 4904–12, 2003.
8. Helaleh, M.I.H., Al-Omair, A., Tanaka, K., and Mori, M., Ion chromatography of common anions by use of a reversed-phase column dynamically coated with fluorine-containing surfactant, *Acta Chromatographia,* 15, 247–257, 2005.
9. West, C., and Lesellier, E., Effects of modifiers in subcriticalfluid chromatography on retention with porous graphitic carbon, *J. Chromatogr. A,* 1087, 64–76, 2005.
10. West, C. and Lesellier, E., A unified classification of stationary phases for packed column supercritical fluid chromatography, *J. Chromatogr. A,* 1191, 21–39, 2009.
11. Shibukawa, M., Unno, A., Miura, T., Nagoya, A., and Oguma, K., On-column derivatization using redox activity of porous graphitic carbon stationary phase: an approach to enhancement of separation selectivity of liquid chromatography, *Anal. Chem.* 75, 2775–2783, 2003.
12. Saitoh, K., Yamada, N., Ishikawa, E., Nakajima, H., and Shibukawa, M., On-line redox derivatization liquid chromatography using double separation columns and one derivatization unit, *J. Sep. Sci.,* 29, 49–56, 2006.
13. Balard, H., Maafa, D., Santini, A., and Donnet, J.B., Study by inverse gas chromatography of the surface properties of milled graphites, *J. Chromatogr. A,* 1198–1199, 173–180, 2009.

14. Perez-Mendoza, M., Almazan-Almazan, M.C., Mendez-Linan, L., Domingo-Garcia, M., and Lopez-Garzon, F.J., Evaluation of the dispersive component of the surface energy of active carbons as determinated by inverse gas chromatography at zero surface coverage, *J. Chromatogr. A*, 1214, 121–127, 2008.

15. Polyakova, Y., and Row, K.H., HPLC of some polar compounds on a porous graphitized carbon Hypercarb column, *J. Liq. Chromatogr. Rel. Technol.*, 28, 3157–3168, 2005.

16. Koivisto, P., and Stefansson, M., Retention studies of sulphated glycosaminoglycan disaccharides on porous graphitic carbon capillary columns, *Chromatographia*, 57, 37–45, 2003.

17. Kulikov, N.S. and Bobyleva, M.S., Molecular modeling in chromatoscopy as a new tool in the structure elucidation of novel isomers by GC/MS, *Struct. Chem.*, 15, 51–64, 2004.

18. Hanai, T., Htano, H., Nimura, N., and Kinoshita, T., Molecular recognition in chromatography aided by computational chemistry, *Supramol. Chem.* 3, 243–247, 1994.

19. Hosoya, H., Application of graph theory for chemical physics, *Kotaibuturi*, 33, 181, 1998.

20. Hanai, T., Inamoto, Y., and Inamoto, S., Chromatography of guanidino compounds, *J. Chromatogr. B*, 747, 123–138, 2000.

21. Supelco data figure 37 in *Supelco GC catalogue* (Sigma-Aldrich 1999), 1999, p. 29.

22. Supelco data figure 39 in *Supelco Bulletin 890A, Packed column GC application guide,* 1999 (Sigma-Aldrich 1999).

23. Supelco data figure 94 in *Supelco GC catalogue* (Sigma-Aldrich 1999), 1999, p. 48.

24. Hanai, T., Quantitative *in silico* analysis of the selectivity of graphitic carbon synthesized by different methods, *Anal. Bioanal. Chem.* 390, 369–375, 2008.

25. Kaliszan, R., Osmialowski, K., Bassler, B.J., and Hartwick, R.A., Mechanism of retention in high-performance liquid chromatography on porous graphitic carbon as revealed by principal component analysis of structural descriptors of solutes, *J. Chromatogr.* 499, 333–344, 1990.

26. Hanai, T., Chromatography *in silico*, Basic concept in reversed-phase liquid chromatography and computational chemistry, *Anal. Bioanal. Chem.* 382, 708–717, 2005.

27. Hanai, T., Quantitative *in silico* analysis of ion-exchange from chromatography to protein, *J. Liq. Chromatogr. Rel. Technol.* 30, 1251–1275, 2007.

28. Tazerouti, F., Badjah-Hadj-Ahmed, A.Y., and Hanai, T., Analysis of the mechanism of retention on a modified β-cyclodextrin/silica chiral stationary phase using a computational chemical method. *J. Liq. Chromatogr. Rel. Technol.* 30, 3043–3057, 2007.

29. Knox, J.H., and Gilbert, M.T., Preparation of porous carbon, *UK Patent* 2035282, (1978).

30. Hanai, T., and Homma, H., Quantitative *in silico* analysis of the specificity of a graphitic carbon column, *J. Liq. Chromatogr. Rel. Technol.* 32, 647–655, 2009.

31. Kiselev, A.V., and Poshkus, D.P., Chromtostructural analysis (Chromatoscopy). A new method of determination of molecular structure, *Faraday Symp. Chem. Soc.* 15, 13–24, 1980.

32. Lesellier, E., Retention mechanisms in super/subsuper fluid chromatography on packed columns, *J. Chromatogr. A*, 1216, 1881–1890, 2009.

33. Gaudin, K., Hanai, T., Chaninade, P., and Baillet, A., Retention behavior of polyunsaturated fatty acid methyl esters on porous graphitic carbon, *J. Chromatogr. A*, 1157, 56–64, 2007.

34. Petro, M., Belliardo, F., Novak, I., and Berek, D., Use of porous pyrolytic carbon for analytical and microscale high-performance liquid chromatographic bioseparation, *J. Chromatogr. B*, 718, 187–192, 1998.

35. Knox, J.H., Kaur, B., and Millward, G.R., Structure and performance of porous graphitic carbon in liquid chromatography, *J. Chromatogr.* 352, 3–25, 1986.

36. Obayashi, T., Ozawa, M., and Kawase, T., Carbon beads, process of producing the same and chromatography containing the same, Tonen Corporation, *European Patent*, 0458548A (1990).
37. Ichikawa, Yokoyama, A., Kawai, T., Moryta, H., Komiya, K., and Kato, Y., Packing material for liquid chromatography and method of manufacturing thereof, Nippon Carbon Co. Ltd. and Tosoh Corporation, *European Patent*, 0484176A (1991).
38. Rittenhouse, C.T., and Olesik, S.V., High performance liquid chromatographic evaluation of a low-temperature glassy carbon stationary phase, *J. Liq. Chromatogr. Rel. Technol.* 19, 2997–3022, 1996.
39. Engell, T.M., Olesik, S.V., Callstrom, M.R., and Diener, M., Liquid chromatography based on a low-temperature method of producing glassy carbon, *Anal. Chem.* 65, 3691–3700, 1993.
40. Hirayama, C., Nagaoka, S., Matsumoto, T., Ihara, H., Nonbo, T., Kurisaki, H., and Ikegami, S., Liquid chromatographic separataion of geometrical isomers using spherical carbon packings prepared from spherical cellulose particles, *J. Liq. Chromatogr.* 18, 1509–1520, 1995.
41. Barro, R., Regueiro, J., Llompart, M., and Garcia-Jares, C., Analysis of industrial contaminants in indoor air: Part 1. Volatile organic compounds, carbonyl compounds, polycyclic aromatic hydrocarbons and polychlorinated biphenyls, *J. Chromatogr. A*, 1216, 540–566, 2009.
42. Lara-Gonzalo, A., Sanchex-Uria, J.E., Segovia-Garcia, E., and Sanz-Medel, A., Critical comparison of automated purge and trap and solid-phase microextraction for routine determination of volatile organic compounds in drinking waters by GC-MS, *Talanta*, 74, 1455–1462, 2008.
43. Kimball, B.A., Russell, J.H., Griffin, D.L., and Johnston, J.J., Response factor consideration for the quantitative analysis of western red cedar (*Thuja plicata*) foliar monoterpens, *J. Chromatogr. Sci.* 43, 253–258, 2005.
44. Razborsek, M.I., Voncina, D.B., Dolecek, V., and Voncina, E., Determination of major phenolic acids, phenolic diterpenes and triterpenes in rosemary (*Rosmarinus officinals* L.) by gas chromatography and mass spectrometry, *Acta Chim. Slovenica*, 54, 60–67, 2007.
45. Herrero, M., Ibanez, E., Cifuentes, A., and Bernal, J., Multidimensional chromatography in food analysis, *J. Chromatogr. A*, 1216, 7110–7129, 2009.
46. Krynitsky, A.J., Niemann, R.A., and Nortrup, D.A., Determination of perchlorate anion in foods by ion chromatography-tandem mass spectrometry, *Anal. Chem.* 76, 5518–5522, 2004.
47. Niemann, R.A., Krynitsky, A.J., and Nortrup, D.A., Chromatographic determination of perchlorate in foods by on-line enrichment and suppressed conductivity detection, *J. Agr. Food Chem.* 54, 1137–1143, 2006.
48. Ueno, E., Oshima, H., Matsumoto, H., Saito, I., and Tamura, H., Determination of spinosad in vegetables and fruits by high-performance liquid chromatography with UV and mass spectrometric detection after gel permeation chromatography and solid-phase extraction cleanup on a 2-layered column, *J. AOAC Int*. 89, 1641–1649, 2006.
49. Ueno, E., Oshima, H., Saito, I., Matsumoto, H., Yoshimura, Y., and Nakazawa, H., Multiresidue analysis of pesticides in vegetables and fruits by gas chromatography/mass spectrometry after gel permeation chromatography and graphitized carbon column cleanup, *J. AOAC Int.* 87, 1003–1015, 2004.
50. Saito, Y., Kodama, S., Matsunaga, A., and Yamamoto, A., Multiresidue determination of pesticides in agricultural products by gas chromatography/mass spectrometry with large volume injection, *J. AOAC Int.* 87, 1356–1367, 2004.

51. Ueno, E., Oshima, H., Saito, I., Matsumoto, H., and Nakazawa, H., Multiresidue analysis of pesticides in agricultural products by GC-ECD after GPC and graphitized carbon column cleanup, *J. Food Hygiene Soc. Jap.* 45, 212–223, 2004.
52. Okihashi, M., Kitagawa, Y., Akutsu, K., Obana, H., and Tanaka, Y., Rapid method for the determination of 180 pesticide residues in foods by gas chromatography/mass spectrometry and flame photometric detection, *J. Pestic. Sci.* (Tokyo, Japan) 30, 368–377, 2005.
53. Tanaka, T., Hori, T., Asada, T., Oikawa, K., and Kawata, K., Simple one-step extraction and cleanup by pressurized liquid extraction for gas chromatographic-mass spectrometric determination of pesticides in green leafy vegetables, *J. Chromatogr. A*, 1175, 181–186, 2007.
54. Peng, C.F., Kuang, H., Li, X.Q., and Xu, C.L., Evaluation and interlaboratory validation of a GC-MS method for analysis of pesticide residues in teas, *Chemical Papers*, 61, 1–5, 2007.
55. Song, S., Ma, X., and Li, C., Rapid multiresidue determination method for 100 pesticides in vegetables by one injection using gas chromatography/mass spectrometry with selective ion storage technology, *Anal. Letter*, 40, 183–197, 2007.
56. Mol, H.G.J., Rooseboom, A., Van Dam, R., Roding, M., Arondeus, K., and Sunarto, S., Modification and re-validation of the ethyl acetate-based multi-residue method for pesticides in produce, *Anal. Bioanal. Chem.*, 389, 1715–1754, 389.
57. Balinova, A., Miadenova, R., and Shtereva, D., Solid-phase extraction on sorbents of different retention mechanisms followed by determination by gas chromatography-mass spectrometric and gas chromatography-electron capture detection of pesticide residues in crops, *J. Chromatogr. A*, 1150, 136–144, 2007.
58. Conrad, A., Couderchet, M., and Biagianti-Risbourg, S., Analysis of terbumeton and its major metabolites by SPE and DAD-HPLC in soil bulk water, Chromatographia, 65, 155–161, 2007.
59. Toba, T., Kaji, N., Takahata, J., and Nakamura, R., Effects of acetone volume on purification of pyrifenox in powdered tea, *Sendai-shi Eisei Kenkyushoho* Volume Date 2001, 31, 123–127, 2003.
60. Gonzalez-Rodriguez, R.M., Rial-Otero, R., Cancho-Grande, B., and Simal-Gandara, J., Determination of 23 pesticide residues in leafy vegetables using gas chromatography-mass spectrometry and analyte protectants, *J. Chromatogr. A*, 1196–1197, 100–109, 2008.
61. Watanabe, E., Baba, K., and Eun, H., Simultaneous determination of neonicotinoid insecticides in agricultural samples by solid-phase extraction cleanup and liquid chromatography equipped with diode-array detection, *J. Agric. Food Chem.* 55, 3798–3804, 2007.
62. Gonzalez-Rodriguez, R.M., Cancho-Grande, B., Simal-Gandara, J., Multiresidue determination of 11 new fungicides in grapes and wines by liquid-liquid extraction/clean-up and programmable temperature vaporization injection with analyte protectants/gas chromatography/ion trap mass spectrometry, *J. Chromatogr. A*, 1216, 6033–6042, 2009.
63. Esteve-Turrillas, F.A., Caupos, E., Liorca, I., Pastor, A., and De la Guardia, M., Optimization of large-volume injection for the determination of polychlorinated biphenyls in children's fast-food menus by low-resolution mass spectrometry, *J. Agric. Food Chem.*, 56, 1797–1803, 2008.
64. Moretton, C., Cretier, G., and Rocca, J.-L., Hear-cutting two-dimensional liquid chromatography methods for qualification of 2-acetyl-4-(1,2,3,4-tetrahydroxybutyl)imidazole in Class III caramel colors, *J. Chromatogr. A*, 1198–1199, 73–79, 2009.
65. Roukos, J., Plaisance, H., Leonardis, T., Bates, M., and Locoge, N., Development and validation of an automated monitoring system for oxygenated volatile organic compounds and nitrile compounds in ambient air, *J. Chromatogr. A*, 1216, 8642–8651, 2009.

66. Esteve-Turrillas, F.A., Ly-Verdu, A., Pastor, A., and De la Guardia, M., Development of a versatile, easy and rapid atmospheric monitor for benzene, toluene, ethylbenzene and xylenes determination in air, *J. Chromatogr. A*, 1216, 8549–8556, 2009.
67. Kovalova, L., McArdell, C.S., and Hollender, J., Challenge of high polarity and low concentrations in analysis of cytostatics and metabolites in wastewater by hydrophilic interaction chromatography/tandem mass spectrometry, *J. Chromatogr. A*, 1216, 1100–1108, 2009.
68. Van Leeuwen, S.P.J., and De Boer, J., Extraction and clean-up strategies for the analysis of poly- and perfluoroalkyl substances in environmental and human matrices, *J. Chromatogr. A*, 1153, 172–185, 2007.
69. Al-Degs, Y.S., and Al-Ghouti, M.A., Preconcentration and determination of high leachable pesticides residues in water using solid-phase extraction coupled with high-performance liquid chromatography, *Int. J. Environ. Anal. Chem.*, 88, 2008, 487–498.
70. D'Archivio, A.A., Fanelli, M., Mazzeo, R., and Ruggieri, F., Comparison of different sorbents for multiresidue solid-phase extraction of 16 pesticides from groundwater coupled with high-performance liquid chromatography, *Talanta*, 71, 25–30, 2007.
71. Castells, P., Parera, J., Santos, F.J., and Galceran, M.T., Occurrence of polychlorinated naphthalenes, polychlorinated biphenyls and short-chain chlorinated paraffins in marine sediments from Barcelona (Spain), *Chemosphere*, 70, 1552–1562, 2008.
72. Al-Haddad, A., A selective method for the determination of benzo[α]pyrene in soil using porous graphitic carbon liquid chromatography columns, *Talanta*, 59, 845–848, 2003.
73. Xu, L., and Hian, K., Novel approach to micro-wave assisted extraction and micro-solid-phase extraction from soil using graphite fibers as sorbent, *J. Chromatog. A*, 1192, 203–207, 2008.
74. Zheng, W., Yates, S.R., and Bradford, S.A., Analysis of steroid hormones in a typical dairy waste disposal system, *Environ. Sci. Technol.*, 42, 530–535, 2008.
75. Banos, C.E., and Silva, M., Comparison of several sorbents for continuous *in situ* derivatization and preconcentration of low-molecular mass aldehyde prior to liquid chromatography-tandem mass spectrometric determination in water samples, *J. Chromatogr. A*, 1216, 6554–6559, 2009.
76. Dagnac, T., Garcia-Chao, M., Pulleiro, P., Garcia-Jares, C., and Llompart, M., Dispersive solid-phase extraction followed by liquid-chromatography-tandem mass spectrometry for multi-residue analysis of pesticides in raw bovine milk, *J. Chromatogr. A*, 1216, 3702–3709, 2009.
77. Patil, S.H., Banerjee, K., Dasgupta, S., Oulkar, D.P., Patis, S.B., Jadhav, M.R., Savant, R.H., Adsule, P.G., and Deshmukh, M.B., Multiresidue analysis of 83 pesticides and 12 dioxin-like polychlorinated biphenyls in wine by gas chromatography-time-of-flight mass spectrometry, *J. Chromatogr. A*, 1216, 2307–2319, 2009.
78. Bacaloni, A., Callipo, L., Corradini, E., Giananti, P., Gubbiotti, R., Samperi, R., and Lagana, A., Liquid chromatography-negative ion atmospheric pressure photoionization tandem mass spectrometry for the determination of brominated flame retardants in environmental water and industrial effluents, *J. Chromatogr. A*, 1216, 6400–6409, 2009.
79. Rosen, J., Nyman, A., and Hellenaes, K.-E., Retention studies of acrylamide for the design of a robust liquid chromatography-tandem mass spectrometry method for food analysis, *J. Chromatogr. A*, 1172, 19–24, 2007.
80. Bessard, J., Saviuc, P., Chane-Yene, Y., Monnet, S., and Bessard, G., Mass spectrometric determination of acromelic acid A from a new poisonous mushroom: Clitocybe amoenolens, *J. Chromatogr. A*, 1055, 99–107, 2004.
81. Monser, L., Liquid chromatographic determination of four purine bases using porous graphitic carbon column, *Chromatographia*, 59, 455–459, 2004.

82. Vilaro, F., Canela-Xandri, A., and Canela, R., Quantification of abscisic acid in grape-vine leaf (Vitis vinifera) by isotope-dilution liquid chromatography-mass spectrometry, *Anal. Bioanal. Chem.* 386, 306–312, 2006.

83. Antonio, C., Pinheiro, C., Chaves, M.M., Ricardo, C.P., Ortuno, M.F., and Thomas-Oates, J., Analysis of carbohydrates in Lupinus albus stems on imposition of water defi-cit, using porous graphitic carbon liquid chromatography-electrospray ionization mass spectrometry, *J. Chromatogr. A*, 1187, 111–118, 2008.

84. Kawasaki, N., Itoh, S., Ohta, M., and Hayakawa, T., Microanalysis of *N*-linked oligosaccharides in a glycoprotein by capillary liquid chromatography/mass spec-trometry and liquid chromatography/tandem mass spectrometry, *Anal. Biochem.* 316, 15–22, 2003.

85. Hashii, N., Kawasaki, N., Itoh, S., Hyuga, M., Kawanishi, T., and Hayakawa, T., Glycomic/glycoproteomic analysis by liquid chromatography/ mass spectrometry: anal-ysis of glycan structural alteration in cells, *Proteomics*, 5, 4665–4672, 2005.

86. Zhang, J., Xie, Y., Hedrick, J.L., and Lebrilla, C.B., Profiling the morphological distri-bution of *O*-linked oligosaccharides, *Anal. Biochem.*, 334, 20–35, 2004.

87. Itoh, S., Kawasaki, N., Hashii, N., Harazono, A., Matsuishi, Y., Hayakawa, T., and Kawanishi, T., *N*-linked oligosaccharide analysis of rat brain Thy-1 by liquid chroma-tography with graphitized carbon column/ion trap-fourier transform ion cyclotron reso-nance mass spectrometry in positive and negative ion modes, *J. Chromatogr. A*, 1103, 296–306, 2006.

88. Tanabe, K., and Ikenaka, K., In-column removal of hydrazine and *N*-acetylation of oli-gosaccharides released by hydrazionolysis, *Anal. Biochem.* 348, 324–326, 2006.

89. Robinson, S., Bergstroem, E., Seymour, M., and Thomas-Oates, J., Screening of underivatized oligosaccharides extracted from the stems of Triticum aestivum using porous graphitized carbon liquid chromatography-mass spectrometry, *Anal. Chem.* 79, 2437–2445, 2007.

90. Hashii, N., Kawasaki, N., Itoh, S., Harazono, A., Matsushi, Y., Hayakawa, T., and Kawanishi, T., Specific detection of Lewis x-carbohydrates in biological samples using liquid chromatography/multiple-sage tandem mass spectrometry, *Rapid Commun. in Mass Spectr.*, 19, 3315–3321, 2005.

91. Antonopoulos, A., Herbreteau, B., Lafosse, M., Helbert, W., Comparative analysis of enzymatically digested κ-carrageenans, using liquid chromatography on ion-exchange and porous graphitic carbon columns coupled to an evaporative light scattering detector, *J. Chromatogr. A*, 1023, 231–238, 2004.

92. Bohnstedt, K.C., Karlberg, B., Basun, H., and Schmidt, S., Porous graphitic carbon chromatography-tandem mass spectrometry for the study of isoprostanes in human cere-brospinal fluid, *J. Chromatogr. B*, 827, 39–43, 2005.

93. Bohnstedt, K.C., Karlberg, B., Wahlund, L.-O., Jonhagen, M.E., Basun, H., and Schmidt, S., Determination of isoprostanes in urine samples from Alzheimer patients using porous graphitic carbon liquid chromatography-tandem mass spectrometry, *J. Chromatogr. B*, 796, 11–19, 2003.

94. Barroso, B., Didraga, M., and Bischoff, R., Analysis of proteoglycans derived sulphated disaccharides by liquid chromatography/mass spectrometry, *J. Chromatogr. A*, 1080, 43–48, 2005.

95. Yuan, J., Hashii, N., Kawasaki, N., Itoh, S., Kawanishi, T., and Hayakawa, T., Isotope tag method for quantitative analysis of carbohydrates by liquid chromatography-mass spectrometry, *J. Chromatogr. A*, 1067, 145–152, 2005.

96. Jacquet, R., Pennanec, R., Elfakir, C., and Lafosse, M., Liquid chromatography analysis of monosubstituted sufobutyl ether-β-cyclodextrin isomers on porous graphitic carbon, *J. Sep. Sci.* 27, 1221–1228, 2004.

97. Tachon, R., Pichon, V., Le Borgne, M.B., and Minet, J.-J., Use of porous graphitic carbon for the analysis of nitrate ester, nitramine and nitroaromatic explosives and by-products by liquid chromatography-atmospheric pressure chemical ionization-mass spectrometry, *J. Chromatogr. A*, 1154, 174–181, 2007.

98. Crescenzi, C., Albinana, J., Carisson, H., Holmgren, E., and Ratlle, R., On-line strategies for detemination of trace levels of nitroaromatic explosives and related compounds in water, *J. Chromatogr. A*, 1153, 186–193, 2007.

99. Zhou, T., and Lucy, C.A., Selective preconcentration of chemical warfare agent degradation products using a zirconia preconcentration column, *J. Chromatogr. A*, 1213, 8–13, 2008.

100. Dufour, A.P., Evans, O., Behymer, T.D., and Cantu, R., Water ingestion during swimming activities in a pool: a pilot study, *J. Water and Health*, 4, 425–430, 2006.

101. Delepee, R., Chaimbault, P., Antignac, J.-P., and Lafosse, M., Validation of a real-time monitoring method for aniline in freshwater by high-performance liquid chromatography on porous graphitic carbon/electrospray ionization tandem mass spectrometry, *Rapid Commun. in Mass Spectr.*, 18, 1548–1552, 2004.

102. Moberg, M., Holmstrom, S.J.M., Lundstrom, U.S., and Markides, K.E., Novel approach to the determination of structurally similar hydroxamate siderophores by column-switching capillary liquid chromatography coupled to mass spectrometry, *J. Chromatogr. A*, 1020, 91–98, 2003.

103. Martens-Lobenhoffer, J., and Bode-Boger, S.M., Determination of levetiracetam in human plasma with minimal sample pretreatment, *J. Chromatogr. B*, 819, 197–200, 2005.

104. Vonaparti, A., Kazanis, M., and Panderi, I., Development and validation of a liquid chromatographic/electrospray ionization mass spectrometric method for the determination of benazepril, benazeprilat and hydrochlorothiazide in human plasma, *J. Mass Spectr.* 41, 593–605, 2006.

105. Thiebaut, D., Vial, J., Michel, M., Hennion, M.-C., and Greibrokk, T., Evaluation of reversed-phase columns designed for polar compounds and porous graphitic carbon in "trapping" and separating neurotransmitters, *J. Chromatogr. A*, 1122, 97–104, 2006.

106. Sakhi, A.K., Russnes, K.M., Smeland, S., Blomhoff, R., and Gundersen, T.E., Simultaneous quantification of reduced and oxidized glutathione in plasma using a two-dimensional chromatographic system with parallel porous graphitized carbon columns coupled with fluorescence and coulometric electrochemical detection, *J. Chromatogr. A*, 1104, 179–189, 2006.

107. Tornkvist, A., Sjoberg, P.J.R., Markides, K.E., and Bergquist, J., Analysis of catecholamines and related substances using porous graphitic carbon as separation media in liquid chromatography-tandem mass spectrometry, *J. Chromatogr. B*, 801, 323–329, 2004.

108. Forgacs, E., Cserhati, T., Miksik, I., Echardt, A., and Deyl, Z., Simultaneous effect of organic modifier and physicochemical parameters of barbiturates on their retention on a narrow-bore PGC column, *J. Chromatogr. B*, 800, 259–262, 2004.

109. Monser, L., and Trabelsi, H., A rapid LC method for the determination of haloperidol and its degradation products in pharmaceuticals using a porous graphitic carbon column, *J. Liq. Chromatogr. Rel. Technol.*, 26, 261–271, 2003.

110. Xu, J.Q., and Aubry, A.-F., Impurity profiling of non-nucleoside reverse transcriptase inhibitors by HPLC using a porous graphitic carbon stationary phase, *Chromatographia*, 57, 67–71, 2003.

111. Jansen, R., Rosing, H., Schellens, J.H.M., and Beijnen, J.H., Retention studies of 2′-2′-difluorodeoxyuridine nucleosides and nucleotides on porous graphitic carbon: Development of a liquid chromatography-tandem mass spectrometry method, *J. Chromatogr. A*, 1216, 3168–3174, 2009.

112. Wang, J., Lin, T., Lai, J., Cai, Z., and Yang, M.S., Analysis of adenosine phosphates in HepG-2 cell by a HPLC-ESI-MS system with porous graphitic carbon as stationary phase, *J. Chromatogr. B*, 877, 2019–2024, 2009.
113. Crauste, C., Lefebvre, I., Hovaneissian, M., Puy, J.Y., Roy, B., Peyrottes, S., Cohen, S., Guitton, J., Dumomtet, C., and Perigaud, C., Development of a sensitive and selective LC/MS/MS method for the simultaneous determination of intraceller 1-β-D-arabinofuranosylcytosine (araCTP) and deoxycytidine triphosphate (dCTP) in a human follicular lymphoma cell line, *J. Chromatogr. B*, 877, 1417–1425, 2009.
114. Krijt, H., Duta, A., and KoZich, V., Determination of S-adenosylmethionine and S-adenosylhomocysteine by LC-MS/MS and evaluation of their stability in mice tissues, *J. Chromatogr. B*, 877, 2061–2066, 2009.
115. Cohen, S., Megherbi, M., Jordheim, L.P., Lefebvre, I., Perigaud, C., Dumontet, C., and Guitton, J., Simultaneous analysis of eight nucleoside triphosphates in cell lines by liquid chromatography coupled with tandem mass spectrometry, *J. Chromatogr. B*, 877, 3831–3840, 2009.
116. Duy, S.V., Lefebvre Tournier, I., Pichon, V., Hugon-Chapuis, F., Puy, J.-Y., and Perigaud, C., Molecularly imprinted polymer for analysis of zidovudine and stavudine in human serum by liquid chromatography-mass spectrometry, *J. Chromatogr. B*, 877, 1101–1108, 2009.
117. Abbas-Ghaleb, K., Gilon, N., Cretier, G., and Mermet, J.M., Preconcentration of selenium compounds on a porous graphitic carbon column in view of HPLC-ICP-AED speciation analysis, *Anal. Bioanal. Chem.*, 377, 1026–31, 2003.
118. Issaq, H., Chan, K.C., Blonder, J., Ye, X., and Veenstra, T.D., Separation, detection and quantitation of peptides by liquid chromatography and capillary electrochromatography, *J. Chromatogr. A*, 1216, 1825–1837, 2009.
119. Xia, Y.-Q., Jemal, M., Zheng, N., and Shen, X., Utility of porous graphitic carbon stationary phase in quantitative liquid chromatography/tandem mass spectrometry bioanalysis: quantitation of diastereomers in plasma, *Rapid Commun. in Mass Spectr.*, 20, 1831–1837, 2006.
120. Lecoeur-Lorin, M., Delepee, R., Adamczyk, M., and Morin, P., Simultaneous determination of optical and chemical purities of a drug with two chiral centers, *J. Chromatogr. A*, 1206, 123–130, 2008.
121. Quinton, L., Gaudin, K., Baillet, A., and Chaminade, P., Microanalytical systems for separations of stratum corneum ceramides, *J. Sep. Sci.* 29, 390–398, 2006.
122. Bieri, S., Varesio, E., Munoz, O., Veuthey, J.-L., and Christen, P., Use of porous graphitic carbon column for the separation of natural isomeric tropane alkaloids by capillary LC and mass spectrometry, *J. Pharm. Biomed. Anal.*, 40, 545–551, 2006.
123. Holmgren, E., Carlsson, H., Goede, P., and Crescenzi, C., Determination and characterization of organic explosives using porous graphitic carbon and liquid chromatography-atmospheric pressure chemical ionization mass spectrometry, *J. Chromatogr. A*, 1099, 127–135, 2005.
124. Essen, S.A., Bylund, D., Holmstrom, S.J.M., Moberg, M., and Lundstrom, U.S., Quantification of hydroxamate siderophores in soil solutions of podzolic soil profiles in Sweden, *Biometals: An Internation J. on the role of metal ions in Bio. Biochem. Med.*, 19, 269–82, 2006.
125. Monser, L., and Darghouth, F., Simultaneous determination of naproxen and related compounds by HPLC using porous graphitic carbon column, *J. Pharm. Biomed. Anal.* 32, 1087–1092, 2003.
126. Sanchez, C., Carlsson, H., Colmsjo, A., Crescenzi, C., and Batlle, R., Determination of nitroaromatic compounds in air samples at fentogram level using C18 membrane sampling and on-line extraction with LC-MS, *Anal. Chem.*, 75, 4639–45, 2003.

127. Urte, L.-von V., Streck, G., and Brack, W., Automated fractionation procedure for poly-cyclic aromatic compounds in sediment extracts on three coupled normal-phase high-performance liquid chromatography columns, *J. Chromatogr. A*, 1185, 31–42, 2008.
128. Wang, J.-L., Din, G.-Z., Chan, C.-C., Validation of a laboratory-constructed automated gas chromatograph for the measurement of ozone precursors through comparison with a commercial analogy, *J. Chromatogr. A*, 1027, 11–18, 2004.
129. Cohen, A., Schagerlof, H., Nilsson, C., Melander, C., Tjerneld, F., and Gorton, L., Liquid chromatography-mass spectrometry analysis of enzyme-hydrolyzed carboxymethylcel-lulose for investigation of enzyme selectivity and substituent pattern, *J. Chromatogr. A*, 1029, 87–95, 2004.
130. Riddle, L.A., and Guiochon, G., Separation of free sterols by high temperature liquid chromatography, *J. Chromatogr. A*, 1137, 173–179, 2006.
131. Heinisch, S., and Rocca, J.-L., Sense and nonsense of high-temperature liquid chroma-tography, *J. Chromatogr. A*, 1216, 642–658, 2009.

7 Application of Column-Switching Methods in HPLC for Evaluating Pharmacokinetic Parameters[*]

Hwa Jeong Lee and Kyu-Bong Kim

CONTENTS

[*] The authors would like to thank So Young Um (Korea Food and Drug Administration), You Jin Kim, Song Hwa Chae, and Hyun Joo Kim (Ewha Womans University, Korea) for providing support with data, tables, and figures.

291

7.1   INTRODUCTION

7.1.1   BRIEF HISTORY OF COLUMN-SWITCHING TECHNIQUE IN HPLC METHOD FOR ANALYSIS

High-performance liquid chromatography (HPLC) is a powerful technique to separate chemicals from matrices in modern analytical sciences. For more powerful analytical capacity and to increase resolution and sensitivity, HPLC has been further developed. For example, to improve resolution for the separation of chemicals, ultra-high pressure was adapted into a very small, particles-packed (<5 μm) column, such as the ultra-performance liquid chromatography (UPLC) system. Also, mass spectrometry (MS), such as HPLC-MS, has been used to enhance the sensitivity of the HPLC system. In general, HPLC uses one separation column with a simple flow-through course. When additional columns are branched and switching devices (valves) are adapted to control or adjust the flow of a chromatographic network, the HPLC system is able to increase its already powerful separation ability. In 1982, Riley and his colleague [1] reported analysis of platinum in urine using HPLC automated column switching. They used platinum complexes retained on anion exchangers, and automated column switching was applied for recovery of purified cisplatin fractions. Werkhoven-Goewie and his colleague [2] also used the column-switching technique to automatically separate the drug secoverine from blood by enzymatic hydrolysis of blood samples with subtilisin-A. The column-switching technique was applied to decrease the analysis time. In 1984, Little et al. [3] explained the role of column switching in analyzing complex samples. They proposed the column-switching technique for sample and column clean-up followed by column reequilibration to minimize analysis time.

7.1.2   ADVANTAGES AND DISADVANTAGES OF COLUMN-SWITCHING TECHNIQUE

The column-switching techniques possess various advantages over conventional techniques in an analytical system. A trace level of compound in biological samples is able to be analyzed by enrichment or preconcentration using the column-switching techniques. The advantage of sample cleanup is the major motivation behind the advent of the column-switching techniques. The column-switching techniques minimize sample handling and reduce analytical time and processing. Online sample processing and automation are additional advantages of the column-switching techniques, which increases reproducibility of analyses in biological samples. Sequential connection of analytical columns enhances precision and accuracy. An internal standard is not necessary in the column-switching techniques

TABLE 7.1

Advantages and Disadvantages of Column-Switching Techniques

Advantages	Disadvantages
Trace enrichment	Increasing risk of carryover effect
Improvement of sensitivity	Additional devices of switching valves, analytical columns, and pumping systems
	Requirement of compatible mobile phase
Sample cleanup	Periodic change of precolumn
Simpler sample handling	
Online sample processing and full automation	
Saving of considerable time and labor	
Increasing accuracy and precision	
Improvement of selectivity of analytes	
Prevention of an internal standard	
Minimal consumption of organic solvents	

because of improvement on accuracy and precision. The selectivity of analytes is able to be improved by combining different chromatographic modes. Control of flow of the mobile phase enables minor consumption of organic solvents, which are contaminants or toxicants to the environment and human health. On the other hand, there are disadvantages to the column-switching techniques. One of the significant problems of the HPLC system with highly sensitive detectors in the analysis of biological samples is a carryover effect, and the major source exists in the autosampler. An online sample pretreatment of column-switching HPLC definitely increases the risk of the carryover effect due to the several switching valves and complex connecting lines. The column-switching techniques for sample cleanup require switching valves, additional columns, and pumping systems. These additional devices increase the costs of the analytical system and necessitate relatively trained personnel to be familiar with or to troubleshoot the system. Compatible mobile phases should be selected and required for each analytical column. The precolumn for online sample preparations should be replaced periodically to prevent overload of intact biological samples. Table 7.1 summarizes the advantages and disadvantages of column-switching techniques.

Since column-switching techniques were adopted, they have been widely used as powerful tools to analyze chemicals, especially in biological samples such as blood and urine. In this chapter, column-switching techniques will be discussed for their versatile applications and functions.

7.2   APPLICATIONS OF COLUMN SWITCHING

7.2.1   Sample Cleanup

It is generally recognized that biological samples need some preparation before injection into a HPLC system to decrease interfering components from the sample. The aim of sample preparation is to selectively extract, concentrate, and purify the target analyte from interfering components. However, this method of preparation can expend much time in analysis of biological samples and is recognized as a time-limiting step. Sample preparation often determines whether the analysis of biological samples can be done effectively or rapidly.

There are a lot of biomaterials for drug or chemical pharmacokinetic analysis. Materials can include not only blood, plasma, serum, and urine but also saliva, hair, sweat, cerebrospinal fluid (CSF), tears, and homogenized organs. However, separation is not easy due to their low content as well as their complex components, which can include various salts, acid, base, proteins, and organic materials. Although the sensitivity of currently used analytical equipment is excellent, in today's world it remains the biggest challenge and most important step to extract materials efficiently without interference from other components.

In analysis, many problems are inevitable when analytes are injected into the HPLC system directly. For example, when a protein-like matrix is injected directly into an HPLC system, which uses the reverse stationary phase, the protein is irreversibly absorbed on the reverse stationary phase, and protein denaturalization occurs. Accordingly, the efficiency of HPLC columns and reproducibility are suddenly decreased.

To solve these problems, it is essential to use appropriate sample preparation. Consequently, biological samples should be applied into analytical instruments like graphitized carbon (GC), HPLC, GC/MS, and HPLC/MS for quantitative and qualitative analysis only after interferences or matrices are removed. Several factors should be considered in biological sample preparation. Target components should be extracted efficiently and selectively by minimizing interfering ingredients. The recovery and reproducibility of the target components should be maintained in the processes of filtration, partition, and purification. In addition, the sample preparation process should be rapid, simple, and economical, and it should not be harmful to the analyst or to the environment.

Generalized methods of extracting biological samples include liquid–liquid extraction (LLE) and solid-phase extraction (SPE), as discussed subsequently. LLE is the most classical method. The advantages of LLE are the variable choice of solvents, with a wide range of polarity and applications. However, LLE may induce environmental problems due to consumption of large quantities of organic solvents and have a low efficacy of separation. To overcome this weak point, various advanced methods have been developed, such as solid-phase extraction (SPE), solid-phase microextraction (SPME), liquid-phase microextraction (LPME), and supercritical fluid extraction (SFE). A recent trend has been to substitute LLE with SPE, which uses various stationary phases and generally requires fewer solvents than LLE.

Table 7.2 compares methods (LLE, SPE, and column switching) of cleanup of biological samples and lists the advantages and disadvantages of each method. Classical sample preparation with LLE and SPE can involve numerous steps, which can lead to the loss of analytes and thus reduce the selectivity or sensitivity of the samples. However, column-switching techniques improve the conventional sample cleanup process. Lindenblatt et al. [4] compared LLE with automated online column-switching (SPE) technique for the quantification of psilocin, a hallucinogen, in human plasma samples. They concluded that automated online column-switching (SPE) technique

TABLE 7.2

Advantages and Disadvantages of Sample Cleanup by LLE, SPE, and Column Switching

Liquid-Liquid Extraction (LLE)	Solid-Phase Extraction (SPE)	Column-Switching
	Advantage	
Improvement of sensitivity by no limit of sample amounts	Simple and fast method	Simple and fast method
Various choice of extracting solvent according to the analytes	Reduction of solvent consumption	Reduction of solvent consumption
Increase in the concentrating factor of non-volatile analytes	Automation of process	Automation of process
The higher selectivity of organic solvent than protein precipitation	High recovery of the analytes	The highest recovery of the analytes
Unnecessariness of internal standard		
	Decrease in sample preparation time	Decrease in sample preparation time
	Disadvantage	
Large quantities of organic solvent for extraction, separation, and purification of analytes	Variation between different cartridges	Complex online system and need of well-trained operator
Laborious and time consuming	Expensive commercial cartridge	Limited column lifetime
Risk of loss of analytes during sample preparation		
Analytical variation due to many preparation steps		
Low recovery and reproducibility		
Induction of health problem due to consumption of much organic solvent		
Induction of environmental problem due to organic solvent waste		

had better selectivity, less manual effort, smaller plasma volume, and better recovery (less loss) of psilocin compared with LLE. Tables 7.3 and 7.4 briefly summarize the plasma sample preparations and time program for the online SPE method to compare the processes of LLE with the online SPE method. Figure 7.1 demonstrates and compares HPLC chromatograms of the high and differing selectivity from two different sample preparation methods of LLE and online SPE. Psilocin (peak 1) was identified in C and D by comparison with blank human plasma (A and B).

The switching-column valves enable the off-line multiple sample preparation steps of SPE to be a single step with online. The principle of column-switching sample cleanup is to clean and analyze the smallest fraction of biological samples and then to discard the others with automation. In a primary or precolumn, the biological sample is eluted, and the fraction of samples containing the target analytes is trapped. Interfering compounds of the biological sample are discarded, and the

TABLE 7.3
Processing Scheme for Psilocin in Human Plasma

LLE Method	Online SPE Method
Pipette 2 mL plasma	Pipette 400 μL plasma
Add 1.2 mL methanol, including I.S.$_1$ 10ng	Add 20 μL I.S.$_2$ solution (2.4 ng)
Add 50 μL 0.5M Na$_2$CO$_3$ until pH 8.5	Add 400 μL PEG 6000 solution 20%
Cool 5 min. (ice bath)	Cool 5 min. (ice bath)
Centrifuge at 2875 g/10 min/20°	Centrifuge at 2875 g/3 min/20°
Give supernatant on column	
Wait 5 min.	
Elute with 6mL CH$_2$Cl$_2$ (5 min.)	
Elute a second time with 6mL CH$_2$Cl$_2$ (10 min.)	
Evaporate under nitrogen, 40°	
Add 100 μL CH$_2$Cl$_2$	Load vials into autosampler rack
Inject 10 μL	Inject 410 μL

Notes: I.S.$_1$, 5-hydroxyindole. I.S.$_2$, bufotenine.
Source: Adapted from Lindenblatt, H. et al., *J. Chromatogr. B,* 709, 255–263, 1998 (with permission).

fraction of target analytes is cut into the secondary or analytical column for separation to identify or quantify the target analytes. In column-switching cleanup technique, "zone cutting" is the most useful technique, which may elute the fraction of analytes at the front (front-cut), in the middle (heart-cut), or at the end (end-cut) of the chromatogram of the primary precolumn. In this "cut" technique, a narrow-retention time region or fraction containing target components is "cut" from the chromatogram and transferred onto another analytical column for further separation. In this process, no adsorptive or degradable losses would be expected by the quantitative transfer of the components. Another advantage of the column-switching technique for sample cleanup is the improved selectivity by the judicious choice of two or more HPLC stationary phases.

7.2.1.1 Precolumns

Precolumns are important to clean up the samples in column-switching techniques. Efficiency and separation of the precolumn is the main issue for the sample cleanup in column-switching techniques. Precolumns are discussed by dimensions and packings.

The dimensions of precolumn require (1) a high sample loading capacity to avoid analyte losses by breakthrough during flushing and (2) practically, the shorter the better. To prevent losses of target analytes, the precolumn should be considered according to the sample volume to be injected. In general, the length of the precolumn is shorter than the analytical column because of (1) minimization of the duration of flushing time for removing the interfering components, (2) easy and less packing materials, and (3) column maintenance. However, the length of the precolumn can be increased according to the analytes. When tightly binding to plasma proteins, the

TABLE 7.4
Time Program for the Online SPE Method

Time	% A	% B	% C	Valve 1	Valve 2	Clamp	Move	Flow	Remarks
0	0	0	100	On	Off	Closed	Off	0.5	Injection and washing of sample
6.0	0	0	100	Off	Off	Open	Off	0.5	Clamp open
6.1	0	0	100	Off	Off	Open	On	0.5	Ring moves one position
6.2	0	0	100	Off	Off	Closed	Off	0.5	Clamp close
6.3	100	0	0	Off	Off	Closed	Off	2.0	Start of elution onto the analytical column, start Integrator
8.3	100	0	0	On	Off	Closed	Off	2.0	End of elution, activation of new cartridge
9.8	100	0	0	Off	Off	Closed	Off	2.0	
9.9	0	100	0	Off	Off	Closed	Off	2.0	
11.9	0	100	0	Off	Off	Closed	Off	2.0	
12.0	0	100	0	On	Off	Closed	Off	2.0	Washing of cartridges before reuse
22.0	0	100	0	Off	Off	Closed	Off	2.0	
22.1	0	0	100	Off	Off	Closed	Off	2.0	
24.1	0	0	100	Off	Off	Closed	Off	2.0	
24.2	0	0	100	On	Off	Closed	Off	2.5	Preconditioning of cartridge
29.2	0	0	20	On	Off	Closed	Off	2.5	

Notes: A, Methaol. B, acetic acid. C, lithium acetate solution. pH 6.8. Flow in mL/min. Valve 1, sample enrichment. Valve 2, elution.

Source: Adapted from Lindenblatt, H. et al., *J. Chromatogr. B,* 709, 255–263, 1998 (with permission).

analytes should be partitioned relatively longer as needed, and the precolumn length must be increased to prevent the breakthrough of target analytes [5]. To improve the selectivity of the precolumn, its length can be increased. In one study, a precolumn was three times longer than the analytical column to preseparate different drugs in plasma [6]. The internal diameter of both the analytical and precolumn should be identical to minimize extra-column band broadening [7].

7.2.1.2 Extraction Materials

In sample preparation, special extraction sorbents to be packed in the precolumn play a crucial role in allowing the direct or multiple injections of complex biological

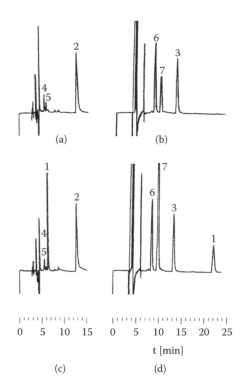

FIGURE 7.1   HPLC chromatograms obtained from real plasma samples of volunteer 6 after 0 min. (a, b) and 70 min. (c, d) after oral administration of psilocybin at a dose of 0.2 mg/kg. (a, c) LLE method; (b, d) online SPE method. 1 = Psilocin; 2 = 5-hydroxyindole (I.S., LLE), 3 = bufotenine (I.S., online SPE), 4, 5, 6, 7 = biomolecules. Retention times in (c): 1 = 6.71 min., 2 = 13.62 min., 4 = 6.00 min., 5 = 6.18 min. Retention times in (d): 1 = 23.07 min., 3 = 13.86 min., 6 = 8.73 min., 7 = 10.13 min. (From Lindenblatt, H. et al., *J. Chromatogr. B,* 709, 255–263, 1998. With permission.)

matrices. The restricted access media (RAMs) were introduced for rapid extraction with direct injection of biological samples. These extractions lead to the automation, simplification, and speeding up of the sample cleanup. RAMs contain the common characteristic of excluding macromolecules, whereas the analytes are generally retained by hydrophobic or electrostatic interactions.

7.2.1.2.1   Restricted Access Media

In a bioanalytical field, sample extraction is focused mainly on the separation of relatively small analytes from big molecules such as proteins. Desilets et al. [8] suggested the term *restricted access media* in 1991. RAMs allow direct injection of biological samples into the HPLC system by limiting the accessibility of interaction sites within the pores to small analytes. Macromolecules are excluded by a physical barrier made by the pore size (diameter) or by a chemical diffusion barrier made by a protein at the outer surface of the particle. RAMs have been further subdivided according to surface chemistry and classified as bimodal or unimodal [9]. The authors considered

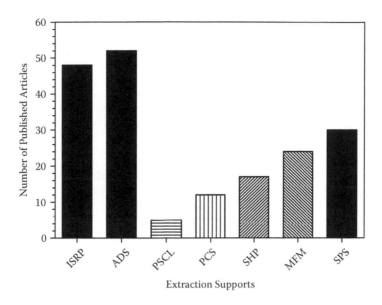

FIGURE 7.2   Articles published on the use of RAM supports for the online sample preparation of biological samples. Extraction supports are internal surface reversed-phase (ISRP), alkyl-diol-silica material (ADS), porous silica covered by a combined ligand (PSCL), protein-coated silica (PCS), shielded hydrophobic phase (SHP), mixed-functional material (MFM), and semipermeable surface (SPS).

phases with different types of bonding in internal and external surface to be bimodal phases, and phases with a specific bonding to both surfaces to be unimodal phases. Commercially available RAM sorbents are reviewed (Figure 7.2).

7.2.1.2.1.1 Internal Surface Reversed Phase (ISRP)   Internal surface reversed phase consists of porous silica particles, which have the outer surface covered with a hydrophilic moiety excluding the adsorption of protein (diol-glycine groups) (Figure 7.3). The internal surface is covered with a hydrophobic tripeptide partitioning phase (glycine-phenylalanine-phenylalanine, GFF). When biological samples are directly injected onto this sorbent, large molecules are limited from the internal surface area by a size exclusion mechanism (physical barrier). The particle pore size of GFF is approximately 8 nm, which is able to exclude proteins larger than approximately 20,000 Da [10]. Blood proteins such as albumin (65,600 Da), immunoglobulin (150,000–900,000 Da), and fibrinogen (340,000 Da) can be directly eluted from the GFF supports, whereas the target small molecules may be retained in the pores by the reverse tripeptide phase [11]. Nakagawa et al. [12] suggested that the mechanism of retention in the pores is mainly through the π-electron interactions and a weak cation exchange because of the free carboxyl group of the terminal phenylalanine. Perry et al. [13] provided a new GFF support to improve performances in efficiency, retentivity, and reproducibility. ISRP allows several hundred biological sample injections without losing performance ability [14]. GFF material has been assumed to be an acceptable approach for the direct injection of biological samples. GFF support

TABLE 7.5
Compounds Analyzed by Direct Injection of Biological Fluids onto RAM Using Column-Switching HPLC System

Compounds (Matrix)	RAM Sorbent	References
Aceclofenac (plasma)	Capcell Pak MF Ph-1	70
Aloesin (plasma)	Capcell Pak MF Ph-1	77
Amoxicillin (serum)	SPS C18	104
Anxiolytic agent CP-93 393 (plasma)	Bio Trap C18	60
Arachidonic acid (urine)	LiChrospher RP-18 ADS	24
Artemisinin (plasma, saliva)	LiChrospher RP-18 ADS	27
Atenolol (serum)	BioTrap C18	61
Atropine (plasma)	LiChrospher XDS	105
Azole pesticides (urine)	GFF	106
Barbiturates (plasma)	LiChrospher RP-18 ADS	26
Barbiturates (serum)	Capcell Pak MF Ph-1	107
Baicalin (plasma)	Capcell Pak MF C(8)	85
Benzodiazepines (plasma, serum)	BioTrap MS	62
Benzodiazepines (plasma, urine)	Lichrospher RP-18 ADS	108
Berberin (plasma)	Capcell Pak MF C(8)	85
Beta-blockers (plasma)	BioTrap C18	61
Beta-blockers (plasma)	LiChrospher XDS	105
Beta-blockers (plasma, microdialysate)	LiChrospher RP-8 ADS	109
Beta-blockers (microdialysate)	Lichrospher RP-18 ADS	132
Beta-blockers (serum)	BioTrap C18	61
Beta-blockers (urine)	Lichrospher RP-18 ADS	44
Biphenyldimethyl dicarboxylate (plasma)	Capcell Pak MF Ph-1	68
Carbamazepine (plasma)	Lichrospher RP-18 ADS	34
Ceftazidime (bronchial secretions, serum)	LiChrospher RP-8 ADS	39
Cisapride (serum)	Capcell Pak MF Ph-1	69
Citalopram (plasma)	LiChrospher RP-4 ADS	18
Citalopram (serum)	LiChrospher CN-20 ADS	134
Clomipramine (plasma)	Capcell Pak MF Ph-1	110
Cocaine (plasma)	Lichrospher RP-18 ADS	32
Creatinine (serum)	GFF	111
Di-(2-ethylhexyl)phthalate (DEHP) and metabolites (urine)	Lichrospher RP-8 ADS	133
Diclofenac (plasma)	Capcell Pak MF Ph-1	70
Difluprednate (aqueous humor)	Capcell Pak MF Ph-1	81
Dopamine agonist (plasma)	GFF	112
Epirubicin (plasma, liver homogenate, liver tumour homogenate)	LiChrospher RP-4 ADS	16
Entacapone glucuronide (plasma)	Lichrospher RP-18 ADS	19
Escitalopram (serum)	LiChrospher CN-20 ADS	134

TABLE 7.5 (continued)
Compounds Analyzed by Direct Injection of Biological Fluids onto RAM Using Column-Switching HPLC System

Compounds (Matrix)	RAM Sorbent	References
Felodipine (plasma, tissue)	Lichrospher RP-18 ADS	29
Fenoterol (plasma)	LiChrospher XDS	105
Flunitrazepam (plasma)	BioTrap MS LiChrospher RP-18 ADS	20
Fluvastatin (plasma)	Capcell Pak MF Ph-1	67
Granisetron (plasma)	GFF	14
Heroin (urine)	Capcell Pak MF SCX	84
Ibuprofen (serum)	BioTrap C18	61
Ipratropium (plasma)	LiChrospher XDS	105
Ketoprofen (plasma)	Lichrospher RP-18 ADS	17
Linezolid (serum,urine)	LiChrospher RP-8 ADS	39
Local anaesthetics (plasma)	SPS C8	22
Local anaesthetics (plasma)	Lichrospher RP-18 ADS	23
Lonazolac and metabolites (cell culture media)	BioTrap 500 MS	63
Matrix metalloprotease inhibitors (plasma)	SPS C18	113
Meloxicam (plasma)	Lichrospher RP-18 ADS	33
Meropenem (bronchial secretions, serum)	LiChrospher RP-18 ADS	38
Methadone (serum)	GFF ⊠, LiChrospher RP-4 ADS	114
Methotrexate (plasma)	LiChrospher RP-8 ADS	25, 51
8-Methoxysporalen (plasma)	LiChrospher RP-8 ADS	15
Montelukast sodium (plasma)	Capcell Pak MF Ph-1	72
Mycophenolic acid (serum)	Capcell Pak MF Ph-1	83
Neuropeptide Y (plasma)	LiChrospher XDS	115
Nitrendipine (plasma,tissue)	Lichrospher RP-18 ADS	29
2-(2-Nitro-4-trifluoromethylbenzoyl)-1,3-cyclohexanedione (plasma)	BioTrap C18	59
Omeprazole (plasma)	Capcell Pak MF Ph-1	80
Paclitaxel (plasma)	LiChrospher RP-4 ADS	21
Parabens (milk)	LiChrospher RP-18 ADS	138
Paraquat (plasma)	Lichrospher RP-18 ADS	31
Phenols (milk)	LiChrospher RP-18 ADS	138
Pirlindol (plasma)	LiChrospher RP-4 ADS	28
Procaine (plasma)	LiChrospher XDS	105
Propafenone (serum)	Lichrospher RP-18 ADS	41
Propentofylline (serum)	Hisep	96
Propiverine (plasma)	Capcell Pak MF Ph-1	141
Propranolol (serum)	BioTrap C18	61
Proton pump inhibitor (KR60436) (plasma)	Capcell Pak MF Ph-1	116
Rhein (plasma)	Capcell Pak MF C(8)	85

(continued on next page)

TABLE 7.5 (continued)
Compounds Analyzed by Direct Injection of Biological Fluids onto RAM Using Column-Switching HPLC System

Compounds (Matrix)	RAM Sorbent	References
Ropivacaine and bupivacaine (plasma)	SPS C18	117
Sibutramine (serum)	Capcell Pak MF Ph-1	190
Steroid compounds (plasma)	Lichrospher RP-18 ADS	118
Steroid compounds (cell culture)	LiChrospher RP-4 ADS	46
Steroid compounds (hepatocytes)	BioTrap 500 MS	64
Tamoxifen (plasma)	SPS CN	50
Terbutaline (plasma)	LiChrospher XDS	105
Tetracyclines (urine)	ChromSpher BioMatrix	119
Tofisopam (serum)	Capcell Pak MF Ph-1	82
Triclosan (milk)	LiChrospher RP-18 ADS	138
Trimethoprim (milk)	Lichrospher RP-18 ADS	120
Verapamil (serum)	SPS C8	90
Verapamil (cell culture, plasma)	LiChrospher RP-8 ADS	30
YM087, YM440 (plasma)	Lichrospher RP-18 ADS	121
Zaltoprofen (plasma)	Capcell Pak MF Ph-1	66

also implies a potential for the direct injection of endogenous substances in serum and peptides from complex extracts.

7.2.1.2.1.2 Alkyl-Diol-Silica Material (ADS)   Alkyl-diol-silica material is the most popular RAM support for sample extraction (Figure 7.3). Vielhauer et al. [15] suggested ADS as the new family of restricted-access materials containing a hydrophilic, electroneutral outer particle surface and a hydrophobic internal pore surface for online sample processing and analysis of the photoreactive drug 8-methoxypsoralen (8-MOP) in plasma. Rudolphi et al. [16] also reported the use of porous ADS as the new RAMs in sample processing and analysis of epirubicin and four metabolites in human plasma, liver homogenate, and liver tumor homogenate in 1995. In that study, they used a hydrophilic and electroneutral external particle surface (glyceryl-residues) and a hydrophobic reversed-phase internal surface (butyryl-, octanoyl-, or octadecyl-residues) for the precolumn. These bimodal chromatographic properties allow retention of low molecular analytes by classical RP-chromatography exclusively at the lipophilic pore surface. Macromolecular constituents of the sample matrix (e.g. proteins or nucleic acids) are size excluded by 5 nm pores and quantitatively eliminated in the interstitial void volume, whereas the adsorption centers for small analytes or target molecules with a molecular mass below 15,000 are localized exclusively at the inner pore space with alkyl chains (e.g., n-alkyl groups, ion-exchange groups). A very efficient method to get a hydrophilic and nonparticipating particle surface is to derivatize diol groups. The subsequent alkyl-diol-silica (ADS,

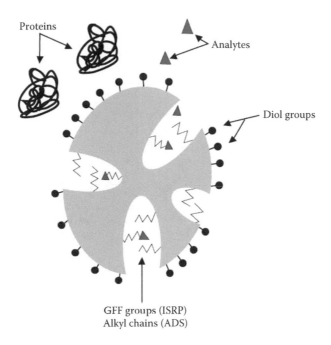

FIGURE 7.3 Internal surface reversed-phase (ISRP) with GEF and alkyl-diol-silica (ADS) with alkyl chains.

e.g., LiChrospher RP-ADS) has nondenaturing properties with proteins or nucleic acids. A RAM sorbent with RP-4, RP-8, or RP-18 modification is available to properly match the polarity of the target analytes. Through this method, the interfering macromolecules can be removed from a biological fluid, avoiding the requirement of pH- or solvent-induced precipitation of proteins. The precolumn (e.g., 25 × 4 or 25 × 2 mm for LC-MS) filled with the RAM sorbent can be integrated into the existing HPLC setup via a simple valve-switching procedure and is available for up to 2000 fractionation cycles under optimal conditions. LiChrospher RP-ADS, the commercially available ADS, was popularly used to achieve the direct analysis in biological matrices such as plasma [15–37], serum [35,38–43], urine [24,39,40,44], saliva [27], liver homogenates [16,45], intestinal aspirates [40], cell cultures [30,46], bronchial secretions [38], milk [45], and tissues [29].

7.2.1.2.1.3 Porous Silica Covered by a Combined Ligand (PSCL)   Varian introduced ChromSpher BioMatrix as a new RAM for direct loading of biological samples [47]. The porous silica (pore size 13 nm) is able to physically exclude macromolecules of proteins or nucleic acids with a ligand possessing similar internal and external surface chemistries of hydrophobic and hydrophilic properties (Figure 7.4). The ligand consists of a combination of diol and hydrophobic groups. Only small or target analytes can access the hydrophobic surface coated with phenyl groups, whereas macromolecules can interact only with diol groups such as alkanolic (poly-

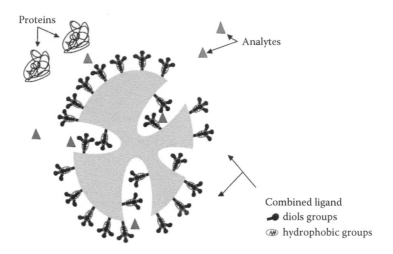

FIGURE 7.4   Porous silica covered with a combined ligand.

glycidol). ChromSpher Biomatrix supports have been applied to the direct injection and analysis of small molecules from complex matrices such as plasma [48].

7.2.1.2.1.4 Semipermeable Surface (SPS)   Polyoxyethylene was first adopted as a RAM by its characteristics of hydrophobic adsorption and covalent bonding to selected C_8 and C_{18} HPLC packings, thereby providing a semipermeable hydrophilic layer over the alkylsilane surface [8]. Brewster et al. [49] compared three commercially available HPLC columns (internal surface reversed phase, shield hydrophobic phase, and semipermeable surface) for suitability of the direct injection of bovine serum to analyze sulfonamide antibiotic residues. They concluded that the semipermeable surface column (C_{18}) provided good separation of analytes and matrix peaks with significantly more retention and selectivity. The semipermeable hydrophilic layer is able to restrict protein access to the hydrophobic stationary phase (Figure 7.5). The trade name of SPS was developed for a new RAM sorbent. SPS material was composed of independent inner and outer chemistries synthesized and covalently grafted to the surface of the silica particles. The hydrophilic outer surfaces (e.g., polyoxyethylene, polyethylene glycol, cyano group) restrict large molecules in matrices, whereas the hydrophobic inner surfaces (e.g., nitrile, phenyl, C_8, C_{18}) provide good retention of target analytes that get through the polymer surface layer. Yu et al. [21] evaluated SPS column tolerability of up to 50 mL of plasma sample. SPS material supports have been applied to the direct injection and analysis of small molecules from complex matrices such as plasma [50,51], serum [49,52,53], and urine [54].

7.2.1.2.1.5 Protein-Coated Silica (PCS)   Hermansson and Grahn [55] introduced α1-acid glycoprotein bonded to the outer surface of a reversed-phase chromatographic packing to exclude protein molecules. They used the column based on particles with a biocompatible outer surface and a hydrophobic inner surface, which

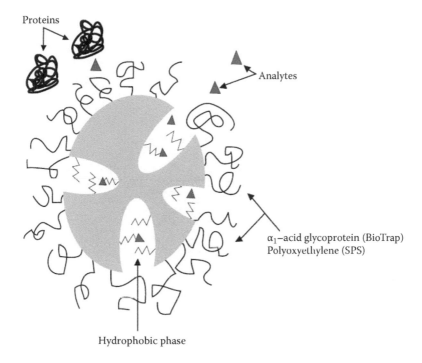

Proteins

Analytes

α_1–acid glycoprotein (BioTrap)
Polyoxyethylene (SPS)

Hydrophobic phase

FIGURE 7.5   Semipermeable surface and protein-coated silica supports.

enables the pore to exclude large molecules and retain small analytes. BioTrap is a trade name of commercially available PCS material [56]. Based on porous silica and human plasma protein, the PCS material has α1-acid glycoprotein (AGP) attached to the outer surface of the packing, and the inner surface is derivatized with a C_8- or C_{18}-ligand, which is able to interact with small analytes (Figure 7.5). PCS material has showed good pretreatment for direct analysis of compounds in biological fluids [57–60]. BioTrap tolerates up to 30 mL of biological fluids without any remarkable deterioration [61]. BioTrap MS, which possesses a hydrophobic polymer, provides the advantage of use in the wider pH working range (2 to 10) than that of a silica material (pH range 2.5 to 7.5). This new RAM support has been used for the direct analysis of chemicals in biological samples [20,62–64].

7.2.1.2.1.6 Mixed-Functional Material (MFM)   Kanda et al. [65] developed and suggested silicone polymer-coated mixed-functional (PCMF) silica packing materials for direct injection of biological samples such as plasma and serum. They used two hydrophilic groups of polyoxyethylene and oligoglyceryl groups attached to porous silica and showed that polyoxyethylene groups possessed higher recoveries of injected proteins and a greater retention of target molecules than oligoglyceryl groups. Figure 7.6 represents MFM, which devises the external and internal surfaces with a mixture of hydrophilic polyoxyethylene groups, which exclude the large molecules and retain the small molecules, and hydrophobic styrene groups on silicon

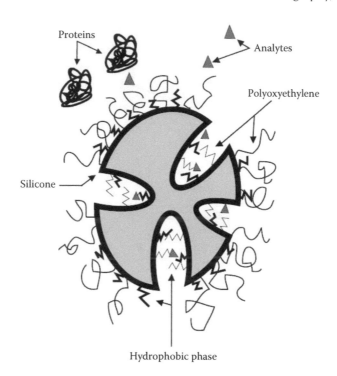

Proteins

Analytes

Polyoxyethylene

Silicone

Hydrophobic phase

FIGURE 7.6   Mixed-functional material.

polymer coated with porous silica gel (8 nm). Instead of styrene groups, phenyl, C_8, and strong cation exchanger (SCX) are also available as hydrophobic groups. Small or target analytes are retained by the adsorption or interaction with hydrophobic or ion exchanger groups. Capcell Pak MF and Capcell Pak SCX materials are commercially available as RAMs and are used for direct determination of drugs in biofluids, although these new materials have a limited lifetime compared with other RAMs. Choi et al. [66] and Um et al. [67] used Capcell Pak MF for direct analysis of zaltoprofen and fluvastatin in plasma using the column-switching HPLC method. There are many other applications for direct injection of biological fluids to analyze drugs using MFM supports [68–87].

7.2.1.2.1.7 Shielded Hydrophobic Phase (SHP)   A new type of RAM, termed shield hydrophobic phase, was proposed by Gisch et al. [88]. SHP is a stationary phase with a chemical barrier excluding proteins from accessing the functional groups responsible for separation of small molecular analytes. A hydrophilic network of polyethylene oxide or polyethylene glycol possesses the embedded hydrophobic phenyl group within the polymer network. In addition, this whole unit is covalently bound to surface silica-based materials (Figure 7.7). The hydrophilic network with embedded hydrophobic groups prevents protein penetration (hydrophilic shielding), whereas small molecules and analytes can penetrate through the polymer layer and adhere to the hydrophobic groups. This extraction material was introduced with the

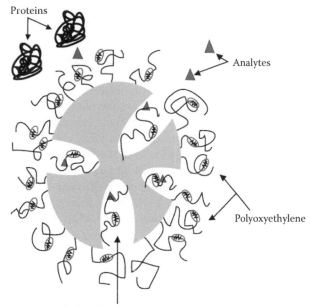

Proteins

Analytes

Polyoxyethylene

Embedded Hydrophobic phase

FIGURE 7.7   Shielded hydrophobic phase.

name Hisep SHP (Supelco, USA). The Hisep column can tolerate 16 mL of serum sample without significant loss of performance [89]. Hisep material has been used for the direct analysis of chemicals in biological fluids [89–103]. Table 7.5 shows RAM sorbents applied for biological matrices in online column-switching analysis [14–34,38–39,41,44,46,50–51,59,61–64,66–70,72,77–78,80–85,90,96,103–120].

7.2.2   TRACE ENRICHMENT

In general, interesting compounds or target molecules often exist at very low concentrations in biological fluids. Therefore, many efforts have been made to enrich target molecules. The main idea of trace enrichment by the column-switching method is based on the trapping of target analytes in the narrow zone of the precolumn after a large volume of sample is loaded. The eluent is then changed to flush out and move the enriched molecules to the analytical column. However, many other compounds can also be concentrated in the precolumn, and there is some limitation to the degree of trace enrichment. The choice of precolumn (dimensions and packing materials) and eluent is important to optimize the chromatographic conditions. To obtain good recovery and reproducibility, the column is not overloaded, and column capacity should not be exceeded. Sample dilution can prevent overloading, and pertinent choice of column can prevent the exceeding the column capacity. An additional factor to consider is to back-flush the target analytes from the precolumn to the analytical column. Nielsen [122] showed that direct injection of biological samples (fluids, feces, and tissues) of up to 50 mL into the various precolumns was performed and then the compounds were back-flushed on to the analytical column to obtain the enrichment factors of about 200. Nielsen [122] reported that the best results were achieved from LiChroprep RP-18 among different packing materials for the precolumn.

In 1981, Roth et al. [123] reported that application of column-switching technique for sample enrichment. They designed an alternating precolumn for a sample enrichment device with a programmable automatic sample unit, which was connected from two alternating precolumns to an analytical column. After injection of biological fluids to precolumn 1, the precolumn was washed with buffer. The substances to be detected were selectively adsorbed and enriched in the reverse-phase precolumn. Then another precolumn 2 was eluted in the back-flushing mode onto the analytical column while the precolumn 1 was reconditioned. A Cardiotonic agent, an antiplatelet drug, and a dipyridamole were successfully analyzed with this new method in plasma, urine, and saliva. Figure 7.8 shows the column-switching system using the alternating precolumn sample enrichment. Juergens [124] proposed the determinations of eight common antiepileptic drugs and metabolites in serum using a similar system of alternating precolumn sample enrichment. Juergens [124] used a precolumn consisting of very short cartridges (length of 0.5 cm) filled with spherical octadesylsilane (ODS) silica gel (particle size 30 μm).

Koenigbauer et al. [125] analyzed diazepam and its metabolites in serum using sample enrichment in a column-switching system. They used a microbore precolumn packed with 5 μm Spherisorb ODS-2. They evaluated varying precolumn dimensions, precolumn loading time, and sample volume.

Online column-switching technique has been successfully applied to sample enrichment to analyze trace level of analytes in biological samples [63–64,126–128].

7.2.3 GROUP SEPARATION

Groups of components can be preliminarily fractionated to reduce the number of components transferred onto the secondary column according to chromatographic characteristics. The groups of components can be selected by similar molecular size and similar adsorption characteristics (e.g., anion, cation). The results of fewer peaks and increased resolution through group separation provide good application for the direct separation of the whole biological samples.

Online group separation was applied for industrial wastewater samples using small precolumns packed with C_{18}, PRP-1 (a polystyrene-divinylbenzene phase), and cation-exchange materials [129]. They divided samples into three analyzing groups of a fraction containing nonpolar compounds, a fraction containing medium polarity compounds, and a fraction containing polar anilines and other polar bases. C_{18} (10 μm) adsorbed and fractionated nonpolar dye stuffs and nonpolar solutes and PRP-1 (10 μm) trapped the more polar aromatics such as p-nitrophenol and p-chloroaniline. Cation exchanger (Aminex A5, 13 μm) adsorbed the polar phenylenediamines and other anilines. Each fraction was subsequently transferred onto a C_{18} analytical column using the column-switching technique.

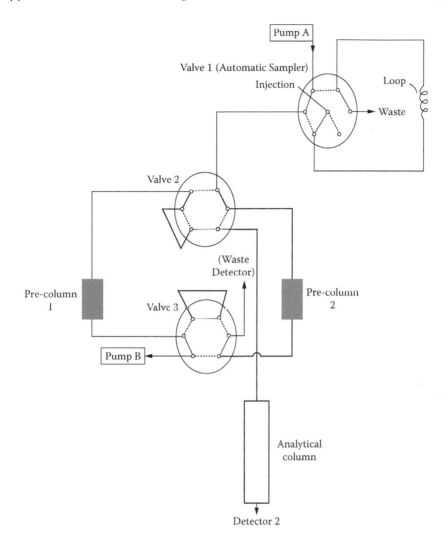

FIGURE 7.8 Column-switching configuration for alternating precolumn sample enrichment.

7.3 PRINCIPLES OF COLUMN-SWITCHING TECHNIQUE

The column-switching technique can be a precolumn or postcolumn procedure. A valve-switching procedure can be used to select a particular column on which to separate a particular sample (e.g. it may be used to select a polar column, a dispersive column, or an ion exchange column) in a precolumn procedure. Alternatively, the valve-switching system may select a precolumn on which to concentrate a very low amount of sample from which it can be displaced onto an analytical column. In a postcolumn procedure, a column-switching system consisting of valve selection can be used to divert the column eluent from the detector to another column with different selectivity. For example, if the sample is primarily separated on a precolumn packed with RAMs, then a group of solutes containing a particular molecular size range can be diverted to a dispersive column and be separated on the basis of dispersive selectivity. The switching valves and any detector situated between columns (e.g., to determine the switching time) must tolerate the high pressures that may develop. Column-switching technique significantly increases the versatility of a chromatographic system.

7.4 COLUMN-SWITCHING HPLC CONFIGURATION

In general, the simple column-switching HPLC configuration is composed of a precolumn, an analytical column, and a six-port valve for a direct analysis of biological samples. Figure 7.9 shows the basic schematic diagram of the automated column-switching technique. When the valve is in the off position, the precolumn is washed with mobile phase, and large molecules like plasma protein or endogenous interferences are supposed to be removed (Figure 7.9a). When the valve is in the on position, the precolumn is back-flushed with different mobile phase, and the target analytes are eluted from precolumn to analytical column (Figure 7.9b). Many biological sample analyses have used this simple column-switching HPLC configuration [130–141].

To improve the resolution and automation, additional precolumn (trap column) can be applicable for column-switching technique [66–67]. Figure 7.10 depicts a schematic diagram. When a biological sample is directly injected onto the precolumn, a mobile phase washes the precolumn and removes the large endogenous molecules (Figure 7.10a). After the valve is switched to the on position, the enriched

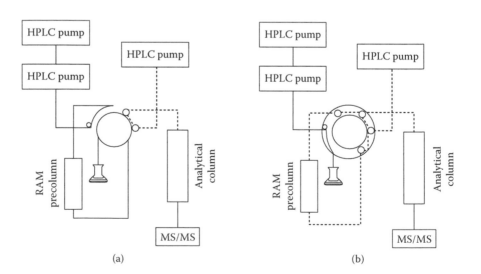

FIGURE 7.9 Simple column-switching configuration.

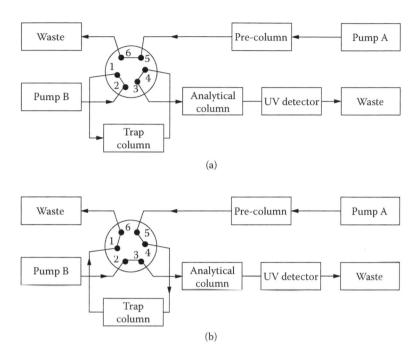

(a)

(b)

FIGURE 7.10 Schematic diagram of the column switching HPLC system using a six-port switching valve. Step1, valve A. Step2, valve B. Step3, valve A.

compounds are eluted from precolumn to the other precolumn to trap the analytes in the back-flush mode (Figure 7.10b). Afterward, the mobile phase is returned to the initial valve condition (off position of valve), and the enriched compounds in trap column are eluted into the analytical column to separate the analyte (Figure 7.10a). This column-switching HPLC configuration can be modified and variable according to physicochemical properties of analytes, RAMs, and polarity of mobile phases.

Another column-switching HPLC configuration is to use a two valves system (Figure 7.11). When two valves are in the off position, a mobile phase washes the first precolumn (Figure 7.11a). After this wash period, the left valve is positioned to on, and the first precolumn is flowed in the reverse (Figure 7.11b). The flow to the second precolumn is the same as the off position of the right valve. The analytes enriched in the first precolumn are transferred to the second precolumn to clean up or to separate. When the fraction of target analytes begins to elute from the second precolumn, the right valve is rotated to the on position, and the left valve is rotated to the off position (Figure 7.11c). After the whole fraction elutes through the second precolumn, the right valve is restored (off position), and the separation on the third analytical column starts (Figure 7.11a). In this column-switching HPLC system, we can manage various analytical conditions using three columns and three different mobile phases [142–144].

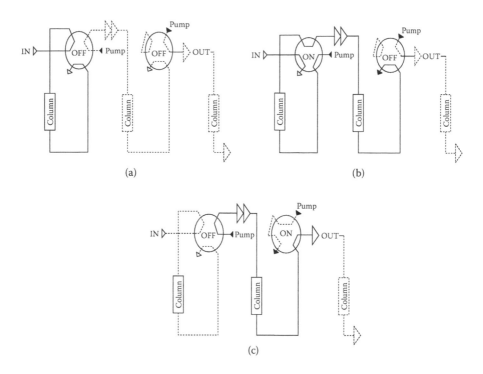

FIGURE 7.11 Schematic diagram of the column switching HPLC system using two valves and three columns.

7.5 APPLICATION OF COLUMN-SWITCHING HPLC METHOD FOR ANALYSIS OF COMPOUNDS IN BIOLOGICAL MATRICES

A column-switching HPLC method has been applied for the quantification of drugs, chemicals, and their metabolites in various biological matrices including plasma, urine, bile, and brain of humans and animals. The HPLC analysis of drug components in herbal medicines has also been performed using a column-switching technique. Table 7.6 summarizes column-switching HPLC methods applied for the analysis of compounds in biological samples to evaluate their pharmacokinetics.

7.5.1 DETERMINATION OF A DRUG IN BIOLOGICAL FLUIDS OF HUMANS

A potent urokinase-type plasminogen activator, UK-356,202, investigated for the topical treatment of chronic ulcerous wounds, has been determined in human plasma by column-switching HPLC with fluorescence detection. The HPLC system was composed of three columns with the injection cycle time of 9.5 min. per each sample. The compound was extracted from the plasma by simple protein precipitation with acetonitrile before injection into the HPLC system. Young male and elderly volunteers were involved in a single-blind and parallel group study to investigate the safety, tolerance, and systemic pharmacokinetics of UK-356,202. It was found

TABLE 7.6
Application of Column-Switching HPLC Methods in Evaluating Pharmacokinetics of Compounds

Precolumn	Trap Column	Analytical Column	Analyte	Sample	Detector	Reference
		Determination of a Drug or a Chemical in Biological Fluids				
Kromasil 100C1	Spherisorb S5CX	SUPELCOSIL ABZ+ Plus	UK-356,202,	Human plasma	Fluoresence	145
Shim-pak MAYI-ODS	×	Symmetry C18	Voriconazole	Human plasma	UV	146
Capcell Pak MF C8 SG 80	Capcell Pak C18 UG 120V	Capcell Pak C18 UG 120V	Cetirizine	Human plasma	UV	149
Capcell Pak MF pH-1	Capcell Pak C18 UG 120U	Capcell pak C18 MG	Glimepiride	Human plasma	UV	150
AC-ODS	×	Luna 2 C18	Clarithromycin	Human plasma	ECD	151
Ultrasphere C18	×	CSC-nucleosil C8	Docetaxel	Human plasma	UV	153
Brownlee Spheri-5 cyano	×	NovaPak C8	Paclitaxel	Human plasma	UV	160
Manually packed with protein-coated uBondapak phenyl silica	×	ODS/TM	Propofol	Human serum	UV	152
BondapakC18 Corasil	×	Superspher 100 RP-18	Mofarotene, arotinoid	Human, rat & dog plasma	UV	161
X-Terra RP18	×	Synergi polar RP	Risedronate	Human urine	UV	163
Spheri 5 amino	×	Nucleosil C18	Cefepime	Human urine, dialysis fluid	UV	164
Shim-pak MAYI-ODS(G)	×	L-column ODS	Fluvoxamine	Rat plasma	UV	165
Capcell Pak MF pH-1	Capcell Pak C18 UG120	Capcell Pak C18 UG120	Verapamil	Rat plasma	Fluoresence	166
Capcell Pak MF pH-1	Capcell pak C18	Capcell pak C18	Zaltoprofen	Rat plasma	UV	167

(continued on next page)

TABLE 7.6 (continued)
Application of Column-Switching HPLC Methods in Evaluating Pharmacokinetics of Compounds

Precolumn	Trap Column	Analytical Column	Analyte	Sample	Detector	Reference
Capcell Pak MF pH-1	Capcell Pak C18 UG120	Capcell Pak C18 UG120	Asiaticoside	Rat plasma and bile	UV	169
Wakosil-II 5 C18	×	Develosil ODS-5	Bisphenol A	Rat brain microdialysate	Fluoresence	172
Capcell Pak C18 MG II	Inertsil NH2	TSKgel ODS-80TsQA	Kynurenic acid	Rat brain	Fluoresence	173
Bondesil C2	×	Shim-pack C18	Terbutaline	Dog plasma	UV	168
Simultaneous Determination of Drugs, Chemicals, and Their Metabolites in Biological Fluids						
TSK BSA-ODS/S	×	Chiralcel OD-RH	Warfarin enatiomers, 7-hydroxywarfarin enantiomers	Human plasma	UV	174
Phenomenex silica	Inertsil silica guard	Chiralcel OJ-H	(+)-(R)-bevantolol, (-)-(S)-bevantolol	Human plasma	UV	179
TSK BSA-ODS/S	×	Develosil C8-5	Itraconazole, hydroxyitraconazole	Human plasma	UV	180
TSK gel PW	×	Inertsil ODS-80A	Omeprazole, 5-hydroxtomeprazole, omeprazole sulfone	Human plasma	UV	181
TSK gel PW	×	C18 STR ODS-II	Fluvoxamine, fluvoxamino acid	Human plasma	UV	183
TSK gel PW	×	C18 Grand ODS-80TM TS	Rabeprazole, rabeprazole thioether	Human plasma	UV	185
Capcell Pak MF pH-1	Capcell Pak C18 UG120	Luna 2 Phenyl-hexyl	Aceclofenac, diclofenac	Human plasma	UV	70

Shim-pak MAYI-ODS	×	Shim-pak VP-ODS	Loxoprofen, trans- and cis-alcohol metabolite	Human serum	UV	182
Pinkerton GFF2 RAM		Micra RP-18 non-porous silica	lidocaine, pindolol, metoprolol, oxprenolol, diltiazem, verapamil	Human serum	UV	186
Inertsil ODS-2		L-column ODS	N-(trans-4-isopropylcyclohexanecarbonyl)-D-phenylalanine and its seven metabolites	Human plasma and urine	UV	188
Phenomenex silica	Phenomenex silica guard	Sumichiral OA-4900	(S)-(+)-terbutaline; (R)-(-)-terbutaline	Human urine	Fluoresence	189
MC-SCX	×	L-column ODA	Atenolol, Sulpiride	Rat plasma	UV	191
MF Ph-1 pretreatment cartridge	Capcell pak C18 MG	Capcell pak C13 MG II	Metabolites of sibutramine	Rat serum	UV	190
TSKgel Super-Phenyl	TSKgurdgel ODS-80Ts	TSKgel ODS-80Ts	2,7,8-trimethyl-2-(β-carboxyethyl)-6-hydroxy chroman, LLU-α	Rat plasma	Fluoresence	192
TSK gel ODS-80Ts	×	A tandem series of two chiral columns (Sumichiral OA-2500(S)&(R)	D- and L-Serine	Rat brain microdialysate	Fluoresence	193
Determination of Drug Components in Herbal Medicine						
Capsell Pak MF C$_8$	×	Kromasil C$_{18}$	Baicalin, rhein, berberine	Rat plasma	UV	85
C$_{18}$ SPE	×	Pinnacle 11 C$_{18}$	Andrographolide, dehydroandrographolide	Rabbit plasma	UV	196

that the extraction, trace enrichment, and analytical columns for analysis lasted for approximately 2000, 500, and 1000 clinical plasma samples, respectively, suggesting robustness of this column-switching system [145].

The plasma concentration of voriconazole, a wide-spectrum triazole antifungal drug, was quantified using a fully automated column-switching HPLC system after oral administration of a voriconazole tablet in healthy volunteers and patients. Therapeutic monitoring of plasma levels of voriconazole is important to minimize adverse reactions, such as skin rashes, liver dysfunction, and photopsia, which can occur in patients who are poor metabolizers of the CYP2C19 isozyme, a major metabolizing enzyme of voriconazole. In the study, plasma samples were prepared by addition of borate buffer, followed by vortex mixing and filtration. The filtered plasma sample was then directly injected onto a Shim-pack MAYI-ODS precolumn. This system was able to analyze more than 300 plasma samples with acceptable precision and accuracy and appeared to be useful for clinical pharmacokinetic studies [146].

The sensitivity for the analysis of cetirizine, an antihistamine, has been enhanced more than two times using the column-switching technique, compared with the GC-NPD and HPLC-ultraviolet (UV) methods [147,148], by introducing a large volume of plasma sample extracted with methylene chloride into the pretreatment column without an additional concentration process. This column-switching method was successfully applied to determine cetirizine in human plasma, providing pharmacokinetic parameters of the drug following oral administration to human subjects [149].

The semimicrobore HPLC method with column-switching was sensitive enough for the quantification of glimepiride, an antidiabetic drug, in human plasma following a single oral dosing of the Amaryl (glimepiride) tablet. In this study, plasma proteins were removed by simple extraction with ethanol and acetonitrile, and three columns (Capcell Pak MF Ph-1 precolumn, intermediate C_{18} UG120U column, and analytical C_{18} MG column) were found to be stable for more than 300 injections of plasma samples [150].

Clarithromycin, a semisynthetic macrolide antibiotic, has been analyzed in human plasma by a simple and reliable HPLC method using the column-switching technique and electrochemical detection. Among several bonded stationary phases such as octyl, octadecyl, cyanopropyl, or phenyl, the octadecyl column was shown to be more suitable than others for the quantification of clarithromycin in plasma with respect to the capacity factor, selectivity, and sensitivity. In addition, AC-ODS precolumn packing was found to be more effective in removing interferences from plasma samples and eluting the adsorbed drug than LiChroprep RP-8 packing. Based on a hydrodynamic voltammogram of clarithromycin, 0.87 V was chosen as the optimum potential for the oxidation of the drug. The suitability of the proposed method was proven by a pharmacokinetic study of the drug in human subjects following a single oral administration [151].

Propofol, a rapid-acting intravenous (IV) anesthetic agent, has been assayed in human serum by the column-switching HPLC method using a precolumn manually packed with protein-coated µBondapak phenyl silica (particle size 20 ~ 30 µm) and an ODS/TM analytical column. Propofol was well retained onto the precolumn due to its high hydrophobicity, whereas the more polar components in serum samples passed through the precolumn with good recovery of the drug. The precolumn used for sample cleanup before elution onto the analytical column lasted for about 400 injections of serum samples. By using this method, propofol was rapidly and quantitatively analyzed directly from human serum samples obtained from women undergoing caesarean section after an IV bolus injection of propofol without sample pretreatment or use of an internal standard, suggesting its applicability for pharmacokinetic studies as well as for its adjustment as a dosage regimen in clinical situations [152].

A rapid and sensitive column-switching HPLC method has been reported for the determination of docetaxel, an anticancer drug, in human plasma [153]. By using a column-switching technique and a 7.5 cm long precolumn, the retention time of docetaxel and total run time of the analytical method were reduced in the absence of late eluting peaks, which increased the analysis time of docetaxel in biological fluids [154,155]. In addition, C_8 nucleosil analytical column with a less polar mobile phase (high percentage of acetonitrile) allowed for a remarkable reduction in retention time and increase in detection of the drug. The suitability of the method for quantification of docetaxel in plasma was proven by clinical pharmacokinetic studies [153].

Because endogenous constituents in plasma are more strongly retained than paclitaxel, an anticancer drug, onto the stationary phase under reversed-phase conditions [156–158], manual sample preparations by LLE or SP prior to chromatographic analysis has not satisfactorily removed these lipophilic plasma substances [159]. Therefore, a fully automated column-switching HPLC has been developed for specific and sensitive determination of paclitaxel in human plasma by preventing the introduction of nonpolar plasma constituents onto the analytical column and simplifying the sample manipulation procedure. The online sample cleanup was successfully achieved by using a 3 cm short Brownlee Spheri-5 cyano cartridge precolumn after extraction of plasma samples with tert-butyl methyl ether. The chromatogram of human plasma determined by an automated column-switching HPLC method developed in this study shows a substantial improvement in selectivity compared with that of a conventional HPLC method with solid phase extraction (Figure 7.12). The chromatographic data was consistent over several weeks, indicating infrequent replacement of the precolumn and analytical column. This assay method was applicable for the monitoring of plasma concentrations of paclitaxel after a 96 h IV infusion in combination with other anticancer drugs to cancer patients [160].

Mofarotene, a retinoid containing an amine group, has been analyzed in human plasma by column-switching HPLC. It has been known that HPLC with a column-switching technique is suitable for the determination of retinoids, which are sensitive to photoisomerization and oxidation. In this analysis, the supernatant of a plasma sample was directly injected onto a precolumn after a simple protein precipitation using ethanol to overcome low recovery of the highly protein-bound retinoid. Approximately 1500 plasma samples were assayed by this method in a clinical pharmacokinetic study of mofarotene, replacing the precolumn after 100 injections [161].

Although bisphosphates have been reported to be difficult to chromatograph due to their strong chelating effect interacting with metals in HPLC systems [162], Vallano et al. [163] developed a column-switching ion-pair HPLC method for the quantification of risedronate in human urine. After precipitation of risedronate and endogenous phosphates such as calcium salts in urine, the precipitates were dissolved

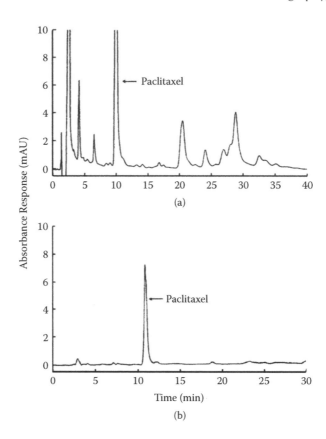

FIGURE 7.12 Chromatograms of human plasma spiked with 1.0 μg/mL paclitaxel prepared for conventional HPLC analysis with solid phase extraction (a) and with 190 ng/mL paclitaxel prepared for an automated column-switching HPLC analysis with a cyano precolumn and an octyl silica analytical column. Numerous endogenous peaks that existed between 20 and 35 min. by conventional HPLC method were made to disappear by an automated column-switching HPLC method. (Adapted from Supko, J.G.. et al., *J. Pharm. Biomed. Anal., 21,* 1025–1036, 1999. With permission.)

in ethylene glycol tetraacetic acid (EGTA) and then extracted using Waters HLB cartilage with 1-octyltriethylammonium phosphate as the ion-pair reagent to eliminate Ca^{2+}. The analytes were then injected onto a Waters X-terra RP18 column to be separated from coextracted endogenous substances followed by injection onto a Phenomenex Synergi Polar RP column for final separation. A Phenomenex Synergi Polar RP column enhanced retention time of risedronate by π-π interactions, improving selectivity and providing good peak shape. The addition of etidronate, a competing biphosphonate, in the mobile phase improved chromatographic reproducibility and peak symmetry by masking potential adsorption sites within the HPLC system.

Cherti et al. [164] reported HPLC determination of cefepime, a parenteral cephalosporin antibiotic, in urine and dialysis fluid using a column-switching technique for therapeutic drug monitoring in patients treated by intermittent venovenous hemodiafiltration. Cefepime in urine and dialysate was rapidly assayed within 10 min. by the present analytical method following intravenous infusion in patients with acute renal failure undergoing hemodiafiltration.

7.5.2 DETERMINATION OF A DRUG OR A CHEMICAL IN BIOLOGICAL FLUIDS OF ANIMALS

For the analysis of fluvoxamine, a polar base with ionizable amine groups and high pKa (8.7), a column-switching and ion-pair HPLC method was developed to improve sensitivity, precision, accuracy, and recovery of the assays compared with traditional methods and a simple online method. An ion-pair mobile phase used for enhancing the selectivity and improving separation of fluvoxamine was composed of acetonitrile-0.1% phosphoric acid (36:64, v/v) containing 2 mM 1-octanesulfonic acid sodium salt. This analytical method was applicable to the pharmacokinetic study of fluvoxamine following oral administration in rats [165].

An automated column-switching HPLC method with fluorescence detection increased speed and work capacity for quantifying verapamil, a calcium channel blocker, in rat plasma by direct injection of a small quantity of sample (10 µL) without a sample pretreatment. The present method was applied to measure plasma concentrations of verapamil in rats after intravenous injection to control and hepatofibrotic rats, suggesting the usefulness of the method for pharmacokinetic study of verapamil in experimental and clinical situations [166].

To compare the pharmacokinetic parameters of zaltoprofen and its sodium salt, a column-switching HPLC method was developed and validated for the determination of zaltoprofen in rat plasma. In this study, 50 samples could be analyzed on a daily basis and the precolumn lasted for about 200 plasma samples. It was observed that zaltoprofen sodium absorbed much faster than zaltoprofen, suggesting that it may have a rapid onset of action in the treatment of inflammatory diseases [167].

Terbutaline, a selective β2-receptor agonist for the treatment of bronchial asthma, was quantified in dog plasma by a column-switching HPLC method with simple deproteinization using acetonitrile. This analytical method was found to be adequate for the pharmacokinetic study of terbutaline after oral administration of pulsatile and immediate-release tablets containing terbutaline to dogs. A short C_2 precolumn (50 × 4.6 mm, 40 µm particle size) used in this assay retained terbutaline but not other matrix components and was stable even after more than 250 plasma sample analyses [168].

Because asiaticoside, a terpenoid used for wound healing, is not easily extracted by liquid–liquid extraction, a column-switching HPLC method with a simple protein precipitation has been developed for the sufficient separation of the compound from interfering substances in rat plasma and bile. The Capcell Pak MF Ph-1 precolumn possessing a mixture of hydrophobic and hydrophilic phases did not retain the majority of endogenous components such as proteins in the presence of washing solvents containing 9% (plasma) or 6.5% (bile) acetonitrile in 10 mM phosphate buffer. The fraction of asiaticoside retained strongly on the precolumn was concentrated on the enrichment column and then transferred to the analytical column for further separation. The proposed analytical method was found to be useful for the pharma-cokinetic study of asiaticoside in rats [169].

Bisphenol A (BPA), a component of polycarbonate plastics, has been reported to influence adult behavior by exposure during the sexual differentiation period of the brain [170] and shows weak estrogenic activity [171]. Sun et al. [172] developed an effective and sensitive HPLC method for the quantification of BPA in rat brain microdialysis samples by using a fluorescent derivatization with 4-(4,5-diphenyl-1H-imidazole-2-yl)benzoyl chloride and a column-switching technique to elucidate whether BPA can be transported across the blood–brain barrier (BBB). Following a single oral or IV administration, BPA was detected for several hours in brain microdialysis sample, indicating its capability of penetration through the BBB.

For the investigation of the alteration of kynurenic acid (KYNA), an endogenous antagonist of N-methyl-D-aspartate and α7 nicotinic receptors in the brain, a precise and accurate column-switching HPLC method with fluorescence detection has been developed for the determination of KYNA in rat brain tissues [173]. After extraction of KYNA from rat brain homogenates with acetone, the residue reconstituted with mobile phase was injected onto the first ODS column (Capcell Pak C_{18} MG II) for the efficient removal of interfering endogenous components eluted with the acidic mobile phase, 0.1% acetic acid in water-acetonitrile (95:5). Then, KYNA entrapped in an Inertsil NH_2 column was separated on the second ODS column (TSKgel ODS-80TsQA), eluted with 50 mM ammonium acetate in water-acetonitrile (95:5, pH 7.0), giving an intense fluorescence of KYNA with zinc ion. The KYNA concentrations in rat plasma, cerebrum, cerebellum, and brainstem were determined by the proposed method, suggesting that this analytical method can be applicable for pharmacokinetic and pharmacological studies of KYNA.

7.5.3 SIMULTANEOUS DETERMINATION OF DRUGS AND THEIR METABOLITES IN BIOLOGICAL FLUIDS OF HUMANS

For the analysis of enantiomers of warfarin and its metabolite, 7-hydroxywarfarin, in human plasma, the analytes were injected onto the TSK BSA-ODS/S precol-umn for sample cleanup and then transferred to a Chiralcel OD-RH analytical column for separation after extraction using diethyl ether-chloroform (80:20, v/v). For clinical application of warfarin, an anticoagulant, individualized dose adjustment is recommended to maintain the international normalized ratio (INR) within an optimal range. In addition, (S)-warfarin is known to be more potent than (R)-warfarin. Therefore, the measurements of therapeutic concentrations of warfarin enantiomers and 7-hydroxywarfarin enantiomers using this analytical method may be informative for clinical application and applicable to pharmacokinetic studies in humans [174].

Because considerable pharmacokinetic, pharmacodynamic, and metabolic differ-ences have been reported between the enantiomers of some β-adrenergic blockers [175–178], a coupled achiral-chiral column-switching HPLC method was applied for the quantification of (+)-(R) and (-)-(S)-bevantolol in plasma from healthy volunteers dosed with 200 mg of racemic bevantolol·HCl tablets. After solid-phase extraction of plasma samples with Sep-Pak Plus C_{18} cartilage, bevantolol enantiomers were separated from plasma substances on a Phenomenex silica precolumn and trapped in an Inertsil silica guard column for 2 min. After that step, the analytes were transferred to a Chiralcel OJ-H chiral column for separation between $(+)$-(R) and $(-)$-(S)-bevantolol. Both bevantolol enantiomers were rapidly absorbed, but their disposition was stereoselective, resulting in a greater maximum plasma concentration and a faster elimination rate of $(+)$-(R)-bevantolol relative to those of $(-)$-(S)-bevantolol after oral administration of racemic bevantolol [179].

The simultaneous determination of itraconazole and its active metabolite, 14-hydroxyitraconazole, using a sensitive column-switching HPLC method with UV detection is very useful for therapeutic drug monitoring and pharmacokinetic studies, because plasma concentrations of itraconazole plus its active metabolite are important to obtain effective antifungal activity. For this, n-heptane-chloroform (60:40, v/v) was used as an extraction solvent, and the extract was injected onto a hydrophilic meta-acrylate polymer column (TSK BSA-ODS/S precolumn) for sample cleanup and a Develosil C8-5 analytical column for separation [180].

For the simultaneous determination of omeprazole and its two main metabolites, 5-hydroxyomeprazole and omeprazole sulfone, in human plasma by column-switching HPLC, a mobile phase of pH values of < 8.0 was used and did not induce decomposition of the compounds during chromatographic separation. This method was suitable for evaluating precise pharmacokinetics of omeprazole and its two metabolites with reference to *CYP2C19* genotypes following intravenous injection and oral administration to human subjects [181].

For the direct and simultaneous analysis of loxoprofen and its *trans-* and *cis-* alcohol metabolites in human serum using a column-switching technique, an acidic mobile phase (acetonitrile-water (45:55, v/v) containing 0.1% formic acid) was used and resulted in short chromatographic run times (18 min.) with enhanced sensitivity and specificity. In addition, the precolumn was stable during online extraction by a simple deproteinization using 0.1% formic acid even after the analysis of more than 600 serum samples. Pharmacokinetic parameters of loxoprofen and its metabolites following a single oral dosing of the drug in human subjects were successfully determined by using this analytical method [182].

Yasui-Furukori et al. [183] reported a column-switching HPLC method for the simultaneous determination of fluvoxamine and its active metabolite in human plasma, because the monitoring of fluvoxamine and fluvoxamino acid is required for clinical treatment due to correlation between the combined steady-state plasma concentration of both compounds and therapeutic outcome. The sensitivity of the analysis was improved by using a simple liquid–liquid extraction (toluene-chloroform, 85:15, v/v), which was superior to a previous HPLC method [184]. This simple and sensitive analytical method resulted in precise measurement of plasma concentrations of fluvoxamine and fluvoamino acid following a single oral administration of fluvoxamine to human volunteers [183].

A column-switching HPLC method developed for the simultaneous measurements of rabeprazole and rabeprazole thioether in human plasma was made sensitive and specific by using a diethyl ether-dichloromethane (90:10, v/v) as an extraction solvent and a longer C_{18} analytical column (250 mm × 4.6 mm), enabling the monitoring of plasma concentrations of both compounds up to 24 h following a single oral dosing of rabeprazole in human subjects [185].

Aceclofenac, a potent nonsteroidal antiinflammatory and analgesic drug, and its active metabolite, diclofenac, have been simultaneously determined from human plasma by using narrow-bore HPLC columns, an MF Ph-1 phase consisting of poly-oxyethylene groups and phenyl groups bonded on the surface of silica, and an automated column-switching technique. In the present method, the use of intermediate C_{18} column reduced analysis time and protected the MF Ph-1 precolumn from high pressure. The narrow-bore phenyl-hexyl analytical column provided excellent resolution, selectivity, and good sensitivity. In addition, the intermediate and analytical columns lasted for more than 250 plasma sample injections with no change in column efficiency. The proposed method proved to be suitable for pharmacokinetic study of aceclofenac in human subjects [70].

An automated column-switching technique has been used for direct HPLC analysis of six cardiovascular drugs: lidocaine, pindolol, metoprolol, oxprenolol, diltiazem, and verapamil in human serum. Online sample cleanup was efficiently carried out using a Pinkerton GFF2 restricted access precolumn. The separation among six compounds in human serum samples was completed using a microsphere nonporous silica C_{18} analytical column with minimal organic solvent, suggesting that this method can be applied to pharmacokinetic studies involved in therapeutic drug monitoring [186].

A column-switching HPLC with amperometric detection using a carbon fibre microelectrode procedure has been used for monitoring plasma and urine concentrations of levodopa, carbidopa, and their metabolites. By using this analytical method, the pharmacokinetics and bioavailability of levodopa, carbidopa, and their metabolites were evaluated after oral administration of two different controlled-release dosage forms as well as a conventional tablet in healthy subjects [187].

N-(trans-4-isopropylcyclohexanecarbonyl)-D-phenylalanine and its seven metabolites in human plasma and urine have been determined simultaneously by two column-switching HPLC methods. Plasma samples were purified by solid-phase extraction, and urinary samples were directly diluted with mobile phase before injection onto the Inertsil ODS-2 precolumn. According to their different retention times, these compounds were divided into two groups and analyzed by using two different mobile phases: acetonitrile-0.05 M sodium phosphate buffer (20:80, v/v, pH 6.6) for M1, M2, and M3; and acetonitrile-ethanol-0.05 M sodium phosphate buffer (32:6:62, v/v, pH 6.6) for N-(trans-4-isopropylcyclohexanecarbonyl)-D-phenylalanine, M4, M5, M6, and M7. The present method was applicable to measure N-(trans-4-isopropylcyclohexanecarbonyl)-D-phenylalanine and its metabolites in plasma and urine after administration of the compound to humans [188].

A coupled achiral–chiral HPLC method using a column-switching technique has been developed for the separation of (S)-(+) and (R)-(−)-terbutaline in human urine. Terbutaline was separated from urinary endogenous interferences by injecting urine samples extracted using a Sep-pak silica cartridge into an achiral Phenomenex silica precolumn. After that, the fraction containing terbutaline was concentrated in a trap column for 2 min., and the compounds were transferred to a Sumichiral OA-4900 column for the complete separation of terbutaline enantiomers. The peak of (S)-(+)-terbutaline was eluted before that of (R)-(−)- terbutaline with a resolution factor (R_s) value of 1.65. The present method was successfully applied to the stereoselective pharmacokinetic study of terbutaline in a human subject dosed orally with racemic mixture of terbutaline [189].

7.5.4 SIMULTANEOUS DETERMINATION OF DRUGS, CHEMICALS, AND THEIR METABOLITES IN BIOLOGICAL FLUIDS OF ANIMALS

For the analysis of two active metabolites of sibutramine (an antiobesity drug)—N-mono-desmethyl sibutramine (M1) and N-di-desmethyl sibutramine (M2) in rat serum—the serum samples were directly injected onto the precolumn after dilution with mobile phase A, methanol-acetonitrile-20 mM ammonium phosphate buffer (pH 6.0) at the same volume. This simple and rapid method appeared to be suitable for the evaluation of pharmacokinetic parameters and bioavailability of M2 in animals and humans [190].

The methylcellulose-immobilized strong cation-exchange (MC-SCX) precolumn was proved to be effective in online enrichment of cationic drugs such as atenolol, sulpiride, and pyridoxine in rat plasma by using a low ionic strength extraction mobile phase. The precolumn stability was maintained for 300 injections of plasma samples. A column-switching HPLC method with the MC-SCX precolumn and L-column ODS was applicable for the direct analysis of diverse cationic drugs having different physicochemical properties (hydrophilicity and pKa) as well as for pharmacokinetic evaluation of atenolol and sulpiride [191].

A HPLC method for the enantiomeric determination of 2,7,8-trimethyl-2-(β-carboxyethyl)-6-hydroxy chroman (LLU-α) in rat plasma has been developed using three nonchiral columns: a phenyl precolumn; octadecyl silica trap and analytical columns; and a chiral CHIRALCEL OD-RH column connected via two column-switching valves with a fluorometric derivatization. The present method was applied to measure the concentrations of S-LLU-α and R-LLU-α in plasma after IV injection of racemic LLU-α to rats to elucidate the pharmacokinetics of LLU-α enantiomers [192].

D- and L-serine in rat brain microdialysates have been simultaneously determined by a column-switching HPLC method following fluorescence derivatization using 4-fluoro-7-nitro-2,1,3-benzoxadiazole (NBD-F) as an achiral fluorogenic reagent [193]. Because microdialysis can simplify the sample cleanup procedure and minimize biological fluid loss during sampling, it is widely applied to the monitoring of drugs in body organs including brain tissues [194]. By applying this method, the D-serine concentration in extracellular fluid as a function of time was examined in rat brain after intraperitoneal administration [193].

7.5.5 DETERMINATION OF DRUG COMPONENTS IN HERBAL MEDICINES

For the clear separation of baicalin, rhein, and berberine in rat plasma following oral dosing of Xiexin-Tang decoction at a dose of 20 mg/kg to rats, a long MF C_8 column

(150 mm × 4.6 mm) was used as a precolumn because the short column is more efficient for removal of plasma proteins. Compared with previous assay methods reported for the simultaneous analysis of these compounds in biological fluids, this column-switching HPLC method was simpler due to direct injection of centrifuged plasma samples (20 μL) with similar precision and accuracy but higher stability and recovery [85].

Andrographolide and dehydroandrographolide, major medicinal constituents of *Andrographis panicula* (*A. panicula*), are commonly used to treat common cold and respiratory inflammations [195]. An online solid-phase extraction method of HPLC using a column-switching technique was developed to quantify andrographolide and dehydroandrographolide in rabbit plasma following oral administration of *A. panicula* extract at a dose of 2 mL/kg. The dose of 2 mL extract was found to contain 35.2 mg andrographolide and 20.7 mg dehydroandrographolide. By using this analytical method, a relatively clean chromatogram was obtained following injection of a plasma sample containing the herb medicine extract. The SPE column used for concentration of the analytes lasted for more than 200 plasma samples, suggesting it can be used repetitively [196].

7.6 PHARMACOKINETIC DATA ANALYSIS

The main objectives for pharmacokinetic data analysis are to describe the pharmacokinetics of the drug and to make pharmacokinetic predictions. Pharmacokinetic analysis is carried out by compartmental or noncompartmental methods.

7.6.1 COMPARTMENTAL ANALYSIS

Because the movement of the drug within the body is inherently and infinitely complex, mathematical pharmacokinetic models are useful to describe and predict drug concentrations in the body as a function of time with any dosage regimen. In addition, pharmacokinetic models can be used to calculate the individualized optimum dosage regimen, to correlate drug concentrations with drug actions, to evaluate bioequivalence between two formulations, to describe how pathological conditions and physiological alterations can change pharmacokinetics of drugs, and to estimate drug accumulations. In pharmacokinetic models, assumptions should be simplified to describe a complex biological system concerning the passage of the drug through the body.

Compartmental models are very simple and useful tools in pharmacokinetics, based on linear assumptions using linear differential equations. In this model, the body is simply divided into one or more homogeneous compartments or tanks in which the drug is uniformly distributed. A compartment is considered to be a group of tissues having similar blood flow and drug affinity but not a real anatomical region. In particular, the compartmental models are useful when limited information is available about the tissues.

Various compartment models provide a visual depiction of the rate processes and exhibit how many pharmacokinetic constants are required to express the process properly (Figure 7.13). The difficulty in developing and validating an adequate model is the main disadvantage of the compartmental pharmacokinetic analysis [197].

A) One compartment model with IV bolus injection

B) One compartment model with first-order absorption.

C) Two compartment model with IV bolus injection

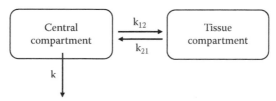

D) Two compartment model with first-order absorption.

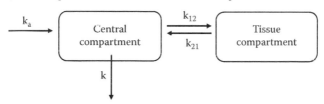

FIGURE 7.13 Several types of compartment models.

7.6.2 NONCOMPARTMENTAL ANALYSIS

Noncompartmental analysis is based on the estimation of the area under the plasma concentration–time curve (AUC) and the first-moment curve (AUMC). The major advantage of noncompartmental analysis is that it does not require the assumptions of a specific compartment model for a drug. Noncompartmental model approaches have been used to estimate certain pharmacokinetic parameters such as bioavailability, clearance, apparent volume of distribution at steady state, and mean residence time [197,198].

7.6.3 PHARMACOKINETIC PARAMETERS

Pharmacokinetic parameters are estimated from a plasma drug concentration–time curve by compartmental analysis or noncompartmental analysis. The maximum plasma drug concentration (C_{max}) and the time required to reach C_{max} (T_{max}) are determined by a visual inspection of individual plasma drug concentration–time curves. The area under the plasma drug concentration–time curve from 0 h to time t (AUC_{0-t}) is calculated by the trapezoidal rule. The area under the plasma drug concentration–time curve from 0 h to infinity ($AUC_{0-\infty}$) is calculated by the trapezoidal rule until the last sampling time point, and the final segment of the AUC is calculated as C_p^*/k, where C_p^* is the plasma concentration determined at the last sampling time point, and k is the elimination rate constant estimated from the slope of the line by log-linear least squares regression analysis. The elimination half-life ($t_{1/2}$), or the time required for the amount of drug in the body to decrease by one-half, is calculated by using the formula: 0.693/k. The apparent volume of distribution (V_d), the volume of the body in which the drug is dissolved but not a true physiological anatomic space, is calculated by A_B/C_p, where A_B is the amount of drug in the body, and C_p is the plasma concentration of drug. Total body clearance (Cl_t), expressed as volume per unit time, is a constant describing drug elimination from the body without identifying the mechanism or process and is calculated by drug elimination rate divided by the plasma drug concentration or dose divided by $AUC_{0-\infty}$ [197,198].

Figure 7.14 shows the mean plasma concentration–time profiles of zaltoprofen obtained after the oral administration of zaltoprofen or its sodium salt. The pharmacokinetic parameters for zaltoprofen obtained following the oral administration of zaltoprofen or its sodium salt are shown in Table 7.7. Plasma concentrations of zaltoprofen were determined by an automated column-switching HPLC method reported by Yang et al. [167]. Once drug concentrations in biological fluids such as plasma, serum, and urine are measured by an analytical method, pharmacokinetic parameters can be calculated from the drug concentration in biological fluids and time profile using compartmental or noncompartmental analysis.

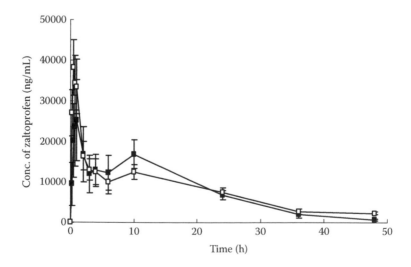

FIGURE 7.14 Mean plasma zaltoprofen concentration-time profiles following the oral administration of (■) zaltoprofen minicapsules or (□) zaltoprofen sodium minicapsules to rats (mean ± SE; n = 6–7). (Adapted from Yang, H.K. et al., *Biomed. Chromatogr.*, 23, 537–542, 2009. With permission.)

TABLE 7.7
Pharmacokinetic Parameters of Zaltoprofen and Its Sodium Salt Observed Following the Oral Administration to Rats

Parameters	Zaltoprofen	Zaltoprofen Sodium
C_{max} (µg/mL)	26.8 ± 8.77	42.6 ± 7.02
T_{max} (h)	10.1 ± 4.63	0.536 ± 0.101*
$t_{1/2}$ (h)	7.48 ± 0.368	15.7 ± 4.21
V_d/F (mL/kg)	667 ± 92.0	1050 ± 287
Cl_t/F (mL/h/kg)	62.4 ± 8.33	50.6 ± 6.87
AUC_{0-48} (µg × h/mL)	311 ± 22.9	348 ± 40.4
$AUC_{0-\infty}$ (µg × h/mL)	372 ± 61.5	465 ± 74.5

Notes: All the values presented in the table are means ± SE. The symbol * indicates a significant statistical difference between the two groups.

Source: Adapted from Yang, H.K. et al., *Biomed. Chromatogr.,* 23, 537–542, 2009.

7.7 CONCLUSIONS

Column switching is a powerful technique for the separation and cleanup of multi-compartment mixtures, possessing the following advantages: minimal sample manipulation, automated sample processing, considerable reduction of analytical time and labor, improved selectivity by combining different chromatographic modes, prevention of an internal standard, protection of photosensitive analytes, and minor consumption of extraction solvents. Column-switching techniques can simplify the HPLC analysis of drugs, and chemicals in biological matrices by facilitating total automation of the chromatographic process, resulting in the increased speed and work capacity. Therefore, the column-switching HPLC method has been widely applied for the determination of drugs, chemicals, and their metabolites in various biological samples of humans and animals to evaluate their pharmacokinetics or toxicokinetics.

REFERENCES

1. Riley, C.M., Sternson, L.A., Repta, A.J. and Siegler, R.W. High-performance liquid chromatography of platinum complexes on solvent generated anion exchangers. III. Application to the analysis of cisplatin in urine using automated column switching. *J. Chromatogr.,* 1992, 229, 373–386.
2. Werkhoven-Goewie, C.E., de Ruiter, C., Brinkman, U.A., Frei, R.W., de Jong, G.J., Little, C.J. and Stahel, O. Automated determination of drugs in blood samples after enzymatic hydrolysis using precolumn switching and post-column reaction detection. *J. Chromatogr.,* 1983, 255, 79–90.
3. Little, C.J., Stahel, O. and Hales, K. The role of column switching in analysing complex samples. *Int. J. Environ. Anal. Chem.,* 1984, 18, 11–23.

4. Lindenblatt, H., Kramer, E., Holzmann-Erens, P., Gouzoulis-Mayfrank, E. and Kovar, K.A. Quantitation of psilocin in human plasma by high-performance liquid chromatography and electrochemical detection: comparison of liquid-liquid extraction with automated online solid-phase extraction. *J. Chromatogr. B*, 1998, 709, 255–263.

5. Taylor, R.B. and Alexander, C. A study of precolumn and analytical column dimensions for online sample pretreatment in HPLC. *J. Liq. Chromatogr.*, 1994, 17, 1479–1495.

6. Takahagi, H., Inoue, K. and Horiguchi, M. Drug monitoring by a fully automated high-performance liquid chromatographic technique, involving direct injection of plasma. *J. Chromatogr.*, 1986, 352, 369–379.

7. Goewie, C., Nielen, M., Frei, R. and Brinkman, U. Optimization of precolumn design in liquid chromatography. *J. Chromatogr.*, 1984, 301, 325–334.

8. Desilets, C.P., Rounds, M.A. and Regnier, F.E. Semipermeable-surface reversed-phase media for high-performance liquid chromatography. *J. Chromatogr.*, 1991, 544, 25–39.

9. Boos, K.S. and Rudolphi, A. The use of restricted-access media in HPLC, part I—classification and review. *LC GC*, 1997, 15, 602–611.

10. Hagestam, I.H. and Pinkerton, T.C. Internal surface reversed-phase silica supports for liquid chromatography. *Anal. Chem.*, 1985, 57, 1757–1763.

11. Cook, S.E. and Pinkerton, T.C. Characterization of internal surface reversed-phase silica supports for liquid chromatography. *J. Chromatogr.*, 1986, 368, 233–248.

12. Nakagawa, T., Shibukawa, A., Shimono, N., Kawashima, T., Tanaka, H. and Haginaka, J. Retention properties of internal-surface reversed-phase silica packing and recovery of drugs from human plasma. *J. Chromatogr.*, 1987, 420, 297–311.

13. Perry, J., Invergo, B., Wagner, H., Szczerba, T. and Rateike, J. An improved internal surface reversed phase. *J. Liq. Chromatogr.*, 1992, 15, 3343–3352.

14. Boppana, V.K., Miller-Stein, C. and Schaefer, W.H. Direct plasma liquid chromatographic-tandem mass spectrometric analysis of granisetron and its 7-hydroxy metabolite utilizing internal surface reversed-phase guard columns and automated column switching devices. *J. Chromatogr. B*, 1996, 678, 227–236.

15. Vielhauer, S., Rudolphi, A., Boos, K.S. and Seidel, D. Evaluation and routine application of the novel restricted-access precolumn packing material Alkyl-Diol Silica: coupled-column high-performance liquid chromatographic analysis of the photoreactive drug 8-methoxypsoralen in plasma. *J. Chromatogr. B*, 1995, 666, 315–322.

16. Rudolphi, A., Vielhauer, S., Boos, K.S., Seidel, D., Bathge, I.M. and Berger, H. Coupled-column liquid chromatographic analysis of epirubicin and metabolites in biological material and its application to optimization of liver cancer therapy. *J. Pharm. Biomed. Anal.*, 1995, 13, 615–623.

17. Baeyens, W.R., Van der Weken, G., Haustraete, J., Aboul-Enein, H.Y., Corveleyn, S., Remon, J.P., Garcia-Campana, A.M. and Deprez, P. Application of the restricted-access precolumn packing material alkyl-diol silica in a column-switching system for the determination of ketoprofen enantiomers in horse plasma. *J. Chromatogr. A*, 2000, 871, 153–161.

18. Ohman, D., Carlsson, B. and Norlander, B. Online extraction using an alkyl-diol silica precolumn for racemic citalopram and its metabolites in plasma. Results compared with solid-phase extraction methodology. *J. Chromatogr. B*, 2001, 753, 365–373.

19. Keski-Hynnila, H., Raanaa, K., Forsberg, M., Mannisto, P., Taskinen, J. and Kostiainen, R. Quantitation of entacapone glucuronide in rat plasma by online coupled restricted access media column and liquid chromatography-tandem mass spectrometry. *J. Chromatogr. B*, 2001, 759, 227–236.

20. El Mahjoub, A. and Staub, C. High-performance liquid chromatography determination of flunitrazepam and its metabolites in plasma by use of column-switching technique: comparison of two extraction columns. *J. Chromatogr. B*, 2001, 754, 271–283.

21. Mader, R.M., Rizovski, B. and Steger, G.G. Online solid-phase extraction and determination of paclitaxel in human plasma. *J. Chromatogr. B*, 2002, 769, 357–361.

22. Yu, Z. and Westerlund, D. Direct injection of large volumes of plasma in a column-switching system for the analysis of local anaesthetics. I. Optimization of semipermeable surface precolumns in the system and characterization of some interference peaks. *J. Chromatogr. A*, 1996, 725, 137–147.

23. Yu, Z. and Westerlund, D. Direct injection of large volumes of plasma in a column-switching system for the analysis of local anaesthetics. II. Determination of bupivacaine in human plasma with an alkyldiol silica precolumn. *J. Chromatogr. A*, 1996, 725, 149–155.

24. Van der Hoeven, R.A., Hofte, A.J., Frenay, M., Irth, H., Tjaden, U.R., van der Greef, J., Rudolphi, A., Boos, K.S., Marko Varga, G. and Edholm, L.E. Liquid chromatography-mass spectrometry with online solid-phase extraction by a restricted-access C18 precolumn for direct plasma and urine injection. *J. Chromatogr. A*, 1997, 762, 193–200.

25. Yu, Z., Westerlund, D. and Boos, K.S. Determination of methotrexate and its metabolite 7-hydroxymethotrexate by direct injection of human plasma into a column-switching liquid chromatographic system using post-column photochemical reaction with fluorimetric detection. *J. Chromatogr. B*, 1997, 689, 379–386.

26. Ceccato, A., Boulanger, B., Chiap, P., Hubert, P. and Crommen, J. Simultaneous determination of methylphenobarbital enantiomers and phenobarbital in human plasma by online coupling of an achiral precolumn to a chiral liquid chromatographic column. *J. Chromatogr. A*, 1998, 819, 143–153.

27. Gordi, T., Nielsen, E., Yu, Z., Westerlund, D. and Ashton, M. Direct analysis of artemisinin in plasma and saliva using coupled-column high-performance liquid chromatography with a restricted-access material precolumn. *J. Chromatogr. B*, 2000, 742, 155–162.

28. Chiap, P., Ceccato, A., Gora, R., Hubert, P., Geczy, J. and Crommen, J. Automated determination of pirlindole enantiomers in plasma by online coupling of a precolumn packed with restricted access material to a chiral liquid chromatographic column. *J. Pharm. Biomed. Anal.*, 2002, 27, 447–455.

29. Heinig, K. and Bucheli, F. Application of column-switching liquid chromatography-tandem mass spectrometry for the determination of pharmaceutical compounds in tissue samples. *J. Chromatogr. B*, 2002, 769, 9–26.

30. Walles, M., Borlak, J. and Levsen, K. Application of restricted access material (RAM) with precolumn-switching and matrix solid-phase dispersion (MSPD) to the study of the metabolism and pharmacokinetics of verapamil. *Anal. Bioanal. Chem.*, 2002, 374, 1179–1186.

31. Brunetto, M.R., Morales, A.R., Gallignani, M., Burguera, J.L. and Burguera, M. Determination of paraquat in human blood plasma using reversed-phase ion-pair high-performance liquid chromatography with direct sample injection. *Talanta*, 2003, 59, 913–921.

32. Brunetto, R., Gutierrez, L., Delgado, Y., Gallignani, M., Burguera, J.L. and Burguera, M. High-performance liquid chromatographic determination of cocaine and benzoylecgonine by direct injection of human blood plasma sample into an alkyl-diol-silica (ADS) precolumn. *Anal. Bioanal. Chem.*, 2003, 375, 534–538.

33. Baeyens, W.R., Van der Weken, G., D'haeninck, E., Garcia-Campana, A.M., Vankeirsbilck, T., Vercauteren, A. and Deprez, P. Application of an alkyl-diol silica precolumn in a column-switching system for the determination of meloxicam in plasma. *J. Pharm. Biomed. Anal.*, 2003, 32, 839–846.

34. Brunetto, M.R., Obando, M.A., Fernandez, A., Gallignani, M., Burguera, J.L. and Burguera, M. Column-switching high-performance liquid chromatographic analysis of carbamazepine and its principal metabolite in human plasma with direct sample injection using an alkyl-diol silica (ADS) precolumn. *Talanta*, 2002, 58, 535–542.

35. Chiap, P., Piette, M., Evrard, B., Frankenne, F., Christiaens, B., Piel, G., Cataldo, D., Foidart, J.M., Delattre, L., Crommen, J. and Hubert, P. Automated method for the determination of a new matrix metalloproteinase inhibitor in ovine plasma and serum by coupling of restricted access material for online sample clean-up to liquid chromatography. *J. Chromatogr. B*, 2005, 817, 109–117.

36. Christiaens, B., Chiap, P., Rbeida, O., Cello, D., Crommen, J. and Hubert, P. Fully automated method for the liquid chromatographic determination of cyproterone acetate in plasma using restricted access material for sample pre-treatment. *J. Chromatogr. B*, 2003, 795, 73–82.

37. Christiaens, B., Fillet, M., Chiap, P., Rbeida, O., Ceccato, A., Streel, B., De Graeve, J., Crommen, J. and Hubert, P. Fully automated method for the liquid chromatographic-tandem mass spectrometric determination of cyproterone acetate in human plasma using restricted access material for online sample clean-up. *J. Chromatogr. A*, 2004, 1056, 105–110.

38. Ehrlich, M., Daschner, F.D. and Kummerer, K. Rapid antibiotic drug monitoring: meropenem and ceftazidime determination in serum and bronchial secretions by high-performance liquid chromatography-integrated sample preparation. *J. Chromatogr. B*, 2001, 751, 357–363.

39. Ehrlich, M., Trittler, R., Daschner, F.D. and Kummerer, K. A new and rapid method for monitoring the new oxazolidinone antibiotic linezolid in serum and urine by high performance liquid chromatography-integrated sample preparation. *J. Chromatogr. B*, 2001, 755, 373–377.

40. Oertel, R., Richter, K., Gramatte, T. and Kirch, W. Determination of drugs in biological fluids by high-performance liquid chromatography with online sample processing. *J. Chromatogr. A*, 1998, 797, 203–209.

41. Kubalec, P. and Brandsteterova, E. Determination of propafenone and its main metabolite 5-hydroxypropafenone in human serum with direct injection into a column-switching chromatographic system. *J. Chromatogr. B*, 1999, 726, 211–218.

42. Egle, H., Trittler, R., Konig, A. and Kummerer, K. Fast, fully automated analysis of voriconazole from serum by LC-LC-ESI-MS-MS with parallel column-switching technique. *J. Chromatogr. B*, 2005, 814, 361–367.

43. Egle, H., Trittler, R. and Kummerer, K. A new, rapid, fully automated method for determination of fluconazole in serum by column-switching liquid chromatography. *Ther. Drug Monit.*, 2004, 26, 425–431.

44. Lamprecht, G., Kraushofer, T., Stoschitzky, K. and Lindner, W. Enantioselective analysis of (R)- and (S)-atenolol in urine samples by a high-performance liquid chromatography column-switching setup. *J. Chromatogr. B*, 2000, 740, 219–226.

45. Decolin, D., Leroy, P., Nicolas, A. and Archimbault, P. Hyphenated liquid chromatographic method for the determination of colistin residues in bovine tissues. *J. Chromatogr. Sci.*, 1997, 35, 557–564.

46. Chang, Y.C., Li, C.M., Li, L.A., Jong, S.B., Liao, P.C. and Chang, L.W. Quantitative measurement of male steroid hormones using automated online solid phase extraction-liquid chromatography-tandem mass spectrometry and comparison with radioimmunoassay. *Analyst.*, 2003, 128, 363–368.

47. Liu, L., Cheng, H., Zhao, J.J. and Rogers, J.D. Determination of montelukast (MK-0476) and its S-enantiomer in human plasma by stereoselective high-performance liquid chromatography with column-switching. *J. Pharm. Biomed. Anal.*, 1997, 15, 631–638.

48. Cassiano, N.M., Lima, V.V., Oliveira, R.V., de Pietro, A.C. and Cass, Q.B. Development of restricted-access media supports and their application to the direct analysis of biological fluid samples via high-performance liquid chromatography. *Anal. Bioanal. Chem.*, 2006, 384, 1462–1469.

49. Brewster, J.D., Lightfield, A.R. and Barford, R.A. Evaluation of restricted access media for high-performance liquid chromatographic analysis of sulfonamide antibiotic residues in bovine serum. *J. Chromatogr.,* 1992, 598, 23–31.

50. Fried, K.M. and Wainer, I.W. Direct determination of tamoxifen and its four major metabolites in plasma using coupled column high-performance liquid chromatography. *J. Chromatogr. B,* 1994, 655, 261–268.

51. Yu, Z. and Westerlund, D. Ion-pair chromatography of methotrexate in a column-switching system using an alkyl-diol silica precolumn for direct injection of plasma. *J. Chromatogr. A,* 1996, 742, 113–120.

52. Buchberger, W., Malissa, H. and Mulleder, E. Direct serum injection in ion chromatography on packing materials with a semi-permeable surface. *J. Chromatogr.,* 1992, 602, 51–55.

53. Haque, A. and Stewart, J.T. Direct injection HPLC analysis of some non-steroidal anti-inflammatory drugs on restricted access media columns. *Biomed. Chromatogr.,* 1999, 13, 51–56.

54. Marrubini, G., Hogendoorn, E.A., Coccini, T. and Manzo, L. Improved coupled column liquid chromatographic method for high-speed direct analysis of urinary trans, trans-muconic acid, as a biomarker of exposure to benzene. *J. Chromatogr. B,* 2001, 751, 331–339.

55. Hermansson, J. and Grahn, A. Determination of drugs by direct injection of plasma into a biocompatible extraction column based on a protein-entrapped hydrophobic phase. *J. Chromatogr. A,* 1994, 660, 119–129.

56. Yu, Z. and Westerlund, D. Characterization of the precolumn bio trap 500 C 18 for direct injection of plasma samples in a column-switching system. *Chromatographia,* 1998, 47, 299–304.

57. Emara, S., El-Gindy, A., Mesbah, M.K. and Hadad, G.M. Direct injection liquid chromatographic technique for simultaneous determination of two antihistaminic drugs and their main metabolites in serum. *J. AOAC Int.,* 2007, 90, 384–390.

58. Emara, S., Saleh, G., Fathy, M. and Bakr, M.A. Chromatographic assay and pharmacokinetic studies of propofol in human serum. *Biomed. Chromatogr.* 1999, 13, 299–303.

59. Bielenstein, M., Astner, L. and Ekberg, S. Determination of 2-(2-nitro-4-trifluoromethylbenzoyl)-1,3-cyclohexanedione in plasma by direct injection into a coupled column liquid chromatographic system. *J. Chromatogr. B,* 1999, 730, 177–182.

60. Needham, S.R., Cole, M.J. and Fouda, H.G. Direct plasma injection for high-performance liquid chromatographic-mass spectrometric quantitation of the anxiolytic agent CP-93 393. *J. Chromatogr. B,* 1998, 718, 87–94.

61. Hermansson, J., Grahn, A. and Hermansson, I. Direct injection of large volumes of plasma/serum on a new biocompatible extraction column for the determination of atenolol, propranolol and ibuprofen. Mechanisms for the improvement of chromatographic performance. *J. Chromatogr. A,* 1998, 797, 251–263.

62. El Mahjoub, A. and Staub, C. High-performance liquid chromatographic method for the determination of benzodiazepines in plasma or serum using the column-switching technique. *J. Chromatogr. B,* 2000, 742, 381–390.

63. Friedrich, G., Rose, T. and Rissler, K. Determination of lonazolac and its hydroxy and O-sulfated metabolites by online sample preparation liquid chromatography with fluorescence detection. *J. Chromatogr. B,* 2002, 766, 295–305.

64. Friedrich, G., Rose, T. and Rissler, K. Determination of testosterone metabolites in human hepatocytes. I. Development of an online sample preparation liquid chromatography technique and mass spectroscopic detection of 6beta-hydroxytestosterone. *J. Chromatogr. B,* 2003, 784, 49–61.

65. Kanda, T., Kutsuna, H., Ohtsu, Y. and Yamaguchi, M. Synthesis of polymer-coated mixed-functional packing materials for direct analysis of drug-containing serum and plasma by high-performance liquid chromatography. *J. Chromatogr. A,* 1994, 672, 51–57.

66. Choi, S.O., Um, S.Y., Jung, S.H., Jung, S.J., Kim, J.I., Lee, H.J. and Chung, S.Y. Column-switching high-performance liquid chromatographic method for the determination of zaltoprofen in rat plasma. *J. Chromatogr. B*, 2006, 830, 301–305.
67. Um, S.Y., Jung, S.H., Jung, S.J., Kim, J.I., Chung, S.Y., Lee, H.J., Han, S.B. and Choi, S.O. Column-switching high-performance liquid chromatographic analysis of fluvastatin in rat plasma by direct injection. *J. Pharm. Biomed. Anal.*, 2006, 41, 1458–1462.
68. Jeong, C.K., Kim, S.B., Choi, S.J., Sohn, D.H., Ko, G.I. and Lee, H.S. Rapid microbore liquid chromatographic analysis of biphenyldimethyl dicarboxylate in human plasma with online column switching. *J. Chromatogr. B*, 2000, 738, 175–179.
69. Lee, H.M., Choi, S.J., Jeong, C.K., Kim, Y.S., Lee, K.C. and Lee, H.S. Microbore high-performance liquid chromatographic determination of cisapride in rat serum samples using column switching. *J. Chromatogr. B*, 1999, 727, 213–217.
70. Lee, H.S., Jeong, C.K., Choi, S.J., Kim, S.B., Lee, M.H., Ko, G.I. and Sohn, D.H. Simultaneous determination of aceclofenac and diclofenac in human plasma by narrowbore HPLC using column-switching. *J. Pharm. Biomed. Anal.*, 2000, 23, 775–781.
71. Song, D. and Au, J.L. Direct injection isocratic high-performance liquid chromatographic analysis of mitomycin C in plasma. *J. Chromatogr. B*, 1996, 676, 165–168.
72. Ochiai, H., Uchiyama, N., Takano, T., Hara, K. and Kamei, T. Determination of montelukast sodium in human plasma by column-switching high-performance liquid chromatography with fluorescence detection. *J. Chromatogr. B*, 1998, 713, 409–414.
73. Baek, M., Rho, Y.S. and Kim, D.H. Column-switching high-performance liquid chromatographic assay for determination of asiaticoside in rat plasma and bile with ultraviolet absorbance detection. *J. Chromatogr. B*, 1999, 732, 357–363.
74. Hsieh, Y., Brisson, J.M., Ng, K. and Korfmacher, W.A. Direct simultaneous determination of drug discovery compounds in monkey plasma using mixed-function column liquid chromatography/tandem mass spectrometry. *J. Pharm. Biomed. Anal.*, 2002, 27, 285–293.
75. Hsieh, Y., Bryant, M.S., Brisson, J.M., Ng, K. and Korfmacher, W.A. Direct cocktail analysis of drug discovery compounds in pooled plasma samples using liquid chromatography-tandem mass spectrometry. *J. Chromatogr. B*, 2002, 767, 353–362.
76. Yamaguchi, J., Matsuno, Y., Hachiuma, K., Ogawa, N. and Higuchi, S. A strategy for quantitative bioanalysis of non-polar neutral compounds by liquid chromatography/ electrospray ionization tandem mass spectrometry: determination of TS-962, a novel acyl-CoA:cholesterol acyltransferase inhibitor, in rabbit aorta and liver tissues. *Rapid Commun. Mass Spectrom.*, 2001, 15, 629–636.
77. Baek, M., Jeong, J.H. and Kim, D.H. Determination of aloesin in rat plasma using a column-switching high-performance liquid chromatographic assay. *J. Chromatogr. B*, 2001, 754, 121–126.
78. Jeong, C.K., Lee, H.Y., Jang, M.S., Kim, W.B. and Lee, H.S. Narrowbore high-performance liquid chromatography for the simultaneous determination of sildenafil and its metabolite UK-103,320 in human plasma using column switching. *J. Chromatogr. B*, 2001, 752, 141–147.
79. Song, D., Wientjes, M.G. and Au, J.L. Isocratic high-performance liquid chromatographic determination of thiacetazone by direct injection of plasma into an internal surface reversed-phase column. *J. Chromatogr. B*, 1997, 690, 289–294.
80. Yim, D.S., Jeong, J.E. and Park, J.Y. Assay of omeprazole and omeprazole sulfone by semimicrocolumn liquid chromatography with mixed-function precolumn. *J. Chromatogr. B*, 2001, 754, 487–493.
81. Yasueda, S., Kimura, M., Ohtori, A. and Kakehi, K. Analysis of an anti-inflammatory steroidal drug, difluprednate, in aqueous humor by combination of semi-micro HPLC and column switching method. *J. Pharm. Biomed. Anal.*, 2003, 30, 1735–1742.

82. Baek, S.K., Choi, S.J., Kim, J.S., Park, E.J., Sohn, D.H., Lee, H.Y. and Lee, H.S. Analysis of tofisopam in human serum by column-switching semi-micro high-performance liquid chromatography and evaluation of tofisopam bioequivalency. *Biomed. Chromatogr.*, 2002, 16, 277–281.

83. Teshima, D., Kitagawa, N., Otsubo, K., Makino, K., Itoh, Y. and Oishi, R. Simple determination of mycophenolic acid in human serum by column-switching high-performance liquid chromatography. *J. Chromatogr. B*, 2002, 780, 21–26.

84. Katagi, M., Nishikawa, M., Tatsuno, M., Miki, A. and Tsuchihashi, H. Column-switching high-performance liquid chromatography-electrospray ionization mass spectrometry for identification of heroin metabolites in human urine. *J. Chromatogr. B*, 2001, 751, 177–185.

85. Yi, L., Jian-Ping, G., Xu, X. and Lixin, D. Simultaneous determination of baicalin, rhein and berberine in rat plasma by column-switching high-performance liquid chromatography. *J. Chromatogr. B*, 2006, 838, 50–55.

86. Kim, C.K., Yeon, K.J., Ban, E., Hyun, M.J., Kim, J.K., Kim, M.K., Jin, S.E. and Park, J.S. Narrow-bore high performance liquid chromatographic method for the determination of cetirizine in human plasma using column switching. *J. Pharm. Biomed. Anal.*, 2005, 37, 603–609.

87. Jhee, O.H., Hong, J.W., Om, A.S., Lee, M.H., Lee, W.S., Shaw, L.M., Lee, J.W. and Kang, J.S. Direct determination of verapamil in rat plasma by coupled column microbore-HPLC method. *J. Pharm. Biomed. Anal.*, 2005, 37, 405–410.

88. Gisch, D.J., Hunter, B.T. and Feibush, B. Shielded hydrophobic phase: a new concept for direct injection analysis of biological fluids by high-performance liquid chromatography. *J. Chromatogr.*, 1988, 433, 264–268.

89. Uno, K. and Maeda, I. Simultaneous determination of sulphamonomethoxine and its N4-acetyl metabolite in blood serum by high-performance liquid chromatography with direct injection. *J. Chromatogr. B*, 1995, 663, 177–180.

90. Brandsteterova, E. and Wainer, I.W. Achiral and chiral high-performance liquid chromatography of verapamil and its metabolites in serum samples. *J. Chromatogr. B*, 1999, 732, 395–404.

91. Ueno, R. and Aoki, T. High-performance liquid chromatographic method for the rapid and simultaneous determination of sulfamonomethoxine, miloxacin and oxolinic acid in serum and muscle of cultured fish. *J. Chromatogr. B*, 1996, 682, 179–181.

92. Riva, E., Merati, R. and Cavenaghi, L. High-performance liquid chromatographic determination of rifapentine and its metabolite in human plasma by direct injection into a shielded hydrophobic phase column. *J. Chromatogr.*, 1991, 553, 35–40.

93. Djurdjevic, P., Jelikic-Stankov, M. and Laban, A. High-performance liquid chromatographic assay of fleroxacin in human serum using fluorescence detection. *Talanta*, 2001, 55, 631–638.

94. Djurdjevic, P., Laban, A. and Jelikic-Stankov, M. Validation of an HPLC method for the determination of fleroxacin and its photo-degradation products in pharmaceutical forms. *Ann. Chim.*, 2004, 94, 71–83.

95. Lockemeyer, M.R. and Smith, C.V. Analysis of pentoxifylline in rabbit plasma using a Hisep high-performance liquid chromatography column. *J. Chromatogr.*, 1990, 532, 162–167.

96. Kuroda, N., Hamachi, Y., Aoki, N., Wada, M., Tanigawa, M. and Nakashima, K. Simple and rapid high-performance liquid chromatography analysis of propentofylline and its main metabolites in serum using a direct injection technique. *Biomed. Chromatogr.*, 1999, 13, 340–343.

97. Ueno, R., Uno, K. and Aoki, T. Determination of oxytetracycline in blood serum by high-performance liquid chromatography with direct injection. *J. Chromatogr.*, 1992, 573, 333–335.

98. Ma, J., Liu, C.L., Zhu, P.L., Jia, Z.P., Xu, L.T. and Wang, R. Simultaneous determination of the carboxylate and lactone forms of 10-hydroxycamptothecin in human serum by restricted-access media high-performance liquid chromatography. *J. Chromatogr. B,* 2002, 772, 197–204.

99. Rao, R.N., Shinde, D.D. and Agawane, S.B. Rapid determination of rifaximin in rat serum and urine by direct injection on to a shielded hydrophobic stationary phase by HPLC. *Biomed. Chromatogr.,* 2009, 23, 563–567.

100. Laban-Djurdjevic, A., Jelikic-Stankov, M. and Djurdjevic, P. Optimization and validation of the direct HPLC method for the determination of moxifloxacin in plasma. *J. Chromatogr. B,* 2006, 844, 104–111.

101. Kishida, K. and Furusawa, N. Application of shielded column liquid chromatography for determination of sulfamonomethoxine, sulfadimethoxine, and their N4-acetyl metabolites in milk. *J. Chromatogr. A,* 2004, 1028, 175–177.

102. Pistos, C. and Stewart, J.T. Direct injection HPLC method for the determination of selected benzodiazepines in plasma using a Hisep column. *J. Pharm. Biomed. Anal.,* 2003, 33, 1135–1142.

103. Pistos, C. and Stewart, J.T. Direct injection HPLC method for the determination of selected phenothiazines in plasma using a Hisep column. *Biomed. Chromatogr.,* 2003, 17, 465–470.

104. Van Zijtveld, J. and Van Hoogdalem, E. Application of a semipermeable surface column for the determination of amoxicillin in human blood serum. *J. Chromatogr. B,* 1999, 726, 169–174.

105. Chiap, P., Rbeida, O., Christiaens, B., Hubert, P., Lubda, D., Boos, K.S. and Crommen, J. Use of a novel cation-exchange restricted-access material for automated sample clean-up prior to the determination of basic drugs in plasma by liquid chromatography. *J. Chromatogr. A,* 2002, 975, 145–155.

106. Martínez Fernández, J., Martínez Vidal, J.L., Parrilla Vázquez, P. and Garrido Frenich, A. Application of restricted-access media column in coupled-column RPLC with UV detection and electrospray mass spectrometry for determination of azole pesticides in urine. *Chromatographia,* 2001, 53, 503–509.

107. Shirota, O., Suzuki, A., Kanda, T., Ohtsu, Y. and Yamaguchi, M. Low concentration drug analysis by semi-microcolumn liquid chromatography with a polymer-coated mixed-function precolumn. *J. Microcolumn Sep.,* 1995, 7, 29–36.

108. Mullett, W.M. and Pawliszyn, J. Direct LC analysis of five benzodiazepines in human urine and plasma using an ADS restricted access extraction column. *J. Pharm. Biomed. Anal.,* 2001, 26, 899–908.

109. Mišl'anová, C. and Hutta, M. Influence of various biological matrices (plasma, blood microdialysate) on chromatographic performance in the determination of β-blockers using an alkyl-diol silica precolumn for sample clean-up. *J. Chromatogr. B,* 2001, 765, 167–177.

110. Lee, H.M., Jeong, C.K., Choi, S.J., Yoon, B.M., Na, D.H., Lee, K.C. and Lee, H.S. Direct analysis of clomipramine in human plasma by microbore high performance liquid chromatography with column-switching. *Chromatographia,* 2000, 51, 353–356.

111. Puhlmann, A., Dulffer, T. and Kobold, U. Multidimensional high-performance liquid chromatography on pinkerton ISRP and RP18 columns: Direct serum injection to quantify creatinine. *J. Chromatogr.,* 1992, 581, 129–133.

112. Ruckmick, S.C. and Hench, B.D. Direct analysis of the dopamine agonist (-)-2-(N-propyl-N-2-thienylethylamino)-5-hydroxytetralin hydrochloride in plasma by high-performance liquid chromatography using two-dimensional column switching. *J. Chromatogr.,* 1991, 565, 277–295.

113. Peng, S.X., Strojnowski, M.J. and Bornes, D.M. Direct determination of stability of protease inhibitors in plasma by HPLC with automated column-switching. *J. Pharm. Biomed. Anal.*, 1999, 19, 343–349.

114. Ortelli, D., Rudaz, S., Souverain, S. and Veuthey, J.L. Restricted access materials for fast analysis of methadone in serum with liquid chromatography-mass spectrometry. *J. High Resolut Chromatogr.*, 2002, 25, 222–228.

115. Racaityte, K., Lutz, E.S.M., Unger, K.K., Lubda, D. and Boos, K.S. Analysis of neuro-peptide Y and its metabolites by high-performance liquid chromatography-electrospray ionization mass spectrometry and integrated sample clean-up with a novel restricted-access sulphonic acid cation exchanger. *J. Chromatogr. A*, 2000, 890, 135–144.

116. Lee, H.M., Lee, H.Y., Choi, J.K. and Lee, H.S. High performance liquid chromato-graphic analysis of a new proton pump inhibitor KR60436 and its active metabolite O-demethyl-KR60436 in rat plasma samples using column-switching. *Arch. Pharm. Res.*, 2001, 24, 207–210.

117. Yu, Z., Abdel-Rehim, M. and Westerlund, D. Determination of amide-type local anaes-thetics by direct injection of plasma in a column-switching high-performance liquid chro-matographic system using a precolumn with a semipermeable surface. *J. Chromatogr. B*, 1994, 654, 221–230.

118. Van der Hoeven, R., Hofte, A., Frenay, M., Irth, H., Tjaden, U., Van der Greef, J., Rudolphi, A., Boos, K.S., Varga, G.M. and Edholm, L. Liquid chromatography—mass spectrometry with online solid-phase extraction by a restricted-access C18 precolumn for direct plasma and urine injection. *J. Chromatogr. A*, 1997, 762, 193–200.

119. Weimann, A. and Bojesen, G. Analysis of tetracyclines in raw urine by column-switching high-performance liquid chromatography and tandem mass spectrometry. *J. Chromatogr. B*, 1999, 721, 47–54.

120. Blahova, E., Bovanova, L. and Brandsteterova, E. Direct HPLC analysis of trimethop-rim in milk. *J. LIQ. CHROM. & REL. TECHNOL.*, 2001, 24(19), 3027–3035.

121. van Zijtveld, J., van den Berg, S. and Swart, P. Development and validation of a direct plasma injection LC-MS method for YM087 and YM440 and metabolites in human plasma. *Chromatographia*, 2003, 57, 23–27.

122. Nielsen, H. Use of enzymatic solubilization of tissues and direct injection on precol-umns of large volumes for analysing biological samples by high-performance liquid chromatography. *J. Chromatogr.*, 1986, 381, 63–74.

123. Roth, W., Beschke, K., Jauch, R., Zimmer, A. and Koss, F.W. Fully automated high-performance liquid chromatography. A new chromatograph for pharmacoki-netic drug monitoring by direct injection of body fluids. *J. Chromatogr.*, 1981, 222, 13–22.

124. Juergens, U. Routine determination of eight common anti-epileptic drugs and metabo-lites by high-performance liquid chromatography using a column-switching system for direct injection of serum samples. *J. Chromatogr.*, 1984, 310, 97–106.

125. Koenigbauer, M.J., Assenza, S.P., Willoughby, R.C. and Curtis, M.A. Trace analysis of diazepam in serum using microbore high-performance liquid chromatography and online preconcentration. *J. Chromatogr.*, 1987, 413, 161–169.

126. Hubert, P., Renson, M. and Crommen, J. A fully automated high-performance liquid chromatographic method for the determination of indomethacin in plasma. *J. Pharm. Biomed. Anal.*, 1989, 7, 1819–1827.

127. Palmerini, C.A., Cantelmi, M.G., Minelli, A., Fini, C., Zampino, M. and Floridi, A. Determination of plasma serotonin by high-performance liquid chromatography with precolumn sample enrichment and fluorimetric detection. *J. Chromatogr.*, 1987, 417, 378–384.

128. Jenner, D.A. and Richards, J. Determination of cortisol and cortisone in urine using high-performance liquid chromatography with UV detection. *J. Pharm. Biomed. Anal.,* 1985, 3, 251–257.

129. Nielen, M., Brinkman, U. and Frei, R. Industrial wastewater analysis by liquid chromatography with precolumn technology and diode-array detection. *Anal. Chem.,* 1985, 57, 806–810.

130. Yamamoto, E., Kato, T., Mano, N. and Asakawa, N. Effective online extraction of drugs from plasma using a restricted–access media column in column-switching HPLC equipped with a dilution system: application to the simultaneous determination of ER-118585 and its metabolites in canine plasma. *J. Pharm. Biomed. Anal.,* 2009, 49, 1250–1255.

131. Yamamoto, E., Igarashi, H., Sato, Y., Kushida, I., Kato, T., Kajima, T. and Asakawa, N. Reliable online sample preparation of basic compounds from plasma using a reversed phase restricted access media in column-switching LC. *J. Pharm. Biomed. Anal.,* 2006, 42, 587–592.

132. Mislanova, C., Stefancova, A., Oravcova, J., Horecky, J., Trnovec, T. and Lindner, W. Direct high-performance liquid chromatographic determination of (R)- and (S)-propranolol in rat microdialysate using online column switching procedures. *J. Chromatogr. B,* 2000, 739, 151–161.

133. Preuss, R., Koch, H.M. and Angerer, J. Biological monitoring of the five major metabolites of di-(2-ethylhexyl)phthalate (DEHP) in human urine using column-switching liquid chromatography-tandem mass spectrometry. *J. Chromatogr. B,* 2005, 816, 269–280.

134. Greiner, C., Hiemke, C., Bader, W. and Haen, E. Determination of citalopram and escitalopram together with their active main metabolites desmethyl(es-)citalopram in human serum by column-switching high performance liquid chromatography (HPLC) and spectrophotometric detection. *J. Chromatogr. B,* 2007, 848, 391–394.

135. Greiner, C. and Haen, E. Development of a simple column-switching high-performance liquid chromatography (HPLC) method for rapid and simultaneous routine serum monitoring of lamotrigine, oxcarbazepine and 10-monohydroxycarbazepine (MHD). *J. Chromatogr. B,* 2007, 854, 338–344.

136. Hamase, K., Morikawa, A., Ohgusu, T., Lindner, W. and Zaitsu, K. Comprehensive analysis of branched aliphatic D-amino acids in mammals using an integrated multi-loop two-dimensional column-switching high-performance liquid chromatographic system combining reversed-phase and enantioselective columns. *J. Chromatogr. A,* 2007, 1143, 105–111.

137. Kenk, M., Greene, M., Lortie, M., Dekemp, R.A., Beanlands, R.S. and Dasilva, J.N. Use of a Column-switching high-performance liquid chromatography method to assess the presence of specific binding of (R)- and (S)-[(11)C]rolipram and their labeled metabolites to the phosphodiesterase-4 enzyme in rat plasma and tissues. *Nucl. Med. Biol.,* 2008, 35, 515–521.

138. Ye, X., Bishop, A.M., Needham, L.L. and Calafat, A.M. Automated online column-switching HPLC-MS/MS method with peak focusing for measuring parabens, triclosan, and other environmental phenols in human milk. *Anal. Chim. Acta,* 2008, 622, 150–156.

139. Ye, X., Kuklenyik, Z., Needham, L.L. and Calafat, A.M. Measuring environmental phenols and chlorinated organic chemicals in breast milk using automated online column-switching-high performance liquid chromatography-isotope dilution tandem mass spectrometry. *J. Chromatogr. B,* 2006, 831, 110–115.

140. Ye, X., Tao, L.J., Needham, L.L. and Calafat, A.M. Automated online column-switching HPLC-MS/MS method for measuring environmental phenols and parabens in serum. *Talanta,* 2008, 76, 865–871.

141. Ban, E., Maeng, J.E., Woo, J.S. and Kim, C.K. Sensitive column-switching high-performance liquid chromatography method for determination of propiverine in human plasma. *J. Chromatogr. B,* 2006, 831, 230–235.

142. Preuss, R. and Angerer, J. Simultaneous determination of 1- and 2-naphthol in human urine using online clean-up column-switching liquid chromatography-fluorescence detection. *J. Chromatogr. B,* 2004, 801, 307–316.

143. Champmartin, C., Simon, P., Delsaut, P., Dorotte, M. and Bianchi, B. Routine determination of benzo[a]pyrene at part-per-billion in complex industrial matrices by multi-dimensional liquid chromatography. *J. Chromatogr. A,* 2007, 1142, 164–171.

144. Wang, Q., Li, X., Dai, S., Ou, L., Sun, X., Zhu, B., Chen, F., Shang, M. and Song, H. Quantification of puerarin in plasma by online solid-phase extraction column switching liquid chromatography-tandem mass spectrometry and its applications to a pharmacokinetic study. *J. Chromatogr. B,* 2008, 863, 55–63.

145. Bayliss, M.A.J., Venn, R.F., Edgington, A.M., Webster, R. and Walker, D.K. Determination of a potent utokinase-type plasminogen activator, UK-356,202, in plasma at pg/ml levels using column-switching HPLC and fluorescence detection. *J. Chromatogr. B,* 2009, 877, 121–126.

146. Nakagawa, S., Suzuki, R., Yamazaki, R., Kusuhara, Y., Mitsumoto, S., Kobayashi, H., Shimoeda, S., Ohta, S. and Yamato, S. Determination of the antifungal agent voriconazole in human plasma using a simple column switching high-performance liquid chromatography and its application to a pharmacokinetic study. *Chem. Pharm. Bull.,* 2008, 56, 328–331.

147. Baltes, E., Coupez, R., Brouwers, L. and Gobert, J. Gas chromatographic method for the determination of cetirizine in plasma. *J. Chromatogr.,* 1988, 340, 149–155.

148. Moncrieff, J. Determination of cetirizine in serum using reversed-phase high-performance liquid chromatography with ultraviolet spectrophotometric detection. *J. Chromatogr.,* 1992, 583, 128–130.

149. Kim, C.-K., Yeon, K. J., Ban, E., Hyun, M.-J., Kim, J.-K., Kim, M.-K., Jin, S.-E. and Park, J.-S. Narrow-bore high performance liquid chromatographic method for the determination of cetirizine in human plasma using column switching. *J. Pharm. Biomed. Anal.,* 2005, 37, 603–609.

150. Song, Y.-K., Maeng, J.-E., Hwang, H.-R., Park, J.-S., Kim, B.-C., Kim, J.-K. and Kim, C.-K. Determination of glimepiride in human plasma using semi-microbore high performance liquid chromatography with column-switching. *J. Chromatogr. B,* 2004, 810, 143–149.

151. Choi, S.J., Kim, S.B., Lee, H.-Y., Na, D.H., Yoon, Y.S., Lee, S.S., Kim, J.H., Lee, K.C. and Lee, H.S. Column-switching high-performance liquid chromatographic determination of clarithromycin in human plasma with electrochemical detection. *Talanta,* 2001, 54, 377–382.

152. Emara, S., Saleh, G., Fathy, M. and Bakr, M.A. Chromatographic assay and pharmacokinetic studies of ptopofol in human serum. *Biomed. Chromatogr.,* 1999, 13, 299–303.

153. Rouini, M.R., Lotfolahi, A., Stewart, D.J., Molepo, J.M., Shirazi, F.H., Vergniol, J.C., Tomiak, E., Delorme, F., Vernillet, L., Giguere, M. and Goel, R. A rapid reversed phase high performance liquid chromatographic method for the determination of docetaxel (Taxoterer®) in human plasma using a column switching technique. *J. Pharm. Biomed. Anal.,* 1998, 17, 1243–1247.

154. Song, D. and Au, J.L.S. Isocratic high-performance liquid chromatographic assay of taxol in biological fluids and tissues using automated column switching. *J. Chromatogr. B,* 1995, 663, 337–344.

155. Richheimer, L.S., Tinnermeier, D.M. and Timmons, D.W. High-performance liquid chromatographic assay of taxol. *Anal. Chem.* 1992, 64, 2323–2326.

156. Jamis-Dow, C.A., Klecker, R.W., Sarosy, G., Reed, E. and Collins, J.M. Steady-state plasma concentrations and effects of taxol for a 250 mg/m2 dose in combination with granulocyte-colony stimulating factor in patients with ovarian cancer. *Cancer Chemother. Pharmacol.,* 1993, 33, 48–52.

157. Longnecker, S.M., Donehower, R.C., Cates, A.E., Chen, T.-L., Brundett, R.B., Grochow, L.B., Ettinger, D.S. and Colvin, M. High-performance liquid chromatographic assay for taxol in human plasma and urine and pharmacokinetics in a phase I trial. *Cancer Treat. Rep.,* 1987, 71, 53–59.

158. Rizzo, J., Riley, C., von Hoff, D., Kuhn, J. Phillips, J. and Brown, T. Analysis of anticancer drugs in biological fluids: determination of taxol with application to clinical pharmacokinetics. *J. Pharm. Biomed. Anal.,* 1990, 8, 159–164.

159. Sonnischsen, D.S., Hurwitz, C.A., Pratt, C.B., Shuster, J.J. and Relling, M.V. Saturable pharmacokinetics and paclitaxel pharmacodynamics in children with solid tumors. *J. Clin. Oncol.,* 1994, 12, 532–538.

160. Supko, J.G., Nair, R.V., Seiden, M.V. and Lu, H. Adaptation of solid phase extraction to an automated column switching method for online sample cleanup as the basis of a facile and sensitive high-performance liquid chromatographic assay for paclitaxel in human plasma. *J. Pharm. Biomed. Anal.,* 1999, 21, 1025–1036.

161. Wyss, R., Bucheli, F. and Hess, B. Determination of the arotinoid mofarotene in human, rat and dog plasma by high-performance liquid chromatography with automated column switching and ultraviolet detection. *J. Chromatogr. A,* 1996, 729, 315–322.

162. Matuszewski, B.K. in: Bijvoet, O., Fleish, H.A., Canfield, R.E. and Russel, G. Bisphosphonate on bones, Elsevier Science, Amsterdam, 1995, Chapter 16.

163. Vallano, P.T., Shugart, S.B., Kline, W.F., Woolf, E.J. and Matuszewski, B.K. Determination of risedronate in human urine by column-switching ion-pair high-performance liquid chromatography with ultraviolet detection. *J. Chromatogr. B,* 2003, 794, 23–33.

164. Cherti, N. Kinowski, J.-M., Lefrant, J.Y. and Bressolle, F. High-performance liquid chromatographic determination of cefepime in human plasma and in urine and dialysis fluid using a column-switching technique. *J. Chromatogr. B,* 2001, 754, 377–386.

165. Liu, S., Shinkai, N., Kakubari, I., Saitoh, H., Noguchi, K., Saitoh, T. and Yamauchi, H. Automated analysis of fluvoxamine in rat plasma using a column-switching system and ion-pair high-performance liquid chromatography. *Biomed. Chromatogr.,* 2008, 22, 1442–1449.

166. Jhee, O. H., Hong, J. W., Om, A. S., Lee, M. H., Lee, W. S., Shaw, L. M., Lee, J.-W. and Kang, J. S. Direct determination of verapamil in rat plasma by coupled column microbore-HPLC method. *J. Pharm. Biomed. Anal.,* 2005, 37, 405–410.

167. Yang, H.K., Kim, S.Y., Kim, J.S., Sah, H. and Lee, H.J. Application of column-switching HPLC method in evaluating pharmacokinetic parameters of zaltoprofen and its salt. *Biomed. Chromatogr.,* 2009, 23, 537–542.

168. Zhang, Y. and Zhang, Z.R. Simple determination of terbutaline in dog plasma by column-switching liquid chromatography. *J. Chromatogr. B,* 2004, 805, 211–214.

169. Baek, M., Rho, Y.-S. and Kim, D.-H. Column-switching high-performance liquid chromatographic assay for determination of asiaticoside in rat plasma and bile with ultraviolet absorbance detection. *J. Chromatogr. B,* 1999, 732, 357–363.

170. Farabollini, F., Porrini, S. and Dessi-Fulgherit, F. Perinatal exposure to estrogenic pollutant bisphenol A affects behavior in male and female rats. *Pharmacol. Biochem. Behav.,* 1999, 64, 687–694.

171. Matthews, J.B., Twomey, K. and Zacharewski, T.R. *In vitro* and *in vivo* interactions of bisphenol A and its metabolite, bisphenol A glucuronide, with estrogen receptors α and β. *Chem. Res. Toxicol.,* 2001, 14, 149–157.

172. Sun, Y., Nakashima, M.N., Takahashi, M. Kuroda, N. and Nakashima, K. Determination of bisphenol A in rat brain by microdialysis and column switching high-performance liquid chromatography with fluorescence detection. *Biomed. Chromatogr.*, 2002, 16, 319–326.

173. Fukushima, T., Misuhashi, S., Tomiya, M., Kawai, J., Hashimoto, K. and Toyo'oka, T. Determination of rat brain kynurenic acid by column-switching HPLC with fluorescence detection. *Biomed. Chromatogr.*, 2007, 21, 514–519.

174. Uno, T., Niioka, T., Hayakari, M., Sugawara, K. and Tateishi, T. Simultaneous determination of warfarin enantiomers and its metabolite in human plasma by column-switching high-performance liquid chromatography with chiral separation. *Ther. Drug Monit.*, 2007, 29, 333–339.

175. Barrett, A.M. and Cullum, V.A. The biological properties of the optical isomers of propranolol and their effects on cardiac arrythmias. *Br. J. Pharmacol.*, 1968, 1, 43–55.

176. Lennard, M.S., Tucker, G.T., Silas, J.H., Freestone, S., Ramsay, L.E. and Woods, H.F. Differential stereoselective metabolism of metoprolol in extensive and poor debrisoquin metabolizer. *Clin. Pharmacol. Ther.*, 1983, 34, 732–737.

177. Hsyu, P.H. and Giacomini, K.M. Stereoselective renal clearance of pindolol in humans. *J. Clin. Invest.*, 1985, 76, 1720–1726.

178. Wille, U.K., Walle, T., Bai, S.A. and Olanoff, L.S. Stereoselective binding of propranolol to human plasma, α_1-acid glycoprotein, and albumin. *Clin. Pharmacol. Ther.*, 1983, 34, 718–723.

179. Oh, J.W., Trung, T.Q., Sin, K.S., Kang, J.S. and Kim, K.H. Determination of bevantolol enantiomers in human plasma by coupled achiral-chiral high performance-liquid chromatography. *Chirality*, 2007, 19, 528–535.

180. Uno, T., Shimizu, M. Sugawara, K. and Tateishi, T. Sensitive determination of itraconazole and its active metabolite in human plasma by column-switching high-performance liquid chromatography with ultraviolet detection. *Ther. Drug Monit.*, ,2006, 526–531.

181. Shimizu, M., Uno, T., Niioka, T., Yaui-Furukori, N., Takahata, T., Sugawara, K. and Tateishi, T. Sensitive determination of omeprazole and its two main metabolites in human plasma by column-switching high-performance liquid chromatography: Application to pharmacokinetic study in relation to CYP2C19 genotypes. *J. Chromatogr. B*, 2006, 832, 241–248.

182. Cho, H.-Y., Park, C.-H. and Lee, Y.-B. Direct and simultaneous analysis of loxoprofen and its diastereometric alcohol metabolites in human serum by online column switching liquid chromatography and its application to a pharmacokinetic study. *J. Chromatogr. B*, 2006, 835, 27–34.

183. Yasui-Furukori, N., Inoue, Y., Kaneko, S. and Otani, K. Determination of fluvoxamine and its metabolite fluvoxamine acid by liquid-liquid extraction and column-switching high-performance liquid chromatography. *J. Pharm. Biomed. Anal.*, 2005, 37, 121–125.

184. Ohkubo, T., Shimoyama, R., Otani, K., Yoshida, K., Higuchi, H. and Shimizu, T. High-performance liquid chromatographic determination of fluvoxamine and fluvoxamino acid in human plasma. *Anal. Sci.*, 2003, 859–864.

185. Uno, T., Yasui-Furukori, N., Shimizu, M., Sugawara, K. and Tateishi, T. Determination of rabeprazole and its active metabolite, rebeprazole thioether in human plasma by column-switching high-performance liquid chromatography and its application to pharmacokinetic study. *J. Chromatogr. B*, 2005, 824, 238–243.

186. Mangani, F., Luck, G., Fraudeau, C. and Verette, E. Online column-switching high-performance liquid chromatography analysis of cardiovascular drugs in serum with automated sample clean-up and zone-cutting technique to perform chiral separation. *J. Chromatogr. B*, 1997, 762, 235–241.

187. Sagar, K.A. and Smyth, M.R. Bioavailability studies of oral dosage forms containing levodopa and carbidopa using column-switching chromatography followed by electrochemical detection. *Analyst*, 2000, 125, 439–445.

188. Ono, I., Matsuda, K. and Kanno, S. Determination of *N*-(*trans*-4-isopropylcyclohexanecarbonyl)-D-phenylalanine and its metabolites in human plasma and urine by column-switching high-performance liquid chromatography with ultraviolet detection. *J. Chromatogr. B*, 1997, 692, 397–404.

189. Kim, K.H., Kim, H.J., Kim, J.H. and Shin, S.D. Determination of terbutaline enetiomers in human urine by coupled achiral-chiral high-performance liquid chromatography with fluorescence detection. *J. Chromatogr. B*, 2001, 751, 68–77.

190. Um, S.Y., Kim, K.-B., Kim, S.H., Ju, Y.C., Lee, H.S., Oh, H.Y., Choi, K.H. and Chung, M.W. Determination of the active metabolites of sibutramine in rat serum using column-switching HPLC. *J. Sep. Sci*, 2008, 31, 2820–2826.

191. Yamamota, E., Sakaguchi, T., Kajima, T., Mano, N. and Asakawa, N. Novel methylcellulose-immobilized cationa-exchange precolumn for online enrichment of cationic drugs in plasma. *J. Chromatogr. B*, 2004, 807, 327–334.

192. Hattori, A., Fukushima, T., Hamamura, K., Kato, M. and Imai, K. A fluorimetric, column-switching HPLC and its application to an elimination study of LLU-α enantiomers in rat plasma. *Biomed. Chromatogr.*, 2001, 15, 95–99.

193. Fukushima, T., Kawai, J., Imai, K. and Toyo'oka, T. Simultaneous determination of D- and L-serine in rat brain microdialysis sample using a column-switching HPLC with fluorimetric detection. *Biomed. Chromatogr.*, 2004, 18, 813–819.

194. de Lange, E.C.M., Danhof, M., de Boer, A.G. and Breimer, D.D. Methodological considerations of intracerebral microdialysis in pharmacokinetic studies on drug transport across the blood-brain barrier. *Brain Res. Rev.*, 1997, 25, 27–49.

195. The Pharmacopoeia Commission of the People's Republic of China, in: Pharmacopoeia of PRC, Chemical Industry Press, Beijing, 2005, p. 189.

196. Chen, L., Yu, A., Zhuang, X., Zhang, K., Wang, X., Ding, L. and Zhang, H. Determination of andrographolide and dehydroanfrographolide in rabbit plasma by online solid phase extraction of high-performance liquid chromatography. *Talanta*, 2007, 74, 146–152.

197. Shargel, L., Wu-Pong, S. and Yu, A.B.C. Introduction to biopharmaceutics and pharmacokinetics, *Applied Biopharmaceutics & Pharmacokinetics*, 5ᵗʰ ed., McGraw-Hill, New York, 2005.

198. Gibaldi, M. and Perrier, D. Noncomparmental analysis based on statistical moment theory, *Pharmacokinetics*, 2nd ed., Dekker, New York, 1982.

8 Chromatographic Procedures in a Regulated Environment

Terrence P. Tougas

CONTENTS

8.1 INTRODUCTION

The focus of this chapter is on both scientific and regulatory aspects of developing, validating, transferring, and using chromatographic procedures in the highly regulated pharmaceutical environment. While it is always good scientific practice to understand the capabilities of the chemical measurements used in scientific pursuits, when chromatographic procedures are used in support of the development or distribution of commercial pharmaceuticals many regulatory expectations must also be considered. In large part there is overlap between the scientific and regulatory perspectives on method development, validation, and implementation. However, the scientist working in this environment should not only be a competent chromatographer but should also be aware of the regulatory and compliance aspects of working in the pharmaceutical environment.

8.2 APPLICATION OF CHROMATOGRAPHY IN PHARMACEUTICAL PRODUCT DEVELOPMENT AND COMMERCIALIZATION

Chromatographic methods are pivotal to the development and subsequent quality control of most pharmaceutical products. During the development phase, separation science supports numerous aspects of the development process such as designing the dosage form, elucidating degradation pathways, demonstrating safety and efficacy, characterizing the active pharmaceutical ingredient (API) and drug product, development of the API synthesis, and development of the drug product manufacturing process. Both during the clinical development phase and later in support of commercial distribution, chromatographic methods are key elements of the quality control strategy of pharmaceuticals. This latter role emphasizes the importance of understanding the performance characteristics of the specific chromatographic procedures used for these purposes, but some level of understanding of performance characteristics is important for all applications of chromatography within the pharmaceutical enterprise. Table 8.1 outlines some common uses of chromatography in pharmaceutical development.

8.2.1 REGULATORY ENVIRONMENT

Owing to the implications for public health, governmental regulatory oversight of the entire pharmaceutical enterprise exists worldwide. The development and use of analytical procedures is among the numerous aspects regulated. Due to the historically independent development of drug laws across most countries in the world, the modern pharmaceutical company must deal with myriad drug regulations to develop and market its drugs in a global economy. From both an economic and ethical perspective, most pharmaceutical companies develop products for worldwide registration. This requires adherence to this worldwide network of regulations that can be complex, confusing, and at times contradictory. Since its inception in 1990 the International Conference on Harmonization (ICH) has strived to harmonize drug regulation worldwide [1] through collaboration between industry and regulators from the three members regions: the United States, European Union, and Japan. It accomplishes

TABLE 8.1

Common Uses of Chromatography in Pharmaceutical Development (and Commercial Manufacturing)

API Related Uses	Drug Product Related Uses
Release and retest of API	Release of drug product
Assay	Assay
Process impurities	Content uniformity
Degradation products	Degradation products
Residual solvents	Dissolution
Stereochemical impurities	
Stability testing of API	Stability testing of drug product
Assay	Assay
Degradation product	Degradation product
	Dissolution
Release of API starting materials, reagents, critical intermediates	Release of drug product excipients
Development and optimization of API synthesis process	Development and optimization of drug product manufacturing process
Elucidation of structure of the API	Formulation development and optimization
Identification of API impurities and DP degradation products (GC-MS, LC-MS)	
Characterization of test article (API or toxicology formulation) for toxicology studies	
Drug metabolism and pharmacokinetics studies	

this mission primarily through the generation of regulatory guidelines in three general subject areas (quality, safety, and efficacy) that are recognized officially by the regulatory bodies of the member regions. This effort has made significant progress in simplifying drug regulation, but the worldwide regulatory environment is still a very complex web of laws and expectation.

The ICH guidelines of most interest to chromatographers are those in the quality section. This section encompasses topics related to chemical and pharmaceutical quality assurance (i.e., Module 3 of the ICH Common Technical Dossier [2]) or what is termed in the United States as chemistry, manufacturing, and controls (CMC). Of particular note for this discussion is the ICH Q2 (R1) guidance on method validation [3], which will be discussed in more detail.

It is insightful for this discussion to have a general understanding of how drugs are regulated. While specific regulations differ country to country, the approaches are similar. As an example, consider the U.S. regulatory environment. The overarching legislation in the United States is the Federal Food, Drug and Cosmetic Act (FD&C Act) [4]. This law and related legislation have a rich history that grew originally in response to fraudulent activities by food and drug producers [5]. It is interesting to note that the current emergence of China both with respect to pharmaceutical manufacture and regulation has some startling similarities to the emergence of the U.S. drug regulation (i.e., scandals provoking a public outcry for better regulatory oversight) [6].

In the United States, the drug laws are enforced within the executive branch of government by the Food and Drug Administration (FDA), which in turn is part of Health and Human Services (HHS). The FD&C Act is by its nature a broad and overarching statement of the requirements for developing and marketing pharmaceuticals. More specific expectations are established by FDA through executive rules in the "Code of Federal Regulations" (CFR). The CFR includes all executive rules, with Title 21 specifically reserved for drugs. As a pertinent example of the relationship between the FD&C Act and CFR Title 21, the FD&C Act states:

> A drug or device shall be deemed to be adulterated … if it is a drug and the methods used in, or the facilities or controls used for, its manufacture, processing, packing, or holding do not conform to or are not operated or administered in conformity with current good manufacturing practice.

The so-called current good manufacturing practices (cGMP) are specified in more detail (including requirements for analytical testing) in CFR Title 21 sections 210 [7] and 211 [8].

Besides legislation and executive rules, U.S. regulatory authorities use a number of other tools to set regulatory expectations. The FDA maintains a series of guidance documents on various topics that represent the agency's current thoughts or expectations on a particular topic. The creation and maintenance of these guidance documents are governed by the FDA's good guidance practices [9,10]. While not absolute requirements, there is an expectation of adherence unless acceptable alternatives are proposed. It is usually a good practice to discuss alternative proposals with the FDA prior to submission to avoid unnecessary time delays.

The FDA also sets regulatory expectations through internal policy and inspection guides. Although these are intended as guides to FDA personnel, they are public documents and are used by industry to gain insight into FDA expectations. Another manner in which the FDA sets regulatory expectations is through its review and inspectional practices. Review comments in a product submission or citations connected with a facility inspection allows the FDA to focus on current hot topics and to create awareness within the industry of its concerns. Finally, FDA personnel frequently present at public scientific forums on topics of interest to the agency. While they often have disclaimers indicating that the content should not be construed as official policy, these presentations nonetheless provide insights into the current thinking of agency personnel.

Another element in the pharmaceutical regulatory environment is pharmacopeial standards, which are compilations of officially recognized analytical procedures and other standards. Worldwide there is again a host of country or regional pharmacopeia that offer another opportunity for international harmonization. Though the industry must consider many pharmacopeia, the most prominent are the U.S. Pharmacopeia (USP), European Pharmacopeia (Ph.Eur.), and the Japanese Pharmacopeia (JP).

With the advent of the Internet, extensive information about the regulatory expectations of the various worldwide regulatory authorities are now readily available through actively maintained Web sites. In addition, most of the pharmacopeias are available in online editions (Table 8.2).

TABLE 8.2

Online Resources Related to Regulatory Expectations for Pharmaceuticals

Organization	Country or Region	Web Site
Regulatory Authorities		
Food and Drug Administration (FDA)	United States	http://www.fda.gov/
European Medicines Authority (EMA, formerly EMEA)	European Union	http://www.ema.europa.eu/
Ministry of Health, Labour and Welfare	Japan	http://www.mhlw.go.jp/english/index.html
Pharmaceuticals and Medical Devices Agency	Japan	http://www.pmda.go.jp/english/index.html
State Food and Drug Administration (SFDA)	China	http://eng.sfda.gov.cn/cmsweb/webportal/w43879537/index.html
Therapeutic Goods Administration	Australia	http://www.tga.gov.au/index.htm
Health Canada	Canada	http://www.hc-sc.gc.ca/index-eng.php
Medicines and Healthcare products Regulatory Agency (MHRA)	United Kingdom	http://www.mhra.gov.uk/index.htm
Pharmacopeia		
United States Pharmacopeia	United States	http://www.usp.org/
European Pharmacopeia	European Union	http://online.edqm.eu/entry.htm
Japanese Pharmacopeia	Japan	http://jpdb.nihs.go.jp/jp15e/
British Pharmacopeia	Great Britain	http://www.pharmacopoeia.co.uk/
Other		
International Conference on Harmonisation of Technical Requirements for Registration of Pharmaceuticals for Human Use (ICH)	United States, European Union, Japan	http://www.ich.org/

8.3 VALIDATION OF ANALYTICAL PROCEDURES

The ICH Q2 (R1) Guideline [3] defines validation through the following statement:

> The objective of validation of an analytical procedure is to demonstrate that it is suitable for its intended purpose.

An FDA draft guideline on method validation [11] contains a similar statement and the added expectation that systematic studies are conducted and documented

> Methods validation is the process of demonstrating that analytical procedures are suitable for their intended use. The methods validation process for analytical procedures begins with the planned and systematic collection by the applicant of the validation data to support the analytical procedures.

The common theme in the previous examples and found throughout many other regulatory sources is that method validation involves some set of designed experiments intended to demonstrate that a specific analytical procedure is capable of producing reliable results with respect to whatever chemical or physical measurement the procedure was designed to measure. One consequence of these definitions is that it is fundamental to define the nature and requirements of the intended measurement. Without this starting point it is impossible to design the subsequent validation studies and criteria by which the suitability of the procedure will be judged.

It is important to recognize that, while the focus here is on the validation of chromatographic procedures, it is only one element of a broader quality system that ensures the reliability of analytical results coming from a particular laboratory, operation, or individual procedure. Besides knowing that a particular procedure is capable of producing reliable and meaningful results, elements need to be in place that assure the quality of all results obtained. This means that at a minimum the following are addressed through a formal quality system that governs operations within the particular laboratory:

- Instrumentation is maintained and periodically calibrated as appropriate (e.g., for an HPLC system periodic calibration of mobile phase flow rates, gradient accuracy, column oven temperatures, detector wavelength, lamp intensity might be performed).
- New instrumentation undergoes installation and operational qualifications.
- Periodic performance qualifications are conducted on all critical equipment.
- Critical vendors (e.g., instrumentation, software) are qualified through an auditing program.
- Software used to obtain or store chromatographic results should be validated (for chromatographers the primary software is typically the Laboratory Data System [LDS] used to acquire and process chromatographic data and the Laboratory Information Management System (LIMS) used to archive chromatographic results).
- Analysts should be properly trained prior to generating analytical results.

The emphasis here is on a "formal quality system." By this it is meant that the aforementioned practices are governed by standard operating procedures, the adherence to these practices is assured through an independent auditing program, and all aspects are well documented. Again, both good scientific practice and regulatory expectations drive this formality when chromatographic results are used for critical purposes associated with pharmaceuticals impacting human health. For a more in-depth discussion on this topic see the text by Swartz and Krull [12].

8.3.1 REGULATORY REQUIREMENT TO VALIDATE ANALYTICAL PROCEDURES

While there is probably general agreement among pharmaceutical scientists about the scientific value of validating a chromatographic procedure intended for repeated use by multiple laboratories, it important to understand the regulatory mandates with respect to method validation. These mandates are very clear with respect to analytical procedures used to test commercial products, but the clarity diminishes as one moves back through the life cycle of a pharmaceutical into early development.

In the European Union, the principles and guidelines describing Good Manufacturing Practices are specified in Commission Directives 2003/94/EC, of October 8, 2003; 91/356/EC of June 13, 1991; and 91/412/EEC of July 23, 1991. The requirements pertaining to method validation can be found in the "EU Guidelines to Good Manufacturing Practice Medicinal Products for Human and Veterinary Use" [13]. The specific reference for drug products (commercial or clinical supplies) is as follows:

> § 6.15 Analytical methods should be validated. All testing operations described in the marketing authorization should be carried out according to the approved methods.

In addition, Annex 18 of the EU GMP Guideline contains method validation requirements pertaining to the API and starting materials:

> Annex 18, §12.80 Analytical methods should be validated unless the method employed is included in the relevant pharmacopoeia or other recognized standard reference. The suitability of all testing methods used should nonetheless be verified under actual conditions of use and documented.

> Annex 18, §12.81 Methods should be validated to include consideration of characteristics included within the ICH guidelines on validation of analytical methods. The degree of analytical validation performed should reflect the purpose of the analysis and the stage of the API production process.

The U.S. requirements are discussed in the corresponding U.S. cGMP (i.e., 21 CFR § 210 and § 211). The specific citations are as follows:

> § 211.165 (e) The accuracy, sensitivity, specificity, and reproducibility of test methods employed by the firm shall be established and documented. Such validation and documentation may be accomplished in accordance with § 211.194 (a)(2)

> § 211.194 (a)(2) A statement of each method used in the testing of the sample. The statement shall indicate the location of data that establish that the methods used in the testing of the sample meet proper standards of accuracy and reliability as applied to the product tested. (If the method employed is in the current revision of the United States Pharmacopeia, National Formulary, AOAC INTERNATIONAL, Book of Methods [1], or in other recognized standard references, or is detailed in an approved new drug application and the referenced method is not modified, a statement indicating the method and reference will suffice). The suitability of all testing methods used shall be verified under actual conditions of use.

8.3.2 REGULATORY GUIDELINES ON METHOD VALIDATION

In support of these mandates to perform method validation, regulatory bodies and other consortia (notably ICH) have published various guidelines that provide additional detail on conducting method validation and topics related to method validation. A listing of some of these guidelines follows.

8.3.2.1   International Conference on Harmonization (ICH)

ICH Q2(R1), Validation of Analytical Procedures: Text and Methodology
ICH Q3A(R2), Impurities in New Drug Substances
ICH Q3B(R2), Impurities in New Drug Products
ICH Q3C(R4), Impurities: Guideline for Residual Solvents
ICH Q7, Good Manufacturing Practice Guide for Active Pharmaceutical
  Ingredients

- Online access: http://www.ich.org/cache/compo/276-254-1.html

8.3.2.2   FDA Guidance Documents

Draft Guidance for Industry, Analytical Procedures and Methods Validation,
  August 2000
Reviewer Guidance, Validation of Chromatographic Methods, November 1994
Guideline for Submitting Samples and Analytical Data for Methods Validation,
  February 1987
Guidance for Industry, Current Good Manufacturing Practices for Phase 1
  Investigational Drugs, July 2008
Guidance for Industry, INDs for Phase 2 and 3 Studies, May 2003

- Online access:
    http://www.fda.gov/Drugs/GuidanceComplianceRegulatoryInformation/
    Guidances/ucm064979.htm

8.3.2.3   Other FDA Documents

Guide To Inspections of Pharmaceutical Quality Control Laboratories, July 1993

- Online access: http://www.fda.gov/ICECI/Inspections/InspectionGuides/
    ucm074918.htm

ORA Method Development and Validation Program, Doc. No. III-08, Ver. 23,
  revised Dec. 3, 2008

- Online access:
    http://www.fda.gov/downloads/ScienceResearch/FieldScience/
    UCM092187.pdf

8.3.2.4   Pharmacopeia

Validation of Compendial Procedures, USP 32, <1225>
Chromatography, USP 32, <621>
Analytical Instrument Qualification, USP 32, <1058>
The Dissolution Procedure: Development and Validation, USP 32, <1092>
Validation of Analytical Procedures, Japanese Pharmacopeia XV, <30>, pg. 1760

8.3.3 Designing and Executing the Validation of a Chromatographic Method

The exercise of validating an analytical procedure includes designing a series of planned experiments to characterize important attributes of the procedure and verify that the procedure can be expected to perform to the selected criteria. Much has been written on which methods and to what extent relative to the pharmaceutical product life cycle methods should be validated [14,15]. Many favor a staged approach during product development that culminates with the full validation recommended in ICH Q2 by the time the so-called registration batches are subject to testing. Registration batches are those pivotal batches (API and drug product) that are fully representative of the proposed commercial product and process. Extensive information about these batches is submitted in marketing applications.

The rationale for a phased approach is that it is an optimum use of resources versus the demands on the analytical procedure. Procedures used during early development are frequently revised and have limited use, and the laboratory needs time to gain experience before realistic acceptance criteria can be established.

Even in early development, when the analytical procedure is intended for the analysis of samples falling under the cGMP, it is a regulatory expectation that there is some documented experimental evidence that the procedure is suitable for its intended purpose. At the other end of the spectrum, full validations supporting a marketing authorization application (MAA) are expected to be executed in accordance with ICH Q2 under a written protocol with predefined acceptance criteria. While there is not universal consensus on the details of phased validations, most pharmaceutical firms have a specific articulated quality philosophy on this topic that is at least consistently followed within the particular company.

The potential attributes to be considered in validating a chromatographic procedure are enumerated in the ICH Q2(R1) guideline [3] and include the following:

Method Validation Attributes	
Specificity (selectivity)	Intermediate precision
Range	Reproducibility (ruggedness)
Linearity	Detection limit
Accuracy	Quantitation limit
Precision	Robustness[a]
Repeatability	System suitability testing[a]

[a] Considered by ICH in the discussion of method validation attributes but in reality not specific quality attributes of a chromatographic procedure.

What follows is an item-by-item discussion of each of the elements recommended in ICH Q2 with respect to full method validation of analytical procedures.

8.3.3.1   Specificity

ICH [3] defines specificity as follows: "Specificity is the ability to assess unequivo-cally the analyte in the presence of components which may be expected to be pres-ent. Typically these might include impurities, degradants, matrix, etc."

Chromatographic methods generally fall into three categories: identification, assay, or impurity profile methods. In the case of identification and assays, dem-onstrating specificity translates into demonstrating that other potential components in the sample do not coelute or interfere with the sought after constituent. Impurity procedures have the additional challenge of ensuring that all potential impurities or degradation products are separated. It is not uncommon for multiple separations to be required when the impurity profile is complex.

In the case of API methods, consideration must be given to potential impurities that arise due to the synthesis process (e.g., carryover of starting materials or inter-mediates, undesired side reactions) and potential degradation products. It is common to conduct experiments to stress the API (heat, light, pH, and oxidant) in an effort to identify the major decomposition pathways [16,17]. This information is used to guide the selection of potential degradation products to be considered in method development and validation. The actual specificity experiment is then a demonstra-tion of the separation of all these potential impurities (impurity profile methods) or in the case of identification and assay methods the lack of interference with the API peak. Note that diode array detectors are often employed to demonstrate spectral purity of the API peak in assay methods.

8.3.3.2   Range

The range of an analytical procedure is in itself not a validation attribute but rather a statement of the range of concentrations or amounts over which one can expect reliable results. This is derived from demonstrating an appropriate level of accuracy, precision, and linearity over this range and therefore must be taken into consider-ation in designing the experiments related to these three validation attributes. ICH Q2A (R1) discusses recommended ranges by class of method.

8.3.3.3   Linearity

Per ICH Q2 (R1), "The linearity of an analytical procedure is its ability (within a given range) to obtain test results which are directly proportional to the concentration (amount) of analyte in the sample" [3]. The ICH recommended approach to evaluating linearity is through an examination of a plot of signal versus concentration or content at a minimum of five different levels. The assumption is that if the analytical signal is linear then the test result must also be linearly related to the true amount (concen-tration). There are allowances for the special cases where the analytical signal is by design not linearly related to amount, but through some other nonlinear relationship.

For the typical chromatographic procedure, this is generally interpreted to mean the response of standard preparations prepared at various concentrations versus the known concentration. Some caution should be exercised though since this approach does not account for any nonlinear response that might be introduced by the sample matrix.

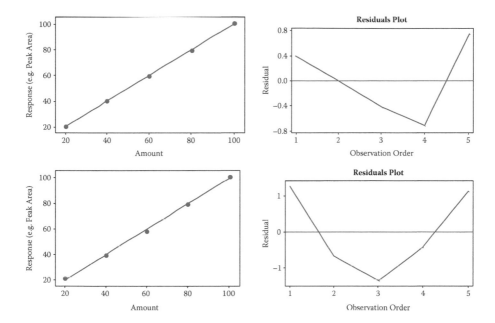

FIGURE 8.1 Example residual plots demonstrating the limited value of constructing such plots for typical method validation results.

The recommended statistical analysis is to perform least squares regression analysis and to report estimates of the slope, y-intercept, correlation coefficient, and residual sum of squares. Optionally, residual analysis is recommended. In practice the latter is of limited value since the numbers of levels typically employed are too few to reliably detect systematic deviations from linearity through residual plots. This is illustrated in Figure 8.1, where a large set of randomly distributed values about a straight line were randomly grouped in sets of five (50 sets of 5). Subsequently linear regression analysis including residual analysis was performed. Since all values are normally distributed about the resulting regression lines, there are no systematic trends, and the corresponding residual plots should not indicate any systematic deviations. The result was an unacceptable number of residual plots that demonstrate an apparent pattern (11 of 50). Two examples of these instances where an apparent residual pattern was observed are illustrated in Figure 8.1.

The correlation coefficient as a goodness-of-fit statistic is frequently the attribute specified in validation protocols. However, care should be taken interpreting this statistic. Mathematically this statistic is the square root of the ratio of the explained sum of squares to the total sum of squares. As such, the fit to a linear model improves as this statistic approaches one, with a value of exactly one indicating perfect correlation with a straight line. Applied to the linear response of a chromatographic procedure, values very close to one are expected, and it is probably best to judge the quality of the fit based on the "number of nines" (i.e., 0.99x, 0.999x, 0.999x). This is illustrated in Figure 8.2 where model calibration curves were generated assuming different variabilities about the linear regression.

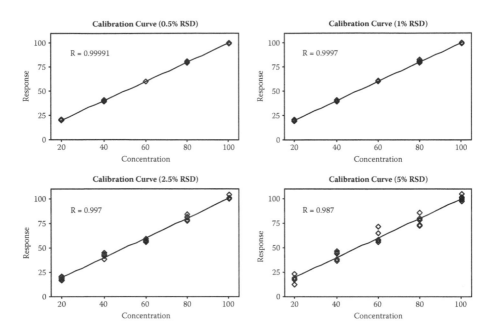

FIGURE 8.2 Influence of increasing variability on the correlation coefficient. The assumed RSD is relative to the highest value (100). Five random normal values at each of the five levels were generated and linear regression analysis was performed on the resulting model calibration curves.

In practice, a correlation coefficient of > 0.99 indicates an acceptable goodness of fit for many common chromatographic procedures. Note that this statistic is unable to diagnose the nature of lack of fit, that is, systematic deviation from linearity versus the degree of random variability about the regression line. This is illustrated in Figure 8.3 where the same correlation coefficient is calculated for both plots, but there is obvious difference in the nature of the lack of fit.

8.3.3.4 Accuracy

The accuracy of a procedure reflects the degree of closeness of the analytical result to the true value. Since the latter is rarely known, measures of accuracy rely on accepted values or well-characterized standards. The typical measure of accuracy for chromatographic procedures is the percent recovery of a standard spiked into the sample matrix. Alternatively, this can be expressed as method bias, that is, the lack of accuracy either in an absolute or relative sense.

ICH [3] recommends assessing accuracy over the operating range of the procedure. In the case of API, this can be accomplished in the following ways:

- Via comparison of results with a second procedure of known accuracy
- By testing of a material of known purity (assay)
- Via inference once specificity, linearity, and precision are established

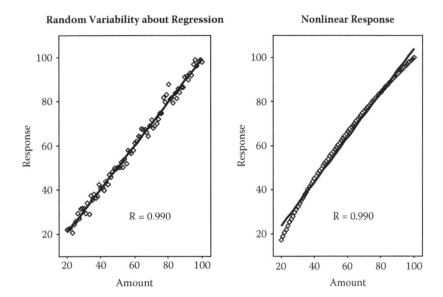

FIGURE 8.3 Limitations of correlation coefficients: inability to distinguish between lack of fit due to the magnitude of the random variability versus some systematic deviation from linearity.

For drug products, accuracy should at least be established by testing samples to which known amounts of analytes have been added. The minimum recommended data is nine determinations spread over three levels—that is, three samples at each of three different concentrations covering the range of the procedure.

8.3.3.5 Precision

The precision of an analytical procedure is the closeness of agreement (degree of scatter) among replicate measurements obtained from multiple sampling of the same homogeneous sample under the prescribed conditions. Traditionally, precision has been subdivided into repeatability and reproducibility [18]. The former is the precision when the entire measurement system (e.g., analyst, instrument, column) is unchanged, and the latter applies when elements are changed. ICH Q2 has altered these definitions to introduce the concept of intermediate precision.

Under ICH, repeatability remains the precision under the same operating conditions over a short interval of time. Intermediate precision reflects reproducibility within laboratories (e.g., variations across different days, different analysts, different equipment). Reproducibility is reserved for estimating the precision among different laboratories (i.e., integral to method transfer).

From a regulatory perspective, the expectation is that results characterizing repeatability and intermediate precision are part of the marketing application supporting the validation of a particular procedure. Method transfer studies (estimating reproducibility) are expected but are typically not included in the marketing application. In the United

States, method transfer studies are subject to review by the FDA during site inspection (i.e., during the preapproval inspection or during a general GMP inspection).

The ICH recommended experimental design for repeatability is to obtain either 9 determinations at three different levels (i.e., three replicates at three different concentrations) or six determinations at 100% test concentration.

Intermediate precision is studied through a designed experiment that considers the major factors pertinent to the specific procedure and the method of analysis of the data (typically analysis of variance [ANOVA]). For a chromatographic procedure, the factors typically considered include analyst, column, day, and chromatographic system.

Caution should be exercised interpreting the results of precision studies. For all precision studies, ICH recommends the following: "The standard deviation, relative standard deviation (coefficient of variation) and confidence interval should be reported for each type of precision investigated." First, repeatability results estimate the ultimate capability of a procedure not the actual performance. Repeatability determinations do not include many indeterminate sources of variability that are difficult to eliminate or control. As such, they are an overly optimistic view of the precision that can be achieved on an ongoing basis. The second consideration rests with the inherent statistical power of the recommended designs. In the case of repeatability, the point estimates of the standard deviations are obtained with either three or six replicates depending on which design is selected. The precision of these estimates of precision is inherently limited by these relatively small numbers of replicates. This arises from the nature of the statistical distributions of sample standard deviations associated with a random normal variable (asymmetrical and a function of sample size).

To put this in perspective, consider the following thought experiment. Consider a hypothetical assay procedure for a 100 mg tablet that has a true standard deviation of 0.5 mg (rsd 0.5%). One hundred sets of n = 3 and n = 6 values randomly distributed ($\mu = 100$, $\sigma = 0.5$) were generated. The 100 sets of six values had estimated standard deviations that ranged from 0.15 to 0.89 mg. The 95% confidence interval about the highest estimate (0.89 mg) was 0.56–2.19 mg. In the case of sets of three values, the 100 estimated standard deviations ranged from 0.04 to 1.08 mg, with the 95% confidence interval about the highest estimate (1.08 mg) of 0.56–6.78. While this one-time experiment may provide some rough estimation of the precision of the particular procedure and satisfies a regulatory expectation, a better estimate could potentially be obtained by building a history of results obtained during the drug development phase.

8.3.3.6   Detection Limit/Quantitation Limit

The detection and quantitation limits of a procedure are closely related validation attributes. The quantitation limit (QL) of an individual analytical procedure is the lowest amount (or concentration) of an analyte in a sample that can be quantitatively determined with suitable precision and accuracy. The quantitation limit is an attribute of quantitative assays for low levels of compounds in sample matrices, and is used particularly for the determination of impurities or degradation products. The detection limit (DL) is the lowest amount (or concentration) of analyte in a sample that can be detected but not necessarily quantitated as an exact value. Both are defined in terms of an amount (or concentration) associated with a user selected signal-to-noise ratio (S/N). Commonly accepted S/N values for pharmaceutical applications are 10 and 3 for QL and DL, respectively. Note, however, that these values may not be applicable in all cases, particularly for ultra-trace methods.

ICH Q2 describes three approaches based on visual examination of the response, deriving signal and noise from a calibration curve and the direct estimation of the baseline noise and signal. While discussions of QL and DL almost always refer to noise and S/N, it is important to understand that noise in this context is synonymous with precision at low levels and is estimated as a standard deviation. With this in mind, it becomes apparent that the same cautions discussed with respect to the estimation of method precision also apply to QL and DL. This leads to the conclusion that the result of any one-time estimation of QL or DL is inherently imprecise. In practice, caution should be exercised interpreting results that fall close to such an estimated QL. In some cases, practitioners claim a more conservative value of QL for regulatory purposes to compensate for this uncertainty. This may be acceptable to regulatory authorities but should be made transparent in the regulatory submissions and communications.

Again, as suggested with respect to precision, multiple estimations of QL and DL obtained over the course of using a chromatographic procedure during the development phase could be an alternative strategy that would lead to a better understanding of these method attributes. In the author's laboratory, this approach has been implemented based on the response of standard solutions used in impurity/degradation methods according to the following:

$$q = kC(s/X)$$

where
   q = DL or QL expressed as a percent
   k = 3 for DL, 10 for QL
   C = Concentration of the low-level standard (typically expressed as a percent relative to the API)
   X = Mean of the areas of the replicate injections of the standard
   s = the standard deviation of the areas of the replicate injections of the standard

This approach has two advantages. First, a better overall estimate of either DL or QL can be obtained as more experience with the particular procedure is gained. Second, this allows an estimation of QL or DL every time the procedure is executed, which may be used as a system suitability criterion.

8.3.3.7 Robustness
USP 32 <1225> defines robustness as follows:

The robustness of an analytical procedure is a measure of its capacity to remain unaffected by small but deliberate variations in procedural parameters listed in the procedure documentation and provides an indication of its suitability during normal usage. Robustness may be determined during development of the analytical procedure.

ICH Q2 recommends that robustness be evaluated during the development phase. The overall goal is to understand which method parameters are critical to method performance and to what degree these experimental conditions need to be controlled to get valid results. The end result is not the estimation of a specific method attribute but rather knowledge about the operating characteristics of the particular chromatographic procedure.

The first step in conducting an evaluation of method robustness is to select the parameters to study. The selection is based on the experience of the method developer with respect to the particular procedure and general knowledge of the type of chromatography. For a typical HPLC procedure this might include the following:

- Sample extraction time and conditions
- Solution or mobile phase pH and composition
- Column age or batch
- Instrumental parameters including flow rate, column temperature, detector wavelength
- Stability of analytical solutions and standards

Once the method parameters are selected, a designed experiment is conducted varying these parameters and measuring the impact on the response of the procedure. Besides gaining insight into the influence of the method parameters, robustness results are part of the basis for selecting system suitability criteria, which in turn ensure at time of use that accurate and reliable results are being obtained.

There are some clear parallels between method robustness studies and method validation overall with the broader topic of quality by design advocated by the FDA and many in industry for pharmaceutical product development. An analytical procedure can be viewed as a process producing a product (an analytical result) analogous to a pharmaceutical manufacturing process that produces a drug product. The concepts of risk assessment and management, evaluating critical process parameters, and quality attributes to gain knowledge of a process or product design space could in principle be applied to method development and validation similar to what is proposed for drug development [19,20].

8.3.3.8 System Suitability

As the name implies, system suitability requirements are intended to verify that the entire analytical system on which the validated chromatographic procedure is being executed is exhibiting acceptable performance at the time of use.

In practice, whenever a chromatographic procedure is executed, results on check samples should be obtained that meet a series of predefined criteria. If these are properly designed, this exercise gives assurance that the entire analytical system (instrumentation, column, mobile phase, sample work-up) is working properly. For chromatographic procedures it is common to set criteria related to the quality of the separation, the precision, and the ability to detect low levels of analyte (applies primarily to impurity and degradation product determinations). The selection of the parameters to be checked and the specific requirements should come from the knowledge gained during the development and validation of the procedure.

Concerning the quality of the separation, a standard preparation containing critical components is often examined for its separation properties. Common parameters examined include resolution between critical pairs of components, tailing factors, plate count, or retention time windows for critical components. To some extent, setting these criteria can be based on the experience from robustness experiments examining different batches and ages of columns. Alternatively or in combination, modeling the loss of separation efficiency can also be a useful and accepted means of establishing these types of system suitability criteria.

Precision is often evaluated by repeated injections of the standard solution and comparison of the resulting estimate of repeatability (rsd of peak area or response factor) to some upper limits (see the section on system suitability within USP 32 <621> Chromatography). Detectability is commonly established through the repeatability of a low concentration standard solution or as the ability to observe a standard solution at a concentration near the procedure QL.

Properly designed system suitability checks are invaluable in detecting many common laboratory errors or system malfunctions and in preventing the reporting of erroneous results. In this context, it is important to understand the regulatory and compliance implications of a failure to meet system suitability criteria.

The failure to meet some system suitability requirement is a signal that there is something wrong with the analytical system at that specific time of use. In some instances, system suitability results can be evaluated prior to obtaining results on actual samples. Where this is the case, the execution of the procedure should be aborted before running the actual samples. With many automated procedures, data on actual samples may be obtained before the system suitability results are evaluated for a particular run. In addition it is common to schedule system suitability samples throughout a particular chromatographic run. In these latter instances, any testing results connected with a system suitability failure should be considered invalid. This means such results have no validity with respect to judging the quality of the drug product, API, or material being tested; that is, they should not be used to judge whether something conforms to specification or is out of specification.

There is a compliance expectation that if the aforementioned scenarios occur, all information, data, and results should be properly documented. Further, a system suitability failure should be investigated for an assignable cause, and corrective action should be taken prior to attempting reanalysis of the original samples. Isolated system suitability failures are probably not of a compliance concern; however, repeated failures may be viewed by authorities as symptomatic of more significant issues. A pattern of failures might lead an inspector to question the training and qualifications of analysts, the suitability of the laboratory facilities, or the suitability of the analytical procedure in question.

Both from a compliance and scientific perspective, establishing and using system suitability requirements appropriately are essential to producing high-quality analytical results. Care should be exercised in establishing an appropriate set of criteria for a particular procedure. Quality systems should be in place that deal effectively with signals arising from failures to meet system suitability criteria.

8.4   METHOD TRANSFER

Within a pharmaceutical product's life cycle, it is often necessary to establish an analytical procedure at another site or laboratory for execution. This typically involves the transfer of the procedure and knowledge about the procedure from one site or laboratory currently competent to execute the procedure to the new laboratory. Common examples include the following:

- Transfer from the original laboratory where the procedure was developed and validated to a quality control (QC) laboratory
- Transfer to an external contract research organization (CRO)
- Transfer among multiple QC sites

This perspective on establishing an analytical procedure at a new site leads to the common term for this process: method transfer. In a holistic sense, method transfer can be defined as an exchange of technical knowledge about an analytical procedure and a systematic process that assesses the ability of additional laboratories or sites to execute the procedure and produce reliable results. Thus, there are two distinct phases to successful method transfers. First, the originating (source) laboratory needs to provide the destination laboratory with enough information concerning the new procedure to enable the destination laboratory to execute the procedure. Second, there should be some data-based exercise that can be used to objectively establish the success of the transfer.

From a scientific or good laboratory practices perspective, it seems logical to conduct this type of two-step process. General guidance concerning method transfer can be found in a number of published sources. The International Society for Pharmaceutical Engineering (ISPE) has issued a good practice guide that extensively covers the concepts and best practices associated with the transfer of analytical procedures [21]. In addition, PhRMA members of the former Analytical Technology Group published in 2002 an acceptable analytical practices paper covering general concepts of analytical method transfer [22].

Beyond a scientific rationale, health authorities expect that some process is conducted and documented that establishes that a particular laboratory is qualified to execute a specific analytical procedure. In practice, these studies are typically part of the technology transfer that occurs between the development site and the commercial site of manufacture or a contract laboratory facility.

However, little regulatory guidance exists on the specific experimental design of these transfer studies. It is expected that the studies are conducted under protocols containing specific acceptance criteria agreed to by both the source and destination laboratories prior to the execution of the study. The results of transfer studies should be well documented and available for inspection by health authorities during facilities inspections but are not generally submitted as part of an MAA. A further expectation is that the organization should have a consistent approach to method transfer, that is, a common quality philosophy articulated formally in the organization's quality system.

These formal method transfer studies should be viewed as a demonstration of the ability of a laboratory to execute a particular chromatographic procedure but are only one element in providing assurance of the reliability of results generated by a laboratory. Several additional requirements are specified in both the United States [7,8] and the European Union (EU) cGMPs [13] pertaining to overall laboratory operations. Deficiencies in any of these elements will cast uncertainty on the reliability of all results generated by a laboratory. These can be expected to the subject of site inspections by health authorities and include the following:

- Adequate facilities suitable for the type of testing conducted
- Competent trained staff
- A calibration program for all critical instruments, apparatus, gauges, and recording devices
- Written procedures, specifications, and sampling plans approved by the quality control unit
- Complete written documentation of all data and information used to generate results

Often the focus of method transfer discussions is on the designed experiment intended to demonstrate suitable performance by the laboratory to be qualified on the chromatographic procedure. While this is an important aspect and contains many technical nuances, other preliminary activities are critical to successfully transferring a procedure to a new laboratory.

First, the procedure to be transferred should be mature and well characterized. The source laboratory should have a body of knowledge about the performance of the procedure that indicates the ruggedness of the procedure and can be used to develop objective criteria to judge the success of interlaboratory studies. Premature attempts to transfer a procedure in most cases turn into an expensive time-consuming exercise in method development (i.e., uncovering and fixing issues that should have been addressed during the method development phase).

Within the pharmaceutical enterprise, the common scenario is that the development organization is developing and transferring procedures to a quality control laboratory responsible for testing of the commercial product. In all cases, it is advantageous to transfer information about procedures to be transferred well in advance of the interlaboratory studies. In the previous scenario, this includes all the way back to the method development phase where design considerations can be developed with the ultimate end user of the procedure: the quality control laboratory.

Prior to executing interlaboratory studies, analysts from the destination laboratory will need to be trained and familiarized with the new procedure. In the author's experience it is advantageous to conduct that training in the source laboratory. This has the advantage of demonstrating to the trainees the successful operation of the method in the exact environment where it was developed under the guidance of the source laboratory personnel.

8.4.1 DEMONSTRATING RELIABLE RESULTS

The primary objective of an interlaboratory study associated with a method transfer is to demonstrate that the destination laboratory is capable of producing reliable results when executing the analytical procedure. Note that this exercise also generates information about the reproducibility of the procedure that should be captured as part of the body of knowledge concerning the procedure. The typical strategy is to compare results obtained at both the source and destination laboratory on the same set of samples over some period of time. An alternative that is sometimes employed is for the destination laboratory to conduct a validation study on the new procedure.

As with any designed experiment, it is important to define the specific questions to be addressed to drive the appropriate statistical design and analysis of the results. In the case of method transfers, several possible questions can be formulated related to the fundamental goal of demonstrating acceptable execution of a procedure by a new laboratory. Hauck et al. [23] discussed this in detail in a recent publication in the context of alternatives to compendial procedures but acknowledge the close relationship to the method transfer problem. They consider the distinctions among demonstrating acceptability, equivalent or better performance, equivalency of results, or equivalency of decision making. These different questions drive differences in the design and analysis of results and reinforce the importance of having statistical input into the creation of a method transfer protocol.

The final decision about the capability of the destination laboratory is made through the comparison of test results to some predefined criteria typically involving some type of statistical hypothesis testing. This type of statistical testing consists of first establishing null (H_0) and alternative (H_1) hypotheses. Results are obtained that allow calculation of a test statistic related to the hypotheses. Finally, the test statistic is compared with some critical value derived from the assumed probability distribution and user selected confidence level to make a decision concerning the hypotheses. The possible decisions are to reject the null hypothesis and therefore to accept the alternative hypothesis or to not reject the null hypothesis. Note that not rejecting the null hypothesis is not the same as accepting the null hypothesis. Depending on the construction of the hypotheses and tests, the focus can be to either detect differences or equivalency with a high degree of confidence.

Historically, conventional two sided t-tests or some other form of difference (point) hypothesis testing were often applied to the mean results from both laboratories to support the method transfer. In these instances the null and alternative hypotheses can be expressed as

$$H_0: \mu_1 = \mu_2 \text{ (means are the same)}$$

versus

$$H_1: \mu_1 \neq \mu_2 \text{ (means are different)}$$

With these assumed hypotheses, the design and sample size are determined with the expectation of detecting a significant difference. The design and sample size combine to establish what the statisticians term the power [24,25] which is a measure of the ability to detect a difference.

More than 10 years ago it began to become apparent that this was not necessarily the best approach to the method transfer problem. Hartmann et al. [26] questioned the use of a simple t-test based approach and is one of the early advocates for interval testing over point testing (difference testing). They recognized the similarities of the method transfer problem to the bioequivalence testing pioneered by Schuirmann [27].

The issue with applying difference testing to the method transfer situation is that one is highly confident of the decision only when a difference is detected (i.e., the rejection of the null hypothesis). When the null hypothesis is not rejected it is not appropriate to assume that the means are equivalent (i.e., the acceptance of the null hypothesis).

Since the goal of method transfer studies is to demonstrate that the destination laboratory can execute the procedure as intended by the source laboratory, more recent work has focused on the notion of equivalence testing [28–34]. In a general sense, this means that the testing hypotheses need to be restructured to reflect a strategy of equivalence. This can be expressed conceptually for the accuracy of the method as

$$H_0: \mu_1 \neq \mu_2 \text{ versus } H_1: \mu_1 = \mu_2$$

or equivalently

$$H_0: \mu_1 - \mu_2 \neq 0 \text{ versus } H_1: \mu_1 - \mu_2 = 0$$

A direct interpretation of these hypotheses is unrealistic since an infinite sample size would be required to test the null hypothesis. The realistic alternative is to define some finite range of differences in the means that are of no practical difference or consequence. This allows definition of an equivalence region and the formulation of testable hypotheses. For example, Schwenke and O'Connor [32] discuss a two one-sided test (TOST) procedure based on the following hypotheses:

$$H_0: \mu_1 - \mu_2 \leq -\Delta \text{ or } \mu_1 - \mu_2 \geq +\Delta \text{ (difference statement)}$$

versus

$$H_1: -\Delta < \mu_1 - \mu_2 < +\Delta, \text{ where } \Delta > 0 \text{ (equivalence statement)}$$

Here Δ is called the "equivalence delta" and defines the bounds of the equivalence region discussed already. For an α-level equivalence test, one constructs two α-level one-sided confidence intervals for $\mu_1 - \mu_2$. These confidence intervals are compared with the equivalence delta to decide if the laboratories can be judged equivalent. If the confidence intervals are contained within $\pm\Delta$, then the null hypotheses are rejected and the laboratories are judged equivalent (see Figure 8.4).

It is important to recognize that the selection of an equivalence region is not fundamentally a statistical question but rather rests primarily with the intended purpose of the analytical procedure. In keeping with regulatory expectations it should be

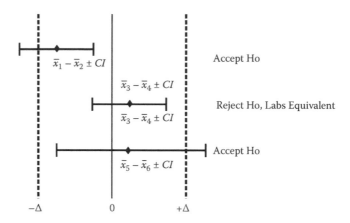

FIGURE 8.4 Illustration of equivalency testing related to method transfer.

predefined and explicitly documented in the study protocol. The selection should encompass a consideration of method performance, specification limits, and characteristics of the drug or component being analyzed. Statistics can provide some support with respect to the following:

- Estimating achievable level of significance
- Estimating the power of being able to detect equivalence

This power estimation relies on information about the experimental design, sample sizes, and estimates of variance components.

This overview of method transfer equivalence testing has focused on demonstrating the equivalence of the accuracy of different laboratories. In some cases, it may be important to test the equivalence of precision. There may also be interest in obtaining information about other factors (e.g., analysts, instruments, day to day). All these factors should be considered in the actual design of experiments to be employed. Several general experimental designs for method transfer have been proposed [32], but this continues to be an evolving area of investigation.

From the preceding discussion, it should be apparent that the design, execution, and analysis of these interlaboratory method transfer studies are complex and require close collaboration between the laboratory scientist and statistician.

REFERENCES

1. Official Web Site of the ICH, http://www.ich.org/cache/compo/276-254-1.html (accessed October 15, 2009).
2. CTD, The Common Technical Document, http://www.ich.org/cache/compo/276-254-1.html (accessed October 15, 2009).
3. ICH Q2 (R1) (2005), "Validation of Analytical Procedures: Text and Methodology," http://www.ich.org/LOB/media/MEDIA417.pdf (accessed October 13, 2009).

4. Federal Food, Drug & Cosmetic Act (FD&C Act), http://www.fda.gov/RegulatoryInformation/Legislation/FederalFoodDrugandCosmeticActFDCAct/default.htm (accessed October 13, 2009).

5. Swann JP, FDA's Origin, http://www.fda.gov/AboutFDA/WhatWeDo/History/Origin/ucm124403.htm (accessed October 13, 2009).

6. AsiaNews.it, Food and drug scandals: black market trade in blood and sick pigs, July 11, 2007, http://www.asianews.it/index.php?l=en&art=9795&size=A (accessed January 21, 2010).

7. 21CFR §210, Current Good Manufacturing Practice In Manufacturing, Processing, Packing, Or Holding Of Drugs; General, Revised as of April 1, 2009.

8. 21CFR §211, Current Good Manufacturing Practice For Finished Pharmaceuticals, Revised as of April 1, 2009.

9. 21CFR §10.115, Good guidance practices, Revised as of April 1, 2009.

10. Manual of Policies and Procedures, Developing and Issuing Guidance, FDA MAPP 4000.2, issued September 13, 2005.

11. FDA Guidance for Industry, "Analytical Procedures and Methods Validation," Draft FDA Guidance, August 2000, http://www.fda.gov/downloads/Drugs/GuidanceComplianceRegulatoryInformation/Guidances/ucm122858.pdf (accessed October 15, 2009).

12. Swartz ME, Krull IS, *Analytical Method Development and Validation*, Marcel Dekker, 1997.

13. EudraLex, The Rules Governing Medicinal Products in the European Union, Volume 4, EU Guidelines to Good Manufacturing Practice Medicinal Products for Human and Veterinary Use, Part I, Chapter 6 Quality Control, http://ec.europa.eu/enterprise/pharmaceuticals/eudralex/vol-4/pdfs-en/2005_10_chapter_6.pdf

14. Boudreau SP, McElvain JS, Martin LD, Dowling T, Fields SM, Method Validation by Phase of Development: An Acceptable Analytical Practice, *Pharm. Tech.,* November 2004, 54–66.

15. Bruce P, Minkkinen P, Riekkola M-L, Practical Method Validation: Validation Sufficient for an Analysis Method, *Mikrochimica Acta*, 128: 93–106 (1998).

16. Reynolds DW et al., Available Guidance and Best Practices for Conducting Forced Degradation Studies. *Pharm. Tech.,* February 1, 2002.

17. Kats M, Forced Degradation Studies: Regulatory Considerations and Implementation, BioPharm International, July 1, 2005, http://biopharminternational.findpharma.com/biopharm/article/articleDetail.jsp?id=170505&pageID=1, (accessed October 19, 2009).

18. McNaught AD, Wilkinson A, *IUPAC. Compendium of Chemical Terminology, 2nd ed. (the "Gold Book")*. Blackwell Scientific Publications, Oxford (1997). XML on-line corrected version: http://goldbook.iupac.org (2006-) created by M. Nic, J. Jirat, B. Kosata; updates compiled by A. Jenkins. ISBN 0-9678550-9-8. doi:10.1351/goldbook.R05305 and doi:10.1351/goldbook.R05293, (accessed, January 13, 2010).

19. Rignall A, Christopher D, Crumpton A, Hawkins K, Lyapustina S, Memmesheimer, H., Parkinson A, Smith M, Wyka B, Kraerger S, Quality by Design for analytical methods for use with orally inhaled and nasal drug products. *Pharm. Tech. Europe* 20(10), (2008).

20. Borman P, Nethercote P, Chatfield M, Thompson D, Truman K, The Application of Quality by Design to Analytical Methods, *Pharm. Tech,* October 2, 2007.

21. ISPE, *Good Practice Guide for Technology Transfer,* (2003), available at http://www.ispe.org

22. Scypinski S, Roberts D, Oates M, Etse J, Acceptable Analytical Practice for Analytical Method Transfer. *Pharm. Tech.* 28: 84–88 (2004).

23. Hauck W, DeStefano A, Cecil T, Abernethy D, Koch W, Williams R, Acceptable, Equivalent, or Better: Approaches for Alternatives to Official Compendial Procedures. *Pharmacopieal Forum,* 35(3), (2009).
24. Park, HM, Hypothesis Testing and Statistical Power of a Test. Working Paper. The University Information Technology Services (UITS) Center for Statistical and Mathematical Computing, Indiana University, 2008, http://www.indiana.edu/~statmath/stat/all/power/index.html
25. Cohen J, *Statistical Power Analysis for the Behavioral Sciences,* Psychology Press 2nd Ed. (1988).
26. Hartmann C, Smeyers-Verbeke J, Penninckx W, Vander Heyden Y, Vankeerberghen P, Massart D, Reappraisal of Hypothesis Testing for Method Validation: Detection of Systematic Error by Comparing the Means of Two Methods or of Two Laboratories. *Anal. Chem.* 67: 4491–4499 (1995).
27. Schuirmann DJ, A Comparison of the Two One-Sided Tests Procedure and the Power Approach for Assessing the Equivalence of Average Bioavailability. *J. Pharmacokinet Biopharmaceutics* 15: 657–680 (1987).
28. Limentani GB, Ringo MC, Ye F, Bergquist ML, McSorley EO. Beyond the *t*-Test: Statistical Equivalence Testing. *Anal Chem.* 77(11): 221A–226A (2005).
29. Schepers U, Wätig H, Application of the Equivalence Test According to a Concept for Analytical Method Transfers from the International Society for Pharmaceutical Engineering (ISPE), *J. Pharm. Biomed. Anal.* 39: 310–314 (2005).
30. Hoffman D, Kringle R, Two-Sided Tolerance Intervals for Balanced and Unbalanced Random Effects Models, *J Biopharm Stat.,* 15(2): 283–293 (2005).
31. Dewé W, Govaerts B, Boulanger B, Rozet E, Chiap P, Hubert Ph, Using Total Error as Decision Criterion in Analytical Method Transfer, *Chem. Int. Lab. Sys.* 85: 262–268 (2007).
32. Zhong J, Lee K, Tsong Y, Statistical Assessment of Analytical Method Transfer, *J Biopharm Stat.,* 18(5): 1005–1012 (2008).
33. Schwenke JR, O'Connor DK, Design and Analysis of Analytical Method Transfer Studies, *J Biopharm Stat.,* 18(5): 1013–1033 (2008).
34. Rozet E, Mertens B, Dewe W, Ceccato A, Govaerts B, Boulanger B, Chiap P, Streel B, Crommen J, Hubert Ph, The Transfer of a LC-UV Method for the Determination of Fenofibrate and Fenofibric Acid in Lidoses: Use of Total Error as Decision Criterion, *J. Pharm. Biomed. Anal.,* 42: 64–70 (2006).

9 The Use of Novel Materials as Solid-Phase Extractors for Chromatographic Analysis

Krystyna Pyrzynska

CONTENTS

9.1 INTRODUCTION

Sample preparation has been one of the most investigated steps in analytical procedures. These investigations are focused on problems such as isolation of analytes from the sample matrix, matrix simplification, analyte enrichment, and the removal of interferences to improve the final quantification. Modern trends in analytical chemistry are toward the simplification and miniaturization of sample preparation as well as minimization of organic solvent used. In particular, the reduction or complete elimination of solvent consumption in analytical procedures is very important according to the rules of green chemistry.

High-performance liquid chromatography (HPLC) is a mature technique commonly used in routine analysis. However, despite the performance of modern chromatographic systems, sample preparation is often considered to be a fundamental step, because it helps not only to achieve detection limits that are as low as legislation requires but also to clean up the sample matrix. In the determination of traces of the analytes, especially in biological fluids, macromolecular compounds have to be removed from a sample prior to HPLC analysis, as they could precipitate by the large amounts of organic solvents. Nonspecific or irreversible bounding on the surface of the chromatographic support could also occur, resulting in a limited lifetime of the column. Moreover, since analyte concentration is usually low, some kind of preconcentration or derivatization is often required to improve analyte detectability.

Solid-phase extraction (SPE) is the most popular sample preparation technique of environmental, food, and biological samples, and it already replaced the classic liquid–liquid extraction [1–4]. The main goal for application of solid-phase extraction is to achieve isolation, preconcentration, and cleanup of the sample in a single step. This can be realized by an appropriate selection of the type of sorbent or their combination. For this reason the properties of the analytes, nature of matrix, the required trace-level concentration, and the type of chromatography involved later in the separation step should be taken into consideration. The strategy of sample pretreatment in SPE-HPLC system is also guided by the method of final detection after chromatographic separation. Application of a simple detection mode (e.g., diode array ultraviolet [UV]) requires more selective isolation and enrichment. When the more specific quantification is used, such as fluorescence, mass spectrometry, or electrochemical methods, application of SPE sample pretreatment can improve the limit of detection.

The extraction process depends on the type of sorbent used, and retention is due to reversible hydrophobic, polar, and ionic interactions between the analyte and the sorptive material. Sorption can be nonspecific, in which case weak dispersive interactions such as van der Waals forces will dominate. However, sorbents using specific interactions resulting from analyte polarity, ionic nature, or the presence of specific functional groups are preferred. The classical sorbents in SPE are silica-based [5,6], carbonaceous materials [7,8] or polymeric, primarily styrene-divinylbenzene copolymers (PS-DVB) [4,9]. The novel sorbents with improved selectivity toward the particular groups of compounds or even individual compounds includes immunosorbents [10,11] and molecularly imprinted polymers (MIPs) [12–15]. Restricted access materials (RAMs) are a class of biocompatible sorbent particles enabling the direct extraction of analytes from biological fluids [16,17]. The objective of this chapter is to present the recent advances in the area of novel materials as solid-phase extractors for chromatographic analysis. The papers published over the last five years are discussed in more detail. Emphasis is also given to the application of several SPE systems for automated preparation of environmental, food, and biological samples.

9.2 BONDED-PHASE SILICA SORBENTS

Silica chemically bonded with various groups has been the most common material for SPE. This sorbent can be classified as reversed-phase sorbents with octadecyl (C_{18}), octacyl (C_8), ethyl (C_2) and phenyl (Ph) or as normal-phase sorbents with cyanopropyl (CN), aminopropyl (NH_2) and diol functional groups. Their interaction mechanisms are mainly based on hydrophobic interaction (van der Waals forces); thus, these SPE packings provide high recoveries for nonpolar analytes. New types of alkyl silicas have also been prepared by bonding with an alkylchain reagent containing an embedded polar carbamate or amide functionality [18,19]. Nevertheless, silica-based sorbents are unstable at pH extremes (2 > pH > 8) and have relatively low capacity and low recovery for basic analytes.

Several kinds of modifications were used to immobilize different compounds on the surface of classical silica-base sorbents to increase their selectivity [20,21]. Jardim and coworker [22,23] have reported immobilization of different polysiloxane on silica and their use for extraction of some pesticides. β-Cyclodextrin bonded silica particles have been applied as a selective solid-phase extraction medium for some phenolic compounds [24,25]. Selective extraction and enrichment of polyunsaturated fatty acid methyl esters from fish oils was achieved by using silica-supported ionic liquid phase coated with silver salts [26]. Liu [27] has used silica gel coated with gold nanoparticles self-assembled with alkanethiols for the extraction of neutral steroidal compounds. The use of this SPE material not only effectively concentrated the analytes through hydrophobic interactions but also removed the interfering signals from urinary proteins through their interactions with residual Au metal surfaces.

The bonded-silica sorbent may be packed in different formats: filled microcolumns, cartridges, or discs. A variety of bonded-silica phases are commercially available in the format of cartridges. Extraction could be also performed with membrane disks containing C_{18}-bonded silica (8 μm particles) on polytetrafluoroethylene or glass fiber supports. The typical composition of the disc is 80–90% (w/w) C_{18} or C_8 and 10–20% support; the disc dimensions are 47 mm in diameter and 0.5 mm thick [28]. Discs provide shorter sample processing time on account of their larger cross sectional area and decreased pressure drop, allowing higher sample flow rates. This is important for environmental samples, where larger sample volumes are usually employed to achieve adequate detection limits.

9.3 POLYMERIC MATERIALS

The polymeric sorbents based on styrene-divinylbenzene exhibit higher capacity and better chemical stability over the whole pH range in comparison with bonded silicas. Due to the specific π–π interactions they are relatively selective for analytes with aromatic rings. Most polymeric sorbents, similar to silica-based sorbents, require a pretreatment with methanol to make the hydrophobic surface more compatible with aqueous samples. The use of highly cross-linked polymeric sorbents with their specific surface up to 800 m^2/g or hypercross-linked polymeric sorbents (over 1000 m^2/g) could improve the retention of analytes as more π–π sites in the aromatic rings will then be accessible to interact with the analytes [29,30]. These sorbents have been applied successfully to the SPE of several groups of compounds from different sample matrices [31–34]. Some studies have shown that these sorbents provided better recoveries than sorbents with a lower cross-linking degree and therefore a lower specific surface area. For example Hyperphere-SH (surface area > 1000 m^2/g) gave better recoveries for phenol derivatives than conventional macroporous resin PRLP-S with its surface area of 500 m^2/g [31, 32]. In spite of their having excellent sorption properties, hypercross-linked sorbents are normally available only with a large particle size (about 50–200 μm), and the particle size distribution is often rather broad [34]. This heterogeneity in particle size distribution leads to inefficient packing of the sorbent in the sorbent bed and, eventually, to inefficiencies in interaction with the analytes. Moreover, their hydrophobic nature leads to poor retention of polar compounds. To overcome these problems, the research in new SPE materials has recently been focused on the development of new hydrophilic polymeric materials. Such sorbents combine high specific surface area and polar interaction between sorbent and analyte due to introduction of a polar moiety to the polymer structure.

9.3.1 HYDROPHILIC POLYMERIC SORBENTS

The hydrophilic polymeric sorbents can be obtained by chemical modification of the existing hydrophobic materials or by copolymerisation of monomers that contain suitable functional groups. The polar substituents reduce the interfacial tension between the polymer surface and aqueous sample improving the wetting characteristics and increase contact between the analyte and polymeric sorbent.

The resins with sulfonic, acetyl, hydroxyl, hydroxymethyl, or carboxybenzoyl substituents on the benzene rings give a more polar surface that functions well without a methanol pretreatment [4]. The modified sorbents allowed a more efficient extraction of polar compounds than the unmodified analogues—the introduction of polar moieties enhances the polar interactions and provides better mass transfer. The first commercially available chemically modified sorbent was Bond Elut PPL from Varian. Another well-known sorbent is Isolute ENV+ from IST (Hendrognet, UK), which is hydroxylated poly(styrene-divinylbenzene) resin. More recently, Strata-X (styrene skeleton modified with a pyrrolidone group) from Phenomenex was introduced. Its retention mechanisms are hydrophobic, hydrogen bonding, and aromatic. Strata-X has been applied in SPE to clean up biological samples, such as blood [35–37], urine [38], saliva [39], or milk [40] as well as to enrich pesticides [41–43] or pharmaceuticals [44–46] from water samples. Some research studies have made a comparative evaluation of the efficiency of Strata-X with other sorbents [40,42,44–49]. The performance of C_{18} BondElut, Lichrolut ENV (composed of an ethylvinylbenzene-divinylbenzene copolymer) and Strata-X for preconcentration of five selected analgestics compounds that can be purchased without medical prescription (diclofenac, ketoprofen, ibuprofen, naproxen, and clofibric acid) was evaluated by spiking 2.0 L of deionised water with 15 µg/L of each compound [46]. The enrichment on hydrophobic sorbents requires acidic conditions; thus, the pH of the samples was adjusted to 2 with HCL solution. For comparison, the recovery studies were also conducted at neutral pH, where acidic pharmaceuticals are present in their respective ionized forms. More complete retention of investigated compounds were obtained, as it was expected, at lower pH (Figure 9.1). Recovery from acidified samples exceeded 88% for all sorbents. Even clofibric acid, the most hydrophilic compound, was almost quantitatively recovered. Application of PS-DVB sorbents with high surface areas and copolymers composed of both liphophilic and hydrophobic monomers, such as Oasis HLB and Abselut Nexsus, gave much lower recoveries for clorofibric acid [49]. Employing the combination of SPE and HPLC analysis, a sample pretreatment procedure was developed allowing the enrichment of water sample by a factor of 1000 [46].

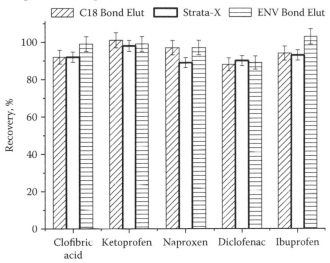

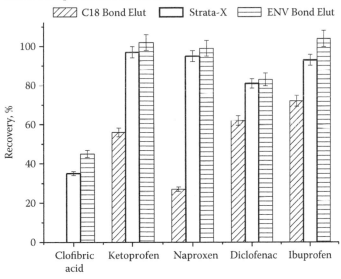

FIGURE 9.1 Recoveries of acidic pharmaceuticals using different solid sorbents from 2.0 L of water was spiked with 15 μg/L of each compound. (From Stafiej, A., Pyrzynska, K., Regan, F., *J. Sep. Sci.* 30: 985–991, 2007. With permission.)

Another approach, which is an alternative to prepare chemically modified functionalised polymers by, is to prepare macroporous copolymers with a hydrophobic monomer. Oasis HLB from Waters is one of the most used hydrophilic sorbents in the extraction of compounds with high polarity. It is a macroporous poly(*N*-vinylpyrrolidone-divinylbenzene) copolymer and has a specific area about

TABLE 9.1

Recent Examples of the Application of Oasis HLB for Enrichment and Isolation of Organic Pollutants

Analytes	Sample	Technique	Reference
Tetracyclines	Muscle	LC-UV	[50]
	Propolis	LC-UV	[51]
Antibiotics	Honey	LC-MS	[52]
	Muscle and liver	GC-MS	[53]
	Serum	MEKC-UV	[54]
Estrogens	Waters	LC-MS	[55]
Pharmaceuticals	Urine	CE-UV	[56]
Pesticides	Waters	LC-MS	[57]
	Blood	GC-MS	[58]
Phenol derivatives	Milk	GC-MS	[59]

Notes: MECK, micellar electrokinetic chromatography. CE, capillary electrophoresis.

800 m^2/g. Oasis HLB has been widely used in SPE, and some recent applications are presented in Table 9.1. Generally, it has been used in SPE to enrich and clean up several types of analytes such as antibiotics [50–54], estrogens [55], pharmaceuticals [56], pesticides [57,58], and phenol derivatives [59]. Most of the studies investigate the performance of Oasis HLB in off-line SPE using different cartridge sizes available (from 30 to 500 mg). Other studies use the direct coupling of online SPE to HPLC with the column-switching technique [60,61] or using 96-well plate [62] to obtain high sample throughput. Abselut Nexus from Varian, the copolymer of methacrylate and divinylbenzene, has been recently commercialized. Its main application is in cleanup of complex samples, such as biological matrices, with the subsequent extraction of analytes [63–67]. No difference between Oasis HLB and Abselut Nexus was observed about extraction performance of methadone enantiomers and some benzodiazepines in serum and urine samples, except that Abselut Nexus cartridge did not need column activation by methanol and water before use [63]. Georga et al. [67] also compared Oasis HLB and Abselut Nexus in the extraction of caffeine-metabolite products from biological samples. The recovery rates of these two sorbents were not significantly different. However, Abselut Nexus was selected for further studies as it has smaller bed volume and therefore provides higher flushing rates. Moreover, it was found that conditioning of the sorbent bed was necessary in spite of the manufacturer's instructions that this step can be omitted.

9.3.1.1 Mixed-Mode Sorbents

Mixed-mode polymeric sorbents combine a polymeric skeleton with ion-exchange groups. Thus, these hybrid materials rely on two types of interaction mechanisms for their performance: reversed phase and ion exchange. Carefully selection of the polymeric skeleton (which enhances the reversed-phase interactions) and the ionic groups (which tune the ion-exchange interactions) could give the combination of two highly desirable properties in SPE (i.e., retentivity and selectivity) in one single material. The benefit of the ion-exchange capacity is that either the analytes, the matrix components in the sample, or even the ionization state of the sorbent (in the case of weak-exchange resins) can be switched during the different steps in SPE procedure. It allows the interference elimination in the washing step and eluting the analytes more selectively, just by suitable pH combination in each step.

Oasis MCX and Oasis MAX from Waters have the same Oasis HLB skeleton (polyvinyl pyrrolidone-divinylbenzene) modified chemically with sulfonic acid and quaternary amine groups, respectively. Recently, the same company commercialized the weak ion-exchange sorbents Oasis WCX and Oasis WAX modified with carboxylic acid and piperazine moieties, respectively. Similarly, Strata-X (from Phenomenex) has been modified with sulfonic groups (Strata-X-C), carboxylic groups (Strata-X-CW), and ethylenediamine (Strata-X-AW). These mixed-mode sorbents are mainly applied for extraction of analytes (charged or not) from complex biological and environmental matrices [68–85]. Some recent applications of mixed-mode polymeric sorbents are presented in Table 9.2.

TABLE 9.2
Some Examples of Recent Applications of Mixed-Mode Polymeric Sorbents

Sorbent	Ionic Mode	Analytes	Sample	Reference
Oasis MCX	Strong cation exchange	Ketamine	Plasma	[68]
		Herbicides	Potatoes	[69]
		Biogenic amines	Wine	[70]
		Metabolites of amphetamine and cocaine	Surface water, urban wastewater	[71]
Oasis WCX	Weak cation exchange	Cocaine and its metabolites	Urine	[72]
Oasis MAX	Strong anion exchange	Antibiotics	Milk	[73]
		Fluoroquinolone drugs	Wastewater	[74]
		Polycyclic aromatic hydrocarbons	Sewage sludge	[75]
		Trehalose-6-phosphate	Tissue of plant	[76]
Oasis WAX	Weak anion exchange	Polysorbate 20	Protein solution	[77]
		Mycophenolic acid	Meat products	[78]
Strata-X-C	Strong anion exchange	Polyphenolic compounds	Urine	[79]
Strata-X-CW	Strong cation exchange	Steroid glucuronides	Urine	[80,81]
		Estrogens	River sediments	[82]
		Tetracycline	Honey	[83]
		Acrylamide	Foodstuffs	[84]
		Rifampicin	Plasma, blood	[85]

Lavén et al. [86] proposed a novel solid phase extraction method whereby 15 basic, neutral, and acidic pharmaceuticals from wastewater were simultaneously extracted and subsequently separated into different fractions. This was achieved using mixed-mode cation- and anion-exchange SPE (Oasis MCX and Oasis MAX) in series. For less complex samples (e.g., the active-sludge-treatment effluent water), Oasis MCX used alone may be an alternative method. With sewage treatment plant influent waters containing high loads of organic compounds the cleanup step using only Oasis MCX was insufficient, leading to unreliable quantitation. Using the ability to separate compounds by mixed-mode SPE according to basic and acidic functionalities should also be very useful in the characterization of unknown water contaminants.

Recently, the synthesis of hypercross-linked polymer resins (surface area 1000–1200 m^2/g) in the form of microspheres produced via precipitation polymerization and modified chemically with 1,2-ethylenediamine and piperazine groups has been reported [87]. Essentially quantitative (~ 100%) recovery of acidic compounds (up to $pK_a < 6.5$) was obtained after a washing step with an organic solvent, which elutes the rest of analytes as well as interferences. The performance of these two resins with weak anion-exchange capacity were superior to the macroporous Oasis-WAX and Strata-X-AW, which was attributed to the selectivity imparted by amine moieties plus the stronger reversed-phase mechanism of the hypercross-linked polymer structures as well as the lower size of the particles (~ 6.5 μm), which ensures analyte retention on cartridge during the application and washing steps. The resin modified with 1,2-ethylenediamine, which displayed the most attractive overall performance characteristics, was tested with 1000 mL of Ebre river water samples with the addition of a 1 μg/L level of the analyte mixture and with effluent wastewater from a treatment plant; in both cases, the resin provided good recoveries.

9.4 CARBONACEOUS MATERIALS

Among carbonaceous materials, activated carbon was certainly one of the first materials applied in SPE. It has been widely used in water and wastewater treatment, primarily as an adsorbent for the removal of organic and inorganic contaminants [88,89]. The high sorptive properties of activated carbon are primarily due to extensive porosity and very large available surface areas in the range of 300–2500 m^2/g [90]. Its adsorption property is strongly affected by its specific surface area, pore size distribution, and content of inorganic elements. The surface structure is highly complex and depends on the raw material used for producing it and on the method of production as well as on pretreatment [91]. Activated carbon fibers have received increasing attention in recent years as a better adsorbent than granular activated carbons, because they normally possess much higher adsorption kinetics and adsorption [88]. Activated carbon fibers have only micropores, which are directly accessible from the external surface of the fiber.

Newer carbon sorbents, such as graphitized carbon blacks (GCB) and porous graphitized carbon (PGC), have a more homogeneous structure and more reproducible properties than activated carbon. GCB are nonspecific and nonporous (surface area 100–210 m^2/g) and are considered to be both reversed-phase sorbents and anion exchangers due to the presence of positively charged groups on their surface [92]. In contrast to GCB, the flat surface of PGC is highly homogeneous, and this property is responsible for its unique selectivity to geometrical isomers. The large layers of carbons containing delocalized π electrons and the high polarizability are responsible for different retention mechanisms such as electron-transfer, ion-pairing, and hydrophobic interactions. This retention effect appears with polar compounds which can either donate or accept electrons when they can polarize the graphite surface [93]. GCB and PGC sorbents have been extensively used for solid-phase extraction of polar organic pollutants from water samples [94–96].

9.4.1 Carbon Nanotubes

The characteristic structures and electronic properties of carbon nanotubes (CNTs) allow them to interact strongly with organic molecules, via noncovalent forces, such as hydrogen bonding, π–π stacking, electrostatic forces, van der Waals forces, and hydrophobic interactions. These interactions as well as hollow and layered nanosized structures make them a good candidate for use as sorbents. The surface, made up of hexagonal arrays of carbon atoms in graphene sheets, interacts particularly strongly with the benzene rings of aromatic compounds. In 2001, Long and Yang [97] observed that dioxins, which have two benzene rings, were strongly adsorbed on CNTs. The amounts of dioxin adsorbed were 10^4 and 10^{17} times greater than that on activated carbon and $\gamma\text{-Al}_2\text{O}_3$, respectively. Dioxins after the enrichment step could be removed from the adsorbent by temperature-programmed desorption.

CNTs possibly contain functional groups such as –OH, –C = O, and –COOH depending on the synthetic procedure and purification process. Oxidation of CNTs with nitric acid is an effective method to remove the amorphous carbon, carbon black, and carbon particles introduced by their preparation process [98]. It is known that oxidation of carbon surface can offer not only a more hydrophilic surface structure but also a larger number of oxygen-containing functional groups, which increase the ion-exchange capability of carbon material. Gas-phase oxidation of activated carbon increases mainly the concentration of hydroxyl and carbonyl surface groups, whereas oxidation in the liquid phase particularly increases the content of carboxylic acids [99]. The amount of carboxyl and lactone groups on the CNTs treated with nitric acid was higher in comparison when the treatment was conducted using H_2O_2 and $KMnO_4$ [100]. Datsyuk et al. [101] found that the nitric acid (65%) treated carbon nanotubes under reflux conditions for 48 h suffered a very high degree of degradation such as nanotube shortening and additional effect generation in the graphitic network. Functional groups can change the wettability of CNTs' surfaces and consequently make them more hydrophilic and suitable for sorption of relatively low molecular weight and polar compounds. On the other hand, functional groups may increase diffusional resistance and reduce the accessibility and affinity of CNTs' surfaces for organic compounds [102].

In several papers, sorption on CNTs has been examined for different aromatic compounds, such as phenol and its derivatives [103–109], herbicides [110–113], pesticides [114–116], insecticides [117], polyhalogenated pollutants [118,119], pharmaceuticals [120], linear alkylbenzene sulfonates [121], and some additives in gasoline [122]. Carbon nanotubes were also applied for trapping volatile organic compounds

[123–125]. The recent applications of CNTs for removal and enrichment of several organic compounds are presented in Table 9.3. Earlier reports were discussed in the review's papers [8,126].

Zhou et al. [127] compared the trapping efficiency of CNTs and C_{18} packed cartridge using sulfonylurea herbicides as the model compounds. When the matrices of the samples were very simple, such as tap water and reservoir water, the enrichment performance between these two adsorbents had no significant difference. However, carbon nanotubes become much more suitable to extract herbicides from complex matrices (seawater and well water). The comparison of carbon nanotubes, activated carbon, and C_{18} silica in terms of analytical performance, application to environmental waters, cartridge reuse, adsorption capacity, and cost of adsorbent has also been made for propoxur, antrazine, and methidation herbicides [114]. The adsorption capacity of CNTs was almost three times higher than that of activated carbon and C_{18}, whereas activated carbon was superior over the other sorbents due to its low cost. It is noteworthy to add that oxidation process of activated carbon with various chemical agents reduced the recovery of some pesticides [114]. A comparative study [116] suggested that carbon nanotubes had a higher extraction efficiency than Oasis HLB for the extraction of methamidophos and acephate, particularly for seawater samples. Figure 9.2 shows the chromatograms of six organophosphorous pesticides in the spiked seawater sample extracted using CNTs and Oasis HLB sorbent. For other tested polar organophosphorous pesticides (dichlorvos, omethoate, monocrotophos, and dimethoate) improvement was not significant; thus, CNTs could supplement Oasis HLB for the extraction of these compounds.

Carbon nanotubes can also remove and preconcentrate volatile organic compounds. Saridara et al. [128] described a mictrotrap operating as a nanoconcentrator and injector for gas chromatography. A thin layer of CNTs were deposited by catalytic chemical vapor deposition on the inside wall of a steel capillary to fabricate the microtrap. The obtained film provides an active surface for fast adsorption and desorption of small organic molecules such as hexane and toluene. Stronger sorption of toluene in comparison with hexane could be explained by the π–π interaction between the sidewall of the carbon nanotubes and the aromatic ring. The purge-and-trap system was used for evaluation of carbon nanotubes as an adsorbent for trapping 16 volatile organic compounds from gaseous mixtures and indirectly from water samples [129]. Due to their porous structure, CNTs were found to have much higher breakthrough volumes than that of graphitized carbon black (Carbopack B) with the same surface area. The carbon nanotubes-bonded fused-silica fiber produced by covalently bonding possesses some special properties such as good stability at high temperature, in organic solvents (polar and nonpolar), acid and alkali solutions, and low detection limits for extracting phenols [107]. Rastkari et al. [122] proposed novel coating for solid-phase microextraction by attaching CNTs onto a stainless steel wire through organic binder. The analytical performance of the proposed fiber was evaluated through determination of the oxygenated ethers in human urine, and the results showed that the CNTs fiber exhibited higher sensitivity and longer life span (over 150 times) than the commercial carboxen/polydimethylsiloxane coating.

TABLE 9.3
Recent Examples for Sorption of Organic Compounds on Carbon Nanotubes

Analytes	Sample	Eluent	Recovery %	Remarks	Reference
Chlorophenols	Aqueous solutions	—	—	Acid refluxing has negative effect on the sorption capacity of CNTs, while the stronger π–π dispersion forces between chlorophenols and sorbent treated with NH_3 produce higher removal of the analytes.	[103]
Resorcinol	Aqueous solutions	—	—	Extraction efficiency for phenolic derivative follows the order: resorcinol > hydroquinone > catechol.	[107]
Phenols	River water and sanitary wastewater	Desorption at 280°C	71–148	Chemically bonded CNTs-fused-silica fibers for solid phase microextraction.	[108]
Phenols and chlorophenols	Tap, river, and waste waters	Methanol	59–99	Disc based on a CNTs sheet enables the use of high flow rates (50 mL/min).	[109]
Chloroacetanilide herbicides	Tap and river water	Ethyl acetate	77–104	pH of 7 was selected as the optimum.	[110]
Sulfonylurea herbicides	Natural waters	Acetonitrile	79–102	The triple layered CNTs disc system showed good extraction efficiency for sample volume up to 3 L.	[111]
Carfentrazone-ethyl	Tap and river water	Ethyl acetate	81–91	Preconcentration factor of 1000 was obtained for 1 L of sample volume.	[112]
Organophosphorous pesticides	Fruit juices	Dichloromethane	73–103	Low amount of sorbent (940 mg) is required.	[114]

(continued on next page)

TABLE 9.3 (continued)
Recent Examples for Sorption of Organic Compounds on Carbon Nanotubes

Analytes	Sample	Eluent	Recovery %	Remarks	Reference
Various pesticides	Tap, reservoir, and stream water	Acetonitrile	81–108	Comparison with C_{18} silica and activated carbon.	[115]
Pesticides	Mineral water	Dichloromethane with 5% (v/v) formic acid	53–94	Optimum pH for preconcentration was 8.0.	[116]
Organophosphorous pesticides	Seawater	Acetone or methanol	79–102	CNTs could supplement Oasis HLB for the extraction of polar organophosphorous pesticides.	[117]
Pyrethroid insectides	Tap, river, reservoir and well water	Acetone	71–118	CNTs showed more powerful sorption properties than C_{18} due to higher capability to extract the analytes in larger volume water solutions.	[118]
Polyhalogenated compounds	River water	Acetone	93–100	The CNTs cartridge could be used for three cycles of sorption/desorption with no loss of efficiency.	[119]
Linear alkylbenzene sulfonates	River waters	Methanol	85–106	CNTs showed stronger retention ability than C_8 and C_{18}.	[121]
Oxygenated ethers	Urine	Desorption at 23°C	90–95	CNTs fiber was prepared by binding nanotubes to the surface of stainless steel wire through organic binder.	[122]

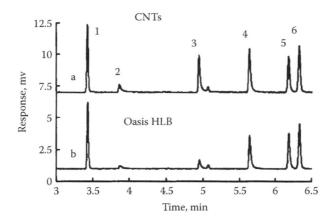

FIGURE 9.2 Chromatograms of organophosphorous pesticides (1.0 μg/L) in the spiked seawater extracted with CNTs and Oasis HLB. Peaks identification: 1-dichlorvos, 2-methamidophos, 3-acephate, 4-omethoate, 5-monocrotophos, 6-dimethoate. (From Li, Q., Wang, X., Yuan, D., *J. Environ. Monit.* 11: 439–444, 2009. With permission.)

Carbon nanotubes could also be used in a format of disc. Incorporating sorbents of small particle size, the disc format possesses a larger surface area than the cartridge, resulting in good mass transfer and fast flow rates [130]. Zheng et al. [131] prepared a CNTs sheet by passing their suspension in toluene (37 mg in 100 mL) through a polytetrafluoroethylene (PTFE) membrane. After drying, a sheet (37 mm in diameter, 60 μm in thickness with a specific surface area of ~ 700 m²/g) was peeled from the membrane and packed into a metal tube for sampling organic vapor containing toluene, methyl ethyl ketone, and dimethyl methylphosphonate. CNTs discs were also prepared with a qualitative paper as support without peeling from the filter to improve their mechanical properties [108,110]. Due to the strong van der Waals forces among the tubes, they could be fixed onto the filter surface without leaking during the processes of filtration or sample loading. To enhance the sorption capacity of the discs, double or even triple disks were used together. A comparison study showed that the double-disc system (comprising two stacked discs with 60 mg of CNTs) exhibited extraction capabilities that were comparable to those of a commercial C_{18} disc with 500 mg sorbent for nonpolar or moderately polar compounds. Moreover, the former system was more powerful than the latter for extracting polar analytes. Chromatograms of phenol derivatives in the spiked wastewater samples preconcentrated with CNTs disks are showed in Figure 9.3.

The proposed μ-SPE device by Sae-Know and Mitra [132] for integrating sampling, analyte enrichment, and sample introduction consists of a syringe attached to a removable capillary probe containing CNTs. Carbon nanotubes were used in self-assembled (open tubular) as well as in packed format. The self-assembled of CNTs were carried in the tube furnace by a chemical vapor deposition, whereas for the packed format 0.3 mg of CNTs was introduced into a 100 mm long capillary plugged with glass wool at both ends. The analytes 92-nitrophenol, 2,6-dichloroaniline, and naphthalene were concentrated by drawing several mililiters of their solutions into

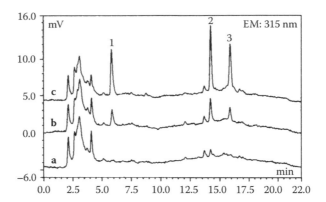

FIGURE 9.3 LC-FID chromatograms for the wastewater samples preconcentrated on CNTs double-disks. (a) Wastewater sample. (b) Wastewater sample spiked with 0.50 ng/mL of each compound. (c) Wastewater sample spiked with 2.0 ng/mL of each compound. Peaks identification: 1- Bisphenol; 2- 4-*tert*-octylphenol; 3- 4-*n*-nonyl-phenol. Volume of sample solution: 1000 mL, volume of methanol for elution: 10 mL. (From Niu, H.Y. et al., *Anal. Bioanal. Chem.* 392: 927–935, 2008. With permission.)

the syringe through the needle and then desorbing them in a few microliters of acetonitrile. However, the first approach was found to be ineffective with an enrichment factor of less than 1, and for packed beds of CNTs an enrichment factor close to 7 was achieved.

Carbon nanotubes could also be readily immobilized into the pore structure of a polymeric membrane for improving the membrane extraction process [133]. The aqueous dispersion of CNTs nanotubes was injected through a polypropylene hollow fiber under pressure and trapped and held within the pores facilitating solute exchange from the donor to the acceptor phase. The effectiveness of this carbon nanotube mediated process was studied by direct solvent enrichment of nonpolar compounds (toluene and naphthalene) as well by selective extraction of trichloroacetic acid and tribromoacetic acid. In both cases, the enrichment factor measured as the ratio of analyte concentrations in the acceptor phase to the donor phase could be increased by more than 200% compared to plain polypropylene membrane.

9.4.2 CARBON-ENCAPSULATED MAGNETIC NANOPARTICLES

Carbon-encapsulated magnetic nanoparticles (CEMNPs) are core-shell materials with similar surface characteristics to carbon nanotubes, and this similarity enables them to be used as solid sorbents. They are composed of the magnetic core (10–100 nm in diameter), which is tightly coated by a carbon coating built from parallel stacked graphitic layers [134–136]. Encapsulation approach primarily protects the nanoparticles against the external environment, hampers aggregation, and also provides the ability for surface functionalization. This process may also improve the dispersion stability of core-shell nanomaterials in a wide range of suspending solvents. A unique and attractive property of CEMNPs is that magnetic nanoparticles can readily be isolated from sample solution by the application of an external magnetic field.

Jin et al. [137] proposed magnetic Fe nanoparticle functionalized CNTs for the removal of benzene and related derivatives. They not only displayed improved water solubility but could also be recovered easily from the solution by magnetic separation. The scheme for their preparation is presented in Figure 9.4. Briefly, CNTs were dispersed in $Fe(NO_3)_3$ solution with the help of an ultrasonic bath. After draining excess of water on a rotary evaporator with a vacuum pump, the resulting materials were reduced using hydrogen at 560°C and 900°C successively. Thus, Fe nanoparticles were deposited inside the inner cavities of carbon nanotubes. Then they were attacked by carbon radicals generated by the thermal decomposition of azodiisobutyronitrile. The produced Fe-CNTs-cyano material was refluxed in sodium hydroxide aqueous solution–methanol mixture. The applicability of the obtained sorbent was studied using four model compounds: benzene, toluene, dimethylbenzene, and styrene. Their mixture was shaken at room temperature with 10 mg of the prepared sorbent for a few minutes to form a homogeneous black dispersion. Then the sorbent was collected from the black dispersion by discarding supernatant liquid with the help of a magnet,

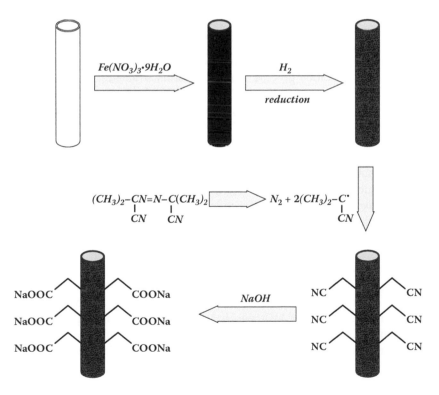

FIGURE 9.4 Scheme for the preparation of magnetic Fe nanoparticles functionalized water-soluble multiwalled carbon nanotubes. (From Jin, J. et al., *Chem. Commun.* 4: 386–38, 2007. With permission.)

and the adsorbed compounds were eluted with 0.2 mL of methanol. The recoveries of studied compounds were higher than 80% [137]. Notably, after being washed several times with methanol and dried in vacuum, this sorbent can be reused.

Magnetic nanoparticles were also applied for preconcentration of middle oxy-ethylated nonylphenols [138] and several typical phenolic compounds such as bis-phenol A, 4-*tert*-octylphenol, and 4-*n*-nonylphenol [139] from environmental water samples. It should be mentioned that silica-encapsulated magnetic nanoparticles have become useful for preconcentration of phenols [140]. To enhance their sorptive tendency toward organic compounds, cetylpyridinium chloride was added, which adsorbed on the surface of nanoparticles and formed mixed hemimicelles. Compared with nonmagnetic nanoparticles, the proposed sorbent material avoids the time-consuming column passing and filtration operation and shows great analytical potential in preconcentration of large volumes of real water samples.

Carbon-encapsulated nanoparticles (10–60 in diameter) showed good sorption properties for removal of heavy metal ions (Cu^{2+}, Co^{2+}, Cu^{2+}) from aqueous solutions [141]. The ions uptake exceeds 80% for cobalt and 95% for cadmium and copper. CEMNPs, due to their excellent magnetic performance, are easily mobile and separable from the solvent under a low magnetic field.

9.5   RESTRICTED ACCESS MATERIALS

To perform a high throughout analysis, efforts have been engaged in developing a faster sample purification process. Among different strategies, the introduction of special extraction sorbents, such as RAMs, allowing the direct and repetitive injection of complex biological matrices, represents a very attractive approach. Integrated in a liquid chromatography system, these extraction supports lead to the automation, simplification, and speeding up of the sample preparation process. RAMs are used mainly for the analysis of compounds with low molecular mass (e.g., drugs, endogenous substances, xenobiotics) in complex matrices containing high-molecular substances (most frequently proteins) [17,142].

RAMs have several different structures, but their mechanism of separation is similar: a hydrophobic barrier enables the small molecules to penetrate through the hydrophobic part of the stationary phase, while at the same time it excludes the macromolecules by physical or chemical means. Glass-fiber filter (GFF) and alkyl-diol-silica (ADS) supports are the most popular RAM materials with a physical barrier (pore size 6–8 nm) that excludes macromolecules. GFF sorbents are constituted of porous silica particles with the outer surface covered by a hydrophilic moiety (diol-glycine groups) and a hydrophobic tripeptide partitioning phase (glycine-L-phenylalanine-L-phenylalanine) on the internal surface (Figure 9.5a). The retention mechanism is mainly due to π–electron interactions. An additional free carboxyl group of phenylalanine shows weak cation-exchange functionality. This material may withstand several hundred plasma or serum injections (total volume of 6–7 mL) without losing performance [17]. The hydrophilic groups (glycerylpropyl, i.e., diol moieties) are bounded at the outer surface of ADS material, while different types of reversed phases (C_4, C_8 or C_{18}) are found on the internal surface. New ADS material (i.e., exchange diol silica, XDS) has been tested for online SPE of basic drugs from

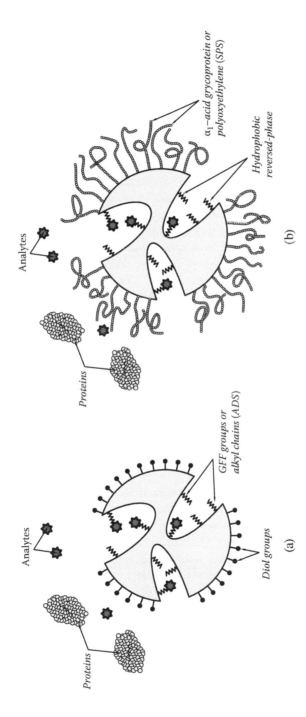

FIGURE 9.5 Schematic drawing of restricted access materials with (a) glycine-L-phenylalanine-L-phenylalanine (GFF) or alkyl-diol-silica (ADS) and (b) semi-permeable surface and protein-coated silica support.

directly injected plasma [143]. The pore of the silica particles is about 6 nm, yielding a molecular mass cutoff of about 15 kDa. Due to this physical barrier and hydrophilic diol groups bounded to the external surface of the silica particles, macromolecules have no access to the inner surface. This sorbent presents the properties of a strong cation exchanger because of the presence of sulfonic acid groups bounded via short alkyl chain to the inner surface. The extraction of the analytes is mainly due to electrostatic interactions. The new XDS sorbent showed higher retention capability towards the basic drugs, selected as model compounds, in comparison with RP-18 ADS, irrespective of the washing liquid pH [144].

RAMs have found several applications, mainly in bioanalysis and environmental analysis [144–153]. In food analysis only a few examples have been reported [154,155]. Certain precautions should be followed when using RAM columns—for instance, the pH and composition of the loading mobile phase. In practice, no more than 20% of organic solvent should be used since higher content leads to protein denaturation and consequently to shorter column lifetime [156]. Moreover, a partial loss of the analytes can occur for those compounds that strongly bind to proteins.

Semipermeable surfaces (SPSs) and protein-coated silica belong to the group of RAM sorbents with a chemical barrier [17,142]. In SPS material the outer surface (hydrophilic polyoxyethylene polymer) repels large molecules, whereas the internal hydrophobic surface retains small analytes that penetrate through the polymer layer (Figure 9.5b). These materials differ mainly in the functional groups forming the reversed phase of the internal surface (e.g., nitrile, phenyl, C_8, C_{18}), and their lifetime is similar to ADS materials. Protein-coated silica uses a protein network (human plasma protein, $\alpha1$-acid glycoprotein) at the outer surface instead of a polymer (Figure 9.5b). It makes the external surface of the particles compatible with a proteinaceous sample that cannot penetrate into small pores (10 nm). An advantage of this material is the wider pH working rage of 2–10 [157].

9.6   IMMUNOSORBENTS

Materials based on antigen–antibody interactions (molecular recognition) allow selective extraction of the analytes from the complex matrix. Antibodies could be linked to an appropriate solid support by covalent bonding, adsorption, or encapsulation to form an immunosorbent. The selected support should have large pore size, should be hydrophilic to avoid any nonspecific interactions, should be chemically and biologically inert, and be pressure resistant for use in online techniques. The last requirement is not obligatory for off-line disposable cartridges. The most common approach involves covalent bonding of the antibodies via carboxyl, amino, or sulfhydryl groups onto activated silica or an agarose gel Sepharose. The synthesis, characteristics, and properties of these molecular recognition materials have been reviewed [158–160].

The binding of antigen to antibody is the result of good spatial complementarities and is a function of intermolecular interactions. This means that antibody can also bind to other analytes with structures similar to the compound, which has induced the immune response. This phenomenon is referred to as the cross-reactivity. It is usually a negative feature of immunosorbents; however, they could be applied for a single analyte, and its metabolites, or for a class of structurally related compounds. Immunosorbents have been tailored by several authors for the extraction of groups of organic compounds including pesticides, herbicides, polyaromatic hydrocarbon metabolites, xenobiotics, and steroid estrogens [161–166].

Preparation of immunosorbent containing antibody against two or more compounds make group-selective extraction possible. An immunosorbent prepared using anti-β-estradiol (E2) antibodies retained more than 90% of E2 but less than 15% and 30% of estrone (E1) and α-ethynylestradiol (EE2) [162]. The mixed immunosorbent containing antibodies against both E1 and E2 extracted 92%, 107%, and 2% of E1, E2, and EE2, respectively, from wastewater effluent. Figure 9.6 shows the total ion chromatogram from an effluent extract without and with purification of sewage effluent using mixed immunosorbents [162]. The estrogens, their surrogate standards, and the internal standards are clearly distinguishable in the chromatogram after sample purification, whereas the chromatogram that was not immunoextraction purified reveals no peaks for any of the analytes or standards due to the presence of coextracted matrix compounds. The selectivity of three immunosorbents based on different monoclonal antitriazine antibodies was evaluated by extraction recovery and stepwise elution measurements [167].

Two parameters can affect the recovery of the analytes: immunosorbent capacity and the affinity of antibodies toward compounds. Insufficient retention induces a low breakthrough volume and an incomplete recovery. The capacities of immunosorbents

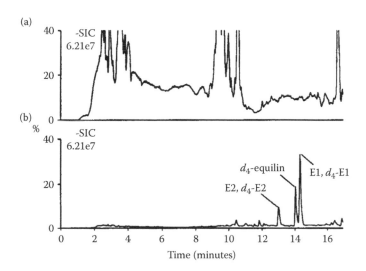

FIGURE 9.6 Total ion chromatograms of (a) raw and (b) extracts of sewage effluent analysed by HPLC-(-)ESI-MS. Peaks corresponding to analytes, surrogate standards, and internal standards are labeled. (From Ferguson, P.L. et al., *Anal. Chem.* 73: 3890–3895, 2001. With permission.)

are rather low compared with those of other reversed-phase sorbents. Thus, it is essential to verify if experimental conditions do not induce an overloading of the capacity. A compromise between cost and sufficient amount of bonded material has to be found.

More developments of immunosorbents are certainly expected in the next years owing to their selectivity and ability for trapping compounds with different polarities. Their important feature is good stability. Immunosorbents can be easily regenerated using phosphate-buffered saline solution, even after being submitted to a high portion of organic solvents applied for the elution step. The high cost and availability are the main limitations in the wider application of SPE immunoaffinity precolumns; they must be prepared for each new analyte, or a class of analytes, and the preparation step often takes a very long time. With the new recombinant technology and the expression of the antibodies in organisms such as yeast, bacteria, and plants it is possible to create antibodies with high affinities for specific antigens in a much faster way and at lower cost [168].

9.7   MOLECULARLY IMPRINTED POLYMERS

The low stability of immunosorbents and the fact that biological antibodies are both difficult to obtain and expensive have led researchers to synthesize antibody mimics such as molecularly imprinted polymers (MIPs) in order to selectively extract the target substances. MIPs are highly cross-linked polymers with specific binding sites for a particular analyte. The print molecule—called the template—is chemically coupled with one or several functional monomers and then spatially fixed in a solid polymer by the polymerisation reaction. After removal of the template, polymers with imprints, which are complementary to the template in terms of size, shape, and functionality, are obtained. These polymers are able to rebind selectively the template molecule or its structural analogues. The right selection of functional monomers is important in molecular imprinting because the interactions with functional groups affect the affinity of MIPs [15].

Two principally different approaches to molecular imprinting may be distinguished. In a noncovalent (or self-assembly) approach, the imprint molecule complexes the monomers by noncovalent or metal ion coordination interactions. The covalent imprinting employs reversible covalent bonds and usually involves a prior chemical synthesis step to link the monomers to the template. The first approach is more flexible in the range of templates that can be used, but covalent imprinting yields better defined and more homogeneous binding sites. Moreover, the former is much easier practically, since complex formation occurs between template and monomers in a solution. Figure 9.7 shows this entire process schematically, and more details on the preparation of imprints can be found elsewhere [169–171]. Proper selection of reagents, reaction medium, and conditions should consider the complexity of formation of selective sites in the polymer structure to obtain a material capable not only of highly selective recognition of target analytes but also of having good kinetic parameters [14].

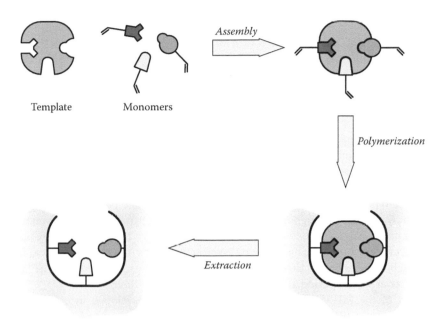

FIGURE 9.7 (*See color insert following page 152.*) Schematic representation of molecular imprinting principle.

MIPs can be obtained in the format of particles, coatings, monolayers of selective compounds bound to the surface of support, monolithic packings, or fibers. Oxelbark et al. [172] presented a comprehensive comparison of five chromatographic stationary phases based on MIPs (e.g., a crushed monolith, microspherical particle material, two silica-based composites, and capillary monolith) in terms of efficiency, imprinting factors, water compatibility, and batch-to-batch reproducibility, all imprinted with the local anesthetic bupivacaine. Except for microparticles, all formats give satisfactory performance, particularly in aqueous solutions.

Depending on the synthetic procedure, there may be some leaching (or so-called bleeding) of the template from the polymer. Since MIPs are made with a large amount of template, a small number of imprint molecules may remain in the resulting polymer, and these may leak during the SPE process. Bleeding is the problem mainly when MIPs have to be applied to extract trace levels of target analytes and may be reduced by thermal treatment of the polymer with extensive washing with strong eluents [15]. The second approach is to use a structural analogue to the analyte of interest as the template for preparing the polymer, thus taking advantage of the cross-reactivity of MIPs [173,174]. Even if leaching of the given analogue occurs during sample recovery, this will not interfere with the analysis.

The selective interactions between the template and the monomers are based on hydrogen bonds and ionic as well as hydrophobic interactions. The different types of interactions involve different levels of specificity, and an increasing number of interacting functional groups in a molecule will increase the selectivity. In organic solvents, imprint recognition is mainly based on interaction with polar functionalities. In aqueous media the recognition of hydrophobic parts of the molecules becomes more significant, and the analytes as well as hydrophobic interfering compounds are nonspecifically adsorbed [175]. Therefore, a selective washing of the MIP sorbent is needed before the elution step to disrupt nonspecific interaction between the analytes and the polymeric matrix. Under these conditions, the specifically retained analytes are not washed out. Since recognition is often best in the solvent used as porogen in the polymerization of the appropriate MIP, this solvent is usually also applied for washing.

The past decade is characterized by the increasing transition of molecular imprinting techniques from a predominantly fundamental experimental and theoretical area to practically useful analytical applications. While biorecognition schemes frequently suffer from degrading bioactivity and long-term stability when applied to real samples, MIPs serving as synthetic antibodies have been successfully applied in environmental monitoring, food and beverage analysis, the pharmaceutical industry, and biological fields [169–171]. Selected application examples of MIP-SPE technique, covering the period 2008–2009, are presented in Table 9.4.

Most of the MIP-SPE reported applications were developed in off-line mode, mainly because of the relatively simple instrumentation required and fairly flexible optimisation of individual steps in SPE procedures. In online systems, the mobile phase desorbs analytes retained on the MIP and transfers the eluate from the extraction cartridge to the injection loop, the content of which is subsequently analyzed on the HPLC column. However, this can be a problem when mobile phase is not powerful enough to disrupt the specific interactions between the analyte and the MIP material [12]. In these cases, a modifier (strong acid or base) is usually added to the mobile phase. Moreover, all the flow parameters in online systems have a pronounced influence on the binding and rebinding in the polymer cavities as well as particle size and packing mode.

TABLE 9.4
Recent Selected Application Examples of MIP-SPE Technique

Template	Analyte	Sample	Functional Monomer/Cross-Linker/ Solvent	Analytical Method	Reference
Ephedrine	Ephedrine, adrenaline, nonadrenaline	Human plasma	MAA/ EGDMA/chloroform	HPLC-UV	[176]
Ursolic acid	Ursolic acid	Extract of herbs	GMA/ EGDMA/ DMSO	HPLC-UV	[177]
Thiabendazol	Benzimidazol	Tap, river and well waters	MMA/ EGDMA/ acetonitrile + toluene	HPLC-UV	[178]
Diethylstilbestrol	Diethylstilbestrol	Fish	TEOS/ APTES/ methanol	HPLC-UV	[179]
Isoxicam	Isoxicam	Pharmaceuticals, human serum	MMA/ EGDMA/ DMF	UV	[180]
Ibuprofen	Ibuprofen	Urine	TEOS/ APTS/ 2-etoxyethanol	HPLC-FLD	[181]
Estrone	Estrone	Biochemical separation of estrone	TEOS/APTES/Fe$_2$O$_3$/ DMSO	HPLC-UV	[182]
Ractopamine	Ractopamine	Pork	TEOS/ APTES/ methanol	HPLC-FLD	[183]
17β-estradiol Atrazine	17β-estradiol Atrazine	Waters	MAA/ EGDMA/ acetonitrile	HPLC-UV	[184]
17β-estradiol	17β-estradiol	Wastewater	MAA/ TRIM/ acetonitrile	HPLC-UV	[185]
2,4,6-trichlorophenol	2,4,6-trichlorophenol	Waters	MAA/ DVB/ toluene + acetonitrile	HPLC-UV	[186]
Polybrominated diphenyl ethers	Polybrominated diphenyl ethers	Wastewater	TEOS/ PTES/ ethanol	GC-MS	[187]

Notes: MAA, methacrylic acid. EGDMA, ethylene glycol dimethacrylate. GMA, glycidyl mtehacrylate. TEOS, tetraethoxysilane. APTES, 3-aminopropyltriethoxysilane. TRIM, trimethylopropane trimethylacrylate. DVB, divinylbenzene. PTES, phenyltrimethoxysilane. DMF, dimethylformamide. DMSO, dimethyl sulfoxide.

REFERENCES

1. Camel, C. 2003. Solid phase extraction of elements. *Spectrochim. Acta Part B* 58: 1177–1233.
2. Pyrzynska, K. 2003. Novel selective sorbents for solid-phase extraction. *Chem. Anal. (Warsaw)* 48: 781–795.
3. Poole, C.F. 2003. New trends in solid-phase extraction. *Trends Anal. Chem.* 22: 362–373.
4. Fontanales, N., Marcé, R.M., Borrull, F. 2007. New materials in sorptive extraction techniques for polar compounds. *J. Chromatogr.* A 1152: 14–31.
5. Jak, P.K., Patel, S., Mishra, B.K. 2004. Chemical modification of silica surface by immobilization of functional groups for extractive concentration of metal ions. *Talanta* 62: 1005–1028.
6. Spivakov, B.Ya., Malofeeva, G.L. Petruskhin, O.M. 2006. Solid-phase extraction on alkyl-bonded silica gels in inorganic analysis. *Anal. Sci.* 22: 503–519.
7. Kyriakopoulos, G., Doulia, D. 2006. Adsorption of pesticides on carbonaceous and polymeric materials from aqueous solutions. *Sep. Purif. Rev.* 35: 97–191.
8. Pyrzynska, K. 2008. Carbon nanotubes as a new solid-phase extraction material for removal and enrichment of organic pollutants in water. *Sep. Purif. Rev.* 37: 375–392.
9. Fontanales, N., Marcé, R.M., Borrull, F. 2004. New hydrophilic materials for solid-phase extraction. *Trends Anal. Chem.* 24: 394–406.
10. Delaunay, N., Pichon, V., Hennion M.C. 2000. Immunoaffinity solid-phase extraction for the trace-analysis of low-molecular-mass analytes in complex sample matrices. *J. Chromatogr.* A, 745: 15–37.
11. Stevenson, D. 2000. Immuno-affinity solid-phase extraction. *J. Chromatogr.* B 745: 39–48.
12. Caro, E., Marcé, R.M., Borrull, F., Cormack, P.G.A., Sherrington, D.C. 2006. Application of molecularly imprinted polymers to solid-phase extraction of compounds from environmental and biological samples. *Trend Anal. Chem.* 25: 143–154.
13. Dias, A.C.B., Fingueiredo, E.C., Grassi, V., Zagatto, E.A.G., Arruda, M.A.Z. 2009. Molecularly imprinted polymer as a solid phase extractor in flow analysis. *Talanta* 76: 988–996.
14. Kloskowski, A., Pilarczyk, M., Przyjazny, A., Namieśnik, J. 2009. Progress in development of molecularly imprinted polymers as sorbents for sample preparation. *Crit. Rev. Anal. Chem.* 39: 43–58.
15. Lasákova, M., Jandera, P. 2009. Molecularly imprinted polymers and their application in solid phase extraction. *J. Sep. Sci.* 32: 799–812.
16. Souverain, S., Rudaz, S., Veuthey, J.-L. 2004. Restricted access materials and large particle supports for on-line sample preparation: an attractive approach for biological fluids analysis. *J. Chromatogr.* B 801: 141–156.
17. Sadilek, P., Šatinsky, D., Solich, P. 2007. Using restricted-access materials and column switching in high-performance liquid chromatography for direct analysis of biologically-active compounds in complex matrices. *Trends Anal. Chem.* 26: 375–384.
18. Kirkland, J.J., Adams, J.B., Van Straten, H.A., Classens H.A. 1998. Bidentate silane stationary phases for reversed-phase high-performance liquid chromatography. *Anal. Chem.* 70: 4344–4351.
19. Buszewski, B., Gadzala-Kopciuch, R., Kaliszan, R., Markuszewski, M., Matyska, M. 1998. Polyfunctional chemically bonded stationary phases for reversed phase high-performance liquid chromatography. *Chromatographia*, 48: 615–622.
20. Parida, S.K., Dash, S., Patel, S., Mishra, B.K. 2006. Adsorption of organic molecules on silica surface. *Adv. Colloid Interfac.* 121: 77–101.

21. Kailasam, K., Fels, A., Müller, K. 2009. Octadecyl grafted MCM-41 silica spheres using trifunctionalsilane precursors—preparation and characterization. *Micropor. Mesopor. Mat.* 117: 136–147.

22. Vigna, C.R.M., Morais, L.S.R., Collins, C.H., Jardim, I.C.S.F. 2006. Poly-(methyloctylsiloxane) immobilized on silica as a sorbent for solid-phase extraction of some pesticides. *J. Chromatogr. A* 1114: 211–215.

23. Faria, A.M., Maldaner, L., Santana, C.C., Jardim, I.C.S.F., Collins, C.H. 2007. Poly-(methyltetradecylsiloxane) immobilized onto silica for extraction of multiclass pesticides from surface waters. *Anal. Chim. Acta* 582: 34–40.

24. Faraji, H. 2005. β-Cyclodextrin-bonded silica particles as the solid-phase extraction medium for the determination of phenol compounds in water samples followed by gas chromatography with flame ionization and mass spectrometry detection. *J. Chromatogr. A* 1087: 283–288.

25. Fan, Y., Feng. Da, S.H. 2003. On-line selective solid-phase extraction of 4-nitrophenol with β-cyclodextrin bonded silica. *Anal. Chim. Acta* 484: 145–153.

26. Li, M., Pham, P.J., Wang T., Pittman Jr., C.U., Li, T. 2009. Selective extraction and enrichment of polyunsaturated fatty acid methyl esters from fish oil by novel π-complexing sorbents. *Sep. Purif. Technol.* 66: 1–8.

27. Liu, F.K. 2008. Preconcentration and separation of neutral steroid analytes using a combination of sweeping micellar electrokinetic chromatography and a Au nanoparticle-coated solid phase exctraction sorbent. *J. Chromatogr. A* 1215: 194–202.

28. Spivakov, B.Y., Malofeeva, G.I., Petrukhin, O.M. 2006. Solid-phase extraction on alkyl-bonded silica gels in inorganic analysis. *Anal. Sci.* 22: 503–519.

29. Davankov, V.A., Tsyurupa, M.P., Ilyin, M.M., Pavlova, L. 2002. Hyper-crosslinked polystyrene and its potential for liquid chromatography. *J. Chromatogr. A* 965: 65–73.

30. Ahn, J.H., Jang, J.E., Oh, C.G., Ihm, S K , Cortez, J , Sherrington D.C. 2006. Rapid generation and control of microporosity bimodal pore size distribution and surface area in Davankov-type hyper-cross-linked resins. *Macromolecules* 39: 627–632.

31. Wissiack, R., Rosenberg, E., Grasserbauer, M. 2000. Comparison of different sorbent materials for on-line solid phase extraction with liquid chromatography—atmospheric pressure chemical ionization mass spectrometry of phenols. *J. Chromatogr. A* 896:159–170.

32. Fontanals, N., Marcé, R.M., Borrull, F. 2006. Improved polymeric materials for more efficient extraction of polar compounds from aqueous samples. *Curr. Anal. Chem.* 2: 171–179.

33. Fontanals, N., Galià, M., Cormack, P.A.G., Marcé, R.M., Sherrington, D.C., Borull, F. 2005. Evaluation of a new hypercrosslinked polymer as a sorbent for solid-phase extraction of polar compounds. *J. Chromatogr. A* 1075: 51–56.

34. Fontanals, N., Marcé, R.M., Cormack, P.A.G., Sherrington, D.C., Borull, F. 2008. Monodisperse, hypercrosslinked polymer microspheres as tailor-made sorbents for highly efficient solid-phase extractions of polar pollutants from water samples. *J. Chromatogr. A* 1191: 118–124.

35. Garces, A., Zerzanova, A., Kucera, R., Barron, D., Barbosa, J. 2006. Determination of a series of quinolones in pig plasma using solid-phase extraction and liquid chromatography coupled with mass spectrometric detection—Application to pharmacokinetic studies. *J. Chromatogr. A* 1137: 22–29.

36. Allanson, A.L., Cotton, M.M., Tettey, J.N.A., Boyter, A.C. 2007. Determination of rifampicin in human plasma and blood spots by high performance liquid chromatography with UV detection: A potential method for therapeutic drug monitoring. *J. Pharm. Biomed. Anal.* 44: 963–969.

37. Higashi, T., Nagahama, A., Otomi, N., Shimada, K. 2007. Studies on neurosteroids XIX—Development and validation of liquid chromatography-tandem mass spectrometric method for determination of 5 alpha-reduced pregnane-type neurosteroids in rat brain and serum. *J. Chromatogr.* B. 848: 188–199.
38. Kuepper, U., Musshoff, F., Madea, B. 2008. A fully validated isotope dilution HPLC-MS/MS method for the simultaneous determination of succinylcholine and succinyl-monocholine in serum and urine samples. *J. Mass Spectrom.* 43: 1344–1352.
39. Higashi, T., Shibayama, Y., Shimada, K. 2007. Determination of salivary dehydroepi-androsterone using liquid chromatography-tandem mass spectrometry combined with charged derivatization. *J. Chromnatogr.* B 846: 195–201.
40. Rodriguez, E., Moreno-Bondi, M.C., Marazuela, M.D. 2008. Development and validation of a solid-phase extraction method coupled to liquid chromatography with fluorescence detection for the determination of fluoroquinolone residues in powdered infant formulae. *J. Chromatogr.* A 1209: 136–144.
41. Tamosiunas, V., Padarauskas, A. 2008. Comparison of LC and UPLC coupled to MS-MS for the determination of sulfonamides in egg and honey. *Chromatographia* 67: 783–788.
42. D'Archivio, A.A., Fanelli, M., Mazzeo, P., Ruggieri, F. 2007. Comparison of different sorbents for multiresidue solid-phase extraction of 16 pesticides from groundwater coupled with high-performance liquid chromatography. *Talanta* 71: 25–30.
43. Hanke, I., Singer, H., Hollender, J. 2008. Ultratrace-level determination of glyphosate, aminomethylphosphonic acid and glufosinate in natural waters by solid-phase extraction followed by liquid chromatography-tandem mass spectrometry: performance tuning of derivatization, enrichment and detection. *Anal. Bioanal. Chem.* 391: 2265–2276.
44. Roberts, P.H., Bersuder, P. 2006. Analysis of OSPAR priority pharmaceuticals using high-performance liquid chromatography-electrospray ionisation tandem mass spectrometry. *J. Chromatogr.* A 1134: 143–150.
45. Melo, S.A.S., Bautitz, I.R., Nogueira, R.F.P. 200. Monitoring pharmaceuticals photo-Fenton degradation process by using solid phase extraction and liquid chromatography. *Anal. Lett.* 41:1682–1690.
46. Stafiej, A., Pyrzynska, K., Regan, F. 2007. Determination of anti-inflammatory drugs and estrogens in water by HPLC with UV detection. *J. Sep. Sci.* 30: 985–991.
47. Michalkiewicz, A., Biesaga, M., Pyrzynska, K. 2008. Solid-phase extraction procedure for determination of phenolic acids and some flavonols in honey. *J. Chromatogr.* A 1187: 18–24.
48. Lajeunesse, A., Gagnon, C. 2007. Determination of acidic pharmaceutical products and carbamazepine in roughly primary-treated wastewater by solid-phase extraction and gas chromatography-tandem mass spectrometry. *Int. J. Environ. An. Ch.* 87: 565–578.
49. Weigel, S., Kallenborn, R., Hühnerfuss, H. 2004. Simultaneous solid-phase extraction of acidic, neutral and basic pharmaceuticals from aqueous samples at ambient (neutral) pH and their determination by gas chromatography-mass spectrometry. *J. Chromatogr.* A 1023: 183–195.
50. Cristofani, E., Antonini, C., Tovo, G., Fioroni, L., Piersanti, A., Galarini, R. 2009. A confirmatory method for the determination of tetracyclines in muscle using high-performance liquid chromatography with diode-array detection. *Anal. Chim. Acta* 637: 40–46.
51. Zhou, J.H., Xue, X.F., Li, Y., Zhang, J.Z., Chen, F., Wu, L.M., Chen, L.Z., Zhao, J. 2009. Multiresidue determination of tetracycline antibiotics in propolis by using HPLC-UV detection with ultrasonic-assisted extraction and two-step solid phase extraction. *Food Chem.* 115: 1074–1080.

52. Vidal, J.L.M., Aguilera-Luiz, M.D., Romero-Gonzalez, R., Frenich, A.G. 2009. Multiclass analysis of antibiotic residues in honey by ultraperformance liquid chromatography-tandem mass spectrometry. *J. Agri. Food Chem.* 57: 1760–1767.

53. Sheen, J., Xia, X., Jiang, H., Li, C., Li, J., Li. X., Ding, S. 2009. Determination of chloramphenicol, thiamphenicol, floorofenicol and florfenicol amine in poultry and porcine muscle and liver by gas chromatography-negative chemical ionization. *J. Chromatogr.* B 877: 1523–1529.

54. Tsai, I.L., Wu, F.L.L., Gau, C.S., Kuo, C.H. 2009. Method development for the determination of teicoplanin in patient serum by solid phase extraction and micellar electrokinetic chromatography. *Talanta* 77: 1208–1216.

55. Pedrouzo, M., Borrull, F., Pocurull, E., Marcé, R.M. 2009. Estrogens and their conjugates: Determination in water samples by solid-phase extraction and liquid chromatography-tandem mass spectrometry. *Talanta* 78: 1327–1331.

56. Lombardo-Agui, M., Cruces-Blanco, C., Garcia-Campana, A.M. 2009. Capillary zone electrophoresis with diode-array detection for analysis of local anaesthetics and opium alkaloids in urine samples. *J. Chromatogr.* B 877: 833–836.

57. Marin, J.M., Gracia-Lor, E., Sancho, J.V., Lopez, F.J., Hernandez, F. 2009. Application of ultra-high-pressure liquid chromatography-tandem mass spectrometry to the determination of multi-class pesticides in environmental and wastewater samples Study of matrix effects. *J. Chromatogr.* A 1216: 1410–1420.

58. Park, M.J., In, S.W., Lee, S.K., Choi, W.K., Park, Y.S., Chung, H.S. 2009. Postmortem blood concentrations of organophosphorus pesticides. *Forensic Sci. Int.* 184: 28–31.

59. Lin, W.C., Wang, S.L., Cheng, C.Y., Ding, W.H. 2009. Determination of alkylphenol residues in breast and commercial milk by solid-phase extraction and gas chromatography-mass spectrometry. *Food Chem.* 114: 753–757.

60. Stoob, K., Singer, H.P., Goetz, C.W., Ruff, M., Mueller, S.R. 2005. Fully automated online solid phase extraction coupled directly to liquid chromatography–tandem mass spectrometry: Quantification of sulfonamide antibiotics, neutral and acidic pesticides at low concentrations in surface waters. *J. Chromatogr.* A 1097: 138–147.

61. Xu, R.N., Fan, L., Rieser, M.J., El-Shourbag, T.A. 2007. Recent advances in high-throughput quantitative bioanalysis by LC-MS/MS. *J. Pharm. Biomed. Anal.* 44: 342–355.

62. Morihisa, H., Fukata, F., Muro, H., Nishimura, K.I., Makini, T. 2008. Determination of indapamide in human serum using 96-well solid phase extraction and high-performance liquid chromatography–tandem mass spectrometry. *J. Chromatogr.* B 870: 126–130.

63. He, H., Sun, C., Wang, X.R., Pham-Huy, C., Chikhi-Chorifi, N., Galons, H., Thevenin, M., Claude, J.R., Warner, J.M. 2005. Solid-phase extraction of methadone enantiomers and benzodiazepines in biological fluids by two polymeric cartridges for liquid chromatographic analysis. *J. Chromatogr.* B 814: 385–391.

64. Samanidou, V.F., Tsochatzis, E.D., Papadoyannis, I.N. 2008. HPLC determination of cefotaxime and cephalexine residues in milk and cephalexine in veterinary formulation. *Microchim. Acta* 160: 471–475.

65. Muniz-Valencia, R., Gonzalo-Lumbreras, R., Santos-Montes, A., Izquierdo-Hornillos, R. 2008. Liquid chromatographic method development for anabolic androgenic steroids using a monolithic column. Application to animal feed samples. *Anal. Chim. Acta* 611:103–112.

66. Samanidou, V.F., Nikolaidou, K.I., Papadoyannis, I.N. 2005. Development and validation of an HPLC confirmatory method for the determination of tetracycline antibiotics residues in bovine muscle according to the European Union regulation 2002/657/EC. *J. Sep. Sci.* 28: 2247–2258.

67. Georga, K.A., Samanidou, V.F., Papadoyannis, I.N. 2001. Use of novel solid-phase extraction sorbent materials for high-performance liquid chromatography quantitation of caffeine metabolism products methlxanthines and methyluric acids in samples of biological origin. *J. Chromatogr.* B 759: 209–218.

68. Legrand, T., Roy, S., Monchaud, C., Grodin, C., Duval, M., Jacqz-Aigrain, E. 2008. Determination of ketamine and norketamine in plasma by micro-liquid chromatography-mass spectrometry. *J. Pharm. Biomed. Anal.* 48: 171–175.

69. Rodriguez-Gonzalo, E., Carabias-Martinez, R., Cruz, E.M., Dominguez-Alvarez, J., Hernandez-Mendez, J. 2009. Ultrasonic solvent extraction and nonaqueous CE for the determination of herbicide residues in potatoes. *J. Sep. Sci.* 32: 575–584.

70. Pena-Gallego, A., Hernandez-Orte, P., Cacho, J., Ferreira, V. 2009. Biogenic amine determination in wines using solid-phase extraction: A comparative study. *J. Chromatogr.* A 1216: 3398–3401.

71. Bijlsma, L., Sancho, J.V., Pitarch, E., Ibanez, M., Hernandez, F. 2009. Simultaneous ultra-high-pressure liquid chromatography-tandem mass spectrometry determination of amphetamine and amphetamine-like stimulants, cocaine and its metabolites, and a cannabis metabolite in surface water and urban wastewater. *J. Chromatogr.* A 1216: 3078–3089.

72. Jagerdeo, E., Abdel-Rehim, M. 2009. Screening of cocaine and its metabolites in human urine samples by direct analysis in real-time source coupled to time-of-flight mass spectrometry after online preconcentration utilizing microextraction by packed sorbent. *J. Am. Soc. Mass Spectr.* 20: 891–899.

73. Kajita, H., Akutsu, C., Hatakeyama, E., Komukai, T. 2008. Simultaneous determination of aminoglycoside antibiotics in milk by liquid chromatography with tandem mass spectrometry. *J. Food Hyg. Soc. Jp.* 49: 189–195.

74. Lee, H.B., Peart, T.E., Svoboda, M.L. 2007. Determination of ofloxacin, norfloxacin, and ciprofloxacin in sewage by selective solid-phase extraction, liquid chromatography with fluorescence detection, and liquid chromatography-tandem mass spectrometry. *J. Chromatogr.* A 1139: 45–52.

75. Pena, M.T., Casais, M.C., Mejuto, M.C., Cela, R. 2008. Development of a matrix solid-phase dispersion method for the determination of polycyclic aromatic hydrocarbons in sewage sludge samples. *Anal. Chim. Acta* 626: 155–165.

76. Delatte, T.L., Selman, M.H.J., Schluepmann, H., Somsen, G.W., Smeekens, S.C.M., de Jong, G.J. 2009. Determination of trehalose-6-phosphate in Arabidopsis seedlings by successive extractions followed by anion exchange chromatography-mass spectrometry. *Anal. Biochem.* 389: 12–17.

77. Hewitt, D., Zhang, T., Kao, Y.H. 2008. Quantitation of polysorbate 20 in protein solutions using mixed-mode chromatography and evaporative light scattering detection. *J. Chromatogr.* A 1215: 156–160.

78. Sorensen, L.M., Nielsen, K.F., Jacobsen, T., Koch, A.G., Nielsen, P.V., Frisvad, J.C. 2008. Determination of mycophenolic acid in meat products using mixed mode reversed phase-anion exchange clean-up and liquid chromatography-high-resolution mass spectrometry. *J. Chromatogr.* A 1205: 103–108.

79. Medina-Remón, A., Barrionuevo-González, A., Zamora-Ros, R., Andres-Lacueva, C., Estruch, R., Martinez-González, M.A., Diez-Espino, J., Lamuela-Raventos, R.M. 2009. Rapid Foli-Ciocalteu method using microtiter 96-well plate cartridges for solid phase extraction to assess urinary total phenolic compounds, as a biomarker of total polyphenols intake. *Anal. Chim. Acta* 634: 54–60.

80. Strahm, E., Rudaz, S., Veuthey, J.L., Saugy, M., Saudan, C. 2009. Profiling of 19-norsteroid sulfoconjugates in human urine by liquid chromatography mass spectrometry. *Anal. Chim. Acta* 613: 228–237.

81. Strahm, E., Kohler, I., Rudaz, S., Martel, S., Carrupt, P.A., Veuthey, J.L., Saugy, M., Saudan, C. 2008. Isolation and quantification by high-performance liquid chromatography-ion-trap mass spectrometry of androgen sulfoconjugates in human urine. *J. Chromatogr.* A 1196: 153–160.

82. Matejicek, D., Houserova, P., Kuban, V. 2007. Combined isolation and purification procedures prior to the high-performance liquid chromatographic-ion-trap tandem mass spectrometric determination of estrogens and their conjugates in river sediments. *J. Chromatogr.* A 1171: 80–89.

83. Huq, S., Garriques, M., Kallury, K.M.R. 2006. Role of zwitterionic structures in the solid-phase extraction based method development for clean-up of tetracycline and oxytetracycline from honey. *J. Chromatogr.* A 1135: 12–18.

84. Bermudo, E., Moyano, E., Puignou, L., Galceran, M.T. 2006. Determination of acrylamide in foodstuffs by liquid chromatography ion-trap tandem mass-spectrometry using an improved clean-up procedure. *Anal. Chim. Acta* 559: 207–214.

85. Ilanson, A..L, Cotton, M.M., Tettey, J.N.A., Boyter, A.C. 2007. Determination of rifampicin in human plasma and blood spots by high performance liquid chromatography with UV detection: A potential method for therapeutic drug monitoring. *J. Pharmaceut. Biomed.* 44: 963–969.

86. Lavén, M., Alsberg, T., Yu, Y., Adolfsson-Erici, M., Sun, H.W. 2009. Serial mixed-mode cation- and anion-exchange solid-phase extraction for separation of basic, neutral and acidic pharmaceuticals in wastewater and analysis by high-performance liquid chromatography-quadrupole time-of-flight mass spectrometry. *J. Chromatogr.* A 1216: 49–62.

87. Fontanals, N., Cormack, P.A.G., Sherrington, D.C. 2008. Hypercrosslinked polymer microspheres with weak anion-exchange character. Preparation of the microspheres and their applications in pH-tuneable, selective extractions of analytes from complex environmental samples. *J. Chromatogr.* A 1215: 21–29.

88. Moreno-Castilla, C.M. 2004. Adsorption of organic molecules from aqueous solutions on carbon materials. *Carbon* 42: 83–94.

89. Kyriakopoilos, G., Doulia, D. 2006. Adsorption of pesticides on carbonaceous and polymeric materials from aqueous solutions. *Sep. Purif. Rev.* 35: 97–191.

90. Efremenko, L., Sheintuch, M. 2006. Predicting solute adsorption on activated carbon: phenol. *Langmuir* 22: 3614–3621.

91. Boehm, H.P. 2002. Surface oxides on carbon and their analysis: a critical assessment. *Carbon* 40: 145–149.

92. Hanai, T. 2003. Separation of polar compounds using carbon columns. *J. Chromatogr.* A 989: 183–196.

93. Hennion, M.C. 2000. Graphitized carbons for solid-phase extraction. *J. Chromatogr.* A 885: 71–95.

94. Forgács, E. 2002. Retention characteristics and practical applications of carbon sorbents. *J. Chromatogr.* A 975: 229–243.

95. Michel, M., Buszewski, B. 2009. Porous graphitic carbon sorbents in biomedical and environmental applications. *Adsorption* 15: 193–202.

96. Li, L., Li, W., Ge, J., Wu, Y., Jiang, S., Liu, F. 2008. Use of graphitic carbon black and primary secondary amine for determination of 17 organophosphorous pesticide residues in spinach. *J. Sep. Sci.* 31: 3588–3594.

97. Long, R.Q., Yang, R.T. 2001. Carbon nanotubes as superior sorbent for dioxin removal. *J. Am. Chem. Soc.* 123: 2058–2059.

98. Yang, K., Zhu, L., Xing, B. 2006. Adsorption of polycyclic aromatic hydrocarbons by carbon nanomaterials. *Environ. Sci. Technol.* 40: 1855–1861.

99. Dastgheib, S.A., Rockstraw, D.A. 2002. Model for the adsorption of single metal ion solutes in aqueous solution onto activated carbon produced from pecan shells. *Carbon* 40: 1843–1851.

100. An, X., Zeng, H. 2003. Functionalization of carbon nanobeads and their use as metal ion adsorbents. *Carbon* 41: 2889–2896.

101. Datsyuk, V., Kalyva, M., Papagelis, K., Parthenios, J., Tasis, D., Siokou, A., Kallitsis, I., Galiiotis, C. 2008. Chemical oxidation of multiwalled carbon nanotubes. *Carbon* 46: 833–840.

102. Cho, H.H., Smith, B.A., Wnuk, J.D., Fairbrother, D.H., Ball, W.P. 2008. Influence of surface oxides on the adsorption of naphthalene onto multiwalled carbon nanotubes. *Environ. Sci. Technol.* 42: 2899–2905.

103. Liao, Q., Sun, J., Gao, L. 2008. Adsorption of chlorophenols by multi-walled carbon nanotubes treated with HNO_3 and NH_3. *Carbon* 46: 544–561.

104. Salam, M.A., Burk, R.C. 2008. Thermodynamics of pentachlorophenol adsorption from aqueous solutions by oxidized multi-walled carbon nanotubes. *App. Surf. Sci.* 255: 1975–1981.

105. Chen, W., Duan, L., Wang, L., Zhu, D. 2008. Adsorption of hydroxyl- and amino-substituted aromatics to carbon nanotubes. *Environ. Sci. Technol.* 42: 6862–6868.

106. Lin, D., Xing, B. 2008. Adsorption of phenolic compounds by carbon nanotubes: role of aromaticity and substitution of hydroxyl groups. *Environ. Sci. Technol.* 42: 7254–7259.

107. Liao, Q., Sun, J., Gao, L. 2008. The adsorption of resorcinol from water using multi-walled carbon nanotubes. *Colloid Surface* A 312: 160–165.

108. Liu, H., Li, J., Liu, X., Jiang, S. 2009. A novel multiwalled carbon nanotubes bonded fused-silica fiber for solid phase microextraction-gas chromatographic analysis of phenols in water samples. *Talanta* 78: 929–935.

109. Niu, H.Y., Cai, Y.Q., Shi, Y.L., Wei, F.S., Liu, J.M., Bin, Jiang, G. 2008. A new solid-phase extraction disk based on a sheet of single-walled carbon nanotubes. *Anal. Bioanal. Chem.* 392: 927–935.

110. Dong, M., Ma, Y., Zhao, E., Qian, C., Han, L., Jiang, S. 2009. Using multiwalled carbon nanotubes as solid phase extraction adsorbents for determination of chloroacetanilide herbicides in water. *Microchim. Acta* 165: 123–128.

111. Niu, H., Shi, Y., Cai, Y., Wei, F., Jiang, G. 2009. Solid-phase extraction of sulfonylurea herbicides from water samples with single-walled carbon nanotube disk. *Microchim. Acta* 164: 431–438.

112. Dong, M., Ma, Y., Liu, F., Qian, C., Han, L., Jiang, S. 2009. Use of multiwalled carbon nanotubes as a SPE adsorbent for analysis of carfentrazone-ethyl in water. *Chromatographia* 69: 73–77.

113. Yan, X.M., Shi, B.Y., Lu, J.J., Feng, C.H., Wang, D.S., Tang, H.X. 2008. Adsorption and desorption of atrazine on carbon nanotubes. *J. Coll. Inter. Sci.* 321: 30–38.

114. Ravelo-Perez, L.M., Hernandez-Borges, J., Rodriguez-Delgado, M.A. 2008. Multiwalled carbon nanotubes as efficient solid-phase extraction materials of organophosphorus pesticides from apple, grape, orange and pineapple fruit juices. *J. Chromatogr.* A 1211: 33–42.

115. El-Skeikh, A.H., Sweileh, J.A., Al-Degs, Y.S., Insisi, A.A., Al-Rabady, N. 2008. Critical evaluation and comparison of enrichment efficiency of multi-walled carbon nanotubes, C18 silica and activated carbon towards some pesticides from environmental waters. *Talanta* 74: 1675–1680.

116. Asensio-Ramos, M., Hernandez-Borges, J., Ravelo-Perez, L.M., Rodriguez-Delgado, M.A. 2008. Simultaneous determination of seven pesticides in waters using multiwalled carbon nanotube SPE and NACE. *Electrophoresis* 29 : 4412–4421.

117. Li, Q., Wang, X., Yuan, D. 2009. Solid-phase extraction of polar organophosphorous pesticides from aqueous samples with oxidized carbon nanotubes. *J. Environ. Monit.* 11: 439–444.
118. Zhou, Q., Xiao, J., Xie, G., Wang, W., Ding, Y., Bai, H. 2009. Enrichment of pyrothroid residues in environmental waters using a multiwalled carbon nanotubes cartridge and analysis in combination with high performance liquid chromatography. *Microchim. Acta* 164: 419–424.
119. Salam, M.A., Burk, R. 2009. Solid phase extraction of polyhalogenated pollutants from freshwater using chemically modified multi-walled carbon nanotubes and their determination by gas chromatography. *J. Sep. Sci.* 32: 10060–1068.
120. Salam, M.A., Burk, R. 2008. Novel application of modified multiwalled carbon nanotubes as a solid phase extraction adsorbent for the determination of polyhalogenated organic pollutant in aqueous solution. *Anal. Bioanal. Chem.* 390: 2159–2170.
121. Pan, B., Lin, D., Maskayekhi, H., Xing, B. 2008. Adsorption and hysteresis of bisphenol A and 17α-ethinyl estradiol on carbon nanomaterials. *Environ. Sci. Technol.* 42: 5480–5485.
122. Guan, Z., Huang, Y.M., Wang, W.D. 2008. Carboxyl modified multi-walled carbon nanotubes as solid-phase extraction adsorbents combined with high-performance liquid chromatography for analysis of linear alkylbenzene sulfonates. *Anal. Chim. Acta* 627: 225–231.
123. Rastkari, N., Ahmadkhaniha, R., Yunesian, M. 2009. Single-walled carbon nanotubes as an effective adsorbent in solid-phase microextraction of low level methyl tert-butyl ether, ethyl tert-butyl ether and methyl tert-amyl ether from human urine. *J. Chromatogr.* B 877: 1568–1574.
124. Sone, H., Fugetsu, B., Tsukada, T., Endo, M. 2008. Affinity-based elimination of aromatic VOCs by highly crystalline multi-walled carbon nanotubes. *Talanta* 74: 1265–1270.
125. Hussain, C.M., Saridara, C., Mitra, S. 2008. Carbon nanotubes as sorbent for the gas phase preconcentration of semivolatile organic in a microtrap. *Analyst* 133: 1076–1082.
126. Shih, Y.H., Li, M.S. 2008. Adsorption of selected volatile organic vapors on multiwalled carbon nanotubes. *J. Hazard. Mater.* 154: 21–28.
127. Pan, B., Xing, B. 2008. Adsorption mechanisms of organic chemicals on carbon nanotubes. *Environ. Sci. Technol.* 42: 9065–9013.
128. Zhou, Q., Xiao, J., Wang, W. 2007. Comparison of multiwalled carbon nanotubes and a conventional absorbent on the enrichment of sulfonylurea herbicides in water samples. *Anal. Sci.* 23: 189–192.
129. Saridara, C., Brukh, R., Iqbal, Z., Mitra, S. 2005. Preconcentration of volatile organics on self-assembled, carbon nanotubes in a microtrap. *Anal. Chem.* 77: 1183–1187.
130. Thurman, E.M., Snavely, K. 2000. Advances in solid-phase extraction disks for environmental chemistry. *Trends Anal. Chem.* 19: 18–26.
131. Zheng, F., Baldwin, D.L., Fifield, L.S., Anheier, N.C., Aardahl, C.L., Grate, J.W. 2006. Single-walled carbon nanotube paper as a sorbent for organic vapor preconcentration. *Anal. Chem.* 78: 2442–2446.
132. Sae-Khow, O., Mitra, S. 2009. Carbon nanotubes as the sorbent for integrating μ-solid phase extraction within the needle of a syringe. *J. Chromatogr.* A 1216: 2270–2274.
133. Hylton, K., Chen, Y., Mitra, S. 2008. Carbon nanotube microscale membrane extraction. *J. Chromatogr.* A 1211: 43–48.
134. Li, Y., Kaneko, T., Ogawa, T., Takahashi, M., Hatakeyama, R. 2007. Magnetic characterization of Fe-nanoparticles encapsulated single-walled carbon nanotubes. *Chem. Comm.* 254–256.
135. Bystrzejewski, M., Lange, H., Huczko, A. 2007. Carbon encapsulation of magnetic nanoparticles. *Fuller. Nanotub. Car. N.* 15: 167–180.

136. Borysiuk, J., Grabias, A., Szczytko, J., Bystrzejewski, M., Twardowski, A., Lange, H. 2008. Structure and magnetic properties of carbon encapsulated Fe nanoparticles obtained by arc plasma and combustion synthesis. *Carbon* 46: 1693–1701.

137. Jin, J., Li, R., Wang, H., Chen, H., Liang, K., Ma, J. 2007. Magnetic Fe nanoparticles functionalized water-soluble multi-walled carbon nanotubes towards the preparation of sorbent for aromatic compounds removal. *Chem. Commun.* 4: 386–388.

138. Safrikova, M., Lunackova, P., Komarek, K., Hubka, T., Safarik, I. 2007. Preconcentration of middle oxyethylated nonylphenols from water samples on magnetic solid phase. *J. Magn. Magn. Mat.* 311: 405–408.

139. Zhao, X., Shi, Y., Cai, Y., Mou, S. 2008. Cetyltrimethylammonium bromide-coated magnetic nanoparticles for the preconcentration of phenolic compounds from environmental water samples. *Emviron. Sci. Technol.* 42: 1201–1206.

140. Zhao, X., Shi, Y., Wang, T., Cai, Y., Jiang, G. 2008. Preparation of silica-magnetite nanoparticles mixed hemimicelle sorbents for extraction of several typical phenolic compounds from environmental water samples. *J. Chromatogr.* A 1188: 140–147.

141. Bystrzejewski, M., Pyrzynska, K., Huczko, A., Lange, H. 2009. Carbon-encapsulated magnetic nanoparticles as separable and mobile sorbents of heavy metal ions from aqueous solutions. *Carbon* 47: 1201–1204.

142. Souverain, S., Rudaz, S., Veuthey, J.-L. 2004. Restricted access materials and large particle supports for on-line sample preparation: an attractive approach for biological fluids analysis. *J. Chromatogr.* B 801: 141–156.

143. Öhman, D., Carlsson, B., Norland, B. 2001. On-line extraction using an alkyl-diol silica precolumn for racemic citalopram and its metabolites in plasma. Results compared with solid-phase extraction methodology. *J. Chromatogr.*, B, 753: 365–373.

144. Brunetto, M.R., Obando, M.A., Fernández, A., Gallignani, M., Burguera, J.L., Burguera, M. 2002. Column-switching high-performance liquid chromatographic analysis of carbamazepine and its principal metabolite in human plasma with direct sample injection using an alkyl-diol silica (ADS) precolumn. *Talanta* 58: 535–542.

145. Mader, R.M., Rizovski, B., Steger, G.G. 2002. On-line solid-phase extraction and determination of paclitaxel in human plasma. *J. Chromatogr.* B 769: 357–361.

146. Meesters, R.J.W., Duisken, M., Jähningen, H., Hollender, J. 2008. Sensitive determination of monoterpene alcohols in urine by HPLC–FLD combined with ESI-MS detection after online-solid phase extraction of the monoterpene-coumarincarbamate derivates. *J. Chromatogr.* B 875: 444–450.

147. Berna, M.J., Ackermann, B.L., Murphy, A.T. 2004. High-throughput chromatographic approaches to liquid chromatographic/tandem mass spectrometric bioanalysis to support drug discovery and development. *Anal. Chim. Acta* 509: 1–9.

148. Huclova, J., Satinsky, D., Maia, T., Karlicek, R. 2005. Sequential injection extraction based on restricted access material for determination of furosemide in serum. *J. Chromatogr.* A 1087: 245–251.

149. Mullet, W.M. 2007. Determination of drugs in biological fluids by direct injection of samples for liquid-chromatographic analysis. *J. Biochem. Biphys. Meth.* 70: 263–273.

150. Santos-Neto, A.J., Bergquist, J., Lancas, F.M., Sjöberg, P.J.R. 2008. Simultaneous analysis of five antidepressant drugs using direct injection of biofluids in a capillary restricted-access media-liquid chromatography–tandem mass spectrometry system. *J. Chromatogr.* A 1189: 514–522.

151. del Rosario Brunetto, M., Contreras, Y., Clavijo, S., Tores, D., Delago, Y. 2009. Determination of losartan, telmisartan, and valsartan by direct injection of human urine into a column-switching liquid chromatographic system with fluorescence detection. *J. Pharm. Biomed. Anal.* 50: 194–199.

152. Winther, B., Moi, P., Nordlund, M.S., Lunder, N., Paus, E., Reubsaet, J.L.E. 2009. Absolute ProGRP quantification in a clinical relevant concentration range using LC–MS/MS and a comprehensive internal standard. *J. Chromatogr.* B 877: 1359–1365.
153. Lien, G.W., Chen, C.Y., Wang, G.S. 2009. Comparison of electrospray ionization, atmospheric pressure chemical ionization and atmospheric pressure photoionization for determining estrogenic chemicals in water by liquid chromatography tandem mass spectrometry with chemical derivatization. *J. Chromatogr.* A 1216: 956–966.
154. Oliveira, R.V., Cass, Q.B. 2006. Evaluation of liquid chromatographic behavior of cephalosporin antibiotics using restricted access medium columns for on-line sample cleanup of bovine milk. *J. Agric. Food. Chem.* 54: 1180–1187.
155. Chico, J., Rubies, A., Centrich, F., Companyo, R., Prat, M.D., Granados, M. 2008. High-throughput multiclass method for antibiotic residue analysis by liquid chromatography-tandem mass spectrometry. *J. Chromatogr.* A 1213: 189–199.
156. Marazuela, M.D., Bogialli, S. 2009. A review of novel strategies of sample preparation for the determination of antibacterial residues in foodstuffs using liquid chromatography-based analytical methods. *Anal. Chim. Acta* 645: 5–17.
157. Friedrich, G., Rose, T., Rissler, K. 2003. Determination of testosterone metabolites in human hepatocytes. Development of an on-line sample preparation liquid chromatography technique and mass spectroscopic detection of 6 beta-hydroxy-testosterone. *J. Chromatogr.* B 784: 49–61.
158. Delaunay, N., Pichon, V., Hennion, M.C. 2000. Immunoaffinity solid-phase extraction for the trace analysis of low-molecular-mass analytes in complex sample matrices. *J. Chromatogr.* B 745: 15–37.
159. Haginaka, J. 2005. Selectivity of affinity media in solid-phase extraction of analytes. *Trends Anal. Chem.* 24: 407–415.
160. Rodriguez-Mozaz, S., de Alda, M.J.L., Barcelo, D. 2007. Advantages and limitations of on-line solid phase extraction coupled to liquid chromatography-mass spectrometry technologies versus biosensors for monitoring of emerging contaminants in water. *J. Chromatogr.* A 1152: 97–115.
161. Vera-Avila, L.E., Vazquez-Lira, J.C., De Llasera, M.G., Covarrubias, R. 2005. Sol-gel immunosorbents doped with polyclonal antibodies for the selective extraction of malathion and triazines from aqueous samples. *Environ. Sci. Technol.* 39: 5421–5426.
162. Ferguson, P.L., Iden, C.R., McElroy, A.E., Brownawell, B.J. 2001. Determination of steroid estrogens in wastewater by munoaffinity extraction coupled with HPLC-electrospray-MS. *Anal. Chem.* 73: 3890–3895.
163. Nevanen, T.K., Simolin, H., Suortti, T., Koivula, A., Soderlund, H. 2005. Development of a high-throughput format for solid-phase extraction of enantiomers using an immunosorbent in 384-well plates. *Anal. Chem.* 77: 3038–3044.
164. Estevez-Alberola, M.C, Marco, M.P. 2004. Immunochemical determination of xenobiotics with endocrine disrupting effects. *Anal. Bioanal. Chem.* 378: 563–575.
165. Chapuis, F., Pichon, V., Lanza, F., Sellergren, B., Hennion, M.C. 2004. Retention mechanism of analytes in the solid-phase extraction process using molecularly imprinted polymers—Application to the extraction of triazines from complex matrices. *J. Chromatogr.* B 804: 93–101.
166. Spoors, J.A., Winger, L.A., Siew, L.K., Dessi, J.L., Jennens, L., Self, C.H. 2002. The firs monoclonal antibody-based, matrix-resistant immunoassay for carbamate herbicide asulam, in water. *J. Environ. Monit.* 4: 917–921.
167. Delaunay-Bertoncini, N., Pichon, V., Hennion, M.C. 2003. Experimental comparison of three monoclonal antibodies for the class-selective immunoextraction of triazines: Correlation with molecular modelling and principal component analysis studies. *J. Chromatogr.* A 999: 3–15.

168. Churchill, R. L.T., Sheedy, C., Yau, Kerrm Y.F., Hall, J. C. 2002. Evolution of antibodies for environmental monitoring: from mice to plants. *Anal. Chim.* Acta 468: 185–197.
169. Mahony, J.O., Nolan, K., Smyth, M.R., Mizaikoff, B. 2005. Molecularly imprinted polymers—potential and challenges in analytical chemistry. *Anal. Chim. Acta* 534: 31:39.
170. Qiao, F., Sun, H., Yan, H., Row, K.H. 2006. Molecularly imprinted polymers for solid phase extraction. *Chromatographia* 64: 625–634.
171. Diaz-Garcia, M.E., Lamo, R.B. 2005. Molecular imprinting in sol-gel materials: recent developments and applications. *Microchim. Acta* 149: 19–36.
172. Oxelbark, J., Legido-Quigley, C., Aureliano, C.S.A., Titirici, M.M., Schillinger, E., Sellergren, B., Courtois, J., Irgum, K., Dambies, L., Cormack, P.A.G., Sherrington, D.C., De Lorenzi, E. 2007. Chromatographic comparison of bupivacaine imprinted-polymers prepared by crushed monolith, microsphere, silica-based composite and capillary monolith formats. *J. Chromatogr.* A 1160: 215–226.
173. Andersson, L.I., Hardenborg, E., Sandberg-Ställ, M., Möller, K., Henriksson, J., Bransby-Sjostrom, I., Olson, L.L., Abdel-Rahim, M. 2004. Development of a molecularly imprinted polymer based solid-phase extraction of local anaesthetics from human plasma. *Anal. Chim. Acta* 526: 147–154.
174. Tamayo, F.G., Titirici, M.M., Martin-Esteban, A., Sellergren, B. 2005. Synthesis and evaluation of new propazine-imprinted polymer formats for use as stationary phases in liquid chromatography. *Anal. Chim. Acta* 542: 38–46.
175. Quaglia, M., Chenon, K., Hall, A.J., De Lorenzi, E., Sellergren, B. 2001. Target analogue imprinted polymers with affinity for folic acid and related compounds. *J. Am. Chem. Soc.* 123: 2146–2154.
176. Laskáková, M., Thiebau, D., Jandeara, P., Pichon, V. 2009. Molecularly imprinted polymer for solid-phase extraction of ephedrine and analogs from human plasma. *J. Sep. Sci.* 32: 1036–1042.
177. Liu, H., Liu, C., Yang, X., Zeng, S., Xiong, Y., Xu, W. 2008. Solid-phase extraction of ursolic acid from herb using β-cyclodexrtin-based molecularly imprinted microspheres. *J. Sep. Sci.* 31: 3573–3560.
178. Cacho, C., Turiel, E., Perez-Conde, C. 2009. Molecularly imprinted polymers: An analytical tool for the determination of benzimidazole compounds in water samples. *Talanta* 78: 1029–1035.
179. Jiang, X., Zhao, C., Jiang, N., Zhang, H., Liu, M. 2008. Selective solid-phase extraction using molecular imprinted polymer for the analysis of diethylstilbestrol. *Food. Chem.* 108: 1061–1067.
180. Rezaei, B., Mallakpour, S., Majidi, N. 2008. Solid-phase molecularly imprinted preconcentration and spectrophotometric determination of isoxicam in pharmaceuticals and human serum. *Talanta* 78: 418:423.
181. Farrington, K., Regan, F. 2009. Molecularly imprinted sol gel for ibuprofen: An analytical study of the factor influencing selectivity. *Talanta* 78: 652–659.
182. Wang, X., Wang, L., He, X., Zhang, Y., Chen, L. 2009. A molecularly imprinted polymer-coated nanocomposite of magnetic nanoparticles for etrone recognition. *Talanta* 78: 327–332.
183. Wang, S., Liu, L., Fang, G., Zhang, C., He, J. 2009. Molecularly imprinted polymer for the determination of trace ractopamine in pork using SPE followed by HPLC with fluorescence detection. *J. Sep. Sci.* 32: 1333–1339.
184. Le Noir, M., Plieva, F.M., Mattiasson, B. 2009. Removal of endocrine-disrupting compounds from water using macroporous molecularly imprinted cryogels in a moving-bed reactor. *J. Sep. Sci.* 32: 1471–1479.
185. Nemulens, O., Mhaka, B., Cukrowska, E., Ramström, O., Tutu, H., Chimuka, L. 2009. Potential of combining of liquid membranes and molecularly imprinted polymers extraction of 17β-estradiol from aqueous samples. *J. Sep. Sci.* 32: 1041–1948.

186. Feng, Q., Zhao, L., Lin, J.M. 2009. Molecularly imprinted polymer as micro-solidphase extraction combined with high performance liquid chromatography to determine phenolic compounds in environmental samples. *Anal. Chim. Acta* doi: 10.1016/j.aca.2009.04.016.
187. Li. M.K.Y., Lei, N.Y., Gong, C., Yu, Y., Lam, K.H., Lam, M.H.W., Yu, H., Lam, P.K.S. 2009. An organically modified silica molecularly imprinted solid-phase microextraction device for the determination of plybrominated dipheyl ethers. *Anal. Chim. Acta* 633: 197–203.

10 Thin-Layer and High-Performance Thin-Layer Chromatographic Analysis of Biological Samples*

Joseph Sherma and Bernard Fried

CONTENTS

---

* We thank Karen E. LeSage for assistance in preparing this manuscript.

10.1 INTRODUCTION

This chapter reviews the use of thin-layer chromatography (TLC) and high-performance TLC (HPTLC) for the analysis of biological samples of particular interest to biologists, biochemists, hematologists, immunologists, medical diagnosticians, and molecular biologists. Determinations of amino acids, drugs, neutral lipids, polar lipids, steroids, gangliosides, glycosides, pigments, phenols, bile acids, oligosaccharides, and other compound classes in a great variety of human and animal sample matrices are included. A limited number of separations and analyses of biologically active compounds such as pesticides, drugs, and phytomedicines from plant samples are also included. The chapter discusses the advantages of using modern TLC for biological applications and summarizes information on stationary and mobile phases and methods used for sample preparation; application of standards and samples; plate development; and zone detection, identification, and quantification. The information in this chapter updates a review article on TLC in biological analysis that covered the literature through 2004 [1].

In TLC and HPTLC, a great number of different sorbents, mobile phases, and development modes are available to provide high-efficiency separations of polar and nonpolar compounds, and use of modern techniques and commercial instruments permits accurate and precise automated quantitative analysis. The off-line nature of TLC allows use of a broad range of selective and universal detection methods in sequence for confirmation of identity, and the ability to simultaneously separate many samples applied on the same plate with corresponding standards leads to high sample throughput, relatively rapid and inexpensive analyses, and accurate and precise quantitative analysis. The advantages of TLC and HPTLC compared with column high-performance liquid chromatography (HPLC) and other analytical methods are described in an introductory book designed to be especially useful for biologists [2].

The following sections briefly describe the steps of biological sample analysis by TLC and HPTLC and offer examples of typical specific applications to a variety of analytes and sample matrices. Additional detailed information on the theory, practice, equipment, instrumentation, and applications of basic and advanced TLC analysis to biological and other samples can be found in a comprehensive handbook of TLC [3] and books on preparative layer chromatography (PLC) [4], TLC in separations and analysis of enantiomers [5], and TLC in phytochemical analysis [6].

10.2 MATERIALS AND TECHNIQUES

TLC, which is sometimes termed *planar chromatography*, is a type of liquid chromatography in which the stationary phase is a thin, uniform layer of a fine-particle sorbent on a glass plate, aluminum foil, or plastic sheet. In the basic TLC procedure, a solution of the sample is applied to the lower part of the plate, and the plate is developed by placing it in a closed chamber whose base is covered with the mobile phase, which is usually a mixture of solvents. After development, the plate is removed from the tank, and the mobile phase front is marked for calculation of R_f (= distance of travel of the zone divided by the distance of the mobile phase front). The zones are detected to visualize the resulting chromatogram of the separated zones. Traditional qualitative or semiquantitative TLC is simple to use and requires only low-cost apparatus (i.e., a development chamber, glass capillary applicator, precoated plate or sheet, and a detection reagent sprayer or dip tank). Modern instruments are available to automate various steps of the procedure, such as sample application, development, and quantification, for improved separations and more accurate and precise quantification.

10.2.1 Sample Preparation

Because TLC plates are not reused, it is often possible to carry out less sample preparation compared with analytical methods such as HPLC in which strongly sorbed impurities from a sample can be eluted slowly and interfere with the analysis of a later sample on the same column. However, most biological samples are complex mixtures (e.g., blood, urine, feces, saliva, cerebrospinal fluid, gastric fluid, body tissues, and other biologics) and may require some degree of sample preparation for purification and concentration prior to TLC analysis.

Liquid–liquid extraction (LLE) [7] is the major sample preparation technique prior to TLC. One of the most used LLE methods for biological samples employs chloroform-methanol (2:1) extraction followed by washing the organic phase with 0.88% potassium chloride (the Folch procedure). This procedure was used to extract neutral lipids from the feces of BALB/c mice infected with *Echinostoma caproni* prior to silica gel-densitometric determination [8]. The following are examples of other reported sample preparation methods prior to TLC:

- Extraction with chloroform-methanol (1:1) followed by LLE and preparative HPLC for gangliosides from tissues (melanoma tumors, sciatic nerves, and brain) [9]
- Ethyl acetate extraction of lamotrigine from serum of epileptic patients [10]
- LLE of concentrated urine samples and TLC of ethanolic extracts for determination of cortisol and cortisone [11]
- Extraction of cardiolipin from mammalian myocardia [12] and guinea pig and bullfrog cardiac and muscle tissue [13] with chloroform-methanol (2:1)
- Extraction of oligosaccharides from pig intestine contents with hot water [14]
- Exhaustive extraction of bioactive substances overnight with methanol [15].

Solvent extraction is improved by vortex mixing for a short period, such as for the extraction of the aldose reductase inhibition drug epalrestat [16] and the fluoroquinolone antibiotic levofloxacin [17] with dichloromethane from human plasma. Column chromatography is often needed for purification of extracts prior to TLC, for example, cleanup over DEAE-sepharose (ion exchange) and silica gel 60 (adsorption chromatography) to obtain a ganglioside GM-3 preparation from Chinese hamster ovary (CHO) cells [18].

Solid-phase extraction (SPE) and pressurized liquid extraction (PLC) are more modern methods that have been used for biological sample preparation prior to TLC analysis. SPE is carried out by passing a liquid sample or sample extract through a small cartridge or column packed with a sorbent. A solvent is passed through the column or cartridge to elute the analyte in a fraction for its TLC analysis. Impurities are eluted first in a separate fraction with a different solvent and discarded, or they remain on the cartridge after elution of the analyte. As an example, internal standard nitazoxanide and HCl were added to a human plasma sample in an Eppendorf tube, and the solution was vortex mixed and loaded into a 3 mL Oasis MCX cartridge previously conditioned with methanol; the cartridge was washed with 0.1 M HCl to remove impurities and then with alkaline methanol to elute the analyte tizoxanide, a broad-spectrum antiparasitic agent; and the eluate was evaporated to dryness and reconstituted for HPTLC-densitometry analysis [19]. Urine samples from abusers of cocaine were submitted to SPE prior to methylation with diazomethane and TLC analysis [20].

PLE with ethanol was used to recover compounds with antioxidant properties from the microalga *Spirulina platensis* [21]. The extraction was carried out at 115°C for 15 min.

10.2.2 STATIONARY PHASES

Normal-phase (NP) or straight-phase adsorption TLC or HPTLC on silica gel with a less polar mobile phase has been used in the majority of reported analyses of biological samples, especially Merck (Darmstad, Germany) silica gel 60. Commercially precoated silica gel layers are usually used as received, but it was shown that activation time and temperature prior to TLC can influence the migration distances and orders of biologically active steroids [22]. Many examples of biological analyses on silica gel are given in the following sections of this chapter. Some other inorganic adsorbents have had limited use, for example, the determination of underivatized aliphatic polyamines in urine samples on calcium sulfate layers [23].

The reversed-phase (RP) TLC of analytes in biological samples has been carried out on octadecylsilyl (C-18 or RP-18) bonded silica gel layers developed with a more polar aqueous mobile phase, such as selected bile acids on RP-18W (water wettable) layers with mobile phases composed of methanol, acetonitrile, acetone, or dioxane mixed with water in different proportions [24] and 12 benzothiazoles on RP-18 developed with mixtures of acetone and Tris(tris(hydroxymethyl)-aminomethane buffer (pH 7.4) [25]. Bonded ethyl (RP-2) [26], cyano (CN), and diol [27] layers were applied to the separation of selected bile acids. Mobile phases for RP-2 were the same as for RP-18W (above), while the optimal mobile phase for CN and diol was n-hexane-ethyl acetate-acetic acid (49:49:2) and (21:21:8), respectively. CN-modified silica gel layers can be used for separations with RP or NP mechanisms, depending on the composition of the mobile phase. Silanized silica gel $60F_{254}$ layers developed with methanol-water (3:2) were used for the RP-TLC of 29 biologically active isomeric chalcones and cyclic chalcone analogs [28].

Manually impregnated (e.g., with paraffin) or chemically bonded RP layers are used for determination of the hydrophobicity of compounds, which is related to biological activity; as examples, RP-18W and CN phases were used to study the hydrophobicity of 22 dihydroxythiobenzanilide fungicides with mobile phases containing methanol or acetone as organic modifiers of aqueous mobile phases [29] and RP-18 phase for 8 N,N-disubstituted-2-phenylacetamides with acetone, tetrahydrofuran, dioxane, methanol, ethanol, 1-propanol, and 2-propanol as organic modifiers [30]. Octylsilyl (C-8), amino, and phenyl bonded layers have also been applied to biological sample analysis, that is, for separation of sterols [31].

Native or microcrystalline cellulose layers provide separations of relatively polar compounds based on the mechanism of normal phase liquid–liquid partitioning, whereas layers containing sodium form sulfonic acid resin (Polygram Ionex-25 SA-Na) are used to separate ionized compounds by the mechanism of ion exchange. Both of these layers were used to separate and determine amino acids in snail digestive gland- gonad (DGG) complex [32].

Preadsorbent TLC and HPTLC plates have a strip of poorly adsorptive diatomaceous earth or wide-pore silicon dioxide adjacent to the main analytical layer. Band-shaped zones are formed, large applied volumes of dilute samples are concentrated, and very strongly attracted impurities are retained in the preadsorbent (also termed a concentration zone). Preadsorbent C-18 plates developed with petroleum ether-acetonitrile-methanol (1:1:2) were used for the HPTLC-densitometry quantification of the chloroplast pigments lutein and beta-carotene in estivated snails [33], and silica gel preadsorbent plates for neutral and polar lipids in estivated and starved snails [34].

Impregnation of layers with buffers, chelating agents such as ethylenediaminetetracetic acid (EDTA), metal ions, or other compounds can increase selectivity for a particular separation. For example, resolution of bile acids was improved by impregnation of silica gel 60 and 60F$_{254}$ plates with 1, 2.5, or 5% solutions of copper, manganese, nickel, or iron(II) sulfate and development with n-hexane-ethyl acetate-acetic acid of different compositions [35]; the optimum separation of all neighboring pairs of the investigated acids was obtained with 5% copper(II) sulfate impregnation and development with n-hexane-ethyl acetate-acetic acid (25:20:2) [36]. Silica gel impregnated with sodium acetate was used for TLC of fructooligosaccharides with butanol-glacial acetic acid-water (2:2:1) mobile phase [14], and silica gel impregnated with silver nitrate (argentation TLC) for separation of sterols from snail bodies [31].

Chiral separation of mandelic acid and derivatives was reported on molecularly imprinted polymers of L-mandelic acid, L-2-chloromandelic acid, and L-4-chlomomandelic acid as chiral stationary phases using mixtures of acetonitrile plus acetic acid in concentrations of 1, 5, and 10% as mobile phases and detection under 254 nm ultraviolet (UV) light [37].

Plates used in methods cited in this chapter were obtained from Merck, Analtech (Newark, DE, USA), Macherey-Nagel (Dueren, Germany), and Whatman (Florham Park, NJ, USA).

10.2.3 Application of Standard and Sample Solutions

Standard and sample solutions are usually manually applied as round spots using a capillary micropipet. It has been shown that initial zones in the form of bands, rather than round spots, allow higher volumes to be applied if the analyte detection limit is low, and bands produce tighter developed zones, higher-resolution separations, and better quantitative results by densitometry. Bands are most conveniently and simply produced when samples are manually applied with a calibrated syringe, such as a 10 or 25 uL Drummond digital dispenser, as diffuse vertical streaks to plates containing a preadsorbent spotting strip below the analytical layer; development with mobile phase automatically produces tight band-shaped initial zones at the sharply defined preadsorbent-analytical sorbent interface [34].

An instrument is often used to apply spots or bands for better results in quantitative analyses by densitometry. The most widely used method is the application of round spots or bands using an automated instrument such as the CAMAG (Wilmington, NC, USA) ATS 4 [17,19], Linomat V [16], or Nanomat [38].

10.2.4 MOBILE PHASES

The mobile phase (developing solvent) is generally selected empirically using prior personal experience and literature reports of similar separations as a guide. It usually consists of two to five components that can include water, organic solvents, an aqueous buffer, and an acid or base. Among the most widely used are the classic Mangold mobile phase for neutral lipid separations on silica gel, consisting of petroleum ether-diethyl ether-glacial acetic acid (80:20:1) [34], and the Wagner et al. system for separation of phospholipids, chloroform-methanol-water (65:25:4) [39]. Examples of other mobile phases that have been reported for silica gel are chloroform-methanol-0.2% aqueous $CaCl_2$ over 60 min. for separation of gangliosides [9], toluene-acetone (5:2) for antiepileptic drugs [40], n-hexane-acetone (3:1) and chloroform-acetone (24:1 and 9:1) for isoflavonoids [41], 1-propanol-chloroform-ethyl acetate-methanol-water (50:50:50:21:18) for cardiolipin [12], and chloroform-methanol-formic acid-water (200:80:20:19) for saponins [42].

A comprehensive study was made of the optimal systems for analysis of 21 biologically important essential and nonessential amino acids in complex mixtures [32]. The five systems found to be most useful were cellulose and silica gel HPTLC plates developed with either 2-butanol-pyridine-glacial acetic acid-water (39:34:10:26) or 2-butanol-pyridine-25% ammonia-water (39:34:10:26), and ion exchange TLC plastic backed sheets developed with pH 3.3 citrate buffer. By means of R_f values in these systems and different colors produced by ninhydrin detection reagent, all acids could be separated and identified except leucine and isoleucine. The analysis of amino acids in the DGG complex of *Biomphalaria glabrata* snails on a cellulose HPTLC plate is shown in Figure 10.1.

Additional mobile phases are specified for particular applications in other sections of this chapter.

10.2.5 LAYER DEVELOPMENT

Isocratic, linear, ascending development has been used primarily for analysis of biological samples by TLC and HPTLC. In this method, the mobile phase is contained in a large volume, covered glass chamber or tank (normal chamber or N-chamber). The twin trough chamber (TTC; CAMAG) is a special N-chamber that is modified with an inverted V-shaped ridge on the bottom dividing the tank

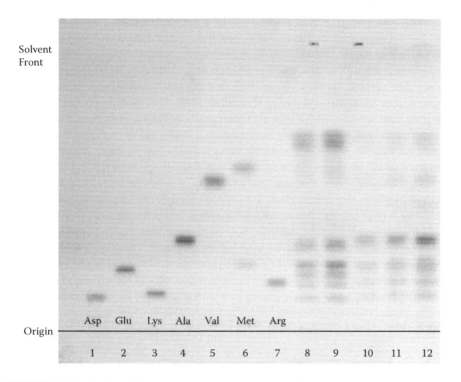

FIGURE 10.1   Amino acid chromatograms on an HPTLC cellulose plate developed with 2-butanol-pyridine-glacial acetic acid-water (39:34:10:36) mobile phase: lanes 1–7, 2 uL aliquots of individual amino acid standards (0.10 ug/uL); lanes 8 and 9, 1 and 2 uL aliquots, respectively, of a mixed amino acid standard (0.10 ug/uL of each acid); lanes 10–12, 1, 2, and 3 uL aliquots, respectively, of the ethanol-water (70:30) amino acid extract from a DGG of the snail *B. glabrata*. Abbreviations: Asp, aspartate; Glu, glutamic acid; Lys, lysine; Ala, alanine; Val, valine; Met, methionine; Arg, arginine. (From Vasta, J.D. et al., *Acta Chromatogr.*, 21, 29–38, 2009. With permission.).

into two sections. The TTC allows development with a very low volume of solvent and easy preequilibration of the layer with vapors of the mobile phase or another conditioning liquid (e.g., a sulfuric acid–water mixture to control humidity) or volatile reagent and has been widely used for biological analyses [19]. The Teflon DS-chamber (Chromdes, Lublin, Poland) has been used in many analyses for horizontal development of plates [25,42]. Development chambers are usually saturated with the mobile phase vapors of for 10–15 min. prior to development [39], but 3 h saturation was required for the silica gel 60 HPTLC of gangliosides with mobile phase composed of chloroform-methanol-water (120:85:20) plus 2 mM calcium chloride [18].

For two-dimensional development, the sample mixture is applied in one corner of the TLC plate. The plate is developed with the first mobile phase, dried, and developed with a second mobile phase in a perpendicular direction, causing the components to be resolved over the entire layer surface instead of on one track as in the usual one dimensional development. Improved separations of complex mixtures (e.g., phospholipids in guinea pig and bullfrog muscle tissues [13]) result because of the doubled separation distance and the use of mobile phases with different selectivities in the two developments.

Automated multiple development (AMD) is a method providing very high efficiency and sample capacity. It involves instrumental incremental gradient development over increasingly longer distances, with the mobile phases becoming progressively weaker. AMD with a 13-step gradient based on methanol (containing 5% ammonia), dichloromethane, and n-hexane was used to screen bioactive compounds in water samples [43] and a 15-step gradient with the same solvents (without ammonia in the methanol) for bioactive compounds in sponges [15].

Overpressured layer chromatography (OPLC) is an instrumental forced flow method in which the mobile phase is pumped through the layer contained in a cassette that is inserted into the separation chamber. It is a planar analog of HPLC that leads to decreased development time, low band spreading, and increased zone resolution and detectability compared with the usual gravity flow TLC. OPLC with an automatic instrument (OPLC-NIT Co. Ltd., Budapest, Hungary) has been applied to study the role of formaldehyde and its reaction products in the effects of antibiotics using the BioArena system. Operating conditions were chloroform-methanol (80:4) mobile phase, external pressure 5.0 MPa, flow rate 250 uL/min, mobile phase volume 8700 uL, and separation time 2680 s [38,44,45].

10.2.6 ZONE VISUALIZATION (DETECTION)

Compounds that are naturally colored, such as chloroplast pigments [33], are viewed directly in daylight or white light with the naked eye. Compounds with native fluorescence are viewed as bright, colored zones on a dark background under 366 nm or 254 nm UV light on layers without fluorescent indicator. Compounds that absorb 254 nm UV light, particularly those with aromatic rings and conjugated double bonds, can be detected on a layer (designated F_{254}) containing a fluorescent indicator or phosphor. When irradiated with 254 nm UV light, absorbing compounds diminish (quench) the uniform layer fluorescence and can be viewed as dark violet spots on a green or pale-blue background (e.g., acetylated kaempferol glycosides [46]).

Universal or selective chromogenic (dyeing) and fluorogenic postchromatographic derivatization reagents are usually applied by spraying onto the layer, dipping the layer into the reagent, exposing the layer to reagent vapors (e.g., iodine [47]), or incorporating reagent into the mobile phase [48]. Examples of detection reagents used in biological analysis are the following:

- Vanillin-sulfuric acid for detection of acetylated kaempferol glycosides as yellow-green zones [46]
- Spraying with 10% aqueous sulfuric acid solution or dipping into 10% ethanolic phosphomolybdic acid (PMA) followed by heating at 120°C for 20 min. for bile acids [24]
- Thionine reagent for cardiolipin prior to densitometry at 600 nm [12]
- 10% sulfuric acid in ethanol or anisaldehyde-sulfuric acid reagent followed by heating at 110°C for 5 min for saponins [42]

- 10% aqueous sulfuric acid in methanol–water (5:85) with the same heating protocol for bioactive compounds [15]
- Dragendorff reagent for Picralima alkaloids [49] and cocaine [20]
- Solutions containing 5% sodium hydroxide and a 1:1 mixture of 2% diphenylamine and 5% formaldehyde for carbaryl as blue-green zones [50]
- 10% cupric sulfate solution followed by heating at 140°C for 10 min for phospholipids [34]
- p-anisaldehyde in 5% sulfuric acid followed by 5% glacial acetic acid in methanol for two serrulatane diterpenes [51]
- Diphenylamine-aniline-phosphoric acid in acetone for fructooligosaccharides as blue-pink zones [14]
- 10% sodium hydroxide and 0.5% aqueous sodium sulfide for the pesticides dichlorvos and diptrex [52]
- Ninhydrin for amino acids [32]
- DPPH (diphenyl-pycrilhydrazyl) radical solution for carotenoid and phenol antioxidants [21]
- 5% sodium hydroxide followed by 1% m-dinitrobenzene for diazepam [53]
- Orcin and primulin reagents for ganglioside GM3 [18]

Optimization studies for PMA detection of cholesterol and bile acids on silica gel and RP-18 layers (application by spraying) indicated the best heating temperature and time were 60°C and >15 min., respectively [54].

Immunostaining has been used to detect gangliosides in tissues. The plate was dipped in polyisobutylmethacrylate solution and bovine serum albumin solution, followed by immersion in antibody-containing supernatant or patient's sera containing antibodies at 4°C overnight. After washing with phosphate buffered saline, detection of Mab binding was by stepwise incubation with biotinylated chain specific antimouse immunoglobulin, followed by steptavidin-horseradish peroxidase complex and visualization of the bound peroxidase with chloro-4-naphthol reagent [9]. Immunostaining was also used to detect Stx1-binding glycosphingolipid bands from human erythrocyte extracts on silica gel plates developed with chloroform-methanol-water (15:30:4) [55].

Competitive protein binding assay was used to measure cortisol and cortisone bands detected under UV light and eluted from the plate with phosphate buffer. Chicken serum was used as the source of corticosteroid binding globulin because it binds both compounds with similar high affinity [11].

The BioArena system has great potential for bioassay guided analysis of biological samples. It consists of regular TLC or OPLC followed by bioautography detection. Formaldehyde (HCHO) is an endogenous component, mainly in the form of hydroxymethyl groups, of all biological systems, and observations have indicated that HCHO is a key molecule in cell proliferation, differentiation, and disease resistance. BioArena was used to study the formaldehydome, which is the complete set of HCHO cycle mediated and nonmediated HCHO pathways of a given biological unit. TLC of trans-resveratrom, semicarbazide, and dimedone (as standards) was performed on silica gel 60 using different mobile phases; OPLC was on sealed silica gel 60 F plates with chloroform-methanol (10:1). After drying, bioautographic detection was by immersion in a bacterial suspension of *Pseudomonas savastanoi* for 25 s and visualization with dye reagent 3-[4,5-dimethylthiazol-2-yl]2,5-diphenyltetrazolium bromide (MTT) either after a short draining period or after overnight incubation [38,44]. Results with the BioArena system showed that not only formaldehyde but also ozone and related bioactive compounds may play a crucial role in the mechanism of activity of antibiotic compounds [45].

Bioluminescent ATP assay was used to optimize conditions for culture of the fungus *Candida albicans* (ATCC 90028) for microbial detection of zones in direct bioautographic TLC. The method allowed differentiation between microbiostatic (bacteriostatic or fungistatic) and microbiocidal (bactericidal or fungicidal) effects on TLC plates. The proposed method can be used for substance specific post-TLC detection, screening for targeted or bioactivity driven isolation processes, detection of antibiotic residues, and standardization of extracts from biologically active materials [56].

The ChromaDex (Boulder, CO, USA) Bioluminex assay [57] and CAMAG BioLuminizer detection device were developed specifically to produce and detect bioluminescence on HPTLC plates. A complex sample containing bioactive substances is separated by HPTLC, and the plate is subsequently immersed in a luminescent bacterial suspension (*Vibrio fischeri*). The reaction takes place quickly to yield a black color for zones with inhibitory or toxic effect and a lighter color for zones with stimulatory effect compared with the background of the plate; chromatograms detected in the BioLuminizer can be digitally imaged by use of a CCD camera. The BioLuminizer was used to screen bioactivity in expressway waste water or landfill leachate [43] and secondary metabolites in sponges [15] after HPTLC separation by AMD.

10.2.7 CHROMATOGRAM DOCUMENTATION

Chromatograms on plates are documented by color digital photography in white or UV light, with a videodensitometer, or by scanning densitometry [39]. The CAMAG DigiStore 2 Documentation System is especially widely used [15].

10.2.8 IDENTIFICATION OF ZONES

R_f values and colors produced by selective detection reagents in comparison with those of standards are used initially to identify unknown sample zones. Identity is more certain if several TLC systems governed by different separation mechanisms are used for these comparisons. Spiking with standard of a presumptive compound in the sample is also used to confirm identity based on the absence of separation.

Qualitative and quantitative analysis of biological samples can be performed by off-line and online coupling of TLC with modern spectrometric methods, such as Fourier transform infrared (FTIR) spectrometry [58].

Surface enhanced Raman spectrometry (SERS) coupled with TLC was tested for analysis of biologically active gibberellic acid (GA), abietic acid, and kaurenoic acid. A special Raman spectrometry plate (Si60F$_{254s}$RAMAN; Merck) was used for TLC, and a typical TLC plate with a modified aluminum backplate foil on one side was used as an interface. Silver colloid was added to the zones on the layer prior to SERS measurements. The strongest signal was obtained for GA [59].

HPTLC was coupled directly to mass spectrometry (MS) for identification of bioactive compounds from sponges. The zone on a plate was positioned in the excited gas stream of a DART (direct analysis in real time) ionization interface from IonSense (Danvers, MA, USA). Mass spectra were acquired in the positive ion full scan mode with an LTQ Orbitrap XL hybrid FT-MS from Thermo Scientific (Madison, WI, USA) as the mass analyzer [15]. HPTLC was also coupled directly to matrix assisted infrared laser desorption/ionization time of flight (IR-MALDI-o-TOF) MS with glycerol matrix for ganglioside GM3 characterization [18].

10.2.9 Quantitative Analysis

Direct quantification using a slit scanning densitometer is the most common method currently used for quantitative TLC and HPTLC biological analysis. A tungsten or halogen lamp is used as the source for scanning colored zones (visible absorption), such as ninhydrin-detected amino acids at 610 nm [32], and a deuterium lamp for scanning UV-absorbing zones directly or as quenched zones on F-layers [60]. The detector is a photomultiplier tube. Plots of scan area versus standard weights are established under the same conditions as for the sample zones separated on the same plate.

Most biological quantitative analyses by densitometry have been validated by determining parameters such as linearity, sensitivity (limits of detection and quantification), selectivity, extraction efficiency, accuracy, and intraday and interday precision [10]. An internal standard is often added to the sample extract to improve quantitative results [19].

An example of HPTLC biological quantitative analyses is determination of epalrestat in human plasma with nitrofurantoin as internal standard on silica gel with ethyl acetate-toluene-acetic acid (30:20:1) mobile phase and densitometric scanning at 290 nm [16].

10.2.10 Preparative Layer Chromatography

PLC is a method used to separate and isolate larger amounts of material (e.g., 1–1000 mg) on thicker layers (0.5–1 mm or greater compared with 0.2 or 0.25 mm analytical layers) [4,61]. The purpose of PLC is to obtain pure compounds for further chromatographic or spectrometric analysis and for investigations of physical, chemical, pharmaceutical, or biological properties. The PLC of (+)-ulein on silica gel with n-hexane-dichloromethane-methanol-diethylamine (25:20:4:1) mobile phase and detection under 254 nm UV light is an example application [62].

10.2.11 Thin-Layer Radiochromatography

Thin-layer radiochromatography (TLRC) involves location and quantification of separated radioisotope-labeled substances on a thin layer by use of x-ray or photographic film autoradiography, zonal analysis with scintillation counting, imaging scanners for in situ measurement, and storage phosphor screen imaging analysis (also termed bioimaging or radioluminography). A recent article described methods and instruments of TLRC and eight applications published in 2006–2008 [63].

Additional papers describing its use for the analysis of biological samples have been
published later [e.g., 64–69].

10.3 APPLICATIONS OF TLC AND HPTLC FOR ANALYSIS OF BIOLOGICAL SAMPLES

This section contains further information on some of the applications mentioned
above and additional representative applications of TLC and HPTLC for analysis
of biological samples, arranged by analyte type. Details of sample collection and
preparation, TLC/HPTLC procedures, and results and discussion of the studies will
be found in the cited references.

10.3.1 AMINO ACIDS

Analyte: Taurine, alanine, threonine, and lysine [32,70]
Sample: Mouse urine extracted with ethanol-water (70:30)
Layer: (A) HPTLC silica gel, (B) HPTLC cellulose
Mobile phase: (A) 2-Butanol-pyridine-glacial acetic acid deionized water
  (39:34:10:26), (B) 2-Butanol-pyridine-25% ammonia-deionized water
  (39:34:10:2.6)
Detection: Ninhydrin reagent
Quantification: Visible mode slit-scanning densitometry at 610 nm

10.3.2 CARBOHYDRATES

Analyte: Simple sugars, glucose and maltose [71]
Sample: Snail tissue, specifically the DGG of *B. glabrata*; extraction with 70%
  aqueous ethanol
Layer: Silica gel preadsorbent TLC plates
Mobile phase: Ethyl acetate-glacial acetic acid-methanol-water (60:15:15:10)
Detection: alpha-Napthol-sulfuric acid reagent
Quantification: Visible mode densitometry

10.3.3 DRUGS

Analyte: Antipyrin and paracetamol [72]
Sample: Human plasma
Layer: Silica gel
Mobile phase: Chloroform-methanol (9:1)
Determination: Use of densitometry in routine clinical analyses to estimate
  amounts of these drugs
Analyte: Colchicine [73]
Sample: Pharmaceutical preparations and vegetal (seeds of meadow saffron)
Layer: Silica gel 60F254

Mobile phase: Chloroform–acetone–diethylamine (5:4:1)

Detection and quantification: Densitometric measurements made at the absorption maximum of colchicine (350 nm) in the reflectance and fluorescence modes

Analyte: Curcuminoids: curcumin, desmethoxycurcumin, and bidesmethoxy-curcumin [74]

Sample: Tumeric (*Curcoma longa*); extraction by sonication with methanol

Layer: Silica gel 60F$_{254}$

Mobile phase: Chloroform-hexane-methanol (1:1:0.1)

Detection and quantification: Image analysis of the scanned TLC plate by Photoshop 7.0 to quantify each of the three curcuminoids

Analyte: Ketorolac tromethamine [75]

Sample: Human plasma; extraction with acetonitrile

Layer: Silica gel 60

Mobile phase: *n*-Butanol-chloroform-acetic acid-ammonium hydroxide-water (9:3:5:1:2)

Quantification: Densitometry at 323 nm

Analyte: Lamotrigine [10]

Sample: Human serum

Layer: Silica gel

Mobile phase: Toluene-acetone-ammonia (14:6:1)

Quantification: Densitometry at 312 nm

Analyte: Minocycline [76]

Sample: Human plasma, saliva, and gingival fluid; extraction with methanol

Layer: Aluminum-backed silica gel 60F$_{254}$

Mobile phase: Methanol-acetonitrile-isopropyl alcohol-water (5:4:0.5:0.5)

Detection and quantification: Densitometry at 345 nm

Analyte: Theophylline [77]

Sample: Postmortem human blood; extraction with methanol-water (80:20 to 30:70)

Layer: Silica gel 60F$_{254}$

Mobile phase: Chloroform-methanol (9:1)

Quantification: UV-densitometry at 277 nm

Analyte: Tinidazole [78]

Sample: Human serum; extraction and protein precipitation with acetonitrile

Layer: Silica gel

Mobile phase: Chloroform-acetonitrile-acetic acid (30:20:1)

Quantification: Densitometry at 320 nm

10.3.4 Lipids

Analyte: Neutral lipids (NL) and phospholipids (PL) [79]

Sample: *B. glabrata* snail whole bodies, DGG, shells, and hemolymph; extraction by the Folch method

Layer: Preadsorbent, channeled HPTLC silica gel

Mobile phase: Petroleum ether-diethyl ether-acetic acid (80:20:1) or hexane-petroleum ether-diethyl ether-acetic acid (50:25:5:1)(NL). Chloroform-methanol-water (65:25:4) (PL)

Detection: 5% Ethanol PMA (NL); 10% cupric sulfate (PL)

Quantification: Absorption-reflectance densitometry at 610 nm (NL) or 370 nm (PL)

Analyte: Sterols [31]

Sample: Snail bodies of *B. glabrata*, *H. trivolvis*, and *P Bridgesii*; extraction by the Folch method

Layer: C-18

Mobile phase: Acetonitrile-chloroform (40:35) or petroleum ether-acetonitrile-methanol

Detection: 5% solution of PMA in ethanol

Qualitative determination: Snails contained desmosterol, camphesterol, brassicasterol, beta-sitosterol, ergosterol, and cholesterol or stigmasterol.

10.3.5 Pigments

Analyte: Bile pigment biliverdin [80]

Sample: The medicinal leech, *Hirudo medicinalis* (whole body); extraction with methanol-12 M HCl (11:1)

Layer: Preadsorbent HPTLC silica gel

Mobile phase: n-Butanol-methanol-deionized water (2:1:1.5)

Detection and quantification: Absorption-reflectance densitometry at 629 nm

Analyte: Carotenoid pigments lutein and beta-carotene [33]

Sample: Snail whole bodies and DGG from *B. glabrata* and *H. trivolvis*; extraction with acetone

Layer: C-18 with concentration zone

Mobile phase: Petroleum ether-acetonitrile-methanol (1:2:2)

Quantification: Densitometry of natural yellow zones at 448 and 455 nm, respectively

The following are additional applications reported using the sample preparation, layers, mobile phases, detection, and quantification cited in Sections 10.3.4 and 10.3.5:

• NL and PL in whole body, DGG, viscera, head–foot, shell, operculum, plasma, and hemocytes of the apple snail *Pomacea bridgesii* [81]

- NL in the urine of humans and BALB/c mice (2′,7′-dichlorofluoroescein, aluminum chloride, and ferric chloride were used as specific reagents to confirm the identity of different NL fractions) [82].
- PL in human and *E. caproni* infected mouse urine [83]
- NL in snail DGG maintained in different salinity environments [84]
- NL and PL in *S. mansoni* infected snails subjected to estivation and starvation [85]
- NL and PL in *B. glabrata* patently infected with *E. caproni* [86]
- PL and sphingolipid in mice that were uninfected and infected with *E. caproni* (alpha-naphthol and ninhydrin spray reagents were used to identify various zones in chromatograms) [87].
- NL in the urine of BALB/c mice infected with *E. caproni* [88]
- NL in the feces of BALB/c mice infected with *E. caproni* [89]
- NL [90] and PL [91] in the livers, abdominal fat bodies, and tails of male and female northern side blotched lizards; NL and PL in the leech *H. medicinalis* and leech conditioned water [92,93]
- NL, PL, and carotenoid pigments in *P. bridgessi* maintained on different diets [94]

10.4   FUTURE PROSPECTS

It is expected that TLC will be applied to a greater range of bioactive analytes and biological sample types in the future as biologists and other life scientists become more aware of its many advantages and opportunities for complementary use with other chromatographic and nonchromatographic analytical methods. This should certainly be true for determination of chiral compounds and phytochemicals in biological samples and use of preparative layer chromatography in addition to analytical TLC because of the recently published books on these topics [4–6].

As have already been implemented more widely in other application areas, such as for foods, drugs, herbal dietary supplements, and environmental samples, biological sample analysis would be improved by greater use of the following materials and methods:

- Modern sample preparation methods such as SPE and supercritical fluid extraction (SFE) in place of LLE because of savings in time, effort, and solvent usage
- HPTLC layers relative to TLC and normal- and reversed-phase bonded layers and impregnated layers compared with plain silica gel to take advantage of enhanced efficiencies and selectivities
- Automated sample application instruments rather than manual application for greater accuracy and precision, especially for quantitative analysis
- More efficient systematic mobile phase selection and optimization methods in place of trial and error
- Forced flow OPLC and automated multiple development to achieve better separations of complex mixtures compared with single development with capillary action mobile phase flow

- Biological detection methods in sequence with chemical and physical methods
- Hyphenated TLC-spectrometry for superior compound identification, especially HPTLC-MS because of the recent availability of a commercial direct, semiautomatic TLC/MS interface (http://www.camag.com/tlc-ms)
- Quantification with video, diode array detector, and flatbed PC densitometers in addition to the slit scanning densitometers employed so far almost exclusively.

Many of these materials and methods will undoubtedly increase in use in the next five years and improve and expand greatly the role of TLC and HPTLC in biological analysis.

REFERENCES

1. Sherma, J. and Fried, B., Thin layer chromatographic analysis of biological samples. A review, *J. Liq. Chromatogr. Relat. Technol.*, 28, 2297–2314, 2005.
2. Fried, B. and Sherma, J., *Thin Layer Chromatography*, 4th ed., Marcel Dekker Inc., New York, 1999.
3. Sherma, J. and Fried, B. (eds.), *Handbook of Thin Layer Chromatography*, 3rd ed., Marcel Dekker Inc., New York, 2003.
4. Kowalska, T. and Sherma, J. (eds.), *Preparative Layer Chromatography*, CRC/Taylor & Francis, Boca Raton, FL, 2006.
5. Kowalska, T. and Sherma, J. (eds.), *Thin Layer Chromatography in Chiral Separations and Analysis*, CRC/Taylor & Francis, Boca Raton, FL, 2007.
6. Waksmundzka-Hajnos, M., Sherma, J., and Kowalska, T. (eds), *Thin Layer Chromatography in Phytochemistry*, CRC/Taylor & Francis, Boca Raton, FL, 2008.
7. Majors, R.E., Practical aspects of solvent extraction, *The Applications Notebook*, Supplement to LC/GC North America, 57–61, February 2009.
8. Bandstra, S.R., Murray, K.E., Fried, B., and Sherma, J., High performance thin layer chromatographic analysis of neutral lipids in the feces of BALB/c mice infected with *Echinostoma caproni*, *J. Liq. Chromatogr. Relat. Technol.*, 30, 1437–1445, 2007.
9. Popa, J. and David, M.-J., Immunoassay detection of gangliosides by specific antibodies, *Camag Bibiography Service (CBS)*, 94, 11–13, 2005.
10. Patil, K.M. and Bodhankar, S.L., High performance thin layer chromatographic determination of lamotrigine in serum, *J. Chromatogr. B*, 823, 152–157, 2005.
11. Fenske, M., Thin layer chromatographic competitive protein binding assay for cortisol and cortisone, and its application to urine samples from healthy men undergoing water diuresis, *Chromatographia*, 63, 383–388, 2006.
12. Helmy, F.M. Cardiolipin, its preferential deacylation in myocardia. Mini review and chromatographic-computational analysis, *Acta Chromatogr.*, 17, 9–19, 2006.
13. Helmy, F., Rothenbacher, F., Nosavanh, L., Lowery, J., and Juracka, A., A comparative study of the phospholipid profiles of guinea pig cardiac muscle and bullfrog cardiac and thigh skeletal muscle, and their in vitro differential deacylation by endogenous phospholipases. Thin layer chromatographic and densitometric analysis, *J. Planar Chromatogr.-Mod. TLC*, 20, 209–215, 2007.
14. Reiffova, K., Podolonovicova, J., Onofrejova, L., Preisler, J., and Nemcova, R., Thin layer chromatography and matrix assisted laser desorption/ionization mass spectrometric analysis of oligosaccharides in biological samples, *J. Planar Chromatogr.-Mod. TLC*, 20, 19–25, 2007.

15. Kloeppel, A., Grasse, W., Bruemmer, F. and Morlock, G.E., HPTLC coupled with bioluminescence and mass spectrometry for bioactivity based analysis of secondary metabolites in sponges, *J. Planar Chromatogr.-Mod. TLC*, 21, 431–436, 2008.

16. Saraf, M.N., Birajdar, P.G., Loya, P., and Mukherjee, S.A., Rapid and sensitive HPTLC method for determination of epalrestat in human plasma, *J. Planar Chromatogr.-Mod. TLC*, 20, 203–207, 2007.

17. Namur, S., Carino, L., and Gonzales-de la Parra, M., Development and validation of a densitometric HPTLC method for quantitative analysis of levofloxacin in human plasma, *J. Planar Chromatogr.-Mod. TLC*, 21, 200–212, 2008.

18. Dreiswerd, K. and Muething, J., Structural characterization of gangliosides by HPTLC/IR-MALDI-o-TOF, *Camag Bibiography Service (CBS)*, 97, 2–5, 2006.

19. Namur, S., Carino, L., and Gonzales-de la Parra, M., Development and validation of a high performance thin layer chromatographic method, with densitometry, for quantitative analysis of tizoxanide (a metabolite of nitazoxanide) in human plasma, *J. Planar Chromatogr.-Mod. TLC*, 20, 331–334, 2007.

20. Yonamine, M. and Cortez, M., A high performance thin layer chromatographic technique to screen cocaine in urine samples, *Leg. Med.* 8, 184–187, 2006.

21. Jaime, L., Mendiola, J.A., Herrero, M., Soler-Rivas, C., Santoyo, S., Senorans, F.J., Cifuentes, A., and Ibanez, E., Separation and characterization of antioxidants from *Spirulina platensis* microalga combining pressurized liquid extraction, TLC, and HPLC-DAD, *J. Sep. Sci.*, 28, 2111–2119, 2005.

22. Pyka, A., Babuska, M., Bober, K., Gurak, D., Klimczok, W., and Miszczyk, M., Influence of temperature of silica gel activation on separation of selected biologically active steroid compounds, *J. Liq. Chromatogr. Relat. Technol.*, 29, 2035–2044, 2006.

23. Khan, H.A., TLC determination of aliphatic polyamines on calcium sulfate layers, *Chromatographia*, 64, 423–427, 2006.

24. Pyka, A. and Dolowy, M. Lipophilicity of selected bile acids as determined by TLC. II. Investigations on RP-18W stationary phases, *J. Liq. Chromatogr. Relat. Technol.*, 28, 297–311, 2005.

25. Brezezinska, E. and Stolarska, J., Determination of the partition and distribution coefficients of biologically active compounds by reversed phase thin layer chromatography, *J. Planar Chromatogr.-Mod. TLC*, 18, 443–449, 2005.

26. Pyka, A. and Dolowy, M. Separation of selected bile acids by TLC. III. Investigations on RP-2 stationary phase, *J. Liq. Chromatogr. Relat. Technol.*, 28, 1765–1775, 2005.

27. Pyka, A. and Dolowy, M. Separation of selected bile acids by TLC. VI. Separation on cyano- and diol-modified silica layers, *J. Liq. Chromatogr. Relat. Technol.*, 28, 1383–1392, 2005.

28. Rozmer, Z., Perjesi, P., and Takacs-Novak, K., Use of RP-TLC for determination of log P of isomeric chalcones and cyclic chalcone analogs, *J. Planar Chromatogr.-Mod. TLC*, 19, 124–128, 2006.

29. Janicka, M., Comparison of different properties—log P, $\log_{KW}$, and phi_0—as descriptors of the hydrophobicity of some fungicides, *J. Planar Chromatogr.-Mod. TLC*, 19, 361–370, 2006.

30. Perisic-Janjic, N., Vastag, G., Tomic, J., and Petrovic, S., Effect of the physicochemical properties of N,N-disubstituted-2-phenylacetamide derivatives on their retention behavior in RP-TLC, *J. Planar Chromatogr.-Mod. TLC*, 20, 353–359, 2007.

31. Jarusiewicz, J., Sherma, J., and Fried, B., Separation of sterols by reversed phase and argentation thin layer chromatography. Their identification in snail bodies, *J. Liq. Chromatogr. Relat. Technol.*, 28, 2607–2617, 2005.

32. Vasta, J.D., Cicchi, M., Sherma, J., and Fried, B., Evaluation of thin layer chromatography systems for analysis of amino acids in complex mixtures, *Acta Chromatogr.*, 21, 29–38, 2009.

33. Arthur, B., Fried, B., and Sherma, J., Effects of estivation on lutein and beta-carotene concentrations in *Biomphalaria glabrata* (NMRI strain) and *Helisoma trivolvis* (Colorado strain) snails as determined by quantitative high performance reversed phase thin layer chromatography, *J. Liq. Chromatogr. Relat. Technol.*, 29, 2159–2165, 2006.

34. White, M.M., Fried, B., and Sherma, J., Determination of the effects of estivation and starvation on neutral lipids and phospholipids in *Biomphalaria glabrata* (NMRI strain) and *Helisoma trivolvis* (Colorado strain) snails by quantitative high performance thin layer chromatography-densitometry, *J. Liq. Chromatogr. Relat. Technol.*, 29, 2167–2180, 2006.

35. Pyka, A., Dolowy, M., and Gurak, D., Separation of silica gel $60F_{254}$ plates impregnated with Cu(II), Ni(II), Fe(II), and Mn(II) cations, *J. Liq. Chromatogr. Relat. Technol.*, 28, 2273–2284, 2005.

36. Dolowy, M., Separation of selected bile acids by TLC. IX. Separation on silica gel 60 and on silica gel $60F_{254}$ aluminum plates impregnated with Cu(II), Ni(II), Fe(II), and Mn(II) cations, *J. Liq. Chromatogr. Relat. Technol.*, 30, 405–418, 2007.

37. Rong, F., Feng, X., Yuan, C., Fu, D., and Li, P., Chiral separation of mandelic acid and its derivatives by thin layer chromatography using molecularly imprinted stationary phases, *J. Liq. Chromatogr. Relat. Technol.*, 29, 2593–2602, 2006.

38. Tyihak, E., Moricz, A., Ott, P.G., Katay, G., and Kiraly-Veghely, Z., The potential of BioArena in the study of the formaldehydome, *J. Planar Chromatogr.-Mod. TLC*, 18, 67–72, 2005.

39. Bandstra, S.R., Fried, B., and Sherma, J., Effects of diet and larval trematode parasitism on lipids in snails as determined by thin layer chromatography, *J. Planar Chromatogr.-Mod. TLC*, 19, 180–186, 2006.

40. Patil, K.M. and Bodhankar, S.L., High performance thin layer chromatography method for therapeutic drug monitoring of antiepilectic drugs in serum, *Indian Drugs*, 42, 665–670, 2005.

41. Ito, C., Murata, T., Itoigawa, M., Nakao, K., Kumagai, M., Kaneda, N., and Furukawa, H., Induction of apoptosis by isoflavonoids from leaves of *Millettia taiwaiana* in human leukemia HL-60 cells, *Planta Med.*, 72, 424–429, 2006.

42. Glensk, M., Wlodarczyk, M., Radom, M., and Cisowski, W., TLC as a rapid and convenient method for saponin investigation, *J. Planar Chromatogr.-Mod. TLC*, 18, 167–170, 2005.

43. Schulz, W., Seitz, W., Weiss, S.C., Weber, W.H., Boehm, M., and Flottmann, D., Use of *Vibrio fischeri* for screening for bioactivity in water samples, *J. Planar Chromatogr.-Mod. TLC*, 21, 427–430, 2008.

44. Tyihak, E., Mincsovics, E., Katay, G., Kiraly-Veghely, Z., Moricz, A.M., and Ott, P.G., BioArena: An unlimited possibility of biochemical interactions in the adsorbent layer after chromatographic separation, *J. Planar Chromatogr.-Mod. TLC*, 21, 15–20, 2008.

45. Tyihak, E., Moricz, A.M., and Ott, P.G., Use of the BioArena system for indirect detection of endogenous ozone in spots after TLC or OPLC separation, *J. Planar Chromatogr.-Mod. TLC*, 21, 77–82, 2008.

46. Kuo, Y.C., Lu, C.K., Huang, L.W., Kuo, Y.H., Chang, C., Hsu, F.L., and Lee, T.H., Inhibitory effects of acylated kaempferol glycosides from leaves of *Linnamomum kotoense* on the proliferation of human peripheral blood mononuclear cells, *Planta Med.*, 71, 412–415, 2005.

47. Khan, M.S.Y. and Akhter, M., Glyceride derivatives as potential drugs: synthesis, biological activity, and kinetic studies of glyceride derivatives of mefenamic acid, *Pharmazie*, 60, 110–114, 2005.

48. Kazmierczak, D., Ciesielski, W., and Zakrzewski, R., Application of the iodine-azide procedure for detection of biogenic amines, *J. Liq. Chromatogr. Relat. Technol.*, 29, 2425–2436, 2006.

49. Okunji, C.O., Iwu, M.M., Ito, Y., and Smith, P.L., Preparative separation of indole alkaloids from the rind of *Picralima nitida* (Staph) T. Durand and H. Durand by pH zone refining countercurrent chromatography, *J. Liq. Chromatogr. Relat. Technol.*, 28, 775–783, 2005.

50. Daundkar, B.B., Mavle, R.R., Malve, M.K., and Krishnamurthy, R., Detection of carbaryl insecticide in biological samples by TLC with a specific chromogenic reagent, *J. Planar Chromatogr.-Mod. TLC*, 19, 467–468, 2006.

51. Smith, J., Tucker, D., Watson, K., and Jones, G., Identification of antibacterial constituents from the indigenous Australian medicinal plant *Eremophila duttonii* F. Muell. (Myoporaceae), *J. Ethnopharmacol.*, 112, 386–393, 2007.

52. Daundkhar, B.B., Malve, R.R., Malve, M.K., and Krishnamurthy, R., Spectrophotometric and TLC detection reagent for the insecticides dichlorvos (DDVP) and diptrex (trichlorfon), and their metabolites, in biological tissue, *J. Planar Chromatogr.-Mod. TLC*, 20, 217–219, 2007.

53. Daundkar, B.B., Malve, M.K., and Krishnamurthy, R., A specific chromogenic reagent for detection of diazepam among other benzodiazepines from biological and nonbiological samples after HPTLC, *J. Planar Chromatogr.-Mod. TLC*, 21, 249–250, 2008.

54. Zarzycki, P.K., Bartosuk, M.A., and Radziwon, A.I., Optimization of TLC detection by phosphomolybdic acid staining for robust quantification of cholesterol and bile acids, *J. Planar Chromatogr.-Mod. TLC*, 19, 52–57, 2006.

55. Meisen, I., Friedrich, A., Karch, H., Witting, U., Peter-Katalinic, J., and Muething, J., Application of combined high performance thin layer chromatography immunostaining and nanospray ionization quadrupole time of flight tandem mass spectrometry to the structural characterization of high and low affinity binding ligands of Shiga Toxin 1, *Rapid Commun. Mass Spectrom.*, 19, 3659–3665, 2005.

56. Nagy, S., Kocsis, B., Koszegi, T., and Botz, L., Optimization of growth conditions for test fungus cultures used in direct bioautographic TLC detection. 3. Test fungus: *Candica albicans*, *J. Planar Chromatogr.-Mod. TLC*, 20, 385–389, 2007.

57. Ikenouye, L., Hickey, S., Verbitski, S., and Gourdin, G., Bioluminex: An effective yet simple tool for screening mixtures, *Camag Bibiography Service (CBS)*, 99, 11–13, 2007.

58. Cimpoiu, C., Qualitative and quantitative analysis by hyphenated (HP)TLC-FTIR technique, *J. Liq. Chromatogr. Relat. Technol.*, 28, 1203–1213, 2005.

59. Orinak, A., Talian, I., Efremov, E.V., Ariese, F., and Orinakova, R., Diterpenoic acids analysis using a coupled TLC-surface enhanced Raman spectroscopy system with TLC, *Chromatographia*, 67, 315–319, 2008.

60. Mennickent, S., Sorbazo, M., Vega, M., Godoy, C., and Diego, M., Quantitative determination of clozapine in serum by instrumental planar chromatography, *J. Sep. Sci.*, 30, 2167–2172, 2007.

61. Sherma, J., in *Handbook of Food Analysis Instruments*, Otles, S., ed., CRC/Taylor & Francis, Boca Raton, FL, 2009, pp. 145–160.

62. Baggio, C.H., De Martini Otofuji, G., De Souza, W.M., De Moraes Santos, C.A., Torres, L.M.B., Rieck, L., De Andrade Marques, M., and Mesia-Vela, S., Gastroprotective mechanisms of indole alkaloids from *Himatanthus lancifolius*, *Planta Med.*, 71, 733–738, 2005.

63. Sherma, J., Thin layer radiochromatography, *J. AOAC Int.*, 92, 29A–35A, 2009.

64. Lakshminarayan, H., Narayanan, S., Bach, H., Sundaram, K.G.P., and Av-Gay, Y. Molecular cloning and biochemical characterization of a serine threonine protein kinase, PknL, from Mycobacterium tuberculosis, *Protein Expres. Purif.* 58, 309–317, 2008.

65. Mandal, C., Srinivasan, G.V., Chowdhury, S., Chandra, S., Mandal, C., Schauer, R., and Mandal, C., High level of sialate-O-acetyltransferase activity in lymphoblasts of childhood acute lymphoblastic leukaemia (ALL): enzyme characterization and correlation with disease status, *Glycoconj. J.* 26, 57–73, 2009.

66. Sadeghpour, H., Jalilian, A.R., Akhlaghi, M., Kamali-Dehghan, M., Mirzaii, M., and Shafiee, A., Preparation and biodistribution of [In-111]-rHuEpo for erythroprotein receptor imaging, *J. Radioanal. Nucl. Chem.*, 278, 117–122, 2008.

67. Jalilian, A.R., Shanehsazzadeh, S., Akhlaghi, M., Garousi, J., Rajabifar, S., and Tavakoli, M.B., Preparation and biodistribution of [Ga-67]-DTPA-gonadorelin in normal rats, *J. Radioanal. Nucl. Chem.*, 278, 123–129, 2008.

68. Jalilian, A.R., Shanehsazzadeh, S., Akhlaghi, M., Garousi, J., Rajabifar, S., and Tavakoli, M.B., Preparation and evaluation of [Ga-67]-DTPA-beta-1-24-corticotrophin in normal rats, *Radiochim. Acta*, 96, 435–439, 2008.

69. Kertesz, V., Van Berkel, G.J., Vavrek, M., Koeplinger, K.A., Schneider, B.B., and Covey, T.R., Comparison of drug distribution images from whole body thin tissue sections obtained using desorption electrospray ionization tandem mass spectrometry and autoradiography, *Anal. Chem.*, 80, 5168–5177, 2008.

70. Vasta, J.D., Fried, B., and Sherma, J., Determination of amino acids in the urine of Balb/c mice infected with *Echinostoma caproni* by high performance thin-layer chromatography-densitometry, *J. Liq. Chromatog. Relat. Technol.*, 32, 1210–1222, 2009.

71. Jarusiewicz, J.A., Sherma, J., and Fried, B., Thin-layer chromatographic analysis of glucose and maltose in estivated *Biomphalaria glabrata* snails and those infected with *Schistosoma mansoni*, *Comp. Biochem. Physiol. B*, 145, 346–349, 2006.

72. Chatterjee, S. and Singh, B.P., A high-performance thin-layer chromatographic (HPTLC) estimation of antipyrin and paracetamol from plasma, *Indian Drugs*, 33, 355–357, 1996.

73. Bodoki, E., Oprean, R., Vlase, L., Tamas, M., and Sandulescu, R., Fast determination of colchicine by TLC-densitometry from pharmaceuticals and vegetal extracts, *J. Pharm. Biomed. Anal.*, 37, 971–977, 2005.

74. Phattanawasin, P., Sotanaphun, U., and Sriphong, L., Validated TLC-image analysis method for simultaneous quantification of curcuminoids in *Curcuma longa*, *Chromatographia*, 69, 397–400, 2009.

75. Lopez-Bojorquez, E., Castaneda-Hermandez, G., Gonzalez-de la Parra, M., and Namur, S., Development and validation of a high performance thin-layer chromatographic method, with densitometry, for quantitative analyses of ketorolac tromethamine in human plasma, *J. AOAC Int.*, 91, 1191–1195, 2008.

76. Jain, N., Jain, G.K., Iqbal, Z., Talegaonkar, S., Ahmad, F.J., and Khar, R.K., An HPTLC method for the determination of minocycline in human plasma, saliva, and gingival fluid after single step liquid extraction, *Anal. Sci.*, 25, 57–62, 2009.

77. Sanganalmath, P.U., Sujatha, K.M., Bhargavi, S., Nayak, V.G., and Mohan, B.M., Simple, accurate and rapid HPTLC method for analysis of theophylline in postmortem blood and validation of the method, *J. Planar Chromatogr.-Mod. TLC*, 22, 29–33, 2009.

78. Guermouche, M.H., Hobel, D., and Guermouche, S., Assay of tinidazole in human serum by high performance thin-layer chromatography-comparison with high performance liquid chromatography, *J. AOAC Int.*, 82, 244–247, 1999.

79. Bandstra, S.R., Fried, B., and Sherma, J., High-performance thin-layer chromatographic analysis of neutral lipids and phospholipids in *Biomphalaria glabrata* patently infected with *Echinostoma caproni*, *Parasitol. Res*, 99, 414–418, 2006.

80. Martin, D.L., Fried, B., and Sherma, J., The absence of beta-carotene and the presence of biliverdin in the medicinal leech *Hirudo medicinalis* as determined by TLC, *J. Planar Chromatogr.-Mod. TLC*, 18, 400–402, 2005.

81. Jarusiewicz, J.A., Fried, B., and Sherma, J., High performance thin layer chromatographic analysis of neutral lipids and phospholipids in the apple snail *Pomacea bridgesii*, *J. Planar Chromatogr.-Mod TLC*, 17, 454–456, 2004.

82. Vasta, J.D., Fried, B., and Sherma, J., High performance thin layer chromatographic analysis of neutral lipids in the urine of humans and BALB/c mice, *J. Planar Chromatogr.-Mod TLC*, 21, 39–42, 2008.

83. Massa, D.R., Vasta, J.D., Fried, B., and Sherma, J., High performance thin layer chromatographic analysis of polar lipid content of human urine and urine from BALB/c mice experimentally infected with *Echinostoma caproni*, *J. Planar Chromatogr.-Mod TLC*, 21, 337–341, 2008.

84. Martin, D.L., Friend, B., and Sherma, J., Effects of increased salinity on survival and lipid composition of *Helisoma trivolvis* (Colorado strain) and *Biomphalaria glabrata* in laboratory cultures, *The Veliger*, 49, 101–104, 2007.

85. White, M.M., Fried, B., and Sherma, J., Effects of aestivation and starvation on the neutral lipid and phospholipid content of *Biomphalarai glabrata* infected with *Schistosoma mansoni*, *J. Parasitol.*, 93, 1–3, 2007.

86. Bandstra, S.R., Fried, B., and Sherma J., High performance thin layer chromatographic analysis of neutral lipids and phospholipids in *Biomphalaria glabrata* patently infected with *Echinostoma caproni*, *Parasitol. Res.*, 99, 414–418, 2006.

87. Murray, K.E., Fried, B., and Sherma, J., Determination of phospholipid and sphingolipid content in the feces of BALB/c mice and those infected with *Echinostoma caproni* by high performance silica gel thin layer chromatography with densitometry, *Acta Chromatogr.*, 18, 190–198, 2007.

88. Vasta, J.D., Fried, B., and Sherma, J., High performance thin layer chromatographic analysis of neutral lipids in the urine of BALB/c mice infected with *Echinostoma caproni*, *Parasitol. Res.*, 102, 625–629, 2008.

89. Massa, D.R., Fried, B., and Sherma, J., Further studies on the neutral lipid content in the feces of BALB/c mice infected with *Echinostoma caproni* as determined by silica gel HPTLC-densitometry, *J. Liq. Chromatogr. Relat. Technol.*, 31, 1871–1880, 2008.

90. Zani, P.A., Counihan, J.L., Vasta, J.D., Fried, B., and Sherma, J., Characterization and quantification of neutral lipids in the lizard *Uta stansburiana* by HPTLC-densitometry, *J. Liq. Chromatogr. Relat. Technol.*, 31, 1881–1891, 2008.

91. Counihan, J.L., Zani, P.A., Fried, B., and Sherma, J., Characterization and quantification of polar lipids in the lizard *Uta stansburiana stansburiana* by HPTLC-densitometry, *J. Liq. Chromatogr. Relat. Technol.*, 32, 1289–1298, 2009.

92. Martin, D.L., Fried, B., and Sherma, J., High performance thin layer chromatographic analysis of neutral lipids and phospholipids in the medicinal leech *Hirudo medicinalis* maintained on different diets, *J. Planar Chromatogr.-Mod TLC*, 19, 167–170, 2006.

93. Martin, D.L., Sherma, J., and Fried, B., High performace thin layer chromatographic analysis of neutral lipids and phospholipids in the medicinal leech *Hirudo medicinalis* and in leech conditioned water, *J. Liq. Chromatogr. Relat. Technol.*, 28, 2597–2606, 2005.

94. Jarusiewicz, J.A., Fried, B., and Sherma, J., Effect of diet on the carotenoid pigment and lipid content of *Pomacea bridgesii* as determined by quantitative high performance thin layer chromatography, *Comp. Biochem. Physiol. B*, 143, 244–248, 2006.

Index

A

Abietic acid, 410
ABO blood group determinants, 170
Abscisic acid (ABA), 55–56, 280
Abselut Nexsus, 368, 370
Accuracy, 352–353
Aceclofenac, 322
Acenocoumarol, 49
Acephate, 374
Acetonitrile organic modifier system, 81
N-Acetylglucosamine, 139
α1-Acid glycoprotein (AGP), 49–53, 148, 151, 304
Activated carbon, 372, 374
Adsorption isotherms, 7–8, 10, 15, 17–18, 114
Affinity chromatography (AFC), 139, 140, 142
Aflatoxins, 199, 227
Alanine, 412
Alanine-based chiral additives, 104
Albumin, 37, 39, 47–49, 322
N-Alkyl amino acid enantioseparation, 81, 88–89, 91, 98, 100–102, 104–105
Alkyl-diol-silica (ADS) material, 302–303, 380
Alprenolol, 52
Amino acid analysis, TLC/HPTLC, 412
Amino acid enantioseparation, See also Chiral ligand exchange chromatography; Enantiomer separation
    chiral mobile phases, 104–112
    cis/trans equilibrium and elution order, 97–98, 100, 117
    coated chiral stationary phases, 91–104
    D- and L- enantiomers, 91–98, 106, 109–111
    factors influencing chromatographic behavior, 80–83
        column temperature, 82–83, 115
        copper salt [Cu(II)] effects, 82, 95, 104, 110, 112
        eluent pH, 78–81, 109
        flow rate, 83
        metal ion concentration, 81
        metal ion type, 82
        organic modifiers, 81–82
    homo-chiral complex, 107
    N-alkyl issues, 81, 88–89, 91, 98, 100–102, 104–105
    preparative scale applications, 113–115
    steric hindrance effects, 104–107

2-Aminobenzamide (2-AB), 141, 152, 156
6-Aminocaproic acid, 152
Aminoindandicarboxylic acid (AIDA), 103
2-Amino-3-methylimidazo[4,5-f] quinoline (IQ), 229
2-Amino-1-methyl-6-phenylimidazo[4,5-b] pyridine (PhIP), 229
5-Amino-2-naphthalenesulfonic acid (ANSA), 152
8-Aminonaphthalene-1,3,6-trisulfonate (ANTS), 141, 152, 156
Aminopropanol-based enantioselector, 95
2-Aminopyridine (2-AP), 141, 152, 155–156
8-Aminopyrene-1,3,6-trisulfonate (APTS), 141, 152
Amino-silica gels, 55
Ammonium acetate buffer, 173
Analytical method transfer, 358–362
Analytical method validation, See Validation of analytical procedures
Andrographolide, 324
Anion vs. cation salt effects, 10, 13, 19, 29
Antibody-based stationary phases, 38, 55–57
Antibody-linked immunosorbents, 382–384
Antipyrin, 412
Antithrombin, 146, 168–169
Antrazine, 374
ANTS, 141, 152, 156
Aromatic amines (AAs), 199
Artificial neural networks (ANNs), 26
Arylamines (AA), 200–201
2-Arylpropionic acid derivatives, 47, 51
Asiaticoside, 319
Asparagine, 119–120
Aspartic acid, 85, 94, 96, 98, 99, 103, 107
Atenolol, 323
Automated multiple development (AMD), 408
Avidin, 38

B

Baicalin, 323
Benzo[a]anthracene (BaA), 200
Benzo[a]pyrene (BaP), 200, 226–228
Benzo[a]pyrene diol epoxide (BPDE), 200, 220–221, 225–229, 234–235
Benzo[b]fluoranthrene (BkF), 200
Benzodiazepines, 47, 51, 370
Benzoin, 52
Benzonase/alkaline phosphatase, 236–237

423

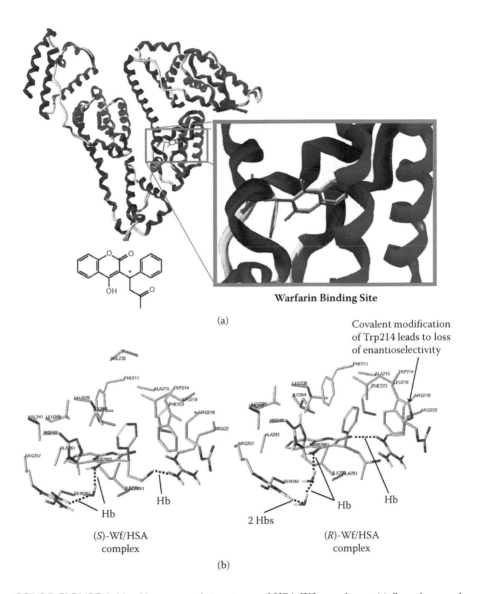

Warfarin Binding Site

(a)

Covalent modification
of Trp214 leads to loss
of enantioselectivity

(S)-Wf/HSA
complex

(R)-Wf/HSA
complex

Hb

Hb

2 Hbs

Hb

Hb

(b)

COLOR FIGURE 2.10 X-ray crystal structures of HSA-Wf complexes. (a) Superimposed complexes, (*R*)-Wf (magenta), (*S*)-Wf (cyan). (b) Active site with binding modes. (From Lämmerhofer, M., *J Chromatogr A*, in press. With permission.)

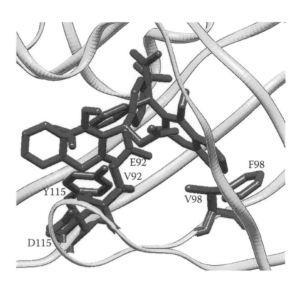

COLOR FIGURE 2.11 Comparison of dockings to ORM 1 and ORM 2 models. ORM 1 side chains and (*S*)-acenocoumarol docked to ORM 1 are indicated with blue color, ORM 2 side chains and (*S*)-acenocoumarol docked to ORM 2 with red color. (From Hazai, E. et al., *Bioorg Med Chem*, 14, 1964, 2006. With permission.)

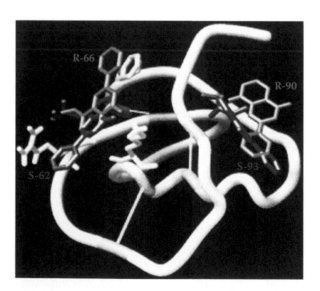

COLOR FIGURE 2.12 Molecular modeling simulation of U-80,413 enantiomers bound to OMTKY3. The white curved graphic represents the protein backbone. Selected protein side chains are also shown in white. Ligands are labeled according to their *R* or *S* chirality and numbered according to their position among the 100 lowest energy-minimized binding orientations. (From Pinkerton, T.C. et al., *Anal Chem*, 67, 2365, 1995. With permission.)

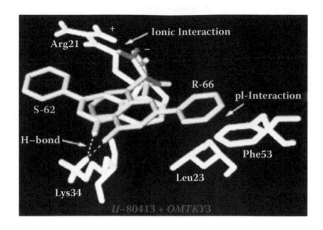

COLOR FIGURE 2.13 U-80413 enantiomers in the enantioselective binding site of OMTKY3. (From Pinkerton, T.C. et al., *Anal Chem*, 67, 2366, 1995. With permission.)

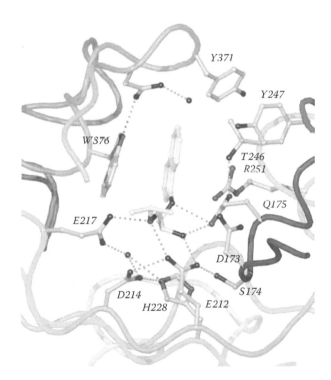

COLOR FIGURE 2.14 Ball and stick representation of the active site with the (*S*)-propranolol molecule (green C atoms), protein residues (light-brown C) with atoms within 4 Å distance from the ligand, and two water molecules (magenta). Probable hydrogen bonds are indicated with blue dots and relevant parts of the C$_\alpha$ backbone are shown, color-ramped from blue at residue 171 to red at the C terminus, residue 434. The main interactions are the hydrophobic stacking of the naphthyl group with Trp376, the bidentate salt-link interaction between the positively charged secondary amine and the catalytic residues Glu212 and Glu217, and hydrogen bonding of the chiral hydroxy group with Gln175 and Glu212. Two residues in contact with the isopropyl moiety, Tyr145 and Tyr171, were omitted for clarity. (From Ståhlberg, J. et al., *J Mol Biol*, 305, 83, 2001. With permission.)

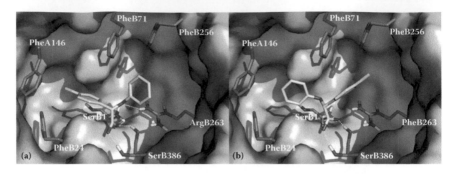

COLOR FIGURE 2.16 Binding mode of (*S*)-2-(4-chlorophenyl)-2-phenoxyacetic acid (a) and (*R*)-2-(4-chlorophenyl)-2-phenoxyacetic acid (b) within PGA. For clarity reasons, only interacting residues are displayed. Hydrogen bonds between ligand and protein are shown as dashed yellow lines. Ligand (white) and interacting key residues (orange) are represented as stick models; the protein is a light gray Connolly surface. (From Lavecchia, A. et al., *J Mol Graph Mod*, 25, 777, 2007. With permission.)

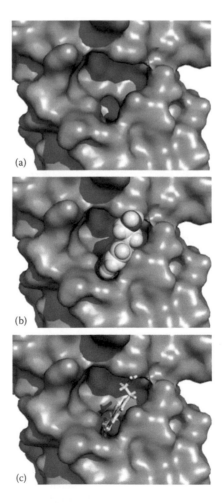

COLOR FIGURE 2.18 Surface contour images of the modeled anti-D-amino acid structure (a) without the ligand, (b) with the docked ligand in a spherical representation, and (c) with the docked ligand in a stick representation. (From Ranieri, D.I. et al., *Chirality*, 20, 565, 2008. With permission.)

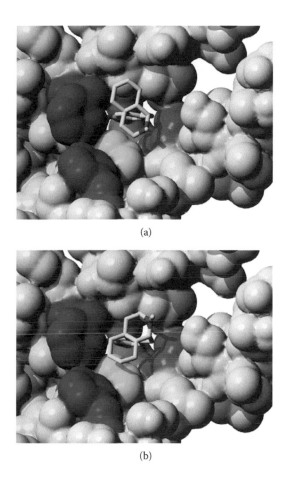

(a)

(b)

COLOR FIGURE 2.21 The most stable docked orientations of (a) DM and (b) LM complexes with the model of the central lumen of the α3β4 nAChR. Hydrophobic clefts formed within the channel are shown in detail. Residues forming the cleft are color coded Phe, blue; Val, green; and Ser, orange. (From Jozwiak, K. et al., *J Chromatogr B*, 797, 203, 2008. With permission.)

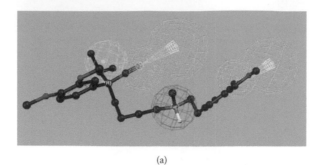

(a)

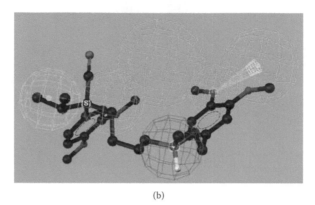

(b)

COLOR FIGURE 2.22 The fit of verapamil enantiomers in the proposed pharmacophore, where (a) the mapping of (*R*)-verapamil and (b) the mapping of (*S*)-verapamil. Nonpolar hydrogen atoms have been omitted for clarity. (From Moaddel, R. et al., *Br J Pharmacol*, 151, 1311, 2007. With permission.)

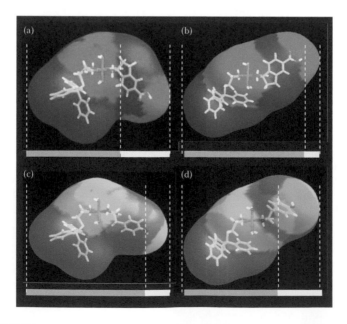

COLOR FIGURE 3.18 The proposed structures for the ternary complexes and surface projections for S-trityl-(*R*)-cysteine as the chiral selector and (a) (*S*)-AIDA, (b) (*R*)-AIDA, (c) (*S*)-Tyr, and (d) (*R*)-Tyr.

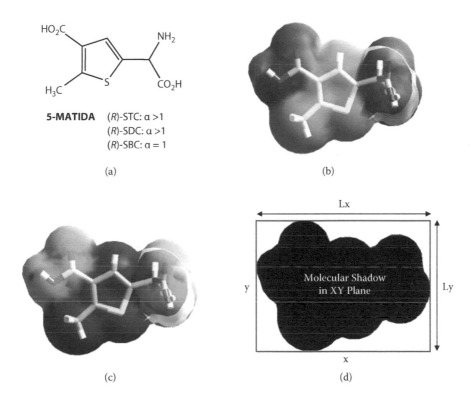

COLOR FIGURE 3.26 Jurs and Shadow descriptors calculated on 5-MATIDA. (a) Chemical structure of 5-MATIDA and diverse separation factor according to the chiral selector used. (b) Representation of some Jurs descriptors on 5-MATIDA: partial positive surface area (PPSA, sum of blue areas) and partial negative surface area (PNSA, sum of red areas). (c) Representation of some Jurs descriptors on 5-MATIDA: total hydrophobic surface area (TASA, sum of brown areas). (d) Representation of some Shadow descriptors on 5-MATIDA: area of the molecular shadow in the xy plane (Sxy, black area). Length of molecule in the x dimension (Lx). Length of molecule in the y dimension (Ly).

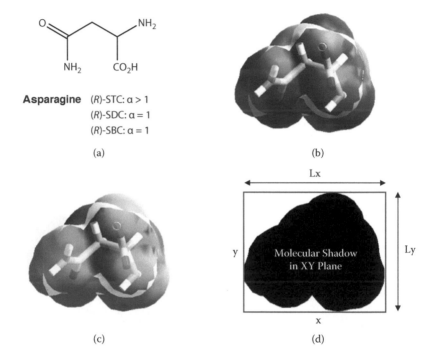

Asparagine (R)-STC: $\alpha > 1$
(R)-SDC: $\alpha = 1$
(R)-SBC: $\alpha = 1$

(a)

(b)

Lx

y   Molecular Shadow in XY Plane   Ly

x

(c)

(d)

COLOR FIGURE 3.27 Jurs and Shadow descriptors calculated on Asparagine. (a) Chemical structure of Asparagine and diverse separation factor according to the chiral selector used. (b) Representation of some Jurs descriptors on Asparagine: partial positive surface area (PPSA, sum of blue areas) and partial negative surface area (PNSA, sum of red areas). (c) Representation of some Jurs descriptors on Asparagine: total hydrophobic surface area (TASA, sum of brown areas). (d) Representation of some Shadow descriptors on Asparagine: area of the molecular shadow in the xy plane (Sxy, black area). Length of molecule in the x dimension (Lx). Length of molecule in the y dimension (Ly).

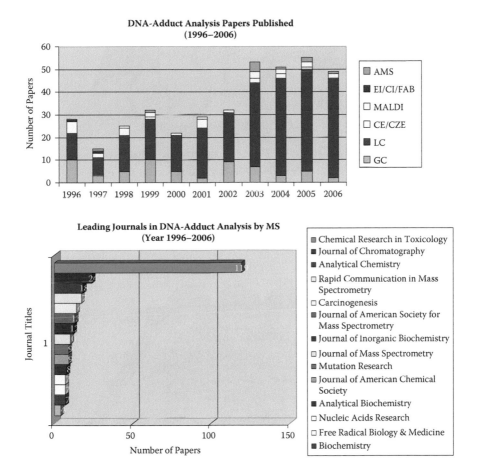

FIGURE 5.2 Number of papers on DNA adducts categorized by the techniques used and leading journals with most publications on the analysis of DNA adducts between years 1996 and 2006, searched through SciFinder.

COLOR FIGURE 6.1 Electron density and atomic partial charge (unit: au) of PAH7.

EonD   0.010

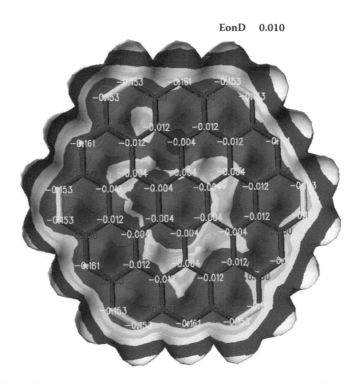

COLOR FIGURE 6.2 Electrostatic potential and atomic partial charge (unit: au) of PAH37.

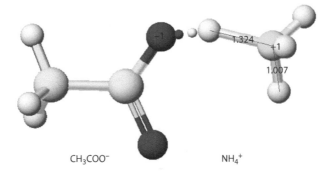

COLOR FIGURE 6.3 Ion pair of ammonium and acetic acid ions. Red balls oxygen, light blue balls nitrogen, yellow balls carbon, white balls hydrogen. Atomic distance unit Å.

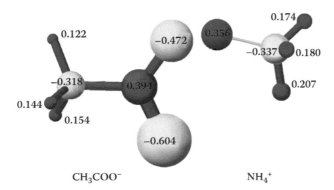

COLOR FIGURE 6.4 Atomic partial charge (unit: au) of ammonium and acetic acid ion pair.

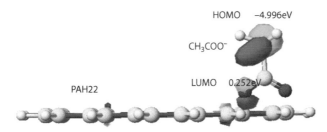

COLOR FIGURE 6.5 Interaction between acetic acid ion and a model graphitized carbon, PAH22, with HOMO and LUMO.

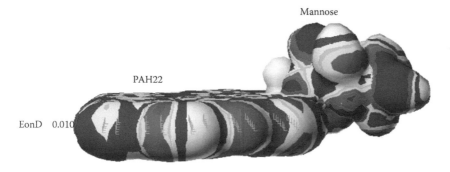

COLOR FIGURE 6.7 Electron density isosurface of mannose and a model graphitized carbon (PAH22) complex.

PAH22

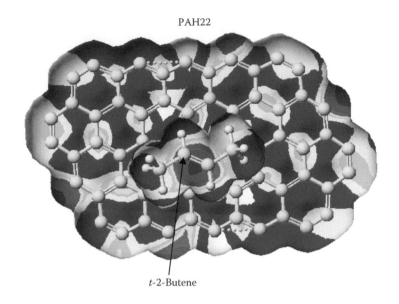

t-2-Butene

COLOR FIGURE 6.8 Retention of *trans*-2-butene on a model graphitized carbon (PAH22).

Carbon (PAH22)

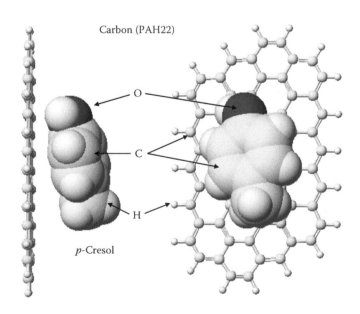

O

C

H

p-Cresol

COLOR FIGURE 6.9 Retention of *p*-cresol on a model graphitized carbon (PAH22). Atomic size of the carbon phase is 20%. Red balls oxygen, yellow balls carbon, and small white balls hydrogen.

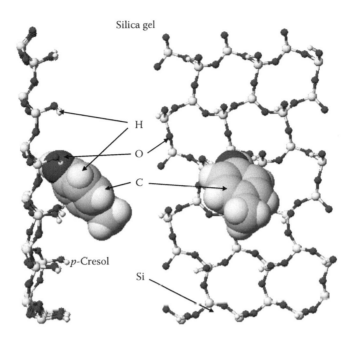

COLOR FIGURE 6.10 Retention of *p*-cresol on a model silica gel. Atomic size of the silica phase is 20%. Red balls oxygen, yellow balls carbon, small white balls hydrogen and large white balls silicone.

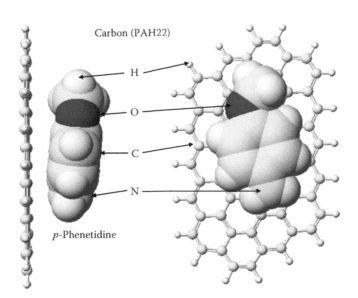

COLOR FIGURE 6.11 Retention of *p*-phenetidine on a model graphitized carbon (PAH22). Atomic size of the carbon phase is 20%. Red balls oxygen, light blue balls nitrogen, yellow balls carbon, white balls hydrogen.

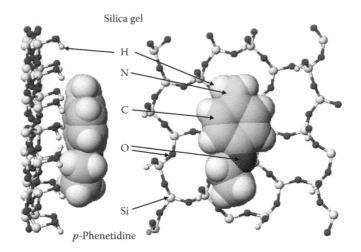

Silica gel

H

N

C

O

Si

p-Phenetidine

COLOR FIGURE 6.12 Retention of *p*-phenetidine on a model silica gel. Atomic size of the carbon phase is 20%. Red balls oxygen, light blue balls nitrogen, yellow balls carbon, small white balls hydrogen and large white balls silicone.

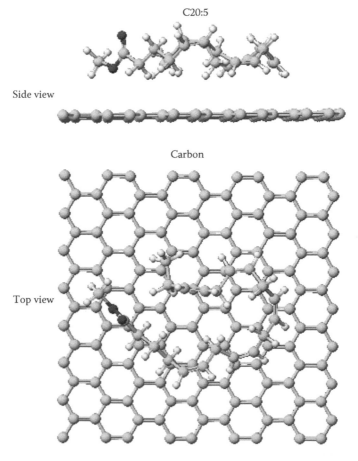

C20:5

Side view

Carbon

Top view

COLOR FIGURE 6.13 Adsorption of C20:5 *cis*-5,8,11,14,17 fatty acid methylester on a model graphitized carbon (PAH195). Red balls oxygen, yellow balls carbon, white balls hydrogen.

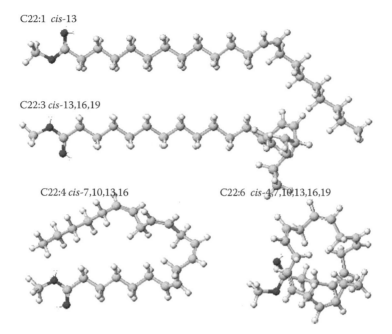

C22:1 *cis*-13

C22:3 *cis*-13,16,19

C22:4 *cis*-7,10,13,16

C22:6 *cis*-4,7,10,13,16,19

COLOR FIGURE 6.14 Stereo structure of 22:1 *cis*-13, 22:3 *cis*-13, 16, 19, 22:4 *cis*-7, 10, 13, 16, and 22:6 *cis*-4, 7, 10, 13, 16, 18 fatty acid methylesters. Red balls oxygen, yellow balls carbon, white balls hydrogen.

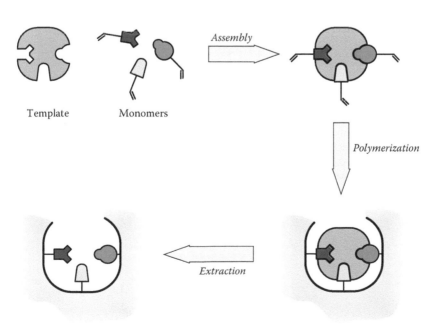

Template Monomers *Assembly*

Polymerization

Extraction

COLOR FIGURE 9.7 Schematic representation of molecular imprinting principle.

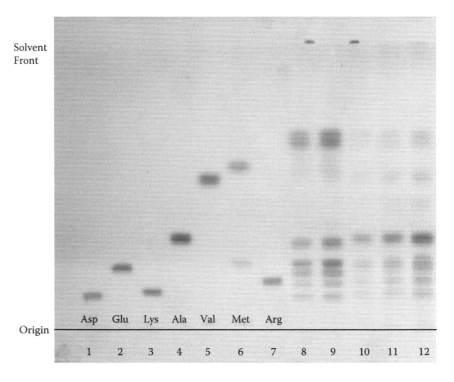

COLOR FIGURE 10.1 Amino acid chromatograms on an HPTLC cellulose plate developed with 2-butanol-pyridine-glacial acetic acid-water (39:34:10:36) mobile phase: lanes 1–7, 2 uL aliquots of individual amino acid standards (0.10 ug/uL); lanes 8 and 9, 1 and 2 uL aliquots, respectively, of a mixed amino acid standard (0.10 ug/uL of each acid); lanes 10–12, 1, 2, and 3 uL aliquots, respectively, of the ethanol-water (70;30) amino acid extract from a DGG of the snail *B. glabrata*. Abbreviations: Asp, aspartate; Glu, glutamic acid; Lys, lysine; Ala, alanine; Val, valine; Met, methionine; Arg, arginine. (From Vasta, J.D. et al., *Acta Chromatogr.*, 21, 29–38, 2009. With permission.).

T - #0333 - 071024 - C16 - 234/156/20 - PB - 9780367383022 - Gloss Lamination